FOOD ADDITIVES

FOOD ADDITIVES

Recent Developments

Edited by J.C. Johnson

NOYES DATA CORPORATION
Park Ridge, New Jersey, U.S.A.
1983

Copyright © 1983 by Noyes Data Corporation
No part of this book may be reproduced in any form
without permission in writing from the Publisher.
Library of Congress Catalog Card Number: 83-2205
ISBN: 0-8155-0935-9
ISSN: 0093-0075
Printed in the United States

Published in the United States of America by
Noyes Data Corporation
Mill Road, Park Ridge, New Jersey 07656

10 9 8 7 6 5 4 3 2 1

Library of Congress Cataloging in Publication Data

Main entry under title:

Food additives, recent developments.

(Food technology review, ISSN 0093-0075 ; no. 58)
Includes indexes.
1. Food additives--Patents. I. Johnson, J. C.
(Jeanne Colbert), 1920–   . II. Noyes Data Corp.
III. Series.
TP455.F65 1983   664'.06'0272   83-2205
ISBN 0-8155-0935-9

FOREWORD

The detailed, descriptive information in this book is based on U.S. patents, issued between January, 1979 and July, 1982 that deal with food additives used for processing, or for preserving shelf life.

This book is a data-based publication, providing information retrieved and made available from the U.S. patent literature. It thus serves a double purpose in that it supplies detailed technical information and can be used as a guide to the patent literature in this field. By indicating all the information that is significant, and eliminating legal jargon and juristic phraseology, this book presents an advanced commercially oriented review of recent developments in the field of food additives.

The U.S. patent literature is the largest and most comprehensive collection of technical information in the world. There is more practical, commercial, timely process information assembled here than is available from any other source. The technical information obtained from a patent is extremely reliable and comprehensive; sufficient information must be included to avoid rejection for "insufficient disclosure." These patents include practically all of those issued on the subject in the United States during the period under review; there has been no bias in the selection of patents for inclusion.

The patent literature covers a substantial amount of information not available in the journal literature. The patent literature is a prime source of basic commercially useful information. This information is overlooked by those who rely primarily on the periodical journal literature. It is realized that there is a lag between a patent application on a new process development and the granting of a patent, but it is felt that this may roughly parallel or even anticipate the lag in putting that development into commercial practice.

Many of these patents are being utilized commercially. Whether used or not, they offer opportunities for technological transfer. Also, a major purpose of this book is to describe the number of technical possibilities available, which may open up profitable areas of research and development. The information contained in this book will allow you to establish a sound background before launching into research in this field.

Advanced composition and production methods developed by Noyes Data are employed to bring these durably bound books to you in a minimum of time. Special techniques are used to close the gap between "manuscript" and "completed book." Industrial technology is progressing so rapidly that time-honored, conventional typesetting, binding and shipping methods are no longer suitable. We have bypassed the delays in the conventional book publishing cycle and provide the user with an effective and convenient means of reviewing up-to-date information in depth.

The table of contents is organized in such a way as to serve as a subject index. Other indexes by company, inventor and patent number help in providing easy access to the information contained in this book.

16 Reasons Why the U.S. Patent Office Literature Is Important to You

1. The U.S. patent literature is the largest and most comprehensive collection of technical information in the world. There is more practical commercial process information assembled here than is available from any other source. Most important technological advances are described in the patent literature.

2. The technical information obtained from the patent literature is extremely comprehensive; sufficient information must be included to avoid rejection for "insufficient disclosure."

3. The patent literature is a prime source of basic commercially utilizable information. This information is overlooked by those who rely primarily on the periodical journal literature.

4. An important feature of the patent literature is that it can serve to avoid duplication of research and development.

5. Patents, unlike periodical literature, are bound by definition to contain new information, data and ideas.

6. It can serve as a source of new ideas in a different but related field, and may be outside the patent protection offered the original invention.

7. Since claims are narrowly defined, much valuable information is included that may be outside the legal protection afforded by the claims.

8. Patents discuss the difficulties associated with previous research, development or production techniques, and offer a specific method of overcoming problems. This gives clues to current process information that has not been published in periodicals or books.

9. Can aid in process design by providing a selection of alternate techniques. A powerful research and engineering tool.

10. Obtain licenses—many U.S. chemical patents have not been developed commercially.

11. Patents provide an excellent starting point for the next investigator.

12. Frequently, innovations derived from research are first disclosed in the patent literature, prior to coverage in the periodical literature.

13. Patents offer a most valuable method of keeping abreast of latest technologies, serving an individual's own "current awareness" program.

14. Identifying potential new competitors.

15. It is a creative source of ideas for those with imagination.

16. Scrutiny of the patent literature has important profit-making potential.

CONTENTS AND SUBJECT INDEX

INTRODUCTION

A food additive may be defined as any substance that becomes part of the final food product whether added intentionally or incorporated accidentally. Approximately 2,800 substances are used by food processors as additives to food products. Many additional materials find their way into food products in small amounts during the course of growing, harvesting, processing, packing and storage.

Sugar, salt and corn syrup are the most widely used additives. These three, plus citric acid, baking soda, colorants, mustard and pepper comprise about 98% by weight of all food additives used in the United States.

This review covers some 247 food additives as disclosed in 258 patents issued since January 1979. The review does not include patents on flavors, colors, sweeteners and nutritive additives. Stabilizers which maintain freshness and retard spoilage are described in 57 patents comprising the first two chapters. Additives which improve the texture or processing of foods are covered in the chapters on vegetable gums, emulsifiers and modified starches by 29, 34 and 27 patents, respectively.

The preparation and use of protein additives and substitutes are disclosed by the 57 patents of the next two chapters. Soy, caseinate, yeast, gluten and other materials serve as the source of the proteins.

The remaining two chapters describe other additives such as fat substitutes, acids and salts, leavening agents, low calorie material, yeast products, etc.

STABILIZERS

ANTIOXIDANTS

Phosphines in Frying Oils and Fats

Poultry, especially chicken, contains a large amount of fatty acids and esters near the surface of the meat, especially in the skin. During frying, the presence of heat and oxygen from the air produces the hydroperoxides of these fatty acids. The hydroperoxides decompose to form polymers, various undefinable gummy materials, aldehydes, acids, ketones, etc., which causes the development of off-flavors and off-odors in the food. The fatty acids and esters which are found in the cooking oil or fat also can be oxidized in the presence of heat and oxygen to similar hydroperoxides of fatty acids which likewise decompose and undesirably affect the flavor of foods (chicken) cooked therein.

B.L. Madison and J.I. Shulman; U.S. Patent 4,164,592; August 14, 1979; assigned to The Procter & Gamble Company has used oligomeric or polymeric phosphine compounds to prevent the formation of hydroperoxides during frying.

Monomeric phosphines, and in particular triphenylphosphine, are toxic and not suitable for use in edible fats and oils. Polymeric phosphine compounds are neither digested nor absorbed after oral ingestion. Preferred for use are the polymeric triaryl- or substituted triarylphosphine compounds with a molecular weight from about 600 to about 3000. Compounds with molecular weights substantially above this range are not effective in preventing hydroperoxide formation. These polymeric phosphine compounds having a molecular weight below about 600 are easily absorbed from the digestive tract of animals and could be metabolized to produce toxic compounds.

Preferred oligomeric and polymeric triarylphosphines and substituted triarylphosphine compounds for use as antioxidants are those derived from polyols. These polyol based triarylphosphine polymers (polyethers) are preferred because they can be made from readily available starting materials. Suitable polyols are glycols, glycerol, sugar alcohols, sugars, including monosaccharides, disaccharides, trisaccharides and tetrasaccharides, among others, and other polyols, for example, pentaerythritol.

Compounds in which the oxygen of the polyether group is replaced with a nitrogen or sulfur are also usable as antioxidants. Hydrocarbyl polymeric phosphine compounds which can be prepared from styryl or alkylene phosphine derivatives are also suitable as antioxidants.

Foods fried in oils containing these polymeric phosphine hydroperoxide inhibiting compounds have an improved flavor over foods fried in oils containing conventional antioxidants, the tocopherols, BHA, BHT, etc. The food in which the fatty acid hydroperoxide formation has been inhibited have a flavor similar to that achieved by frying foods in an oxygen free atmosphere, that is, under a nitrogen or carbon dioxide atmosphere.

These hydroperoxide inhibiting compounds are particularly useful in oils and fats which contain unsaturated fatty acids. Oils which are polyunsaturated, especially safflower oil, sunflower seed oil, soybean oil, and corn oil are particularly susceptible to hydroperoxide formation. Foods fried in these oils, or mixtures thereof, or their corresponding hydrogenated oils, especially need to be protected by these phosphine compounds.

Example: Deodorized safflower oil containing 500 ppm of 1,2,3-tris(p-diphenylphosphino)benzoxypropane is heated in an electric skillet to temperatures of about 325°F. Cut-up frying chicken is fried in this oil until thoroughly cooked. During the frying, the skillet and chicken are open to the atmosphere. When this chicken is eaten, the overall taste is judged to be very good.

When the chicken prepared in this example is compared with chicken fried in deodorized safflower oil not containing the 1,2,3-tris(p-diphenylphosphino)-benzoxypropane, the chicken of this example is judged to have the overall better flavor.

When poly(p-diphenylphosphino)styrene having a degree of polymerization of 8 is used to replace the hydroperoxide inhibiting compound of this example similar results are obtained.

Triarylphosphine Compounds

Oligomeric or polymeric triaryl- and substituted triarylphosphine compounds which are nondigestible and nonabsorbable by animals are disclosed by *J.I. Shulman; U.S. Patent 4,209,468; June 24, 1980; assigned to The Procter & Gamble Company*. These compounds act as antioxidants by inhibiting the buildup of hydroperoxides of unsaturated fatty acids, e.g., in foods.

The polymeric aryl- or substituted arylphosphine compounds are preferably those with a molecular weight from about 600 to about 3000. These compounds are not digested and are not absorbed by the animal ingesting them.

For food use in fats and oils, an amount in the range of about 10 to about 1000 ppm is effective and safe for ingestion by humans and lower animals. The amount used will depend upon several factors, including the amount of unsaturation in the fatty acid fats or oils, and the presence of other antioxidant materials in the fat or oil.

Preferred oligomeric and polymeric triarylphosphines and substituted triarylphosphine compounds for use as antioxidants are those derived from polyols.

These polyol-based triarylphosphine polymers are preferred because they can be made from readily available starting material.

Glycols which are useful are ethylene glycol, 1,2-propylene glycol, 1,3-propylene glycol, the butyl glycols, the pentyl glycols, and others. Ethylene glycol and glycerol are most preferred for use herein.

Methods of preparation and testing of these compounds are shown in the following examples.

Example 1: *Preparation of 1,2,3-tris(p-diphenylphosphino)benzoxypropane* – Step A: A mixture of 114 g of glycerol and 148.5 g of paraformaldehyde in a three-necked flask is stirred mechanically until a thick paste is formed. With stirring and cooling by an external ice bath, excess hydrogen chloride gas is bubbled through the slurry until the mixture becomes very fluid and no more hydrogen chloride is absorbed. The two phases are separated and the lower, cloudy layer is dried overnight over anhydrous $CaCl_2$. Filtration through glass wool and distillation affords 130.5 g of 1,2,3-tris(chloromethoxy)propane, BP 168°-170°C (18 mm).

Step B: To 1.0 mol of n-butyllithium in hexane at -78°C under an inert atmosphere is added dropwise and with efficient stirring 236 g of p-dibromobenzene in dry tetrahydrofuran (total volume 500 ml). The resulting solution is stirred at -78°C for 1 hr, then treated dropwise at this temperature with 220.5 g of chlorodiphenylphosphine. When this addition is complete, the mixture is allowed to warm to room temperature, then solvent is removed under vacuum. The residue is triturated with two 400-ml portions of methanol, then distilled through a Vigreux column. The cut with BP 165°-170°C (0.15 mm) is recrystallized from ethanol to give 100.5 g of p-diphenylphosphinobromobenzene, MP 78°-80°C.

Step C: A mixture of 110 ml of ~1 M tert-butyllithium in hexane and 180 ml of dry tetrahydrofuran is cooled to -70°C under an inert atmosphere. To this is added dropwise a solution of 28 g of the product of Step B in 200 ml of tetrahydrofuran, keeping the reaction temperature below -60°C. Additional small aliquots (1 g) of the product of Step B are added until thin-layer chromatographic analysis shows this compound to be present in excess (in this instance, another 3 g). Then a 6.5 portion of the product of Step A in 5 ml of tetrahydrofuran is added and the solution is warmed to ambient temperature and stirred for 16 hr. The reaction mixture is poured into 1 ℓ of water, the organic layer is separated, and the aqueous layer is extracted with three 100-ml portions of peroxide-free ethyl ether. The combined organic layers are washed with water, then brine, dried over anhydrous $MgSO_4$, and concentrated under vacuum to leave 28.7 g of crude product as a syrup. Preparative layer or column chromatography on silica gel affords high purity 1,2,3-tris(p-diphenylphosphino)benzoxypropane.

When the product of Step C of Example 1 is added at 500 ppm to deodorized safflower oil and the oil is heated at 60°C with air bubbling through it, the initiation time for peroxide build-up is increased. No titratable peroxide can be observed in the oil for at least 51 hr. During this time deodorized safflower oil alone reaches a peroxide value of 5.9 meq/kg.

When the product of Step C of Example 1 is added at 199 ppm to deodorized soybean oil and the oil is heated at 60°C in a loosely capped can for eight days, the peroxide value of the oil reaches 3.8 meq/kg. During this time deodorized soybean oil without addition of the product of Step C reaches a peroxide value of 16.2 meq/kg.

When an antioxidant product made according to this example, but containing a radioactive carbon-14 tracer, is fed to rats, all the radioactivity is excreted in the rats' feces. Less than 0.1% of the fed radioactivity appears in the lymphatic fluid and in expired carbon dioxide, thus indicating that the antioxidant is not absorbed from the gastrointestinal tract.

Example 2: *Preparation of poly(p-diphenylphosphino)styrene* — To 2.42 g (8.4 mmol) of p-diphenylphosphinostyrene in a two-necked round bottom flask is added 6.0 ml of dry benzene. This solution is degassed on a high vacuum line, and, while frozen, the mixture is then placed under nitrogen. The mixture is allowed to partially melt. Then with stirring 0.88 ml (1.4 mmol) of 1.6 M n-butyllithium in hexane is added rapidly via syringe. Within 45 seconds, 0.30 ml of distilled degassed tetrahydrofuran is added to the yellow solution; this serves to activate the anionic initiator. The solution rapidly turns dark. Stirring is continued for 30 minutes, then the reaction is quenched by adding 0.12 ml of methanol. The solution is washed with three 12-ml portions of water and the resulting benzene solution is freeze-dried to leave 2.47 g (100%) of colorless powder.

When the product of Example 2 is added at 500 ppm to deodorized safflower oil and the oil is heated at 60°C with air bubbling through it, the initiation time for peroxide build-up is increased. No titratable peroxide can be observed in the oil for at least 46 hr. After 70 hr, the peroxide value of the oil reaches 1.2 meq/kg. Under these conditions, deodorized safflower oil without the addition of the product of Example 2 reaches a peroxide value of 29.8 meq/kg after 70 hr.

Furfurylidene-bis-Di-tert-Butylphenol

According to *E. Clinton; U.S. Patent 4,222,883; September 16, 1980; assigned to Ethyl Corporation* a compound is disclosed which is a very effective antioxidant in a wide range of organic substrates. This new compound is 4,4'-furfurylidene-bis(2,6-di-tert-butylphenol).

This compound can be made by reacting about 2 mols of 2,6-di-tert-butylphenol with about 1-2 mols of furfural in the presence of a basic catalyst (e.g., KOH) in a lower alkanol solvent (e.g., isopropanol) at reflux temperature. The following example illustrates the preparation of this compound.

Example 1: In a reaction vessel was placed 400 ml ethanol, 6.6 g KOH, and 206 g 2,6-di-tert-butylphenol While stirring at 30°C, 48 g of 2-furfurylidene was added over a 15-minute period and the mixture refluxed (82°-83°C) for 3 hr. The mixture was cooled and after standing 65 hr, solids had precipitated. These were filtered off and washed with cold ethanol yielding 164 g crude product. An additional 57 g of solids was obtained by evaporating the filtrate. A portion of the crude product was recrystallized from ethanol followed by isooctane to yield 4,4'-furfurylidene-bis(2,6-di-tert-butylphenol) (white solid, MP 158°-159°C, analysis 80.1% C, 9.25% H).

The 4,4'-furfurylidene-bis(2,6-di-tert-butylphenol) is an effective stabilizer in a broad range of organic materials including foods of the type normally subject to oxidative deterioration in the presence of oxygen during use over an extended period.

Fats and oils of animal and vegetable origin are protected against gradual deterioration. Examples of these are lard, beef tallow, coconut oil, safflower oil, castor oil, babassu oil, cottonseed oil, corn oil, rapeseed oil, and the like.

Animal feeds such as ground corn, cracked wheat, oats, wheat germ, alfalfa, and the like are protected by mixing a small but effective amount of the additive with these products. Vitamin extracts, especially the fat-soluble vitamins are effectively stabilized against degradation.

The amount of furfurylidene-bis-phenol added is a small antioxidant amount. A useful range is about 0.01-5 weight percent. A more preferred range is about 0.1-3 weight percent.

The antioxidant of this process may be used alone as the sole antioxidant or may be used in combination with other antioxidants or compounds which synergistically affect the effectiveness of the antioxidant.

Methods of incorporating the additive into the substrate are well known. For example, if the substrate is liquid the additive can be merely mixed into the substrate. Frequently the organic substrate is in solution and the additive is added to the solution and the solvent removed. Solid organic substrates can be merely sprayed with a solution of the additive in a volatile solvent. For example, stabilized grain products result from spraying the grain with a toluene solution of the additive.

The following describes organic compositions containing the additive of this process. In these examples "additive" refers to 4,4'-furfurylidene-bis(2,6-di-tert-butylphenol).

Example 2: To 10,000 parts of corn oil is added 15 parts of additive. The mixture is stirred, giving a corn oil highly resistant to normal oxidative degradation.

Example 3: To 10,000 parts of melted lard is added 10 parts of additive and the mixture is stirred until thoroughly blended, resulting in a lard highly resistant to normal oxidative degradation.

Metal Proteinates in Cooking Oils

H.H. Ashmead; U.S. Patent 4,201,793; May 6, 1980 has provided foods cooked in a hot oil bath containing stable metal proteinates.

By fortifying the cooking oil with proper amount of metal proteinates, the oil is stabilized toward oxidation and these metal proteinates will be absorbed into the material being cooked and prolong the life of the oil as well as the product.

A metal proteinate is a chelate of an essential bivalent metal such as iron, zinc, copper, cobalt, manganese, magnesium and calcium with at least two ligands which are amino acids, peptides or polypeptides, all of which are hydrolysis products or proteins. The shorter the chain length of the ligand the stronger the chelation bond will be. As used here the term proteinate and chelate may be used interchangeably.

Protein hydrolysates are prepared by acidic, basic or enzymatic hydrolysis. It is beneficial to first hydrolyze a protein source well so that the subsequent formation of the metal proteinate or chelate can form a product that can be actively transported through the walls of the small intestine. Large protein entities such as metal salts of gelatinates, caseinates or albuminates must be digested before transport can take place. It is believed that unhydrolyzed protein salts, in general, pass through the intestine largely intact with only small amounts being utilized. Therefore, in this process the protein molecules are hydrolyzed to a polypeptide, peptide or amino acid stage prior to mixing with the metal salt to form a proteinate (chelate with at least two protein hydrolysates as the liquid). In order to form a metal proteinate the proper amounts of constituents must be present at the right conditions. The mineral to be chelated must be in soluble form and the protein hydrolysates or amino acids must be free from interfering protons, i.e., nonintact, in the chelation process so that chemical bonds can be formed between the protein ligand and the metal involved.

The products of this process are stable and provide proper fortification of metal proteinates into the products into which they are incorporated. The concentration of metal proteinate in the product will depend upon the concentration of metal proteinate in the cooking oil. The absorptive capacity of the metal proteinates onto the foods is substantially proportionate to the concentration of the metals in the oil and the amount of the oil that penetrates into the cooked food. Concentrations may vary depending upon the metal proteinate or the mixture of proteinates desired. Also the end product may have a lesser propensity for the metal proteinate than the cooking oil. Concentrations may be determined empirically and will depend somewhat upon the recommended daily allowance for each metal. Generally concentrations will vary from about 0.2 to 2 grams of metal (without considering the ligand) per gallon of cooking oil. The resulting deep fried products will contain sufficient metal proteinates to inhibit the onset of rancidity and also to supply a biologically effective amount of essential minerals to the body when the products are ingested.

Example 1: Into a vat equipped with a stirrer containing a mixture of hydrogenated palm and coconut oils maintained at 375°F is added a mixture of iron, zinc and copper proteinates containing about 1.5 g of zinc, 1.8 g of iron, and 0.2 g of copper per gallon of oil. Potatoes sliced on a commercial potato chip slicer are cooked in the hot oil until crisp and a light golden brown whereupon the cooked potato chips are removed from the oil and allowed to drain. Upon analysis the chips are found to contain 6.0 mg of zinc, 7.2 mg of iron and 0.8 mg of copper in the form of proteinates per 2 oz serving of potato chips.

Example 2: The procedure of Example 1 is followed using an iron proteinate only. The oil used is safflower oil containing about 2.0 g of iron in the proteinate form per gallon of oil. Results similar to Example 1 are obtained which upon analysis show a product containing about 8 mg of iron as a proteinate per 2 oz serving.

Example 3: Into a small container equipped with a stirrer containing hydrogenated corn oil is added a mixture of iron, magnesium and zinc proteinates in an amount sufficient to provide 1.5 g of iron, 0.5 g of magnesium and 1.2 g of zinc per gallon of oil. The oil is maintained at a temperature of about 400°F and is constantly agitated. A yeast leavened donut dough is cooked in the hot oil until done and the donuts are allowed to drain. Upon analysis it is shown that the donuts contained about 9.7 mg of iron, 3.2 mg of magnesium and 7.8 mg of zinc for 4 oz serving.

Example 4: Into a vat that constantly strains and recirculates the oil is added a mixture of iron, manganese, copper, and calcium proteinates in an amount sufficient to provide about 0.5 g of each metal in the form of a proteinate in the oil. The oil is maintained at a temperature of about 325°-350°C. Chicken legs are dipped in a batter and placed in the oil until cooked and are then removed and allowed to drain. Upon analysis it is shown that the proteinates are absorbed onto the surface of the batter in an amount of which is approximately proportionate to the concentration of the proteinates in the cooking oil.

Heat Stabilization of Antioxidants with Yeast

Among the most important synthetic antioxidants which are approved for food uses are butylated hydroxyanisol (BHA), butylated hydroxytoluene (BHT), propyl gallate (PG), trihydroxybutyrophenone (THBP), thiodipropionic acid (TDPA), mono-tert-butylhydroquinone (TBHQ), and dilauryl thiodipropionate (DLTDP).

It is desirable for an antioxidant to be able to withstand the thermal conditions present during food processing and still retain its effectiveness. Although natural antioxidants tend to be heat stable, synthetic antioxidants such as BHA, BHT, and TBHQ, for example, volatilize at temperatures below those used during heat processing. Such antioxidants cannot withstand thermoprocessing conditions used for such products as snacks, baked goods, and meats, which use temperatures typically in the range from 200°-400°F.

Hence, for this process a "heat-sensitive antioxidant" is one which is unstable or volatilizes at temperatures in the range of 200°-400°F so as to reduce its effectiveness in foods prepared by thermoprocessing. In such instances the antioxidant must be applied as an additional step after themoprocessing. In the case of snacks, for example, the antioxidant is commonly sprayed onto the snack food after processing. This necessitates additional process equipment which is expensive. It would be desirable in many applications to simply add the antioxidant to the food formulation along with the other ingredients if the antioxidant could retain its effectiveness after being subjected to subsequent processing.

It is therefore an object of *P.G. Schnell; U.S. Patent 4,301,185; November 17, 1981; assigned to Standard Oil Company (Indiana)* to provide a method for stabilizing heat sensitive antioxidants.

Heat labile antioxidants such as butylated hydroxyanisol, butylated hydroxytoluene, propyl gallate and mono-tert-butylhydroquinone are stabilized by incorporation onto an inactive yeast (heat-killed) substrate. The yeast acts to prevent volatilization of the antioxidants at temperatures commonly used in food processes, thus enabling the antioxidants to be added to the food formulation prior to thermoprocessing rather than as a subsequent additional step.

Any inactive yeast (heat-killed) can be used to achieve at least some degree of success in producing the heat-stable properties characteristic of these antioxidant compositions, but *Saccharomyces cerevisiae, Kluyveromyces fragilis,* and *Candida utilis* are particularly advantageous since they are acceptable for food use. The amount of yeast to be added to the antioxidant-containing medium is dependent upon the desired end use of the resulting composition.

As a general guideline, a weight ratio (dry basis) of antioxidant to yeast can advantageously be from about 0.15 to about 1.2 and preferably not more than about

0.8. Weight ratios greater than 1.2 can also be used but with less efficiency. It should be noted that the yeast and antioxidants can be combined in a variety of ways to suit the situation. For instance, during the yeast production, the antioxidant can be added to the live yeast cream, which can be pasteurized and spray-dried to incorporate the antioxidant onto the yeast product. As an alternative method, spray-dried inactive yeast can be slurried in a suitable solvent with the antioxidant(s) and thereafter dried to deposit the antioxidant(s) onto the yeast. In each case, however, the final yeast-antioxidant composition is inactive.

Example: Sugar cookies were prepared which contained varying amounts of two commonly used antioxidants, propyl gallate (PG) and butylated hydroxyanisol (BHA) (Control). Similar cookes were prepared which contained the same initial amounts of the antioxidants, but which antioxidants had first been incorporated onto an inactive yeast substrate (Yeast Test Product) according to this process. After baking, the cookies were analyzed for the presence of the specific antioxidants in question and the Controls were compared with the Yeast Test Products.

In order to prepare the Yeast Test Product, the desired amount of the antioxidant was dissolved at room temperature in 38 ml of ethanol. To this solution was added 100 g of inactive, spray-dried *Candida utilis* yeast. The resulting slurry was subjected to a vacuum at room temperature to evaporate the ethanol.

The weight ratio of the antioxidant to the yeast is not critical and is only determined by the amounts of each which are desired in the final food application. Generally, yeast levels do not exceed about 5 weight percent due to the yeasty flavor imparted by higher levels. Likewise, the levels of antioxidants permissible by FDA regulations limit the amount of antioxidant allowed in the final product.

The control cookies for the comparative tests were prepared using the following ingredients in percent by weight: 26.83 cake flour, 26.83 bread flour, 24.40 shortening, 11.20 sugar, 2.20 whole dry egg, 2.10 corn syrup, 0.69 vanilla, 0.35 salt and 5.40 water. The test cookie recipe was the same except that only 0.59% whole dry egg was used plus 1.61% of yeast test product.

All of the ingredients, except the flour, were blended into a smooth paste. The flour was separately sifted and added to the paste mixture. The entire mixture was folded until smooth and refrigerated until the resulting dough became firm. The dough was then rolled out in a sheet about ¼ inch thick. The cookies were cut out and baked in an oven at 385°F for about 10 minutes or until they were golden brown. The cookies were than analyzed for the specific antioxidant used in the formulation. (The antioxidants in the Control samples were added with the dried egg when making the dough.)

Control cookies containing 50 ppm of PG had no retention of antioxidant after baking. Cookies prepared with the stabilized antioxidant of this process (yeast plus 50 ppm of PG) showed 12 ppm of PG after baking.

These results clearly show the advantage of depositing the antioxidant on a yeast substrate prior to subjecting the antioxidant to high temperature. Similar results can be obtained with the other antioxidants previously mentioned. Such antioxidant-on-yeast compositions can be used in any foodstuff and are not restricted to cookies or baked foods.

Coffee Extract for Stabilizing Coffee Oils

The process by *M. Hamell, R.J. Sims and J.R. Feldman; U.S. Patent 4,156,031; May 22, 1979; assigned to General Foods Corporation* is directed to contacting purified coffee oil which comprises primarily triglycerides and which is essentially free of diterpene esters, with the natural coffee antioxidants contained in soluble coffee solids. Coffee oil may be purified as a result of refining techniques, such as high vacuum distillation, steam vacuum distillation or acid treatment, or chromatographic separation.

Contact between the purified coffee oil and the antioxidants may be effected by directly contacting the oil with soluble coffee solids such as those contained in an aqueous coffee extract. The oil may then be separated from the soluble coffee solids by suitable means such as centrifugation. Alternatively, the natural antioxidants could be first extracted from soluble coffee solids by an organic solvent such as methanol, and then extracted from the organic solvent by the purified oil. This contacting step stabilizes the purified oil to the development of oxidative rancidity by extracting into the oil at least some of the antioxidants naturally found in soluble coffee solids. These antioxidants replace those removed from the oil during the purification process and are not considered to be foreign or synthetic additives.

Crude coffee oil either expressed or solvent extracted from roasted coffee material, such as whole roasted coffee beans or spent coffee grounds, is known to contain significant amounts of nonglyceride materials most notably diterpene esters such as esters of cafestol and kahweol. Methods useful for the purification of crude roasted coffee oil to obtain a material comprised primarily of triglycerides and substantially free of diterpene esters are treatment with strong mineral acids, chromatographic separation, steam vacuum distillation and high vacuum distillation. Various other methods may also be suitable.

Example 1: Whole roasted coffee beans were expressed in a screw or auger type of press at a pressure of at least 5,000 psi to obtain crude coffee oil. The oil recovered had a temperature of about 100°C and the coffee meal residue had a temperature of between 75° and 150°C. The oil was then clarified to remove fines and foots in the oil to less than 0.5%. The volatile aromatic constituents of the expressed oil were then distilled by evaporation from a rapidly moving film of the oil formed on a moving surface at temperatures of about 50°C and a pressure of below 25 mm of mercury. The aromatics were collected as a frost in a liquid nitrogen cold trap (-196°C).

About 2,961 g of the dearomatized coffee oil was placed in a beaker and 148 g (5% by weight) of H_2SO_4 (98% concentration) was added with stirring over a 5-minute period. There was an exotherm to 45°C and a black sludge separated. Stirring was continued for 60 minutes until the oil temperature dropped to 30°C. Then the mixture was diluted with an equal volume of petroleum ether (BP 30°-60°C) and centrifuged for 15 minutes at 5,000 rpm. The ether solution was decanted, the sludge transferred to a beaker, and reslurried with 1 ℓ of petroleum ether. This mixture was then centrifuged and the combined ether solutions were filtered with suction through a diatomaceous earth filter. Solvent was distilled off and the residual oil was stirred for 30 minutes at 90°C with 95 g of bleaching clay. The oil was then steam-vacuum deodorized for 4 hr at 210°C and 0.5 mm pressure. The oil was cooled to 60°C under vacuum with continuous steam stripping before air was readmitted. The yield of purified bland, odorless oil was 1,480 g, or 50% recovery.

Absence of the diterpenoids cafestol and kahweol was shown by analysis of the purified coffee oil via gas-liquid chromatography, thin-layer chromatography, and nuclear magnetic resonance. By all these methods, no diterpenes were found in the H_2SO_4 purified oil.

Example 2: Eighty grams of the purified coffee oil of Example 1 was treated in a flask with 25 g of spray-dried coffee solids dissolved in 75 ml of water. A control example with only the oil and water was also used.

Both flasks were flushed with CO_2, stoppered and agitated with a mechanical shaker overnight. Then the contents were centrifuged to separate the phases. The oil layers were decanted and tested for peroxide values with and without 0.02% BHT addition for stability using the Schaal oven test (60°C).

The control oil (with or without BHT) shows a rapid increase in peroxide value and develops a rancid odor within a few days. The extract oil retains its bland odor even after several days in this accelerated test. BHT gives virtually no additional protection over that obtained by the extraction.

Example 3: Portions of the control oil and the extract oil of Example 2 (both without BHT) were combined with grinder gas aromatics and then injected into glass jars containing spray-dried soluble coffee powder, at a level of 0.4% oil by weight of powder. The glass jars were then sealed under an inert atmosphere. After 8 weeks storage at room temperature and 6 weeks at 95°F, the control oil plated powder was detected as being rancid; whereas, the extract oil plated powder continues to remain stable after 20 weeks at both room temperature and 95°F.

2',6'-Dihydroxy-9-(2,5-Dihydroxyphenyl)Octophenone from Mace

Y. Saito, Y. Kimura, T. Sakamoto, M. Shinbo and S. Kameyama; U.S. Patent 4,195,101; March 25, 1980; assigned to The Lion Dentrifice Co. Ltd., Japan have provided antioxidants for foodstuffs which compare favorably with synthetic antioxidants in antioxidative activities and which are free from the problems of safety and taste.

The compound proposed is 2',6'-dihydroxy-9-(2,5-dihydroxyphenyl)octylphenone, which serves as a powerful antioxidant for foodstuffs, and has not hitherto been known or described in literature.

It has been found that the extraction and separation of mace, or Myristica fragrans Hautt, which is a known spice, successively with petroleum ether, diethylether, n-hexane and carbon tetrachloride, followed by column chromatographic separation, can produce the compound 2',6'-dihydroxy-9-(2,5-dihydroxyphenyl)-octylphenone which is very effective as an antioxidant in foodstuffs, such as lard or the like, exhibiting much higher antioxidative activities than the conventional antioxidant BHA.

The compound may be obtained by the following method. Mace, a dried arillode of nutmeg, is subjected to extraction with petroleum ether to leave a residue, which is further subjected to extraction with diethylether. The diethylether is removed by distillation from the extract, while the residue, after addition to carbon tetrachloride, is heated under reflux and centrifugally separated to give an insoluble matter. The insoluble matter is then subjected to column chromatographic separation with silica gel as the adsorbent and a mixed chloroform-acetone (9:1) solvent and acetone as the eluants. The eluted fractions which exhibit antioxidative activities, are subjected again to column chromatographic separation with silica gel as the adsorbent and a mixed chloroform-acetone-n-hexane (8:1:1) solvent as the eluant to produce fractions exhibiting antioxidative activities. The fractions thus produced are mixed with acetone and n-hexane and subjected to extraction. The resulting solution in n-hexane and acetone is chilled and the precipitated crystalline material is recrystallized from benzene, resulting in obtaining the compound expressed by the structural formula below.

$$\text{(structure): two phenyl rings each bearing two OH groups connected by } -CO-(CH_2)_8-$$

The solvents to be used in the extraction and washing of the compound are not limited to those named above. For example, those in which the compound is readily soluble, such as ethyl acetate and ethyl alcohol, may be used for extraction and those in which the compound is insoluble or hardly soluble, such as ligroin and chloroform may be used for washing.

Alternatively, the compound can be separated from mace by virtue of differences in its solubility in some solvents depending on temperature. That is to say, for example, 1,2-dichloroethane, toluene or benzene in which the compound is insoluble or hardly soluble at room temperature is used for the extraction of the compound direct from mace in heated conditions, the extract being chilled to precipitate the compound and the precipitate being purified by chromatography or other means.

The compound is useful as an antioxidant in order to prevent foodstuffs from oxidative denaturation. The antioxidative activity of the compound is so strong that the oxidative denaturation of lard can be effectively prevented by adding as little as 0.005% by weight of the compound, to the same extent as by adding 0.02% by weight of BHA. When the amount of the compound added is increased to 0.02% by weight, it brings about antioxidative effect approximately twice as high as the same amount of BHA. This is indicative that sufficient antioxidative effects can be obtained by use of the compound of this process in a very small amount.

Different from most of the synthetic antioxidants, this compound which is obtained by extraction and separation from a natural product, has no problem in safety with an LD_{50} value of more than 2,000 mg/kg of mouse by oral administration. The LD_{50} value of the compound is about the same as in BHT, and much better than 1,100 mg/kg of mouse in BHA.

The effective amount of the compound is usually between 0.001 and 0.1% by weight in lard and other foodstuffs. However, any larger amount brings about no problems since the compound is tasteless, odorless, almost colorless and non-irritative. The compound may also be useful as an additive to various organic materials, including plastics, rubbers, pharmaceuticals, cosmetics, paper and the like.

Example: The compound was tested for the rate of oxygen absorption in a corn salad oil to which 0.02% by weight of the compound had been added. For the purpose, 2 g of the corn salad oil sample was taken in the receptacle of Warburg's manometer with thermostat at 60°±0.1°C, which was shaken with 110 vibrations per minute in a vibration amplitude of 60 mm. The rate of oxygen absorption was determined by reading the manometer. In this testing, the receptacle was totally wrapped with an aluminum foil to prevent the photooxidation of the oil, and mercury was used as the manometer fluid in place of Brodie's solution in order to minimize any possible influence caused by fluctuations in the atmospheric pressure.

The oxygen adsorption in $\mu\ell/g$ for this compound was 84.4 at 85 hours, 238.0 at 263 hours and 375.3 at 343 hours. For BHA, the readings at the same time periods were 57.6, 478.9 and 1124.2 respectively. For no additive, the readings were 52.9, 517.5 and 1164.3 respectively.

Antioxidants from Tempeh

It is known that antioxidant properties are possessed by tempeh, a fermented soybean product, obtained by fermenting soybeans with a fungus, either *Rhizopus oligosporus* or *Rhizopus oryzae* and food products containing tempeh, such as fish or fatty meat food products, exhibit improved stability. It has further been found that by extracting tempeh with a mixture of hexane and ethanol, a component tempeh, namely oil of tempeh, can be recovered. Oil of tempeh demonstrates improved antioxidant properties over those of unextracted tempeh. Although the aroma of tempeh and oil of tempeh is an essentially mild and pleasant one, the use of these materials is limited to situations where their basic flavors are desired or at least tolerable.

F.W. Zilliken; U.S. Patents 4,232,122; November 4, 1980; 4,218,489; August 19, 1980 and 4,157,984; June 12, 1979; all assigned to Z-L Limited Partnership has isolated an ergostadientriol and two new isoflavones from the fermented soybean product tempeh. These compounds possess antioxidative properties either alone or in combination.

The isolated ergostadientriol is "Emmerie Engel" positive at the same order of magnitude as vitamin E, i.e., it reduced Fe^{3+} to Fe^{2+} at room temperature, the latter forming a brilliant red complex in the presence of α,α-dipyridil. This ergostadientriol has been identified by UV, IR, and high resolution mass spectrometry to have the formula $C_{28}H_{46}O_3$ and molecular weight 430, with double bonds in the 6-7 and 22-23 positions and hydroxy groups on the 3, 6 and 7 carbon atoms.

It is produced by fermentation of soyban with a fungus, either *Rhizopus oligosporus* or *Rhizopus oryzae.* The ergostadientriol has been produced and recovered from the fungus itself after growth on a suitable culture medium. Suitable fungi for producing this comound are *Rhizopus oligosporus* ATCC No. 22959 and *Rhizopus oryzae* ATCC No. 9363.

Dry, i.e., lyophilized, tempeh powder or fungus is contacted with a 60-70% aqueous methanol solution for an extended period, e.g., overnight, at a relatively low temperature, e.g., about 4°C, thereby extracting methanol-soluble components, including the ergostadientriol from the tempeh powder. The methanol extract solution, after removal of insoluble material, is evaporated to dryness in vacuo at an elevated temperature, e.g., a temperature in the range 40°-60°F, and a solid residue is produced. This solid residue is redissolved in dry methanol. The insoluble portion of this residue is separated from the soluble components by centrifugation, after which the supernatant is extracted with hexane several times, e.g., 2-3X, in order to remove any remaining traces of hexane-soluble impurities, e.g., lipids. Next, the methanol supernatant is evaporated to reduce its volume to a minimal fraction, e.g., 20 ml, and kept at a low temperature, e.g., -20°C, for a brief time, e.g., a time in the range 15-20 minutes. This results in the formation of additional precipitate which is removed. The then-remaining supernatant is subjected to molecular sieve chromatography, e.g., chromatography on Sephadex LH20 using a suitable size column e.g. 2 x 40 cm, and a suitable mobile phase, e.g., n-propanol/ethylacetate/H_2O = 5:5:1.

One of the fractions resulting from this chromatographic separation is fluorescent with emission in the blue range of the visible spectrum. This blue fluorescent fraction is next subjected to adsorption chromatography on a suitable matrix, e.g., silica gel, using an appropriate mobile phase, e.g., ethylacetate/propanol/H_2O = 95:2:3. The resulting blue fluorescent fraction is then subjected to rechromatography on an adsorptive matrix, e.g., silica gel again, employing a different mobile phase, e.g., cyclohexane/dichloromethane/ethyl formate = 35:30:30. The resulting blue fluorescent fraction is next subjected to thin layer chromatography on silica gel using a suitable, mobile phase, e.g., cyclohexane/dichloromethane/ethyl formate/formic acid = 35:30:30:5. The ergostadientriol is recovered in essentially pure form utilizing its differential mobility on the silica gel plate as compared with other components of the blue fluorescent fraction.

In a particularly preferred embodiment, the ergostadientriol is obtained in pure form by preparative high pressure liquid gas chromatography.

In accordance with another embodiment of this process two new isoflavones have been isolated. In admixture, these two new isoflavones provide an antioxidative composition. Mixtures containing the ergostadientriol described above, these two isoflavones and other known isoflavones provide exceptionally effective antioxidative compositions.

One of the two isoflavones isolated is 6-methoxy-7,4'-dihydroxyisoflavone, which has a melting point of 222°C, a molecular weight of 284, emits a bright blue fluorescence after NH_3 treatment when excited with UV-light.

The other isoflavone, emits a bright green fluorescence after NH_3 treatment when excited with UV-light and is an isoflavone having hydroxy or methoxy groups on positions 6, 7 and 4'.

These isoflavones are separately recovered from tempeh by the same process described previously to recover the ergostadientriol. The only difference is in the last step of the recovery process, namely, thin layer chromatography on silica gel. The isoflavones are separated and recovered from each other and from other components of the blue fluorescent fraction on the basis of their unique mobilities.

A mixture of these two isoflavones possesses antioxidative properties and can be used as an antioxidant composition in various applications. In a standard test assay involving the oxidation of lard by exposure to air at 60°C for 72 hr, addition of a mixture of the two isoflavones at a concentration of 0.1% by weight of the mixture results in about 50% protection of the lard against oxidation. Such a mixture can be prepared by mixing the isoflavones after they have been separately recovered according to the procedure described above. Although approximately equimolar mixtures are most effective, mixtures containing as much as 5X the amount of one new isoflavone in comparison with another are still effective antioxidants.

Antioxidant compositions may also be prepared by including two previously known isoflavones the so-called "Murata" compound (6,7,4'-trihydroxy-isoflavone), and Genistein (5,6,4'-trihydroxy-isoflavone) and/or Daidzein (7,4'-dihydroxyisoflavone).

These known isoflavones can be recovered from tempeh either in accordance with known methods or as side products of the procedure set forth hereinabove for recovery of the sterol and the new isoflavones. These known compounds are recovered from the silica gel plate used in the procedure based upon their known mobilities.

Additionally, an antioxidant composition comprising the ergostadientriol and a mixture of the new isoflavones and the three known isoflavones can be recovered from tempeh by the same procedure used for the recovery of the sterol and the isoflavones except that the final blue fluorescent fraction obtained is not subjected to any further treatment. This fraction is comprised of the sterol and the mixture of the five isoflavones.

Alternatively, this particular antioxidant composition can be recovered by contacting tempeh powder with petroleum ether at an elevated temperature, e.g., a temperature in the range 50°-60°C, for a short time, e.g., about 1 hr. The resulting petroleum ether extract containing components of tempeh soluble in petroleum ether is discarded and the residue containing insoluble components is further extracted by adding ether. Insoluble material is discarded, and the extract solution containing the antioxidant composition is evaporated to dryness.

The resulting residue is washed with petroleum ether several times, e.g., three times, to remove any remaining petroleum ether-soluble material, and the remaining residue is dissolved in a mixture of $CHCl_3$:methanol containing equal portions of each. This solution containing the antioxidant composition is subjected to adsorption chromatography on silica gel using a suitable mobile phase, e.g., ether and chloroform respectively, and the antioxidant composition is recovered from the column eluent.

An antioxidant composition comprising the two new isoflavones confers improved stability. An antioxidant composition comprising the ergostadientriol, the two new isoflavones, and the Murata compound confers exceptional stability, as does one which additionally includes Genistein and/or Daidzein.

Such stabilized food compositions can be prepared by addition to food products, such as fish, fatty meat or derivatives thereof, of an effective amount of an antioxidant composition which includes one or more of the new compounds, disclosed herein. Specifically, the antioxidant composition comprising the two new isoflavones confers stability when added in an amount in the range 0.01-1.0% by weight. Antioxidant compositions comprising the sterol and mixtures of isoflavones as described above confer stability when added in an amount in the range 0.005-0.5% by weight, more or less.

Preparation of Isoflavones and Related Compounds

F.W. Zilliken; U.S. Patent 4,264,509; April 28, 1981; assigned to Z-L Limited Partnership has synthesized many of the antioxidant isoflavones isolated from tempeh. In addition other synthetic derivatives of the isoflavones have been prepared and found to be active antioxidants.

Example 1: *Synthesis of 6,7-dihydroxy-4'-methoxyisoflavone [6,7-dihydroxy-3-(4-methoxyphenyl)chromone]* – 35.4 grams of 1,2,4-trihydroxybenzene were suspended in 200 ml of dry ethyl ether containing 30 g of dry zinc-chloride (0.22 mol) and 50 g (0.34 mol) of p-methoxyphenylacetonitrile. The suspension was then exposed for 4 hr at 0°C to a gentle stream of dry hydrogen chloride (HCl), the gas bubbling through the suspension under continuous stirring. Then the reaction mixture was kept for 70 hr at 4°C and thereafter the supernatant was decanted from the heavy oil which had separated. The oil was washed twice with ethyl ether, then one liter of water and a few ml of concentrated hydrochloric acid were added and the mixture boiled for 1 hr under reflux. After cooling to room temperature the precipitate was collected by filtration and recrystallized from ethanol/water. This precipitate was (4-methoxybenzyl)-2,4,5-trihydroxyphenyl ketone.

Four grams (14.6 mmol) of (4-methoxybenzyl)-2,4,5-trihydroxyphenyl ketone were dissolved in 50 ml of dry dimethylformamide. To this solution was added 7.5 g of borontrifluoride-methyletherate, $BF_3 \cdot (CH_3)_2O$, dropwise. Under spontaneous elevation of the temperature the color of the solution turns to yellowish-green. Then the temperature of the reaction is asjusted to 50°C and a solution of 5 g methanesulfonyl chloride (CH_3SO_2Cl) in 25 ml of dry dimethylformamide (DMF) is added dropwise. Thereafter, the solution is heated for 90 minutes at 90°-100°C. After cooling to room temperature, the reaction mixture is poured into 500 ml water and the resulting yellow precipitate is filtered off. After drying in a desiccator the crude product is purified by boiling in 50 ml methanol and then 50 ml of ethyl ether. The resulting white powder can be recrystallized from dioxane or glacial acetic acid. This product is 6,7-dihydroxy-3-(4-methoxyphenyl)-chromone (texasin).

Example 2: *Synthesis of 6,7-dihydroxy-3-(4-methoxyphenyl)chromanon-4* – Six grams of texasin [6,7-dihydroxy-3-(4-methoxyphenyl)chromone] were dissolved and partially suspended in 500 ml of ethanol and hydrogenated at normal pressure and room temperature using 10% palladium/charcoal as a catalyst under addition of 6 drops of triethylamine. The catalytic hydrogenation is continued until no starting material is detectable by means of thin layer chromatography. Thereafter, the catalyst is removed by filtration, an equal amount of water is added to the reaction mixture and the solution is evaporated under reduced pressure.

After removal of the largest part of the ethanol the product precipitates out in pure form. After filtration and drying in a desiccator 5.5 g of a light yellow powder is obtained which may be recrystallized once from ethanol/water. The product has the formula $C_{16}H_{14}O_5$, molecular weight 286, and melting point 215°C. It was characterized by UV, NMR, IR and Mass spectroscopy and determined to be 6,7-dihydroxy-3-(4-methoxyphenyl)-chromanon-4. The product is actually a mixture of optical isomers but since the isomers exhibit the same properties further characterization was not deemed necessary.

Example 3: *Synthesis of 6,7-dihydroxy-3-(4-methoxyphenyl)chroman* – Fifteen grams of texasin [6,7-dihydroxy-3-(methoxyphenyl)chromone] were dissolved and partially suspended in 500 ml of ethanol and hydrogenated at normal pressure and room temperature using 10% palladium/charcoal as a catalyst under addition of 20 drops of concentrated sulfuric acid. The catalytic hydrogenation is continued until neither texasin nor any of the 6,7-dihydroxy-3-(4-methoxyphenyl)-chromanon-4 was detected by thin layer chromatography. The further purification was identical to that described in Example 2. The product has the formula $C_{16}H_{16}O_4$, molecular weight 272, and melting point 160°C. It was characterized by UV, NMR, IR and Mass spectroscopy and determined to be 6,7-dihydroxy-3-(4-methoxyphenyl)chroman. As in Example 2, the product is actually a mixture of optical isomers.

6,7-Dihydroxyisoflavone and its chromanon-4 and chroman hydrogenation products are also prepared in other examples.

Example 4: The compounds 6,7-dihydroxy-3-(4-methoxyphenyl)chromanon-(4) and 6,7-dihydroxy-3-(4-methoxyphenyl)chroman were evaluated as antioxidants using an automated version of the Swift stability test at 100°C.

Tests were also run with BHA and a control with no antioxidant. For the test the lard was heated to 100°C and air bubbled through at 2 ml/min. Every few hours the oil was analyzed for peroxide value using the peroxide value test, AOCS Official Method Cd8-53 (1960). The tests were carried out using two different batches of lard.

According to the results obtained, 6,7-dihydroxy-3-(4-methoxyphenyl)chroman is the most effective of the antioxidants tested. 6,7-dihydroxy-3-(4-methoxyphenyl)chromanon-4 is better than α-tocopherol, but less effective than tert-butylhydroquinone and comparable with BHA and BHT.

ANTISTALING ADDITIVES

Lactose Replacement of Sugar and Fats

S.F. Zenner and D.C. Stanberry; U.S. Patent 4,233,321; November 11, 1980; assigned to Patent Technology, Inc. have prepared white pan bread and rolls with prolonged shelf-life by replacing 10 to 50% of the sugar and fat content of standard dough formulations with lactose.

The process is predicated on the discovery that lactose can be used to replace 10-50% of the sugar content (e.g., sucrose or dextrose) and/or the fat content (e.g., shortening) in standard dough formulations for producing white pan bread and rolls, with unexpected improvements in the tenderness and "freshness" of the resulting bread and roll products as represented by the previously noted significant prolongation of their shelf lives—viz, ranging up to 50-100% longer than normally encountered with products prepared from standard formulations for such products.

When lactose is substituted for the sugar and/or fat content in white bread and roll formulas, the product initially becomes less tender. However, as the lactose level is increased to within the range of the 10 to 50% substitution of the process, the product becomes unexpectedly more tender as reflected by tenderness scores only ½ to ¾ those obtained with the standard formula breads. The result is a significant and unexpected increase in the bread's "freshness" or shelf life over a substantial period of time. Thus if the shelf life of the standard bread is normally two to three days, breads made with approximately 16 to 25% lactose substituted into the formulation will last 4 to 6 days—representing an extraordinary improvement in keeping quality. Substantial improvements in dough qualities, as reflected by improved workability and machining of the dough, and in baking qualities, as represented by significant increases in loaf volumes (10 to 15%) and in external appearance scores (5 to 15%) are also obtained.

An additional benefit is flavor and aroma enhancement, apparently related to an ability of lactose to absorb flavors and aroma and color, and to retard their loss during processing and baking. The foregoing improvements are unexpected, not only because of the lack of sweetness and flavor and the unfermentable character of the lactose, but also because of an inherent lack of plasticity and known volume-depressing characteristic of the lactose ingredient, which would lead a worker in the art away from substituting lactose for either the sugar or the shortening content in the standard formulations.

According to this process, the lactose is incorporated into standard dough formulations for white pan bread or rolls in the manner of any dry subdivided material, such as nonfat milk solids, sugar or salt. Utility and effectiveness of the lactose substitution is indicated for such bread formulations specifying 6 to 10% sugar and 2.5 to 3.5% shortening, and in standard formulations for yeast leavened rolls specifying 10 to 13% sugar and 6 to 8% shortening. The indicated lactose substitution is effective for such formulations as used in the straight dough, sponge-dough, continuous and "no-time" dough procedures.

In general, and assuming a standard formulation of essential dough ingredients including customary proportions of flour, water and leavening, and standard proportions of sugar and shortening as specified above, the improvement comprises replacing from 10 to 50% of the weight of the sugar or shortening (or both) in the dough with lactose, followed by mixing the dough ingredients including the lactose component to form a dough, dividing the dough into units for baking and baking the units to provide bread and roll products having the significant and unexpected characteristics as respects enhanced tenderness or "freshness" and prolonged shelf life, together with the improved characteristics as respects increased volume, external appearance, flavor and aroma.

The following examples are intended to be illustrative of the process with substitution of lactose for 10 to 50% of the sugar and/or shortening content in dough formulations for commercial white pan bread and yeast leavened rolls. In these examples, all concentrations of ingredients are expressed as percent of the flour.

Example 1: In both the standard formulation and the formulation of this process, the white bread contained 100.0 parts flour, 65 parts water, 3.0 parts yeast, 0.5 part yeast food, 0.5 part emulsifier, 2.0 parts salts and 3.0 parts nonfat dry milk. The standard formula contained 6-10 parts sucrose, and 3.0 parts shortening. The formulation of this process uses only 3.0-9.0 parts sucrose, 1.5-2.7 parts shortening and 1.1-6.6 parts lactose. The breads are prepared in the same manner by regular procedures and baked at 400°F for 20 minutes.

Example 2: White pan breads are prepared according to the standard formulation and procedure of Example 1, using 6% sucrose and 3% shortening in the formula. White pan breads are also prepared according to the process employing 4.5% sucrose, 3% lactose, and 2% shortening, and representing a ¼ reduction in sucrose and a ⅓ reduction in shortening. Compressimeter values as to tenderness are recorded over a five day period, as a measure of shelf life or "freshness."

The data show that at one day the tenderness score (11.0) for the lactose-substituted bread is approximately ⅓ less than the score (17.5) for the standard formula. At two days, the value for the lactose-substituted bread is equal to the score for the standard formula bread at one day, representing a shelf life prolongation of 100%. Similar results are indicated for the lactose-substituted breads at three days, and at four days. Thus at four days, the tenderness value for the lactose substituted bread is equal to the value for the standard formula bread at two days, again representing a 100% prolongation of shelf life or freshness.

S.F. Zenner and D.C. Stanberry; U.S. Patent 4,233,330; November 11, 1980; assigned to Patent Technology, Inc. have also produced cakes with prolonged shelf life by replacing sugar and fat with lactose.

Specifically, lactose is used to replace 10 to 35% of the sugar or corn sugar content (sucrose, dextrose) or fat content (shortening) or both, in such standard formulations containing as a norm, 24 to 32% sugar and 8 to 13% fat.

Specifically, it was found that as lactose is substituted for the sugar and/or fat content in commercial yellow, white or devil's food cake mixes, the cake products initially become less tender. However, as the lactose level is increased to within the range of the 10 to 35% substitution, the product becomes unexpectedly more tender as reflected by tenderness scores substantially below those obtained with the standard formula cake mixes. The result is a significant and unexpected increase in the freshness or shelf-life of the cakes over a substantial period of time. For example, if the desired freshness level of a standard commercial cake normally lasts two to three days, the same cakes made with approximately 10 to 15% lactose substituted into the formulation will last 5 to 9 days—representing an extraordinary improvement in keeping quality. Substantial improvements in dough or batter qualities, as reflected by improved workability of the dough, and in baking qualities, as represented by significant increases in cake volumes (10 to 15%) and an internal appearance scores (5 to 15%), are also obtained.

An additional benefit is flavor and aroma enhancement, apparently related to an ability of lactose to absorb flavors and aroma and color, and to retard the loss of these values during processing and baking. The foregoing improvements are unexpected, not only because of the lack of sweetness and flavor of the lactose, but also because of the inherent lack of plasticity and known volume-depressing characteristics of the lactose ingredient, which would lead a worker in the art away from substituting lactose for either the sugar or the shortening content in a standard cake formulation.

The lactose is incorporated into standard cake mixes for yellow, white or devil's food cake in the manner of any dry subdivided material, such as nonfat milk solids, sugar or salt. Utility and effectiveness of the lactose substitution is indicated for such cake formulations specifying between 24 and 32% sugar and 8 to 13% shortening. The indicated lactose substitution is effective for such formulations as used in cake processing based on the old creaming method, the more recent three-stage and two-stage blending methods, and in the continuous cake mixing procedure. In general, and assuming a standard formulation of essential cake mix ingredients, including customary proportions of flour, liquid (water, milk, eggs) and leavening, and standard proportions of sugar and shortening as specified above, the improvement of the process comprises replacing from 10 to 35% of the weight of the sugar or shortening (or both) in the cake mix with lactose, followed by mixing or blending the ingredients including the lactose component to form a batter or dough, forming the dough into a unit (or units) for baking, and baking the same to provide cake products having the significant, unexpected characteristics as respects enhanced tenderness or freshness and prolonged shelf-life, as indicated herein, together with the improved characteristics as respects increased volume, external appearance, flavor and aroma.

In the following examples, all concentrations of ingredients are expressed as a percent of the total formula by weight.

Example 1: The yellow cake mixes used for both the standard formula and the formula for this process include 26.0 parts cake flour, 33.6 parts liquid (water and whole eggs), 2.6 parts nonfat dry milk, 1.3 parts baking powder, and salt to taste. The standard formula included 28.5 parts sucrose, and 8.0 parts shortening. The cake formula of this process contains 18.5-25.6 parts sucrose, 5.2-7.2 parts shortening and 0.8-10.0 parts lactose. The cakes were prepared by standard methods and baked at 375°F for 26 minutes.

Example 2: A series of yellow cakes are prepared according to the standard cake mix formula of Example 1, using 28% sucrose and 9% shortening in the formula. Yellow cakes are also prepared according to this process, employing 24% sucrose, 4% lactose, and 6% shortening, representing an approximate 15% reduction in sucrose and a 33% reduction in shortening. Compressimeter values as to cake tenderness are recorded over a five day period, as a measure of shelf life or freshness.

The data show that at 1 day the tenderness score (7.0) for the lactose substituted cake is approximately $\frac{1}{3}$ less than the score (10.5) for the standard formula cake. At five days, the value for the lactose-substituted cake is equal to the score for the standard mix cake at one day, representing a shelf-life prolongation of 400%. Similar results are indicated for the lactose-substituted cakes at intervening time periods. Thus at each of two, three and four days, the tenderness value for the lactose-substituted cake is substantially less than the value for the standard formula cake at one day, representing a prolongation of shelf life or freshness within the range of at least 100% and ranging up to 400%.

Preservative Sugars in Sweet Baked Goods

It has been discovered that there is a significant relationship between the sweetener composition of sweet baked goods and staling. The process used by *H.F. Zobel and J.L. Maxwell; U.S. Patent 4,291,065; September 22, 1981; assigned to CPC International Inc.* thus provides a means for controlling staling through the use of certain selected sugars as at least part of that composition.

More particularly, it has been discovered that all sugars contribute significant, but different effects to the staling of sweet baked goods. It is therefore possible to reduce staling by employing a sweetener material comprising certain preservative sugars having higher antistaling activities than the sugars heretofore utilized.

The food products of this process need contain only starch and water in addition to sweetener material sugar or sugars. The starch and water are generally present in a bound or gel-like form as a result of baking. These essential components should be finely divided (i.e. not present in individually identifiable aggregates) and lie in essentially homogeneous dispersion in the food product.

The sweetener material of the process generally constitutes between about 15 and 40% of the total weight of the baked goods products. This sweetener material is one which imbues the product with a resistance to staling superior to products heretofore known.

Sucrose has been selected as the sweetener control (or reference for comparison) because of its own relatively high antistaling activity and because it is by far the most common sugar employed in making the baked food products. Sweetener compositions consisting only of sucrose are also believed to be the best antistaling sweeteners (although not known to be such) heretofore employed in baked goods products. In addition, and because it is known that staling is highly temperature dependent, these activities are referenced to 21°C in order to approximate common storage conditions and allow comparability.

The selection of suitable preservative agent sugars having high unit antistaling activities may be made on a simple trial and error basis through substitution into the desired, sweet baked goods product. Measurement of the staling properties of the resultant product over a time of, for example, 3 to 10 days will then reveal the manner in which staling has been affected. These results may then readily be compared to those of an otherwise identical product which contains an equal sweetener weight of sucrose.

An exemplary test model which has proven particularly useful may be utilized and analyzed as follows.

Right cylindrical wheat starch gels (diameter 2.3 cm, height 3.0 cm) may be made by packing an admixture of moistened starch and sweetener into aluminum molds. After heating the packed molds for 2 hr at 100°C, they may be quenched in ice water to cool the gels to 21°C, at which temperature they may be stored until they are analyzed.

For analysis, the cell halves are separated to free the gel cylinder. Rigidity measurements may then be made with an Instron Tensile Tester operating in the compression mode. Compression is desirably limited to about 10% (to avoid exceeding the elastic modulus of the gel and thus permit reuse of the gel specimen after further aging). This analysis will result in a series of data points over an aging period and reflect the progress of staling in the gel.

Mathematical treatment of the data obtained by the above test method will produce two key values, A, the final rigidity of the gel and T the time constant in hours.

In such instance, it has been found that the various sweetener material sugars may be ranked for unit antistaling activity on the basis of decreasing "A" and increasing "T." Thus, for example, sugars having lower A and higher T values than sucrose will provide superior staling-resistant food products.

Preservative agent sugars which are capable of providing higher antistaling activities in baked goods products than equal weights of sucrose most generally are oligosaccharides of from two to three hexose or pentose units. This is not, however, a complete listing and representative sugars include maltose, maltotriose, and xylose. Such sugars of high unit activity preferably constitute between 30 and 100%, most preferably 60-100%, by weight of the total sweetener material employed in the product sweet baked goods.

Of the foregoing high unit antistaling activity sugars, maltose and maltotriose are particularly preferred. Both are relatively inexpensively produced, particularly in corn syrup compositions. Most preferably, the baked food products contain a sweetener material composed of 30 to 100% by weight of at least one of these two sugars.

Example: A gel aging/staling study was performed utilizing various individual sweetener sugars in the Test Model System described above. Wheat starch was utilized and the starch-water-sweetener proportions of the product gel were 1:1:1 by weight.

The initial rigidities of the gels is essentially constant, even in the absence of any sweetener material. The final A and T obtained for the sweeteners used are given in the following format: sweetener (A/T), fructose (19/40), galactose (16.7/38), mannose (16.1/51), dextrose (16/57), isomaltose (12.6/49), sucrose (12.4/63), maltose (9.9/68), xylose (6.9/97), maltotriose (11.4/125) and control, no sweeteners (34.3/64).

All sugars showed substantial improvement in final gel rigidity (i.e. lower A values) and in the time constant (i.e., increased T values) as compared to a gel containing no sweetener. Maltotriose, maltose and xylose far outperformed the sucrose control by providing retarded staling and decreased final product staling.

Stabilized Alpha-Amylase Enzyme

Since bread staling or firming is believed to be caused by crystallization of gelatinized starch, it would appear that these textural changes could be modified or inhibited by using starch hydrolyzing enzymes to fragment the starch polymers, thus reducing the degree of interaction or firming on storage. Attempts to use commercially available enzyme preparations, such as heat stable bacterial alpha-amylase enzyme have not met with success because the degree of enzyme action has been too difficult to control.

A method has been described by *V.A. De Stefanis and E.W. Turner; U.S. Patent 4,299,848; November 10, 1981; assigned to International Telephone and Telegraph Corporation* to inactivate the proteolytic enzyme(s) contained in commercial heat stable bacterial alpha-amylase under conditions which retain full alpha-amylase activity and to use the thus purified alpha-amylase enzyme plus surfactants that are approved for use in bread to inhibit firming and improve the keeping of bread and other bakery products.

This process provides for the inactivation of the proteolytic enzyme(s), which are present in commercially available heat stable bacterial alpha-amylase enzyme preparations obtained from extracts of *Bacillus subtilis, Bacillus sterothermophilis* or other microbial sources. The proteolytic enzyme(s) are inactivated by heating the commercial enzyme preparations in buffer solutions under controlled conditions of time, temperature, pH, and specific ion concentrations. The proteolytic enzyme(s) are inactivated, and alpha-amylase activity is retained by the procedures used.

Example 1: 1.750 grams of commercial bacterial α-amylase, Rhozyme H-39 was diluted to 100 ml with water. Two ml of enzyme (35 mg) was mixed with 10 ml of 0-0.1 M sodium acetate buffer solution adjusted to a pH of 6.00-6.20 using acetic acid. 0.5 M sodium chloride solution was added to the above mixture and heated at 70°C for 30 minutes. The protease activity was reduced to a nondetectable level.

Conditions for the inactivation of all protease activity and the commercial bacterial enzyme preparation is noted above. Studies are conducted to determine whether the alpha-amylase activity is also affected by these conditions. The results show that thermal treatment at 70°C for 30 minutes reduces the activity of both the protease to 0 and the amylase activity to about 10% in the absence of sodium chloride. The alpha-amylase activity is not affected by the thermal treatment when about 0.5 M sodium chloride is present in the enzyme solution. The presence of the sodium chloride during the thermal treatment has a stabilizing effect on the alpha-amylase activity and slightly enhances the heat inactivation of the protease enzyme.

Since calcium sulfate is known to improve the stability of the amylase at 75°C, experiments were conducted to determine the stability as a function of concentration. The results obtained from this experiment show that maximum stability is obtained when 0.0003 to 0.0012 M calcium sulfate was used. Above 0.0012 M calcium sulfate exerts an inhibitory effect on the amylase activity. Therefore, 0.0006 M calcium sulfate appears to be a preferred concentration.

A study was conducted to determine the effect of calcium sulfate and wheat starch on the alpha-amylase activity during heating at 70° and 80°C. 0.500 g of HT concentrate, 4.7 g of sodium chloride, 0.0006 M calcium sulfate and 1.0, 3.0 and 5.0 g of wheat starch were diluted to 100 ml. The enzyme solutions were heated at either 75° or 80°C for 30 minutes.

Maximum enzyme stability was obtained when a commercial enzyme was heated at 75°C together with 0.8 M sodium chloride plus 0.0006 M calcium sulfate plus 1% wheat starch. At 80°C the alpha-amylase activity was slightly reduced when treated in the presence of these 3 components. Increased wheat starch concentrations do not improve enzyme stability.

In summary, the developed method of purification involves the following conditions:

(1)   a. 0.8 M NaCl
      b. 0.0006 M CaSO$_4$·2H$_2$O
      c. 1-3% wheat starch   } adjusted to pH 6.50
      d. enzyme

(2)   Bring the solution to 75°C (without enzyme and starch).

(3)   At 75°C, add the enzyme and wheat starch.

(4)   Hold for 15 to 30 min., depending on the enzyme con-

centration used as well as the initial protease level in the commercial amylase.

(5) Use the same procedure for all commercial preparations. Should the protease be still present after 30 min. at 75°C, then increase the wheat starch and time at 75°C, while keeping the othe conditions constant.

Baking studies were conducted to establish the level of purified bacterial alpha-amylase in combination with 0.5% calcium stearate as emulsifier required to obtain the desired antistaling effect in bread.

Example 2: Breads were baked using 300 DU of the commercial alpha-amylase, Rhozyme H-39, and 0.5% calcium stearate per 800 g of flour (14% MB), and the same enzyme source, same level of alpha-amylase (300 DU) after the Rhozyme H-39 was modified by heat treatment to inactivate the proteolytic enzymes. The results obtained using commercial vs purified alpha-amylase show that the bread made using commercial alpha-amylase has a low loaf volume, open cell structure and a sticky, gummy bread crumb while the bread using purified alpha-amylase had a satisfactory loaf volume and cell structure and the bread crumb was not as gummy as the bread made using the commercial alpha-amylase.

The changing rate of bread firming during a 1 to 6 day storage period using different levels of purified alpha-amylase in combination with 0.5% calcium stearate was tested. The data obtained show that the rate of bread firming was remarkably decreased when using a combination of 54-72 DU of enzyme and 0.5% calcium stearate. When using this combination, bread has as fresh a crumb texture at the end of 6 days as bread made with 0.5% surfactant after 3 days. The combination of the purified bacterial alpha-amylase and calcium stearate is an effective method for control of bread firming. The surfactant system alone promotes greater initial bread softness, however, the rate of staling was the same as bread made without surfactants.

The method described above, using combination of purified heat stable bacterial alpha-amylase enzyme plus calcium stearate as a surfactant is the first effective method to be developed for the control of staling in bread and other baked products.

Stabilized Sugar-Alpha-Amylase

M.S. Cole; U.S. Patent 4,320,151; March 16, 1982 has discovered that the thermal stability of fungal alpha-amylase is substantially increased by dispersing aqueous solutions of the enzyme in concentrated sugar solutions. The syrup-protected fungal alpha-amylase enzyme survives incorporation in a dough and remains active until a temperature is achieved at which starch gelatinization occurs. Partial hydrolysis of starch takes place and the softness of bread is increased. Maximum effectiveness of the protected fungal alpha-amylase enzyme is obtained when it is used concurrently with chemical emulsifiers of the appropriate types. There is a synergistic effect between the protected fungal alpha-amylase enzyme and chemical emulsifiers which causes a reduction in the rate of bread staling. While this reduction in staling primarily relates to yeast-raised bread, it will be apparent that similar results are obtainable in fresh, refrigerated and frozen yeast-raised buns and rolls, yeast-raised sweet doughs, and chemically leavened baked products prepared from doughs, such as muffins, quick breads and biscuits, for example.

Example 1: Aqueous solutions of sucrose were prepared at concentrations of 35 to 65% by weight. Fungal alpha-amylase from *Aspergillus oryzae* (Miles Laboratories, Elkhart, Ind.), containing approximately 5000 SKB (Sandstedt, Kneen & Blish units) per gram of alpha-amylase activity was dissolved in the sucrose solution to obtain an activity of approximately 65 SKB/g of solution. The sugar-enzyme solution was placed in a 170°F water bath. Aliquots of the solution were withdrawn at intervals and observed for their ability to liquify a starch gel as an indication of residual alpha-amylase enzyme activity.

The starch gel liquefaction was determined by preparing a starch gel from a 7.5% dispersion of cornstarch which was cooked to completely gelatinize the starch. The gel was poured into a series of dishes and cooled to room temperature. Aliquots of sugar-enzyme syrup were removed from a 170°F water bath and mixed into the soft starch gel. The time required for liquefaction of the starch gel to occur is measured in minutes. This time period is an indication of the residual alpha-amylase activity, following heating.

Test data showed that significant protection against destruction by heat was provided to the enzyme dissolved in concentrated sucrose solutions at a temperature of 170°F. At least 55% sugar (by weight) was required to provide the protective effect. When the solution was 35% sucrose, there was insufficient protection against thermal denaturation of the enzyme.

Example 2: A series of sugars including sucrose, dextrose, fructose, invert syrup and corn syrup were evaluated for the protection which they afford against thermal denaturation of fungal alpha-amylase. The procedure was the same as that described in Example 1. Sugar concentrations of 40 to 60% (w/w) were tested for the protective effect upon the enzyme at temperatures of 170°F and 180°F.

Measurable protection is afforded the enzyme exposed to the temperature of 180°F. In each instance, the highest tested level of sugar provided the greatest protection. Sucrose provides the highest order of protection to the enzyme exposed to 180°F. It would be anticipated that blends of sucrose and other sugars at a total concentration of at least 60% would provide at least 30 minutes protection against thermal denaturation of the enzyme at 170°F and an intermediate level of protection at 180°F, depending on the relative concentrations of sucrose and other sugars used.

Example 3: A short time bread formulation was employed as a test vehicle to evaluate the combination of a chemical emulsifier and a protected fungal alpha-amylase as an aid for bread softness retention. Short time doughs employ one or more chemical fermentation accelerators as a means for avoiding the long bulk fermentations that are required with conventional dough-making techniques, such as with sponge doughs and straight doughs.

Tests for softness retention showed that the protected enzyme provides greater retention of softness at 72 hr than does unprotected enzyme, in combination with mono- and diglycerides. The rate of decline in bread softness between 24 and 72 hr is significantly less for the combination of a stabilized enzyme plus emulsifier than it is for an unprotected enzyme, plus an emulsifier. Both protected and unprotected fungal alpha-amylase enzyme plus mono- and diglycerides increase bread softness compared to the effect of an emulsifier alone. Higher levels of enzyme yield increased softness at 24 and 72 hr, as compared to the lower level of enzyme. After 72 hr, the combination of mono- and diglycerides plus protected fungal alpha-amylase is clearly superior in softness retention to the combination with the unprotected enzyme.

Emulsifier-Enzyme Compositions

An improved antistaling composition used in the preparation of bakery products has been developed by *F.D. Vidal and A.B. Gerrity; U.S. Patent 4,160,848; July 10, 1979; assigned to Pennwalt Corporation.* This composition which is a combination of glycerol esters, certain fatty acids or certain salts thereof or pentaerythritol monoesters thereof and, preferably certain enzymes, when incorporated in dough used to prepare bakery products, provides unexpectedly improved antistaling characteristics for such products.

In greater detail, this composition comprises a glycerol ester of a C_{10-24} saturated aliphatic fatty acid wherein at least 10 weight percent of the ester is a monoester, and for each part by weight of the ester, from about 0.25 to about 1.0, preferably 0.3 to 0.65, part of a component selected from the group consisting of a free C_{14-20} saturated aliphatic fatty acid, sodium stearate, magnesium stearate, a pentaerythritol monoester of a C_{14-20} saturated aliphatic fatty acid and mixtures thereof.

The above composition preferably contains an enzyme product selected from the group consisting of alpha-amylase, amyloglucosidase, mold derived lipase and mixtures thereof in an amount sufficient to provide increased antistaling characteristics to a bakery product prepared from a dough containing an effective amount of the composition.

From about 30 to about 50, preferably about 35 to about 45 weight percent of the abovedescribed glycerol ester may be a condensation product of ethylene oxide, the ethylene oxide portion consisting of from about 10 up to about 95 weight percent of the condensation product.

A preferred combination of components comprises (1) a mixture of glycerol monostearate and ethoxylated glycerol monostearate, (2) stearic acid and (3) alpha-amylase.

In this process, the range of amount of alpha-amylase beneficially used in the composition is from about 0.006 to about 0.46 SU for each part by weight of the antistaling mixture of glycerol ester and free fatty acid or pentaerythritol monoester, preferably about 0.03 SU. Generally, the foregoing range of amounts based on the antistaling composition will provide from 0.00185 to 0.185 SU, preferably 0.0092 SU for each 100 grams of baking flour when the antistaling composition is added to flour or dough in the prescribed amount.

The antistaling composition is preferably employed in powder, granular or tablet form. Edible fillers or carriers such as cornstarch, microcrystalline cellulose, sucrose, dextrose, dextrins, salt, nonfat milk powder, dicalcium phosphate, calcium sulfate dihydrate and the like may be used in conjunction therewith.

While the antistaling composition is preferably employed as a mixed combination, it may also be incorporated in the dough by mixing part thereof with some dough ingredients and part with other dough ingredients before all are combined in the final dough product.

The composition is incorporated in flour or in a dough preparation for preparing bakery products in amounts ranging from 0.2 to about 1.3 parts, preferably about 0.3 to about 0.6 part, for each 100 parts by weight of flour.

The antistaling compositions may be employed equally as well in either a "straight" dough baking procedure, a "sponge" dough baking procedure or a "continuous" dough baking procedure. While yeast-raised dough is preferred dough for the bakery products used herein, other types of dough may be employed.

In the following examples a standard sponge dough bread recipe and methods of bread preparation were used. Loaves of bread, 520 g each, were baked at 410°F for 18 minutes.

Example 1: 0.3% glycerol monostearate (added to sponge) provided the expected antistaling characteristics for the bakery product.

Example 2: 0.5% of a mixture of 60 parts by weight glycerol monostearate and 40 parts ethoxylated glycerol monostearate consisting of about 55 weight percent ethylene oxide (added to shortening) provided about equivalent antistaling characteristics for the bakery product as did the 0.3% glycerol ester of Example 1.

Example 3: 0.24% of the mixture of Example 2 and 0.1% stearic acid (added to sponge) unexpectedly provided as good antistaling characteristics as 0.5% of the mixture of Example 2.

Example 4: 0.24% of the mixture of Example 2 (added to shortening) and 0.005% (0.0092 SU alpha-amylase per 100 g flour) Rhozyme S (added to sponge) did not perform as favorably with respect to antistaling characteristics as the combination of Example 3.

Example 5: 0.24% of the mixture of Example 2, 0.1% stearic acid (both added to shortening) and 0.005% (0.0092 SU α-amylase per 100 g flour) initially provided the softest bakery product and thereafter provided comparable or better antistaling characteristics than those provided by the additives of the prior examples.

PHOTOPROTECTIVE AGENTS

Stabilization of Hops

A method for stabilizing the alpha acid or humulone content in hops prior to, during, and subsequent to the processing of hops for use in brewing has been devised by *H.L. Grant; U.S. Patent 4,154,865; May 15, 1979; assigned to S.S. Steiner, Inc.*

The lupulin or humulin is the glandular trichomes of the strobiles of hops which are used for making various beverages. The lupulin contains two important constituents which are commonly used in brewing beer. These constituents are humulones, also referred to as alpha acids, and lupulones, sometimes referred to as beta acids. The humulones are the primary bitter constituent of the hops which are utilized in brewing beer.

Humulones or alpha acids tend to undergo isomerization forming isohumulones or iso-alpha acids. These isomerization products are desirable constituents in brewing and thus, isomerization is a preferred process reaction. Isomerization typically is carried out intentionally in the brewing kettle.

Humulones are also subject to certain undesirable phenomena, however. Humulones tend to undergo deterioration and break down with undesirable by-products. For instance, under ambient conditions humulones or alpha acids have a tendency to oxidize and polymerize into hard resin. Moreover, during isomerization by-products may form, e.g., humulinic acids, which waste the humulone or alpha acid content of the hops. Another problem is the interreaction of humulones with other constituents in hops or hop extracts, such as lupulones or beta acids, under the stringent operating conditions of prior art processing methods.

The present method comprises the steps of:

(1) treating hops or hop extracts so that a substantial portion of the alpha acid content is converted to the corresponding isomerized alpha acid; and

(2) contacting the hop extracts containing isomerized alpha acids with a metallic hydride compound suitable for use in foods under suitable conditions to stabilize the isomerized alpha acids.

In addition to stabilizing the hops and alpha acids against deterioration and sensitivity to light, the above process may be used to convert the alpha acids to iso-alpha acids and to reduced isomerized products.

Any of the methods for isomerizing alpha acids, are suitable for this process. In a preferred method of isomerizing the alpha acid constituents of hops, it has been found advantageous to mix the hops or hop extracts with one or more metallic oxides wherein the metal is divalent and suitable for use in food products and wherein the mixing intimately contacts the oxide material with the alpha acids in the hops. Examples of suitable metallic oxides include calcium oxide, magnesium oxide or a mixture of calcium oxide and magnesium oxide.

Although ambient temperatures are adequate, it is also preferred to use elevated temperatures in carrying out the preferred isomerization process, in order to shorten stabilization reaction times and to facilitate stabilization. For example, it has been found that the reaction reaches sufficient conclusion after about 25-30 minutes where the blend of hops and metallic oxides are heated to a temperature in the range from about 70° to about 90°C and the heat source is removed as soon as the hops reach that temperature.

It has been surprisingly found that with the use of the oxide materials in the isomerizing process, the moisture or water naturally present in hops, typically from about 6 to 15% by weight, is sufficient to cause a transformation in most of the alpha acids present to their isomerized form, i.e., iso-alpha acids. If desired, it has been found that the rate of this reaction can be increased by the addition of a small amount of a lower alkanol of from about 1 to about 6 carbon atoms, e.g., methanol, ethanol, propanol, or the like, or a mixture of any of these. The alkanol can be added to the blend before or during mixing. Amounts of from about 5 to about 15% by weight of alkanol, based on the weight of the hops, are preferred.

Example: A hops powder is formed by hammermilling. A magnesium oxide in an amount of 0.75% by weight of hops powder is introduced into a blender and intimately mixed with the hops powder until uniform. The mixing generates some heat which tends to raise the temperature of the mixed product. After a period of 15 minutes, calcium hydride is added in an amount of 0.70% by weight of the hops powder and the mixture is again intimately mixed in the blender until a uniform blend is obtained. Hops pellets were formed directly from (1) the initial hops powder without any additives (control); (2) the isomerized hops powder after being mixed with the magnesium oxide; and (3) the reduced isomerized hops powder after being treated with magnesium oxide and subsequently with calcium hydride. The content of alpha acids, iso-alpha acids and reduced iso-alpha acids was measured in each of the pellets formed.

Hops pellets of (1) above tested for 6.8 alpha acids with no iso-alpha acids or reduced alpha-acids present. Hops pellets from (2) above tested for 0.6 alpha acids and 6.7 iso-alpha acids with no reduced iso-alpha acids. Pellets from (3) tested for 0.4 alpha acid, 0.7 iso-alpha acids and 5.8 reduced iso-alpha acids.

By means of this preferred process, in the order of 80% or more by weight of the alpha acids present in the hops powder prior to treatment are converted to the desired iso-alpha acids, with no appreciable losses of alpha acids or iso-alpha acids during processing. In the order of 85% of the iso-alpha acids formed are converted to the reduced iso-alpha acid derivative.

Prevention of Surface Discoloration of Margarine

Margarines are water-in-oil emulsions in which an aqueous phase is present as droplets distributed within a continuous matrix of oil and fat and are usually manufactured to resemble butter in appearance.

When a food spread is imperfectly wrapped or not hermetically sealed, or is uncovered by the consumer, there will be an exposure of the product to the atmosphere for a period of time, whereupon water evaporates from the surface of the spread, and the surface layer darkens in color. The consumer often confuses this discoloration with product spoilage, which it does not represent. Nevertheless, this surface darkening is considered by the average consumer to be objectionable and unsightly. Two mechanisms are thought to be involved in the discoloration of the food spread (such as oleomargarine) occasioned by its exposure to the atmosphere.

First, evaporation of water from the surface layer of the spread causes an increase in the concentration of the other constituents of that layer, including but not limited to the dye. Thus, in a situation where spreads initially contain large amounts of water, the potential exists for a large increase in dye concentration, and hence, a significant change in color.

The second mechanism thought to be involved in the discoloration process is the loss of water droplets, which serve as light-scattering centers, through evaporation. Their loss causes the food spread to appear darker and glassy. This phenomenon may, in fact, be the dominant mechanism in causing the surface darkening of the product.

H.M. Princen and M.P. Aronson; U.S. Patent 4,176,200; November 27, 1979; assigned to Lever Brothers Company have discovered that the above darkening or discoloration of the surface of a food spread, resulting from a loss of water to the atmosphere, can be reduced by incorporating a substance which makes the optical properties of the spread, when dehydrated, comparable to the optical properties of a freshly made food spread. Such a substance may be incorporated directly into the fat phase of the spread or, alternatively, into the water phase. Typically, such a substance need not cause a substantial or even a perceptible change in the initial color of the food spread when freshly made. Indeed, an appreciable change in the initial color would be undesirable. However, by incorporation of such a substance into the fat or water phase of the spread, the resistance to color change of the end product becomes evident.

In the production of the spreads, according to this process, white pigments have been found to be effective in preventing the discoloration of the spreads. Any food-grade white pigment suitable for food use may be employed, examples being titanium dioxide and calcium carbonate. Titanium dioxide is preferred since its high index of refraction (2.6-2.9) makes possible the use of lower concentrations, for example, as little as 0.05% by weight of the food spread may be used. However, depending on the substance used, an effective amount to prevent the discoloration in question is all that is required. No upper limit can be given in terms of the substances to be used although it is very desirable to keep the amounts of the substance relatively low so that other properties of the spreads, such as, but not limited to, texture, taste, mouth feel, etc., will not be affected. A desirable range for the concentration of the substance to be incorporated into the spread will be from about 0.05-0.1% to about 1% by weight, based on the total weight of the spread.

Food spreads used in the examples were prepared from 59.43% partially hydrated palm and soybean oils, 37.12% water, 2.60% salt and 0.77% of additives (emulsifier, beta-carotene, potassium sorbate and nonfat milk solids).

Example 1: Samples of food spreads as described above, prepared with 0 and 0.4% titanium dioxide incorporated into the aqueous phase, were spread on glass slides and stored at 20°C/20% relative humidity. Microscopic examination showed that water droplets disappeared at about the same rate for samples with and without titanium dioxide. However, samples without titanium dioxide became dark in color and lost their light reflectivity, while those with titanium dioxide retained their light reflectivity and had almost the same appearance as a freshly prepared food spread.

Example 2: The water loss of samples of food spreads, prepared with 0, 0.4% and 1.0% titanium dioxide, was observed using a microbalance. At 20°C/20% relative humidity, the rates of loss of water from samples with titanium dioxide were identical to those of samples without titanium dioxide. The samples without titanium dioxide became dark; those with titanium dioxide remained light in appearance.

Examples 1 and 2 show that titanium dioxide does not significantly affect the rate of water loss from food spreads. The reduction in discoloration, therefore, is not due to retardation of evaporation of water.

Stabilization of Rubrolone Using Quercetin-5'-Sulfonate

A process for reducing the tendency of the pigment rubrolone to fade upon exposure to direct sunlight has been developed by *G.A. Iacobucci and J.G. Sweeny; U.S. Patent 4,285,985; August 25, 1981; assigned to The Coca-Cola Company.* The pigment is combined with quercetin-5'-sulfonate as a protective agent.

Rubrolone is a water-soluble amorphous red solid of microbial origin having the empirical formula $C_{23}H_{23}NO_8$ and with the descriptive name of 8(R),9(R),10(S),-10a(R)-tetrahydro-9,10,10a,11-tetrahydroxy-3,8-dimethyl-1-propyl-6aH(S)-pyrano[2",3":5',4]furo[2',3':5,6]azuleno[2,3-c]pyridine-5,13-dione.

Quercetin-5'-sulfonate, a water-soluble derivative of the flavonol quercetin, has been described many times previously primarily as a reagent for spectrophotometric analysis of zirconium, hafnium, uranium and other elements. It has also been suggested as an ingredient in suntan lotion. For many years it was thought to be the 8-isomer, however, the correct 5'-sulfonate structure was finally determined by NMR [Terpilowski, et al, *Diss. Pharm. Pharmacol.,* 1970 (22), 389-93].

Preparation of quercetin-5'-sulfonate consists generally of dissolving quercetin (1 gram) in concentrated sulfuric acid (4 grams) and allowing the mixture to stand for 2 hours at 80°C. The solution is poured into excess ice water and the solution neutralized with solid $CaCO_3$. The resulting $CaSO_4$ is removed by filtration and the filtrate passed through a strong cation exchange resin in the Na^+ form to remove excess Ca^{2+} ions. The eluent is subsequently freeze-dried to yield the 40-60% pure sulfonate. The pure 5'-sulfonate can then be separated from disulfonate by-products by preparative high pressure liquid chromatography (HPLC).

Alternatively, the quercetin-5'-sulfonate can be recovered in a cruder form (purity: 86%) by pouring the sulfonation reaction mixture into glacial acetic acid, and removing the precipitated product by filtration.

Although it does not appear as if either an upper or lower limit exists on the ratio of the photoprotective agent quercetin-5'-sulfonate to the rubrolone pigment which will result in a reduced tendency of the rubrolone to fade upon exposure to sunlight, practical considerations indicate that a range of molar ratios of photoprotective agent:pigment of from about 1:1 to about 10:1 are preferred, with optimum molar ratios in the range of from about 2:1 to 5:1.

The so-prepared pigment composition may be utilized in any products where coloring with rubrolone, either alone or in admixture with other colorants, is desired. The pigment composition may be prepared as a dry admixture which is suitable for mixing with additional dry components to form a product, e.g., a beverage powder, adapted for dissolution in water. The pigment composition may also be utilized either as a component in preparing water-based concentrates or syrups adapted to be diluted with water to form a single-strength beverage or as a component in single-strength beverages per se.

Example 1: Rubrolone was dissolved at a level of 20 ppm in a 0.01 M citric acid solution containing 200 ppm sodium benzoate as a preservative. A second sample, identically prepared, had added thereto 100 ppm of quercetin-5'-sulfonate. The samples, in 100 ml Pyrex glass flasks, were exposed to direct sunlight and aliquots removed periodically to determine the absorbance at λ_{max}(525 nm). Sunlight incident energy was measured with an Eplex meter.

After receiving 1000 Langleys (cal/cm^2) only 9.5% color remained in the solution containing only rubrolone while 59.2% color was retained by the stabilized solution.

Example 2: A similar experiment was conducted using 30 ppm rubrolone and 100 ppm quercetin-5'-sulfonate in a 0.01 M citric acid solution containing 200 ppm sodium benzoate preservative.

After exposure to 1317 Langleys, the rubrolone alone only retained 13% color while the stabilized solution retained 70% color.

Protective Agents for Anthocyanic Pigments

Anthocyanic pigments, i.e. anthocyanins and anthocyanidins, have been found to account for the natural colors of many fruits, vegetables and flowers. Despite their widespread occurrence in nature, however, these pigments have not been widely used as colorants in foods because of both limited availability and, in many cases, poor stability and resultant color loss.

Such anthocyanic pigments include cyanidin rutinoside, enocianina (grapeskin colorants), and apigeninidin chloride.

It has long been known that the various red to blue shades of flowers are due to anthocyanins either alone or in association with phenolic materials, called "copigments," which are present in the plants along with the anthocyanins. These copigments are known to cause a bathochromic shift in the λ_{max} of the anthocyanin pigment and also an increase in the absorbance at λ_{max}. This copigment effect appears to be at its greatest when the copigment is a flavonol, and the use of flavonol copigments, such as rutin and kaempferol-3-glucoside, has been suggested as a way to enhance the hue and intensity of anthocyanin colorants in foods. The use of nonflavanoid compounds as copigments, however, does not appear to have been shown or suggested previously.

G.A. Iacobucci and J.G. Sweeny; U.S. Patent 4,285,982; August 25, 1981; assigned to The Coca-Cola Company has developed stabilized anthocyanic pigments which may be used in food compositions.

The anthocyanic pigments are combined with a photoprotective agent selected from the group consisting of sulfonated polyhydroxyflavonols, poly(hydroxyalkyl)-flavonols, sulfonated polyhydroxy flavones, sulfonated polyhydroxyisoflavones and sulfonated aurones.

Example 1: In order to test whether the color of anthocyanin-containing food products may be stabilized with respect to sunlight-induced fading through the use of a photoprotective agent, several experimental beverages were prepared containing 30 ppm of the representative anthocyanin cyanidin rutinoside (the main pigment in cherries) and 850 ppm (40 molar excess) of kaempferol-3-glucoside. Beverages containing 30 ppm cyanidin rutinoside and no photoprotective agent were used as controls.

The experimental beverages were prepared from a syrup containing 2.0 kg water, 25 g citric acid, 2.5 kg sugar, and 40 g of commercial grape flavor. The syrup was diluted 4.4:1 with water, the colorant and copigment added, and the resulting beverage carbonated with two volumes of carbon dioxide. Standard flint glass bottles were used as containers. Sunlight exposure was measured with a broad spectrum Langley meter, with samples being removed periodically for analysis.

The use of the natural flavonol kaempferol-3-glucoside greatly enhances the resistance of cyanidin rutinoside to sunlight-induced fading.

After exposure to 5000 Langleys of sunlight, the pigment without being stabilized had no color remaining, while the pigment with the kaempferol-3-glucoside retained 42% of its color.

In another test it was shown that quercetin-5'-sulfonate acts as a photoprotective agent for the representative anthocyanins cyanidin rutinoside and enocianina. Although used at a much lower concentration, the photoprotective effect of quercetin-5'-sulfonate is approximately the same as that of the natural flavonol, kaempferol-3-glucoside as shown in Example 1.

Further tests demonstrated that only certain of the compounds which exhibit a copigment effect also exhibit a photoprotective effect and that, in fact, several of the compounds exhibiting copigment effects enhanced rather than reduced the photodegradation of cyanidin rutinoside. Those copigment species showing promising photoprotective effects are hydroxyethylrutin [a poly(hydroxyalkyl)-flavonol], quercetin disulfonate, quercetin-5'-sulfonate and morin disulfonate [polyhydroxyflavonol sulfonates], biochanin A sulfonate [a polyhydroxyiso-flavone sulfonate], 4'-methoxyaurone sulfonate and 4'-methoxyaurone disul-fonate [aurone sulfonates], and apigenin sulfonates [polyhydroxy flavone sulfo-nates]. Among other copigments tested, quercetin-O-sulfates exhibited a some-what lower degree of photoprotective effect, and certain other copigments in-cluding flavone monosulfonate, flavone disulfonate and 4'-methoxyflavone sul-fonate, actually increased the rate of photodecomposition of cyanidin rutinoside.

MICROBIOLOGICAL STABILIZERS

CONTROLLED WATER-SUGAR-FAT CONTENT

A series of products have been developed by the Rich Products Corporation which are intermediate-moisture foods which remain ready to use at freezer temperatures.

These foods are characterized by a high sugar content, at least equal in weight to the amount of water present in order to provide microbiological stability. The sugars used have a low molecular weight, mainly dextrose and fructose, which comprise together at least about 50% and preferably at least about 75% of the total sugar content. Sucrose has a sweetness between that of fructose and dextrose. Fructose, which is sweeter than dextrose, is preferred since it has a lesser tendency to crystallize and cause apparent hardness. For most foods, particularly where the food comprises an emulsion, it is preferred that the fats used include partially unsaturated fats which tend to provide superior flow properties, and nutritional advantages although less stable than saturated fats. The fat content is usually less than the water content in order to form a stable oil-in-water emulsion; the water content is preferably at least about 25% greater than the fat content.

An important group of foods which have been particularly well-adapted in accordance with this process are oil-in-water emulsions, including butter creams, whipped toppings, low-fat whipped creams, milk mates, nondairy shakes, icings and coffee creamers.

Another class of goods, which forms a unique combination with the foregoing, is bakery products, such as cakes, breads, cookies, pie shells, muffins, turnovers, pancakes, waffles and donuts. The pastries can be filled or topped with the creams and icings of this process.

Many diverse foods can likewise be adapted pursuant to this process, such as dressings, puddings, sauces, gravies, snack spreads, pancake syrups, ice creams, candies, and beverage (soup, tea, juice) concentrates, and meat, fish, fruit or vegetable products.

These foods are generally characterized as microbiologically stable food products comprising about from 15 to 45% water, sugar in a ratio to water of about from 1:1 to 2:1, preferably about from 1–1.5:1, about from 2.5 to 30% fat, and minor but effective amounts of salt, emulsifier, stabilizer and flavoring, provided that the amount of fat is less than the amount of water or equivalent phase, such as nonaqueous water-soluble liquid phase, the solutes content is adequate to provide the product with a water activity of about from 0.8 to 0.9, the amount of dextrose plus fructose is at least about 50% based upon the total sugar content, wherein the foregoing ingredients comprise at least one of fructose and unsaturated fat and the product is spoonable at about 10°F.

The fructose-dextrose syrup used in these products (Isosweet) comprises 29% water and 71% sugars (50% dextrose, 42% fructose, 1.5% maltose, 1.5% isomaltose and 5% higher saccharides).

Since these foods are maintained at freezer temperatures until ready to be used, a water activity (relative vapor pressure) of about from 0.85 to 0.90 is adequate. Freezer temperature, unless stated otherwise, refers to temperatures of about from −5° to +10°F which is a common range for freezers in homes and stores.

Fats high in unsaturation are safflower oil, corn oil, soybean oil, cottonseed oil and sunflower oil, unsaturated fats as used in these products are those having an iodine value of about at least 50 which include partially hydrogenated fats, and the more highly unsaturated fats with an iodine value above about 100. These fats are recommended for dietary purposes, particularly for those with a high plasma cholesterol level which is associated with atherosclerosis.

The saturated fats include the hydrogenated oil products of coconut, cottonseed, corn, soybean, peanut, olive, etc. Fats having a melting point of 90° to 94°F are preferred, i.e., the melting point should be below body temperature.

Emulsifiers are necessary ingredients of those compositions of this process which contain fats and are oil-in-water emulsions. A wide variety of emulsifiers may be employed in amounts on the same order as in previous oil-in-water emulsions for example, about from 0.1 to 5%, preferably about from 0.2 to 1.5%. They induce the formation of a stable emulsion and improve the rate and total aeration obtained. Among the more suitable are: hydroxylated lecithin; mono-, di-, or polyglycerides of fatty acids, such as monostearin and monopalmitin; polyoxyethylene ethers of fatty esters of polyhydric alcohols, such as the polyoxyethylene ethers of sorbitan monostearate (polysorbate 60) or the polyoxyethylene ethers of sorbitan distearate; etc.

Tenderex emulsifier used is a mixture containing polysorbate 60 (11.9%), sorbitan monostearate (31.6%), mono- and diglycerides of fatty acids (2.3%), propylene glycol (9.5%) and water (44.3%).

The emulsion compositions of these products also include one or more stabilizers or hydrophilic colloids to improve the body and texture of toppings, and as an aid in providing freeze-thaw stability. These stabilizers are natural, i.e., vegetable, or synthetic gums and may be, for example, carrageenin, guar gum, alginate, xanthan gum and the like or methylcellulose, carboxymethylcellulose, ethylcellulose, hydroxypropyl methylcellulose (Methocel 65 HG), micro-crystalline cellulose and the like, and mixtures thereof.

Whipped Foods and Toppings

Microbiologically stable whippable or whipped foods which remain soft and ready for use at freezer temperatures have been developed by *M.L. Kahn and K.E. Eapen; U.S. Patent 4,146,652; March 27, 1979; assigned to Rich Products Corporation.*

These microbiologically stable oil-in-water cream-type products, such as butter creams, whipped creams, shakes, nondairy creamers, etc., comprise from 25 to 45% water, sugar in a ratio to water of from 1.5 to 1:1, from 10 to 30% fat, and minor amounts of protein, salt, emulsifier, stabilizer and flavoring. These products have a water activity of from 0.8 to 0.9, wherein the amount of fructose is from 15 to 65% based on the sugar content and the amount of dextrose is at least about 50% based upon the remaining total sugar content, the fat content preferably comprises at least, about 10% unsaturated fat and the foregoing ingredients are adapted to provide a product which will flow at about 10°F. These products have excellent texture and eating properties and are readily whipped to a high volume with a light but firm structure. In addition to microbiological stability these products have physical stability in that they retain a smooth foamed cellular structure without separation of a liquid portion. The products are further characterized by having an overrun of greater than about 150% and a density as low as about 0.3 or 0.4 for a butter cream and whipped cream.

Example 1: A group of useful products made in accordance with this process is the oil-in-water emulsion based material used for preparing butter creams, whipped creams, shakes, coffee lighteners, and the like. Butter creams, which can be used as a topping and/or filling for a confectionery product, is typical in several respects of this class of products and the manner in which the problems raised by this type of product have been overcome can readily be adapted to similar types of products.

The butter cream is an oil-in-water emulsion comprising about from 25 to 45% water, preferably 30 to 40% water, sugar in a ratio to water of about from 1–1.5:1 and about from 10 to 30% fat. At the higher ratios, particularly of fructose, a less firm product is obtained which is less suited as a topping but may be used as a filling, i.e., in an eclair. The sugar preferably comprises some fructose, usually in an amount about from 15 to 65% based on the total sugar used. The remainder of the sugar is at least substantially dextrose, i.e., from at least about 50% up to all of the remaining sugar, preferably the total amount of fructose plus dextrose is about from 75 to 100% of the sugar content. The fat preferably contains about from 10 to 60% unsaturated or partially unsaturated fat. Minor amounts of other ingredients are used in about conventional amounts, i.e., protein concentrate, salt, emulsifier, stabilizer and flavoring.

A process for making a butter cream formulation comprises adding 36.72 parts of dextrose-fructose syrup to 25.32 parts water. Then 0.4 part xanthan gum, 0.26 part sucrose, 0.26 part Methocel 65 HG and 1.67 parts soy protein concentrate were premixed and added to and mixed with the above batch. Heating the batch to 180°F was begun during which 10.57 parts dextrose, 0.14 part salt, 0.28 part polysorbate 60 and 0.1 part hexaglyceryl distearate were added. After 180°F was reached, mixing was continued for 5 minutes. Then 19.2 parts of hard butter and 5.0 parts soybean oil were added. Then 0.1 part lecithin and 0.01 part Tenox 22 antioxidant were dissolved in an additional 0.3 part of hard butter and added. The 0.03 part flavoring was then added to the mixture and homogenized in two steps at 3,000 and 500 psi and the product cooled to 38° to 42°F. The finished product can be packed in suitable containers, and stored in a freezer or refrigerator for whipping later.

The water content of the formulation was 35.97% (including the water in the dextrose-fructose syrup). The formulation also contained 10.95% fructose, 23.61% dextrose and 2.35% higher sugars (36.91% total sugar). The product was whipped and had an overrun value of 286%, with a whipping time of about 4 minutes. The specific gravity of the product was 0.35.

The coli count after five days at room temperature was less than ten and the total plate count at that time was less than one hundred, which shows an excellent room temperature stability. It was found that freshly made samples decreased in coli count upon storage at room temperature and had lower counts than refrigerated samples, which in turn had lower counts than frozen samples, i.e., freshly made samples had a coli count of 152. Three samples were held for fourteen days at various temperatures. At 70°F the coli count was 7, at 40°F the coli count was 53 and at –70°F the coli count was 133.

The product was left standing for ten days at room temperature without any evidence of browning (Maillard reaction).

The water activity of the whipped product was 0.875 at 72°F and its pH was 6.88. It was found that as the sugar/water ratio fell below about one the product quickly lost its microbiological stability and physical integrity. Thus, even at about 45% sugar in the aqueous phase, the coli count and the total plate count increased within two days at room temperature and the butter cream sagged.

The formulation has excellent flow properties at 5°F, the flow test results were: 300 ml after 1 minute, 455 ml after 3 minutes and 570 ml after 6 minutes. The product when whipped was easily applied to cake as a topping and maintained its physical integrity, texture, and appearance in the freezer during a ten-day test and at room temperature during a seven-day test. The butter cream was capable of being whipped at freezer temperatures. It was whipped at a temperature as low as –30°F.

Example 2: A whipped topping made in accordance with this process has the same advantages as the butter cream discussed in connection with the preceding example. The whipped topping has less hard butter and a higher unsaturated fat content than the butter cream formulation; the ingredients are otherwise equivalent. The product retains its texture at freezer temperatures and is microbiologically stable. This product also has the property of being whipped at freezer temperature rather than requiring the expensive and time consuming technique of first taking it to room temperature, whipping it and then cooling it.

Flour-Based Batters

Microbiologically stable flour based batters are also prepared by *M.L. Kahn and K.E. Eapen; U.S. Patent 4,154,863; May 15, 1979; assigned to Rich Products Corporation* by controlling their sugar/fat content.

These batters comprise conventional amounts and types of flour depending on the final product, about from 15 to 40% water, sugar in a ratio of water of about from 1.5 to 1:1, about from 2 to 10% or up to 25% fat, and minor but effective amounts of leavening agent which may be encapsulated, egg products, salt, emulsifier, stabilizer and flavoring, provided that the solutes content is adequate to provide the product with a water activity of about from 0.8 to 0.9, the fructose content of the sugar preferably is about 10 to 40%, the amount of dextrose plus fructose is at least about 50% or from 75 to 100% based upon the total sugar content, and the fat is preferably unsaturated. The batter should have at least one of fructose and unsaturated fat to assist in providing a spoonable and preferably pourable product at about 10°F. The final product made from the batter has a higher penetrometer value than conventional products at 10°F and is edible at that temperature.

Example 1: A microbiologically stable cake batter and cake and other bakery products were made which retain their characteristic texture at freezer temperature. The cake batter is suitable for industrial and home use where stable storage is an important factor. The batter can be kept in a freezer and is always ready for use. The cake of this process is particularly suited for the expanding convenience frozen food market. It can be cut and served promptly upon removal from the freezer. The cake, of course, can be made with fillings and toppings described above which likewise retain a soft texture and are microbiologically stable.

The cake batter comprises about from 20 to 30% water, sugar in a ratio to water of about from 1-1.5:1, and preferably about from 2.5 to 10% fat, and up to 25% fat. The sugar preferably includes fructose in an amount about from 10 to 40% based on the sugar content with the remainder being substantially dextrose (50 to 100%). The type of fat can be varied widely between saturated and unsaturated depending on the type of cake and texture desired. An unsaturated fat will provide superior flow and nutritional properties. Other conventional ingredients are used in their normal proportions such as egg whites, nonfat milk solids, flour, emulsifiers or softeners such as glyceryl monostearate, salt, preservative, coloring and flavoring.

A cake batter is prepared by thoroughly mixing 15.0 parts water, 8.75 parts egg white, 8.75 parts 12X sugar, 18.75 parts dextrose and 15.0 parts dextrose-fructose syrup. Then 1.4 parts Tenderex emulsifier, 1.3 parts baking powder, 2.5 parts vegetable oil and 0.2 part vanilla are added and mixed until uniform. Finally 0.72 part salt, 0.13 part coloring, 2.5 parts nonfat milk solids and 25.0 parts cake flour are mixed in.

The water content of the batter is 27.67% and the sugar content is 38.15% (fructose 4.47%, dextrose 24.08% and other sugars 9.6%).

The batter was frozen and then tested on a penetrometer; it has a value of 19.8 compared to a conventional batter which gave a reading of 4.1. A cake made from this batter was frozen and it gave a penetrometer reading of 6.9 compared to a value of 4.2 for a conventional cake. The cake had a moisture content of 25.2%.

Example 2: A pancake batter can be made in accordance with this process which is sufficiently free-flowing at freezer temperatures to be poured or squeezed from a container. The product can be maintained indefinitely in a freezer and upon removal from the freezer can be poured, without defrosting, onto a griddle to make pancakes in the conventional manner. The pancakes made from the batter can be frozen and stored indefinitely but will remain soft at freezer temperature. The pancakes therefore can be used directly from the freezer by quickly warming them, unlike conventional frozen pancakes which need to be defrosted or subjected to extensive heating to soften them throughout. The pancakes and waffles can be stored at room temperature or at refrigerator temperature for many days without spoilage.

The pancake batter was made by placing the 32.26 parts of liquid egg white into a Norman mixer. Next, 0.58 part salt and 0.82 part sodium acid pyrophosphate were added followed by metering in the 19.42 parts dextrose-fructose syrup with agitation. Then 20.33 parts dextrose and 19.42 parts bread flour were added followed by increasing the mixer setting to high speed and adding the 6.47 parts soybean oil (Type 106) and finally adding 0.60 part sodium bicarbonate and mixing for five minutes. The formulation is then pumped to a cooled hold tank from which it is passed through a votator to cool it to 25° to 28°F, from which it is pumped to another cooled holding tank.

The liquid egg white comprises 87.6% water and this in combination with the 29% water content of the dextrose-fructose syrup gave a total water content of 33.98%. The fructose content of the formulation was 5.79% and the dextrose content was 27.22%, whereas the total sugar content of the batter was 34.11%.

Pancakes were made from this formulation on a greased and covered griddle, frozen and tested on a penetrometer against pancakes made from a conventional batter. The frozen pancake made from the formulation of this process gave a penetrometer reading of 5.1 mm whereas the frozen standard pancake gave a reading of 1.1 mm. The pancake had a moisture content of 25.2%.

Stable Sauce and Soup Concentrates

Sauces and soup concentrates with specific sugar and fat contents are also described by *M.L. Kahn and K.E. Eapen; U.S. Patent 4,220,671; September 2, 1980; assigned to Rich Products Corporation.*

Microbiologically stable soup concentrates and sauces have been made comprising about from 30 to 45% water, sugar in a ratio to water of about from 1.5 to 1:1, about from 5 to 30% fat, and minor but effective amounts of salt, stabilizer and flavoring, wherein the amount of dextrose plus fructose is at least about 50% based upon the total sugar content, the foregoing ingredients comprise at least one of fructose and unsaturated fat and the product is spoonable at about 10°F. Fish, meat and vegetables (which may be infused with solutes) are added to these to provide, for example, a chowder concentrate or Newburg sauce. Such preparation may also comprise about from 5 to 100 ppm, preferably about from 30 to 70 ppm, of a quinine salt.

Example 1: Frozen clam chowder concentrate as marketed is usually defrosted before use. Otherwise, it is difficult to remove from the container and if removed from the container and placed while still solid in boiling water or a hot pot it sinks to the bottom of the pot and may be scorched. In accordance with this process a clam chowder concentrate can be made which will flow at freezer temperature. This product is easily removed from its container and mixed with water or milk to make the final product.

The clam chowder concentrate contains from about 30 to about 45% water, sugar in a ratio to water of about from 1-1.5:1 and about from 5 to 30% fat (saturated or unsaturated). The sugar content preferably includes about from 10 to 40% fructose and the fructose plus dextrose content is about from 75 to 100% of the total sugar. The formulation includes a standard mixture of finely chopped vegetables, a stabilizer like cornstarch, salt, spices and flavorings. Other conventional ingredients can be added, such as milk solids.

A clam chowder concentrate was made by melting 7.32 parts margarine and adding 4.05 parts potato, 5.03 parts celery, 2.81 parts onion, 2.23 parts mushrooms and 0.07 part garlic, all finely chopped.

Alternatively the vegetables can be infused with solutes to control their stability and texture, for example, cooking or subsequently immersing the vegetables in a stabilizing solution having a concentration of water-soluble compounds sufficiently high to effect the desired transfer of solute.

Next 0.25 part salt and 0.03 part black pepper are added. Then 1.31 parts cornstarch are dissolved in 22.68 parts stewed tomatoes and added to the sautéed mixture and simmered until thickened.

Then 0.78 part Worcestershire sauce and 0.78 part sherry are added and simmering is continued for five to seven minutes. Add the desired amount of infused clams (i.e., about 25%) and simmer for five minutes more. Infused clams are those treated to lower their moisture content below 50% and to add solutes comprising sugars, polyhydric alcohols and salts to lower the water activity of the clams to about 0.90 and below, i.e., to 0.75. This can be carried out by cooking or subsequently immersing the clams in a stabilizing solution having a concentration of water-soluble compounds sufficiently high to effect the desired transfer of solute and lowering of water activity, usually under an elevated temperature and pressure.

For example, in one procedure the clams were placed in a solution of 47.4 parts water, 44.3 parts propylene glycol, 7.4 parts salt and 0.9 part potassium sorbate which was brought to boiling temperature and then let stand at room temperature overnight. These clams were placed in the freezer overnight and when removed were soft and ready to eat. An alternative procedure is to use the same technique with a clam-flavored fructose-dextrose syrup having about 5 to 10% salt. Finally add the 20.05 parts fructose-dextrose syrup and 32.68 parts dextrose and mix for ten minutes.

This product had a moisture content of 42.03% and a sugar content of 46.95%. The clam chowder concentrate gave a penetrometer reading of 3.9. A conventional frozen oyster stew semi-condensed soup (Campbell's) tested under the same conditions gave a penetrometer reading of zero, i.e., too hard for penetration.

The product can be frozen until ready to be used. It is then readily dispersed in water or preferably milk and heated. The foregoing formulation and technique can be readily adapted to make other soup concentrates, such as seafood bisque, and cream of chicken, mushroom, cheese and other fish, fowl, meat and vegetables.

Example 2: A Newburg sauce can be made by this process to be sold separately or with shell fish, such as lobster or crab. As pointed out previously the fish may be treated to lower its water content by infusing it with stabilizing solutes to make it microbiologically stable at room temperature. However, since the product is kept frozen and can be used promptly after removal from the freezer, the requirements for microbiological stability are not as rigid as for conventional products.

The procedure for making this product is to dissolve 1.89 parts cornstarch and 6.88 parts nonfat dry milk in 39.07 parts whole milk and add to this 3.06 parts melted margarine and 1.11 parts salt. This is heated and stirred until the mixture thickens. Then 3.33 parts egg yolk and 0.22 part lemon juice concentrate are mixed and stirred into the thickened mixture. Infused crab (about from 30 to 40% of the total formulation) is added, with flavoring as desired, i.e., dry sherry wine and red pepper. The product is cooked for 3 to 4 minutes and 31.11 parts dextrose and 13.33 parts fructose-dextrose syrup are added and mixed well for 10 minutes.

The product contained 36.92% water and 40.57% sugar (3.97% fructose, 35.84% dextrose and 0.75% higher sugars). The milk products would contribute about an additional 5% sugar, but in the form of lactose which contributes comparatively little to lowering the osmotic pressure.

The Newburg sauce had a penetrometer value of 14.9 mm. A conventional frozen Alaska King Crab-Newburg Sauce (Stauffers) tested under the same conditions was too hard to give a measurable penetrometer reading.

Other sauces such as thermidor, bearnaise, hollandaise and cheese may be made by the foregoing technique.

Frozen Puddings and Fillings

Microbiologically stable pudding products and filling materials have also been described by *M.L. Kahn and K.E. Eapen; U.S. Patent 4,234,611; November 18, 1980; assigned to Rich Products Corporation.*

Puddings made in accordance with this process are useful as a ready-to-eat convenience food which can be packaged in any conventionally used container for storage in a freezer; the pudding retains its soft texture at freezer temperature and is microbiologically stable at room temperature. Unlike canned puddings, this pudding does not require sterilization and expensive packaging and unused portions may be left in the refrigerator, or even at room temperature, for subsequent use. Unlike conventional frozen puddings, this pudding does not substantially crystallize with consequent loss of texture, nor is the inconvenience of a long period of thawing necessary before the pudding can be eaten.

The puddings of this process generally comprise an oil-in-water emulsion having about from 30 to 40% water, sugar in a ratio to water of 1-1.5 to 1 and about from 15 to 25% fat. When the sugar does not contain fructose and the fat is saturated, the product tends to a somewhat cohesive consistency and the sugar to water ratio is maintained toward the upper range. The amount of dextrose plus fructose is preferably about 70 to 100% of the sugar content. The use of unsaturated fats, such as soybean oil, would be desirable for flow and nutritional properties. Minor amounts of conventional stabilizers, emulsifiers and flavors are also used. Further, about from 10 to about 30 ppm of a quinine salt may be added.

The process for making microbiologically stable puddings is also applicable for making donut fillings for pastries such as fruit pies, donuts, etc. Some fillings of this process remain in a flowable condition at freezer temperatures so that they can be used directly upon removal from the freezer.

Each of the fillings, whether for fruit pies or for donuts, contains sugar in a ratio to water about from 0.8-1.5 to 1. The sugar is predominantly of low molecular weight, substantially dextrose and/or fructose in an amount totalling from about 50 to 100% of the total sugar content. The amount and form of the fruit used in the filling may vary widely.

The fillings preferably contain from about 40 to 60% fruit, from about 20 to 50% fructose-dextrose syrup, from about 5 to 25% additional sugar, from about 2 to 7% starch and from about 40 to 55% water. The fillings may also contain about from 2 to 30 ppm, preferably about from 3 to 20 ppm of a quinine salt.

Example 1: A pudding emulsion is prepared by heating 31.72 parts water to 140°F and adding 0.20 part polysorbate 60, 0.07 part guar gum, 0.13 part sorbitan monostearate, 0.86 part sodium caseinate, 0.66 part dextrose, 14.72 parts sucrose, 5.30 parts hard butter, 12.60 parts coconut oil and 0.03 part potassium sorbate while heating to 155° to 160°F. This mixture is homogenized in two steps at 7,000 and 500 psi with cooling to 34° to 38°F. This pudding emulsion is a conventional product which is converted to the pudding of this process by adding a premix of 33.14 parts dextrose and 0.23 part sodium alginate at 150°F plus 0.11 part vanilla flavor and 0.23 part 10% calcium chloride solution. Quinine sulfate may be added if desired.

The product had a slightly elastic character and at –7°F a penetrometer value of 29.3 mm, compared to a commercially available pudding (Rich's Chocolate Pudding) which gave a penetrometer reading of 1.3 mm. The water activity of the pudding averaged 0.852 at 72°F.

Example 2: An apple pie filling is prepared from 50 parts frozen apples, 30.95 parts dextrose-fructose syrup, 2.5 parts starch, 0.5 part Avicel (cellulose gum), 5.51 parts dextrose, 10.0 parts sucrose, 0.1 part salt, 0.1 part cinnamon, 0.05 part nutmeg and 0.3 part 50% citric acid. 5 ppm of quinine sulfate may be added to the above composition.

In the foregoing composition, the frozen apples comprise about 75% water, about 16% sucrose and about 8% fructose, and thus the total sugar content is almost 50% and the total water content is almost 47%.

Example 3: An example of a donut filling contains 44.25 parts dextrose-fructose syrup, 5.537 parts starch (Regista), 0.051 part lemon powder, 0.113 part salt, 0.1 part potassium sorbate, 0.0175 part Red Dye No. 40, 49.53 parts cherries (RSP, Grade A, frozen) and 0.4 part 50% citric acid. 5 ppm of quinine sulfate may be added to the above formulation.

The frozen cherries were in conventional packed form containing 5 parts fruit to 1 part sucrose, about 70% water and the remainder almost all fructose.

Soft Ice Creams

M.L. Kahn and K.E. Eapen; U.S. Patent 4,244,977; January 13, 1981; assigned to Rich Products Corporation have also provided an ice cream product which remains soft in the freezer.

Example: The ice cream product comprises about 45 to 60% water, sugar in a ratio to water of about from 0.5–1:1 and from about 8 to 16% fat. The total of fructose and dextrose is from about 75 to 100% of the total sugar content; the amount of fructose preferably is 65 to 100% of the total sugar content. The fat is a butter fat.

For nonregulated ice cream substitutes (where the ingredients can by varied without government regulation) the water content may be about from 40 to 60%, the sugar to water ratio may be about 0.5 to 1.5:1, fat about from 2 to 16%. The amount of fructose plus dextrose equals about 50 to 100% of the sugar content.

A suitable ice cream was made by adding 40 parts whole milk and 24.62 parts heavy cream to a kettle and begin heating. When 140°F was reached 0.10 part polysorbate 60 and 0.10 part sorbitan monostearate were added. While stirring, a premix of 0.70 part sucrose and 0.30 part alginate was added, and then 26.88 parts of an 80% fructose concentrate (90% fructose, 10% dextrose) and 7.0 parts nonfat dried milk solids. Mixing was continued at 160°F for five minutes. The product was then homogenized in a first stage at 3,000 psi and second stage at 500 psi followed by cooling. The product was whipped to an overrun of 100% and removed at 22°F. This ice cream was placed in a freezer at about 0° to 10°F for 72 hours and, during this entire period, it retained a texture suitable for immediate use. The maintenance of this spoonable texture also permits the ice cream to be packaged in a flexible squeeze package (i.e., a Squiggle-Pak) for dispensing in a ribbon form.

The product had a water content of 54.12%, a sugar content of 28.7% (including the sugars in the whole milk, cream and milk solids) and a fat content of 10.26% (from the milk and cream).

Addition of Quinine Salts

The process disclosed by *M.L. Kahn and K.E. Eapen; U.S. Patent 4,199,604; April 22, 1980; assigned to Rich Products Corporation* is directed to the use of quinine salts in intermediate-moisture foods and other products which remain ready to use at freezer temperatures, to modify the taste of the requisite sugar content. The principles and techniques which have been developed for intermediate-moisture foods are applicable to this process, as modified in the manner explained herein below. Most of these foods are maintainable at freezer temperature in a condition ready for immediate use. After removal from the freezer the foods may be held at room temperature or at refrigerator temperature for a considerable period of time without spoilage because of the bacteriostatic effect of the sugar/water ratio.

The high levels of sugar requisite to effect bacteriostasis may render the ultimate food product excessively sweet, however, and thus less palatable to some consumers. In accordance with this process it has been found that addition of a bitterness principle modifies the undesirable sweetness of the sugar. More specifically, addition of a quinine salt such as quinine sulfate, quinine bisulfate or quinine hydrochloride, results in reduced perception of sweetness. In many cases, the perceived sweetness is reduced by more than half the value associated with the unmodified food product. Quinine salts are employed in this regard in amounts up to about 125 ppm per foodstuff. A preferred range of such additive is from 2 to 75 ppm.

The foods of this process are generally characterized as microbiologically stable food products comprising about from 15 to 45% water, sugar in a ratio to water of about from 1:1 to 2:1, preferably about from 1.5–1:1, about from 2.5 to 30% fat, up to about 125 ppm of a quinine salt, and minor but effective amounts of salt, emulsifier, stabilizer and flavoring, provided that the amount of fat is less than the amount of water or equivalent phase, such as nonaqueous water-soluble liquid phase, the solutes content is adequate to provide the product with a water activity of about from 0.8 to 0.9, the amount of dextrose plus fructose is at least about 50% based upon the total sugar content, wherein the foregoing ingredients comprise at least one of fructose and unsaturated fat and the product is spoonable at about 10°F.

Example 1: Puddings made in accordance with this process are useful as ready-to-eat convenience foods which can be packaged in any conventionally used container for storage in a freezer. The pudding retains its soft texture at freezer temperature and is microbiologically stable at room temperature. Unlike canned puddings, the pudding of this process does not require sterilization and expensive packaging, and unused portions may be left in the refrigerator, or even at room temperature, for subsequent use. Unlike conventional frozen puddings, this pudding does not crystallize and harden with consequent loss of texture, nor is the inconvenience of defrosting necessary before the pudding can be eaten.

The puddings of this process comprise oil-in-water emulsions having about from 30 to 40% water, sugar in a ratio to water of 1.5-1:1, about 15 to 25% fat and about from 10 to 30 ppm of a quinine salt. When the sugar does not contain fructose and the fat is saturated, the product tends to a somewhat cohesive consistency and the sugar to water ratio is maintained toward the upper range. The amount of dextrose plus fructose is preferably about 70 to 100% of the sugar content. The use of unsaturated fats, such as soybean oil, would be desirable for flow and nutritional properties. Minor amounts of conventional stabilizers, emulsifiers and flavors are also used.

A pudding emulsion was prepared from 31.72 parts water, 0.20 part polysorbate 60, 0.07 part guar gum, 0.13 part sorbitan monostearate, 0.86 part sodium caseinate, 0.66 part dextrose, 14.72 parts sucrose, 5.30 parts hard butter, 12.60 parts coconut oil and 0.03 part potassium sorbate to a total of 66.29 parts. To this are added 33.14 parts dextose, 0.23 part sodium alginate, 0.11 part vanilla flavor and 0.23 part calcium chloride (10% solution). Quinine sulfate is added to the above composition in the amount of 15 ppm.

The pudding emulsion is a conventional product made by heating the water to 140°F, adding the remaining ingredients, heating the solution to 155° to 160°F, homogenizing in two steps at 7,000 and 500 psi and cooling to 34° to 38°F. The pudding formulation is made by premixing the dextrose and sodium alginate and adding them to the standard pudding emulsion at 150°F. The remaining ingredients, and the quinine sulfate, are then added.

The product had a slightly elastic character and at –10°F a penetrometer value of 29.3 mm, compared to a commercially available pudding (Rich's Chocolate Pudding) which gave a penetrometer reading of 1.3 mm. The water activity of the pudding averaged 0.852 at 73°F.

Example 2: Illustrative of the oil-in-water emulsion based products made in accordance with this process is a nondairy creamer which can be stored in a freezer until ready for use, thawed, and then used or left at room temperature for at least about ten days without spoilage. The product may also be left in a refrigerator for a lengthy period of time without spoilage. This product is useful as a coffee lightener and sweetener.

The coffee lightener contains about from 35 to 45% water, dextrose in a ratio to water of about from 1.5-1:1, about from 10 to 30% fat and about from 3 to 20 ppm of a quinine salt. The fat content preferably comprises from 50 to 100% unsaturated type fats. Other ingredients are included in conventionally minor amounts, such as salts, emulsifiers and a protein concentrate.

An example of a suitable formulation contains 40 parts water, 0.14 part K_2HPO_4, 0.14 part Na_2HPO_4, 0.02 part sodium acid pyrophosphate, 0.50 part soy protein isolate, 0.30 part polysorbate 60, 0.30 part sodium stearoyl lactylate, 0.40 part mono- and diglycerides, 0.10 part potassium sorbate, 40.10 parts dextrose, 16.00 parts soybean and 2.00 parts coconut oil. Quinine sulfate is added to the above composition in the amount of 7 ppm.

The product was made as follows: Heat the coconut oil to 155°F and dissolve in the emulsifiers. The foregoing is then added to the soybean oil. The water is heated to 150°F and the salts and protein are added. The dextrose is added to the aqueous solution, which is then held at 170°F for one minute, after which the oil blend is added. The quinine sulfate is then added. The entire batch is homogenized at 3,000 and then at 500 psi and cooled to 40°F.

This product does not flow, but is a semi-solid when held at 5°F for 3 days. This formulation has a water activity value of 0.9 measured at 72°F and upon storage at 40°F for 32 days, maintained its stability and did not exhibit any off-flavor. In addition, the product maintains its stability at room temperature for many days.

DICARBOXYLIC ACIDS AND DERIVATIVES

Readily Soluble Sorbic Acid Formulations

Since sorbic acid shows a strong antibacterial force in low toxicity, it is widely used as a preservative agent for drinks and foods. However, sorbic acid is sparingly soluble in water and solubility for water is in the order of 0.16 g/100 ml at normal temperature, but it dissolves more in water of which the pH value is high by formation of salts. Sorbic acid, however, shows antibacterial action in a form of a free acid and in the case of using as a preservative agent it is preferred to lower the pH value of foods as much as possible. The amount of sorbic acid added varies according to the kind of foods, but in general, is about 0.05 to 0.3% by weight. This addition amount is close to the saturated solubility for water of sorbic acid and it is extremely difficult to perfectly dissolve in foods at low pH values.

As a means for improving the solubility of sorbic acid it is considered to divide it finely. Powdered sorbic acid, however, shows a strong irritating action on the mucous membranes of human beings and it will harm the working environment to make sorbic acid into finely divided powder high in scatterability.

The mucous membrane irritating action of sorbic acid can be avoided by use of its salts, but the antibacterial action inherent in sorbic acid cannot exhibit itself if it is kept in a salt condition.

Sorbic acid-containing powder or granules free from scatterability and rapidly dissolvable in water are provided by the process developed by *R. Ueno, T. Matsuda and S. Inamine; U.S. Patents 4,172,897; October 30, 1979 and 4,308,281; December 29, 1981; both assigned to KK Ueno Seiyaku Oyo Kenkyujo, Japan.*

This process provides sorbic acid-containing powder or granules, free from scatterability and rapidly dissolvable in water, comprising 5 to 90% by weight, preferably 10 to 80% by weight, of finely divided sorbic acid having particle diameter of 50 μ or less, 10 to 95% by weight, preferably 20 to 90% by weight, of an easily water-soluble substance which is a solid at normal temperature and 0 to 2% by weight, preferably 0.05 to 1% by weight, of a hydrophilic surface-active agent, characterized by having particle diameter of 300 μ or more, preferably 500 to 1,500 μ.

Such sorbic acid-containing powder or granules, are obtained by drying after making powder or granules having particle diameter of 300 μ or more from a mixture consisting of finely divided sorbic acid with particle diameter of 50 μ or less, an easily water-soluble substance which is solid at normal temperature, water and/or an aqueous organic solvent and optionally, a hydrophilic surface active agent, or by making powder or granules having particle diameter of 300 μ or more from the dry mixture after drying the mixture.

The easily water-soluble substance which is a solid at normal temperature is preferably selected from among additives indispensable for the manufacture of foods. As substances of this kind mention is made of sugars, such as cane sugar, grape sugar, fruit sugar and so on; sugar alcohols, such as sorbitol, mannitol and so on; organic acids, such as citric acid, malic acid, tartaric acid, fumaric acid and so on; salts of organic acids, such as the respective sodium salts or potassium salts of acetic acid, citric acid, malic acid, tartaric acid, fumaric acid and sorbic acid, monosodium glutamate, sodium inosinate and so on, as well as sodium primary phosphate, sodium secondary phosphate, sodium tertiary phosphate, sodium pyrophosphate, acid sodium pyrophosphate, sodium metaphosphate and sodium polyphosphate or their corresponding potassium salts. These substances are preferably selected according to the kind of foods and can be used alone or as a mixture of two members or more.

Sorbic acid is used in finely divided form with particle diameter of 50 μ or less so as to be able to immediately dissolve when added to foods.

As the hydrophilic surface active agent, for instance, cane sugar fatty acid esters, preferably those ones which are 11 or more in the HLB, lecithins, preferably high purity lecithins, fatty acid esters of sorbitan (Span-20, for instance), reaction products between sorbitan fatty acid esters and polyoxyethylenes (Tween 20, for instance) and so forth are used alone or as a mixture as the easily, water-soluble substance; in some cases, no surface active agents are required.

Example 1: 4 g of cane sugar fatty acid ester (HLB 15) and 20 g of glycerin were added to 150 ml of water and dissolved by heating at 80°C. The aqueous solution obtained was cooled at about 50°C. Then 1,000 g of finely divided sorbic acid (average particle diameter 30 μ) and 976 g of DL-malic acid were placed in a kneader and uniformly mixed together. While adding the aforesaid aqueous solution the mixture was kneaded for 30 minutes by means of a kneader. Then columnar granules 1 mm in diameter were fabricated from this kneaded substance by means of an extrusion type granulator and dried in hot air at 70°C.

The sorbic acid-containing granulated product so obtained was suitable as one for pickles.

Example 2: 350 g of sodium secondary phosphate dodecahydrate was placed in the kneader with a jacket and melted by heating at 40°C. 100 g of finely divided sorbic acid with average particle diameter of 25 μ was added to this melt and stirred to bring to uniformly suspended condition. After that, with addition of 50 g of anhydrous sodium secondary phosphate the mixture was cooled down to 30°C for solidification with further continued stirring. Solids were further left to stand for 2 hours at 20°C. Then they were pulverized to sieve powder 0.3 to 0.7 mm in particle diameter.

Example 3: 600 g of finely divided sorbic acid (average particle diameter 30 μ) and 389 g of citric acid powder were placed in the kneader. Then, with addition of aqueous solution prepared by dissolving 1 g of Tween-20 and 10 g of glycerin in 100 ml of water the mixture was kneaded for 30 minutes. The kneaded substance was pulverized to sieve particles 0.3 to 0.7 mm in particle diameter.

Example 4: 5 kg of water-washed cucumbers (about 5 to 10 cm long) and 5 ℓ of salt water (10°Bé) were placed in a tub. The tub was covered and a lightweight stone was laid thereon to pickle at 30°C. Two days later common table salt was further added to maintain 10°Bé. It was fermented for three weeks in such condition and checking was made of the condition where the membrane-producing yeast occurred during the fermentation. The occurrence of the membrane-producing yeast is not desirable because it becomes the cause of spoiling flavor and of putrefaction.

When 0.16% of the preservative of this process was added to the salt water, no putrification of the pickles was observed during 21 days of observation. When 0.13% of potassium sorbate was added putrification was noted at 18 days and with no additive, putrification started at 7 days.

Sorbic Acid Stabilization of Cheese

R. Ueno, T. Matsuda and S. Inamine; U.S. Patent 4,207,350; June 10, 1980; assigned to KK Ueno Seiyakuoyo Kenkyujo, Japan have also used a readily soluble sorbic acid in the preservation of processed cheese.

This process comprises admixing with natural cheese during the steps of preparing process cheese from natural cheese sorbic acid-containing powder or granules having particle diameter of 300 microns or more comprising 5 to 90% by weight of finely divided sorbic acid and 10 to 95% by weight of an easily water-soluble substance which is a solid at normal temperature.

The following can be used in the process as the easily water-soluble substances remaining solid at normal temperature: sugars such as cane sugar, grape sugar, milk sugar, fruit sugar and so on; sugar alcohols such as sorbitol and so on; organic acids such as citric acid, tartaric acid, malic acid and so on; phosphates, particularly sodium salts and potassium salts, such as primary phosphate, secondary phosphate, tertiary phosphate, polyphosphate, metaphosphate, hexametaphosphate, acid pyrophosphate and so on; organic acid salts, particularly sodium salts and potassium salts, such as citrate, malate, tartrate, sorbate and so on.

However, it is not preferred to use sorbate in overly great amounts because laws and regulations impose limitations on the amount of sorbic acid added to cheese plus the amount of sorbate added as sorbic acid, and when the amount of sorbate added is made greater, it is required to lower the amount of sorbic acid added, with the result that the pH lowering effect by sorbic acid will be less. These easily water-soluble substances could be used in admixture of two members or more.

The sorbic acid preparation (powder or granule) suited to the purpose of this process cannot be obtained merely by mixing these easily water-soluble substances and finely divided sorbic acid. For instance, it is required to knead together by adding water or an organic solvent or mixed solution of both to a mixture of finely divided sorbic acid and easily water-soluble substances and make powder or granules from the resultant kneaded substance, followed by drying. Or otherwise, the kneaded substance may be first dried and then powder or granules made from the dried substance.

Furthermore, in the case of adding water or an organic solvent, it is also possible to dissolve therein the easily water-soluble substance in advance prior to its addition. As the organic solvent there can be used methanol, ethanol, p-propyl alcohol, isopropyl alcohol, acetone and so forth. If the easily water-soluble substance is the one which is low in its melting temperature (such as sorbitol), powder or granules could be obtained by mixing finely divided sorbic acid to that melt, followed by cooling without drying.

Example 1: 5 kg of finely divided sorbic acid having particle diameter of 50 μ or less and 5 kg of powdered potassium citrate (monohydrate) were mixed together in a kneader. Then the mixture was kneaded with addition of 1,300 ml of water. Columnar granules 0.8 mm in diameter were made from the kneaded substance obtained by means of the extrusion type granulating machine. Granules were dried to give a granulated preparation containing 50% by weight of sorbic acid.

Example 2: A granulated preparation was prepared in the same way as in Example 1 except that there was used sodium citrate (dihydrate) in lieu of potassium citrate (monohydrate).

Example 3: 4 kg of minced Gouda cheese, 1 kg of cheddar cheese, 60 g of sodium secondary phosphate as an emulsifying agent and 60 g of sodium citrate were placed in a kneader heated by passing water held at 80°C through the jacket to mix together for one minute. With further addition of preparations as indicated below the mixture was mixed for another 10 minutes, whereby there was prepared process cheese. After that, it was wrapped in plastic film and left to stand at room temperature for one day. It was cooled down to room temperature and then cut off to a size of 2 x 3 x 0.5 cm. It was placed in the aseptic Schale and preserved at room temperature (20° to 31°C) to observe the condition in which it went rotten.

Samples containing the preparations of Examples 1 or 2 above went at least 13 days before putrefaction occurred while samples containing potassium sorbate powder or sorbic acid powder putrefied in 6 to 8 days.

Botulism-Preventing Sorbic Acid Formulations

The disclosure by *R. Ueno, T. Matsuda, T. Kanayama, K. Tomiyasu, Y. Fujita and S. Inamine; U.S. Patent 4,299,852; November 10, 1981; assigned to Kabushiki Kaishaveno Seiyakuoyo Kenkyojo, Japan* relates to processes for the preparation of botulinal resistant meat products.

"Meat products" as are used in this process generally cover pork products, especially ham, bacon and various sausages. Of these, salted meat products contain as the salting agent table salt and nitrite (sodium or potassium salt).

According to practices in the United States at least 120 ppm to the meat product, sometimes as much as 156 ppm, of nitrite is added for inhibiting the growth of *Clostridium botulinum.*

These processes for the preparation of meat products exhibiting excellent resistance to *Clostridium botulinum* comprise adding to the meat, in the course of its processing, (1) sorbic acid or potassium sorbate, (2) a glycerol monoester or monoesters of a C_{10} and/or C_{12} fatty acid or acids, (3) a nitrous acid compound in the amount corresponding to the minimum nitrite radical necessary for the color development of the product, and/or (4) sodium hexametaphosphate.

According to the processes, (1) sorbic acid or potassium sorbate is added in an amount of 0.05 to 0.2% by weight to the meat. The glycerol monoester of C_{10} and/or C_{12} fatty acids as the component (2) are added, either singly or as a mixture of more than one compound, in the amount of 0.001 to 0.5% by weight to the meat. The component (3), nitrous acid compound, may be nitrous acid, potassium nitrite and the like, preferably sodium nitrite. The minimum amount of nitrite radical (NO_2^-) necessary for the color development is not greater than 30 ppm, preferably 20 to 30 ppm. The component (4), sodium hexametaphosphate, is used in the amount of 0.2 to 0.5% by weight of the starting meat material.

It has been reported that glycerol monoesters of C_{10}–C_{12} fatty acids show antimicrobial activity against bacteria in general. According to these studies, however, the glycerol monoester of C_{10} or C_{12} fatty acid shows no growth inhibiting effect on *Clostridium botulinum,* in meat products, if used alone and at concentrations not detrimental to the taste of meat products. Furthermore, they are again ineffective when used concurrently with a minor amount of nitrous acid compound, for example, that of the amount corresponding to no more than 30 ppm of nitrite radical.

Sodium hexametaphosphate, which also is known to have antimicrobial activity, again is incapable of inhibiting toxin formation of *Clostridium botulinum* in meat products, if used alone or co-used with a minor amount of a nitrous acid compound, for example, that of the amount corresponding to no more than 30 ppm of nitrite radical.

Concurrent use of sodium hexametaphosphate and glycerol monoester of C_{10} and/or C_{12} fatty acid is not effective in meat products.

Furthermore, potassium sorbate shows almost no effectiveness without the concurrent use of nitrite, although it is considerably effective if used with a nitrite.

It is surprising, therefore, that this process using the compounds of each suitable amount, in suitable combination, effectively inhibits the growth of *Clostridium botulinum* in the meat products and prevents toxin formation, when none of the compounds shows such effectiveness if used alone.

In the following example, the additives tested are abbreviated as follows: Sok for potassium sorbate, MC_{12} for lauryl monoglycerides and SHMP for sodium hexametaphosphate. The formula $NaNO_2$ is used for sodium nitrite.

Example: Minced lean pork meat (sausage) was used as the starting material. In each test lot 5 kg of the minced meat was mixed with 2.0% table salt, 1.0% sugar, 0.055% sodium erythorbate, 0.5% blended spice, 0.3% smoke powder and 1.0% sodium caseinate; and the chemicals of the amounts varied for each lot (Lots 1 through 6), cut for 10 minutes with a silent cutter, transferred into a stainless steel cake mix blender, and inoculated with the *Clostridium botulinum* spore suspension at a rate of 10^3 spores/gram, followed by mixing for 5 minutes. The mixture was then filled into a vinylidene chloride film tube (90 mm in outer diameter) (approximately 120 g/tube), heated for 60 minutes in 75°C water and cooled rapidly in ice water.

Lot 1 contained 156 ppm $NaNO_2$; Lot 2 contained 40 ppm $NaNO_2$ and 0.26% Sok; Lot 3 contained 30 ppm $NaNO_2$, 0.20% Sok and 0.05% MC_{12}; Lot 4 contained 30 ppm $NaNO_2$, 0.20% Sok and 0.5% SHMP; Lot 5 contained 30 ppm $NaNO_2$, 0.20% Sok, 0.05% MC_{12} and 0.5% SHMP; and Lot 6 contained only 0.268% Sok, 0.05% MC_{12} and 0.5% SHMP.

The above lots were tested for toxin formation. Lots 1, 2 and 4 showed toxin formation on days 7, 6 and 7 respectively. Lots 3 and 5 which are prepared by this process tested for toxin on days 10 and 15 respectively. Lot 6 which tested for toxin on day 9 proved that according to this process, antibotulinal stability can be improved over that of conventional products, without using nitrite.

Potassium Sorbate plus Nitrite in Bacon

A process for making bacon which shows antibotulinal stability, in spite of the substantially lower level of sodium or potassium nitrite has also been developed by *R. Ueno, T. Matsuda, T. Kanayama, Y. Fujita and S. Inamine; U.S. Patent 4,305,966; December 15, 1981; assigned to KK Ueno Seiyaku Oyo Kenkyujo, Japan.*

This process for making bacon having antibotulinal stability, comprising adding to pork, which has been cured with a curing agent so as to contain no more than 30 ppm of nitrite radical (NO_2^-) and not higher than 0.26% of potassium sorbate based on the weight of meat, an aqueous solution of a water-soluble organic acid and/or inorganic acid, and optionally sodium hexametaphosphate, thereby lowering the average pH of the bacon to approximately 6.0 or below.

This process is based on the discovery that when an aqueous solution of a water-soluble organic acid and/or inorganic acid is injected into the meat, which has been cured with known curing agent so as to contain no more than 0.26% of potassium sorbate and 30 ppm of nitrate radical NO_2^- by weight of the meat, and then the meat is smoked, the pH of the bacon can be lowered to 6.0 or below, without substantially degrading the bacon quality.

When an aqueous solution of a water-soluble organic acid and/or inorganic acid is injected into a thoroughly cured meat lump, in an attempt to lower pH of the latter, surprisingly the water-retaining property of the meat shows little lowering in spite of the pH drop. This is probably due to the fact that salted meat becomes to a certain degree resistant to pH drop.

The lowering of pH bacon brings about conspicuous advantages as follows. (1) Such a low pH as 6.0 or below notably increases the effect of sorbic acid against *Clostridium botulinum*. The antibacterial activity of sorbic acid is derived from its nondissociative molecules. As the pH is lowered, the ratio of its nondissociative molecules increases exponentially in the vicinity of specified pH value. It is therefore extremely useful to lower the pH, for causing the sorbic acid to exhibit its maximum effect. Consequently, the amount of sorbic acid to be added to bacon can be reduced at the lower pH, to make its irritation of mucous membrane less, when the bacon is fried. (2) The nitrite radical remaining in the bacon becomes less as the pH is lowered. Consequently, the amount of N-nitroso pyrrolidine formed when the bacon is fried can also be reduced. (3) The lower pH is inadequate for the growth of *Clostridium botulinum*. This means that the bacon prepared by this process exhibits stronger antibotulinal stability.

According to this process, the antibotulinal stability can be further increased by injecting into the cured meat lumps, hexametaphosphate which is an acidic phosphate, together with the aqueous acid solution. No such antibotulinal stability increasing effect of, for example, sodium hexametaphosphate, is observed with other polyphosphates, for example, sodium pyrophosphate, acid sodium pyrophosphate or sodium tripolyphosphate. This effect is clearly recognizable only with sodium hexametaphosphate. Other phosphates, for example, acid sodium pyrophosphate, show the action to inhibit the color development of bacon pro-. moted by the nitrite, thus adversely affecting the quality of bacon.

In a preferred embodiment, the meat lumps are cured with an ordinarily used pickle solution containing nitrite radicals (NO_2^-) at concentrations not higher than 30 ppm (corresponds to 45 ppm of $NaNO_2$) by weight of the meat (such as table salt, ascorbate or erythorbate, containing if necessary a polyphosphate, sugars, and the like). Then an aqueous solution of a water-soluble organic acid and/or inorganic acid is injected into the meat, which is subsequently smoked to be converted to bacon.

In this process, the nitrite radical level is specified to be not higher than 30 ppm, because such is sufficient for imparting the desired color and taste to the meat, and with which the residual nitrite radical in the bacon becomes substantially zero, by the time the bacon is sold to the consumers through stores so long as the bacon is made under conditions of a pH value of 6.0 or less. Thus, the addition of not higher than 30 ppm of nitrite radical in reality presents no health problem to the consumers.

Examples of the water-soluble organic acids to be used in the process include acetic acid, propionic acid, lactic acid, malic acid, succinic acid, tartaric acid, adipic acid, fumaric acid, citric acid and glucono delta-lactone (GDL), and an example of the inorganic acid is phosphoric acid.

In an aqueous solution of the water-soluble organic acid and/or inorganic acid, the concentration of the acidic substance or substances should be in the range of 1 to 5% by weight. The amount of injection of the aqueous solution should be 2 to 10% by weight of the meat, and 0.05 to 0.15% by weight as the acidic substance or substances. With the injection as above, pH of bacon can be lowered by approximately 0.2 to 0.4. Because the conventional commercial bacon has the pH of 6.2 to 6.3, the specified addition of the acidic substance or substances can lower its pH to 6.0 or below, thus fully accomplishing the intended effect. The pH-lowering in this order can considerably enhance the effect of sorbic acid. For instance comparing the cases wherein the pHs are 6.5 and 6.0, respectively, the concentration of nondissociative molecules of sorbic acid under the pH of 6.0

is approximately double that under the pH of 6.5. Theoretically this means a doubled effect of sorbic acid. Thus, even a minor pH lowering such as about 0.1 shows a substantial merit.

The preferred concentration of sodium hexametaphosphate is normally 0.2 to 0.5% based on the weight of meat. The phosphate is itself weakly acidic, and can lower the pH of the product, although slightly. It is recommended that the sodium hexametaphosphate should be dissolved in the aforesaid aqueous solution of acidic substance or substances, though it is permissible to add it as a separate aqueous solution. Obviously, it is added at the time the aqueous acid solution is injected, i.e., after the meat is cured, whereby the growth of *Clostridium botulinum* and toxin formation can be very effectively inhibited.

Melt Coating Containing Sorbic Acid

K. Sato, A. Asahi and T. Koyama; U.S. Patent 4,267,198; May 12, 1981; assigned to Daicel Ltd., Japan describe a melt-coated preparation having a high sorbic acid content suitable for use as a preservative for meat and fish paste products.

In preparing a sorbic acid preparation by the melt coating method, sorbic acid particles are dispersed in a melt of a coating agent to form a slurry, which is then granulated (i.e., coated sorbic acid particles are collected into grains) by spraying, etc., and solidified by cooling. When the concentration of sorbic acid in the slurry is increased to meet the objects of this process, the slurry becomes very highly viscous, and it is difficult to charge the slurry through a pipe or to granulate it by spraying. It is for this reason that formerly the amount of the coating agent was limited to at least 1.5 times the amount of sorbic acid, and, practically speaking, the amount of the coating agent is about 4 times the amount of sorbic acid, which corresponds to a sorbic acid/coating agent ratio of 0.25.

Also it has previously been found that the high viscosity accompanying a high-concentration slurry could be overcome to some extent by using sorbic acid having a particle diameter in the range of 20 to 80 microns which is larger than the fine powdery sorbic acid used in conventional techniques and a melt-coated preparation containing sorbic acid in an amount of 0.75 to 1.2 times the weight of the coating agent was obtained.

This process has resulted from further investigation and provides the following preparations and process for their production.

One embodiment is a melt-coated preparation comprising sorbic acid particles at least 80% by weight of which have a particle diameter of greater than 80 to about 150 microns and a coating agent composed mainly of solid fat, the amount of the sorbic acid being about 0.9 to 2 times the weight of the coating agent.

Furthermore, in another embodiment the coating agent is composed of a solid fat and an acetomonoglyceride of a higher fatty acid or a monoglyceride of a fatty acid containing 6 to 18 carbon atoms.

This process also provides a method for the preparation of a melt-coated preparation which comprises melting a coating agent composed of a solid fat or both a solid fat and a glyceride, dispersing sorbic acid in the melt, granulating the resulting slurry, and cooling the slurry to solidify it; wherein at least 80% by weight of the sorbic acid has a particle diameter of greater than 80 to about 150 microns and preferably 90 to 130 microns, the amount of the sorbic acid is about 0.9 to about 2 times the weight of the coating agent, and temperature of the slurry to be granulated is about 75° to about 110°C.

Example 1: Sorbic acid was screened through an 80-mesh sieve (the mesh distance being 175 microns) to remove coarse particles. Particles which passed through the 80-mesh sieve were subjected to a 200-mesh sieve (the mesh distance being 74 microns) to remove fine particles. The particle diameter of the resulting sorbic acid particles was found to be mainly 80 to 150 microns when determined by a sedimentation method.

A hardened beef tallow having a melting point of 55° to 60°C was melted, and while stirring 500 parts of the molten beef tallow maintained at 80° to 85°C, 500 parts of the above sorbic acid was added and dispersed fully to obtain a uniform slurry. The slurry had a viscosity, as measured by a B-type viscometer, of about 12 poises. When it was allowed to stand for about 1 hour, little sedimentation was observed.

The slurry was made into fine particles by a granulator equipped with a rotary disc type spraying mechanism. The particles were cooled while they fell through the air at 20° to 30°C to obtain a coated sorbic acid preparation containing 50% of sorbic acid. The particle diameter of the preparation, measured by a sedimentation method was 40 to 350 microns with an average particle size of 180 microns.

One part of the resulting preparation was added to 200 parts of water at 20°C containing 0.1% of a nonionic surface active agent, and the mixture was stirred for 10 minutes. The mixture was then filtered, and the filtrate was titrated with a 0.1 N aqueous solution of sodium hydroxide to determine the amount of the acid eluted. The elution rate of the acid based on the sorbic acid contained in the preparation was only 1.67%.

Then from 450 parts of the same hardened beef tallow and 550 parts of the same sorbic acid, a slurry having a viscosity of 20 poises was prepared. A 55% preparation formed in the same way as above had an elution rate of 1.64%.

Example 2: Coarse particles of sorbic acid were pulverized and sieved to obtain particles which passed through a 100-mesh sieve but remained on a 200-mesh sieve. The particle diameter of these particles was 80 to 150 microns.

600 parts of the resulting sorbic acid particles were added to 400 parts of a hardened beef tallow having a melting point of 55° to 60°C and maintained at 85° to 90°C to form a uniform slurry. The slurry was granulated by a spray granulator, and cooled with cold air. The resulting sorbic acid preparation had a particle diameter of 150 microns on an average, and the elution rate of the preparation measured by the same method as in Example 1 was 3.2%.

C_{13} to C_{16} Dicarboxylic Acids

Antibacterial agents such as monoglyceride caprate or sodium propionate have been used to prevent the food contamination with bacteria or microorganisms during the course of processing or distribution. However, since such conventional antibacterial agents do not have a sufficient antibacterial ability, it has been proposed to add an acid together therewith for maintaining the pH at a lower level to reinforce the antibacterial ability.

However, the use of this prior method has been limited in that it is effective, by its own nature, only to acidic foods because of sourness caused by an acid added thereto. Especially, in the case of monoglyceride caprate, its use is limited from the viewpoint of long-term stability, because an ester bond existing in the molecule thereof renders it readily susceptible to hydrolysis.

Accordingly, an object of the development by *K. Hata, M. Matsukura, S. Hatano, K. Ohsima, I. Kano, H. Umeda and H. Awaji; U.S. Patent 4,247,569; January 27, 1981; assigned to Jujo Paper Co., Ltd., Japan* is to provide an antibacterial agent which has an improved antibacterial ability as well as safety and in which the aforementioned shortcomings of the known antibacterial agents are eliminated.

These antibacterial agents comprise one or more alpha, omega, C_{13-16} dicarboxylic acids or their alkali metal salts. These antibacterial agents are stable in an aqueous solution and a small amount of addition to foods exhibiting an improved antibacterial effect on gram-positive bacteria such as *Staphylococcus aureus* and *Bacillus subtilis.*

The following test method is used in the following examples. One loop was inoculated from slant culture of *Staphylococcus aureus* and *Bacillus subtilis,* respectively, into a liquid culture medium containing 1% glucose, 0.5% meat juice extract, 0.5% peptone and 0.3% sodium chloride and separately subjected to a pre-culture at 37°C for 24 hours. Then, each 0.1 ml of the two resultant cultures from the pre-culture was separately inoculated into 10 ml of each of three culture media having the same composition as that used for pre-culture, to which the dibasic acid antibacterial agent of this process, conventional monoglyceride caprate and sodium propionate were respectively added. Then, these three cultures were subjected to a shake culture in an L-tube at 30°C for 24 hours. Consequently, the concentration of the antibacterial agents at which the growth of each bacterial could be inhibited was determined on the basis of the turbidity of culture due to the growth of bacteria (absorbance at the wavelength of 660 mμ was measured by means of a spectrophotometer).

Example 1: In tests using *S. aureus,* the minimum inhibitory concentration (mg/ℓ) was found to be 10 for the compounds 1,12-dodecamethylene dicarboxylic acid, 1,13-tridecamethylene dicarboxylic acid and 1,14-tetradecamethylene dicarboxylic acid.

The prior art monoglyceride caprate required 500 mg/ℓ and sodium propionate required 100 mg/ℓ.

In tests with *B. subtilis,* the minimum inhibitory concentrations for 1,12-dodecamethylene dicarboxylic acid, 1,13-tridecamethylene dicarboxylic acid and 1,14-tetradecamethylene dicarboxylic acid were 25, 10 and 15 mg/ℓ, respectively. The monoglyceride caprate required 250 mg/ℓ and sodium propionate required 250 mg/ℓ.

Example 2: When equimolar mixtures of dicarboxylic acids or salts of this process were used in the culture media at pH 6, the minimum inhibitory concentrations in mg/ℓ obtained for the following mixtures against *S. aureus* were 70 for a mixture of azelaic acid and brassylic acid, 25 for a mixture of 1,12-dodecamethylene dicarboxylic acid and disodium 1,13-tridecamethylene dicarboxylate while monoglyceride caprate required 500 mg/ℓ and sodium propionate required 1,000 mg/ℓ.

OTHER ACIDS IN MEAT PRODUCTS

Hypophosphorous Acid and Salts in Smoked Fish or Meat Products

The disclosure by *J.S. Thompson and J.F. Jadlocki, Jr.; U.S. Patents 4,277,507; July 7, 1981; and 4,282,260; August 4, 1981; both assigned to FMC Corporation* relates to a composition which inhibits the growth of *Clostridium botulinum* in preserved meat and fish products that have been smoked. For many years, it has been standard practice to add sodium nitrite to preserve meat products such as ham, bacon and other meats, e.g., frankfurters, baloney, Thuringian sausages and salami, having a small particle size. The sodium nitrite is added to inhibit the growth of *Clostridium botulinum* and the production of enterotoxin in the smoked meat products during storage. The addition of sodium nitrite also maintains a pleasing pink color in the meat products.

The presence of sodium nitrite as food additive, and particularly the presence of sodium nitrite in bacon and other smoked meats that are cooked at high temperatures, has become of increasing concern with the knowledge that sodium nitrite can combine with secondary and tertiary amines in cooked meats (particularly bacon) to form nitrosamines. Many nitrosamines have been shown to be carcinogens in animals, and the nitrosamine that is commonly found in fried bacon, nitrosopyrrolidine, is a known carcinogen.

It is quite obvious that reducing the nitrites present in smoked meat and fish is a desirable goal, but it is also necessary to prevent the production of deadly botulinal toxin that may occur on storage.

In accordance with this process, the growth of *Clostridium botulinum* and the production of botulinal toxin is inhibited during storage of smoked meat and fish products by addition thereto of an effective amount of a compound selected from the group consisting of hypophosphorous acid, sodium hypophosphite, potassium hypophosphite, calcium hypophosphite and manganese hypophosphite. Sodium hypophosphite is conveniently used in the form of its monohydrate, $NaH_2PO_2 \cdot H_2O$. The amount (120 ppm) of sodium nitrite that is customarily added to such meat and fish products may be reduced to one-third of that amount or eliminated entirely. The addition of hypophosphorous acid and/or a salt of hypophosphorous acid is believed to suppress or block the formation of N-nitrosamines upon cooking smoked meat and fish products containing sodium nitrite.

The amount of hypophosphite salt (or hypophosphorous acid) that is added may vary with the meat product, the particular salt and the presence or absence of sodium nitrite; and desirably is about 1,000 to 3,000 ppm. It is preferred that about 3,000 ppm of hypophosphite salt be added if all sodium nitrite is eliminated from the meat or fish product. About 1,000 ppm of sodium hypophosphite is effective when 40 ppm of sodium nitrite is also present in the meat composition. As indicated above, the potassium and calcium or manganese hypophosphite salts may be employed in similar amounts with good results. The choice of a particular hypophosphite salt will depend upon its cost and relative effectiveness.

In the practice of this process, the hypophosphorous acid (or its salt) may be added to the pickle (in solution or solid form) and can be added to the meat with the pickle or rubbed into the meat or fish after smoking. While hypophosphorous acid and its salts can be added to smoke cured meat products that are ground or chopped into small particle size, e.g., frankfurters, it is of particular advantage when added to meat products such as ham and bacon either before or after smoking as bacon in particular has the potential of forming nitrosamines if cooked in the presence of substantial amounts of sodium nitrite.

Example 1: Fresh salmon was steeped in a pickle solution containing 1.8 kg salt, 680 g sugar, 16 g saltpeter and 454 g sodium hypophosphite dissolved in 22.7 ℓ water. After 12 hours, the fish is removed from the pickle solution, washed under running water and smoked for 5 hours to an internal temperature of 80°C. The smoked salmon may be stored over a substantial period of time without spoilage.

Example 2: Slices of bacon for testing are randomly picked from 9 sliced bacon samples that have been treated with various quantities of sodium nitrite and sodium hypophosphite. The slices are uniformly inoculated (0.25 ml for 100 g of bacon) with a heat-shocked suspension of *Clostridium botulinum* spores. The inoculum consists of 4 type A strains and 5 type B strains (36A, 52A, 77A, 10755A, ATCC 7949, 41B, 53B, 213B and Lamanna B). The average inoculum level is between 100 and 500 spores per gram of bacon. Vacuum packages (100 grams of bacon per package) of inoculated bacon is prepared for each of the 9 bacon samples.

The packages of each sample incubated at 27°C are tested for botulinum toxin after 0, 7, 10, 14, 24 and 35 days storage.

Four test packages are examined on each of these days for each sample and the day of first toxicity is reported. Testing is terminated when all packages are swollen. Sample 1 with no additive showed first toxicity at 7 days and all samples were swollen at 14 days. In samples which contain no sodium nitrite, sample 4 which contained the most sodium hypophosphite (3,000 ppm) gave the best results, first toxicity at 24 days and total swelling at 35 days. Samples 5, 6, 7 and 8 contained 40 ppm of sodium nitrite with samples 6, 7 and 8 containing various amounts of sodium hypophosphite. All samples containing the sodium hypophosphite showed a delay in toxin formation over sample 5 with no added hypophosphite. Sample 8 with 3,000 ppm of sodium hypophosphite showed no toxin on day 35 while another control sample 9 which contained 120 ppm sodium nitrite tested positive for toxin on day 24.

Lactic Acid in Sausage

The procedure in the preparation of meat sausage is to blend the meats with suitable spices and flavoring and stuff the blended mixture into a suitable casing for further processing. If the ingredients are to be subjected to coagulation by fermentation, fermented sugars such as dextrose are added as an ingredient to the mix along with a bacterial starter. Fermentation is carried out at high humidity while at a temperature of about 80° to 100°F over a long period of time, such as up to 100 hours. In the event that the sausage mix is not to be subjected to coagulation by fermentation, the desired degree of coagulation can be achieved during the smoking and cooking steps. In any event, such coagulation is usually carried out in conjunction with the smoking step, or separate and apart therefrom, and preferably immediately after the smoking step.

The encased sausage mixture is smoked and cooked by suspending in a smokehouse for exposure to smoke, generated directly from hardwood or indirectly from liquid smoke, while at a temperature of about 110°F, until the product reaches the desired flavor.

The smoked or unsmoked sausage is cooked by continuing to heat in the smoke-house while raising the temperature of the smokehouse 15°F per hour until the product reaches an internal temperature of 150° to 160°F, but with the dampers closed to prevent excessive drying.

Thereafter the cooked product is washed, as by showering with hot and/or with cold water, followed with drying and cooling to about 40°F for storage.

D.G. Olson and H.E. Wistreich; U.S. Patent 4,263,329; April 21, 1981; assigned to B. Heller & Company have found that the stability and flavor of the product can be greatly improved when the cooked sausage is exposed, as by immersion, in an aqueous solution of lactic acid, with or without other edible organic acids or preservatives. Such other edible organic acids or preservatives include acetic acid, benzoic acid, sorbic acid, para-hydroxybenzoic acid, and the like and their cor-responding esters such as the methyl, ethyl, propyl and up to heptyl esters. The treatment by immersion can be carried out at 45° to 145°F.

The solution for best treatment of the sausage should be formulated to have a total titratable acidity within the range of 2 to 20% and preferably 4 to 10%, with at least 25% and preferably 40% of the total acidity being in the form of lactic acid. While the total acidity can exceed 20%, little if any benefit is derived from such higher concentration. The desired result can also be achieved with the solution having a total acidity as low as 1% but then a substantially longer expo-sure time is required to achieve the desired result. Immersion for a time sufficient to permit complete penetration is desired. From a practical standpoint, the de-sired results can be achieved by overnight immersion in the acid solution or im-mersion up to a number of weeks at about room temperature.

The post-treatment with lactic and/or lactic plus other edible organic acids operates to lower the pH of the product to less than 5.3 and preferably to within the range of 3.5 to 4.5. The flavor is enhanced by a tangy taste and the product is stable over long periods of time without undesirable increase in bacteria count.

Recipes were given for the production of beef sausage, Polish sausage and Thur-inger by the above method.

Acids in Processing of Dry Sausage

W.E. Kentor; U.S. Patent 4,279,935; July 21, 1981; assigned to Servbest Foods, Inc. has provided a process for preparing a dry sausage in a substantially shorter time period than with heretofore available methods.

Dry sausages are produced in this process by adding a quantity of natural bacteri-cides and bacteriostats to comminuted meats in amounts sufficient to at least inhibit growth of generally encountered pathogenic and nonpathogenic organ-isms, admixing a select amount of at least one acidulating material and relatively promptly (i.e., within about 3 hours after all acidulants are added) forming sausages from such acidulated meats so that a pH of less than about 5.7 is at-tained, thermally treating the so-formed sausages under time-temperature condi-tions sufficient to attain an average internal temperature within each sausage of at least about 58°C and subjecting the thermally treated sausage to a controlled drying environment having a relatively high temperature and a relatively low rela-tive humidity for a period of time sufficient to reduce the moisture level in each of the sausages to a maximum of approximately 35%.

In preferred embodiments of the process bactericides and bacteriostats are selected from potassium nitrite, sodium nitrite, and mixtures thereof and are preferably added to a given mass of comminuted meats in an amount equivalent to at least about 150 ppm and generally in the range of about 100 to 150 ppm. In certain preferred embodiments, the acidulating materials are selected from the group consisting of glucono delta lactone, lactic acid, acetic acid, phosphoric acid, maleic acid, citric acid, fumaric acid, tartaric acid, adipic acid, succinic acid, hydrochloric acid, anhydrides of such acids, edible salts of such acids and mixtures thereof. Preferred acidulating materials from this group comprise glucono delta lactone, lactic acid, acetic acid, hydrochloric acid, and mixtures thereof.

Example: 65% beef, 20% beef tripe, 15% beef fat (flank) were obtained and set aside. A condiment mixture comprised of 3.33% salt, 1.0% cane sugar, 0.4% ground white pepper, 0.2% cracked pepper, 150 ppm sodium nitrite, 0.07% garlic powder, 0.5% glucono delta lactone and 0.3% liquid smoke (Griffith Royal AA) was prepared and set aside.

The beef tripe was cooked in water until soft (water approximately 93°C) and then chilled to about 3°C. Then the beef and beef tripe were ground through a ⅛ inch plate. The beef fat was cooled to about –4°C and ground through a ³⁄₁₆ inch plate. The ground beef and beef tripe were admixed and the pH of the resultant meat mass was determined as being 6.3. The pH of the liquid smoke (which is a complex mixture of food-grade acids) was 2.2.

The ground beef and beef tripe and all dry ingredients were placed in a Hobart paddle mixer and were mixed for about 3 minutes. The temperature of the resultant admixture was about +3°C. The ground fat was then added and mixed for an additional 3 minutes. Temperature of resultant mass was about +2°C.

The ground product so-attained was placed in a piston-type air stuffer (Buffalo Model 300), well tamped, and stuffed into fibrous cellulose casing (Union Carbide 1 SLDS) and clipped to lengths of about 229 mm. The diameter of such casings was approximately 43 mm. The stuffed sausages at 4°C (internal temperature reported) were hung on rods and placed in a smokehouse at 37°C and 65% relative humidity (RH).

The cooking schedule was such that after 2 hours, the smokehouse was at 38°C and sausage recorded 5°C. After another 2 hours, the smokehouse was at 49°C and the sausage was at 26°C while the RH had reduced to 60%. After an additional period of 1¾ hours, the smokehouse registered 60°C and the sausage 43°C. A fourth period of 1¾ hours produced a temperature of 71°C in the smokehouse and 61°C in the sausage. RH had remained at 60%.

Hardwood smoke was introduced after the first hour and maintained for the total process. After the internal temperature of the sausages reached about 61°C and the resultant temperature was maintained for 15 minutes, the sausages were showered briefly with hot water at a temperature of approximately 82°C.

The sausages were then permitted to cool in ambient room temperature of about 26°C for approximately 4 hours. The pH of such sausages was determined to be 5.6 with a Radiometer Meter. The shrinkage of the sausage at this cooked stage was calculated to be 12%. The sausages were then transferred to a drying room. The drying room was operated at 21° to 23°C with the RH varying from 52 to 62%.

Humidity variations were the result of: (a) wet sausages elevating RH for the first 36 hours, and (b) nighttime RH was elevated because external "make-up air" had high RH (approximately 85% vs daytime RH of 60%). After 3 days, RH in the drying room was maintained between 50 and 58%. After 7 full days in the drying room, the total shrinkage calculated from stuffed weight of sausages, was 37%.

After 7 days, the sausages were removed and examined. The sausages were relatively firm and had acceptable color, texture, odor and flavor as evaluated by a taste panel of knowledgeable sausage makers. Further the sausages had a pH of 5.65, a salt level of 5.1%, an Aw (water activity) of 0.86 and a moisture content of 29%.

To confirm that toxins pathogenic to man and those most commonly pathogenic to certain species of animals, do not survive in this process, suitable laboratory tests were conducted on randomly selected sausages from the above example. Negative reports, i.e., absence of toxins frequently encountered in sausages, was obtained in each instance.

Acid-Proteolytic Enzyme Treatment

A process which does not require refrigeration has been developed by *V.P. Cerrillo; U.S. Patent 4,207,344; June 10, 1980* for protecting animal derived foodstuffs against spoilage.

The process has two stages, one the stabilizing stage and the other the recovery stage. In the first stage the foodstuff, an animal ("animal" is used to identify mammals, birds, fishes, selachii, crustaceans, etc.) or part thereof, is immersed in a stabilizing liquid composed of an acid or alkaline buffer solution, a proteolytic enzyme which is active in an acid or alkaline medium, depending upon the pH of the stabilizing liquid, and an antioxidant. Thereafter, the product may be stored at room temperature in either closed or open containers.

The second stage is composed of three steps. In the first step, the pH of the foodstuff is adjusted to a preferred level by immersing the product in an acid or alkaline solution. This results in a rapid change of pH from acid to alkaline or vice versa which has been given the name "ionic blow".

In the second step, carried out after the product has been drained, the product is placed in another receptacle which contains a hypertonic solution at a selected pH to dehydrate the cells of the foodstuff, eliminating the hypotonic solution therefrom.

In the third step the product is placed in yet another receptacle which contains a rehydrating solution. Here, the foodstuff recovers ions it may have lost during preceding steps. This results in the foodstuff being restored, as nearly as possible, to its original, fresh or unpreserved form.

In the stabilizing stage, use is made of the hydrogen ion to protect the proteins, advantage being taken of the inhibiting effect of such ions on enzymatic mechanisms which cause autolysis of cells. The hydrogen ions furnished by the donor also create a bacteriostatic and fungistatic environment.

The component furnishing the hydrogen ions may be a potable organic acid such as acetic, citric, or lactic or an inorganic acid, preferably potable, such as hydrochloric or phosphoric. In any case the acid should be free of pollutants.

The enzymes employed for this process to break down the protein may be of animal or vegetable origin, or they can be produced by different strains of bacteria or molds (fungi). Enzymes successfully used are pepsin, papain, and bromelain.

To guard against autolysis of the foodstuff cells, the selected enzyme are used in a concentration which is approximately one-tenth of that which would result in proteolytic acitivity. While the enzyme concentrations used do not result in breakdown of the cells, they are nevertheless capable of effecting the wanted dissolution of the intercellular cement.

Enzymatic action is insured by using a buffer to adjust the pH of the stabilizing liquid to a level ≤5 which is optimum for the particular enzyme being used. As a buffer, the same acid employed as the hydrogen ion donor or a salt of that acid is used.

It is also necessary to inhibit decomposition of fatty constituents. This is achieved in situ by adding a potable antioxidant to the stabilizing liquid. The amount of antioxidant is correlated to the amount of fat in the product so that the amount of antioxidant will not exceed the limits allowed for the use to which the product will be put.

The concept of treating proteic foodstuffs with a proteolytic enzyme is not unknown. However, heretofore an enzyme has not been employed in a food preservation process or, more particularly, to break down intercellular cement so that an acidic or alkaline material can penetrate through the foodstuff and create an environment which inhibits reactions that would cause decay of the foodstuff.

The product being preserved is immersed seriatim in three different liquids in the second recovery stage of the process. The objective of the first step is to rapidly change the pH of the product. This, which is done by immersing the product in an alkaline solution and is known as ionic blow, produces a beneficial bactericidal and fungicidal effect.

The pH adjustment is facilitated by two characteristics of the stabilized product. First, the opening of the intercellular spaces allows the recovery liquid to penetrate to the interior of the product, insuring a rapid and adequate concentration of the hydroxyl ion in all cells of the product.

The hypotonic alkaline solution employed to deliver the ionic blow is prepared by dissolving a potable base in water in an amount sufficient to produce a pH ≥9. Suitable bases include sodium, potassium and calcium hydroxides and mixtures of the foregoing.

This and the subsequent steps of the recovery stage should be performed under sterile conditions. If the second stage steps are carried out under strict aseptic conditions, the final product may be kept at room temperature in hermetically sealed, sterilized packages. If the conditions are not sufficiently aseptic, it is necessary to refrigerate the reconstituted product to avoid decay.

The second step of the recovery stage involves the immersion of the foodstuff in a hypertonic solution having a pH of 5 to 7 in order to adjust the pH of the foodstuff to the desired level. Because of the previous breakdown of the intercellular cement this step also proceeds rapidly and efficiently.

The composition of the rehydrating solution employed in the third step will depend upon, and can readily be determined for, the particular foodstuff involved. One exemplary composition is shown in the example.

Example: To stabilize 100 kg of sardines with an approximate content of 14% fat, a mixture of 10 ℓ hydrochloric acid (30%), free of contaminants, 7.9 g potassium chloride, 1 g purified pepsin 1/10,000 (Difco), 1 g Ionol [2,6-di-tert-butyl 4-methyl phenol antioxidant (Shell)] and 88 ℓ drinking water were employed.

A buffer solution (Solution A) was prepared by adding the potassium chloride dissolved in 1 ℓ water to 88 ℓ water. Thereafter the hydrochloric acid was added with continual stirring.

The pepsin was dissolved in 400 ml of water and the Ionol was added. This solution called "Solution B" was mixed with Solution A; and the mixture was briefly homogenized before immersing the sardines in it to stabilize them.

The stabilized sardines were placed in perforated plastic boxes or nets to facilitate handling and immersed (in a tank) in a hypotonic solution made by diluting 14 ℓ of a 10 N sodium hydroxide solution (NaOH 10/N) in 86 ℓ drinking water. The liquid was agitated to accelerate the alkalinization process.

The sardines were removed from the hypotonic solution and excess fluid was allowed to drain off. Then the sardines were placed in another washing tank containing a hypertonic solution of sodium chloride to dehydrate the cells, thus accelerating the outflow of the hydroxyl ions. This solution was prepared by dissolving 12 g of sodium chloride per liter in a large quantity of water.

The product was introduced into a third tank containing a rehydrating solution from which the product recovered ions lost in the previous steps.

This solution contained 3,220 g Na^+, 390 g K^+, 100.2 g Ca^{2+}, 36.5 g Mg^{2+} and 3,660 g Cl^- in 1 ℓ water.

Chlorine Dioxide Treatment of Meat

Bacterial contamination of retail meat has been the subject of extensive studies. The microorganisms present in retail portions are derived directly from the initial bacterial load on the carcass surface immediately post-slaughter; thus, meat portions, such as hamburger, having high bacterial counts are traceable to carcasses having high surface contamination.

Several available processes eliminate bacteria from meat by killing them with a contact disinfectant(s) applied in the form of a spray to the carcass surface during chilling. Use of 50 to 200 ppm of aqueous chlorine (hypochlorous acid) or 5 to 50 ppm of aqueous chlorine dioxide reduced bacterial counts during the chill cycle (18 to 24 hours post-slaughter) by killing bacteria introduced onto the carcass during slaughter procedures.

A major problem with use of chlorinated contact disinfectants is reaction of the agent with meat components to produce chloro-organic derivatives such as chloro-substituted lipids and chloro-aromatic compounds. These chlorinated derivates pose a potential health hazard, especially the class of halomethanes (known to be carcinogenic) formed by reaction of bactericidal levels of hypochlorous acid with humic or other organic substances. Reaction of chlorine dioxide at bactericidal concentrations with meat components results in lower but detectable levels of organic chlorine.

K.S. Barta; U.S. Patent 4,244,978; January 13, 1981 has discovered that application to carcass surfaces of aqueous chlorine dioxide at concentrations too low to be effective as a bactericide nevertheless substantially inhibits attachment of spoilage organisms and prevents their growth upon the meat substrate. Chlorine dioxide solutions at concentrations as low as 100-fold lower than heretofore used are effective for inhibiting attachment provided that their application to carcass surfaces commences at a time substantially coincident with contaminating events, as during slaughter procedures and subsequent chilling. At such low concentrations, no detectable organic chlorine is produced by reaction of the chlorine dioxide with meat components.

In this method, aqueous solutions of chlorine dioxide, so weak as to be substantially subtoxic, are first applied substantially coincident with (that is, as close as feasible in point of time with) a significant contaminating event, and thereafter continued by intermittent spraying. Thus, in slaughtering operations, such a solution is first applied as a low pressure (less than 45 psi) spray to meat carcass surface immediately post-slaughter and substantially coincident in time with dehiding and dressing procedures on the kill floor. In addition, shrouds for beef may be soaked in the chlorine dioxide solution prior to draping. Carcasses (customarily in halves or quarters) are then conveyed to a chill room and an aqueous chlorine dioxide solution is intermittently applied to such carcass portions over a conventional 14 to 24 hour chill period. This method is adapted to such processing of fresh meat of domestic animals including but not limited to pork, beef, veal and lamb.

Typically, the chlorine dioxide is generated on site with conventional apparatus and formed into solution with potable water to a concentration of 0.04 to 1.0 ppm (mg/ℓ), preferably less than about 0.1 ppm (mg/ℓ) prior to application. It is critical that the solution be applied to carcass surfaces on the kill floor at a time substantially coincident with contaminating events such as dehiding, disembowelment, etc. Thereafter, the chlorine dioxide solution is applied intermittently during chilling for such intervals and in such volume as prescribed by USDA regulations (generally a maximum total of 0.5 hour spraying time in increments during the entire chill cycle).

Significant contaminating events may occur thereafter, for example, where carcasses are transshipped such as by rail or on ocean-going vessels. Handling during such transshipment may expose the meat to substantial numbers of spoilage organisms. Hence, promptly on any significant contaminating event, the chlorine dioxide solution is applied to their surfaces and is thereafter applied, preferably by intermittent spraying, during subsequent cold storage.

In this process, application of chlorine dioxide solutions at substantially subtoxic concentrations, that is, less than considered bactericidal, and preferably about 0.04 to 1.0 ppm, has been found to prevent attachment of spoilage microorganisms to meat surfaces. Further, application of such substantially nontoxic concentrations will prevent growth of spoilage organisms on the meat surface so as to avoid subsequent spoilage.

Example: Beef cubes (approximately 1.0 g), excised aseptically from the center of a freshly slaughtered beef round, were treated with aqueous chlorine dioxide in various concentrations (zero, trace, 0.1, 0.5, 5.0, 10.0, 50.0 and 100.0). The cubes, seeded with *A. hydrophilis,* were placed spacedly in Petri plates. Molten soft McConky agar (0.2%) supplemented with 2×10^{-4} M glucose was poured into the plates so as to fully immerse the cubes. The plates were covered tightly to avoid evaporative loss of water and incubated for four days at 37°C.

Results then observed were as follows. Those cubes treated with chlorine dioxide at concentrations of 5.0 ppm or greater (i.e., bactericidal concentrations) showed no slime development. At all lesser concentrations a spreading slime front, evidencing bacterial growth, developed outwardly into the soft agar at the edges of the cubes.

At the concentrations of 0.1 and 0.5 ppm, the slime formation was confined to a narrow perimeter (1 to 2 mm) around the cubes, whereas the trace and control cubes showed a spreading slime front wider than 1 cm.

These results indicate that levels of chlorine dioxide insufficient to kill the cells nevertheless significantly retarded production of the slime characteristically secreted by the test organism.

Fatty Acids and Derivatives for Mite Control

The intermediate moisture food components are known to prevent both bacteria and mold growth when the moist food has been stored in any moisture impermeable container. However, on extended periods of time, infestation by minute arachnids or mites may take place. Mites thrive on a soft-moist food especially when it is nutritionally balanced and extensive storage times permit undesirable reproduction and growth.

The intermediate moisture food employs the concept of limiting the amount of unbound water capable of supporting microbiological spoilage. The food contains a sufficient amount of soluble solids which limits the amount of "free" water available for bacterial growth under ambient conditions. In addition to the soluble solids, the food preferably contains an antimycotic agent, since because of the high nutritional content, the food may be susceptible as a host for fungi, yeast or mold.

Additional examples that are water-soluble other than sugar include low molecular compounds such as sorbitol, propylene glycol and sodium chloride.

To prevent mold growth an antimycotic agent is preferably employed at a level sufficient to prevent the growth of such organisms. Illustrative of suitable antimycotic agents are sorbic acid as well as sorbate salts such as potassium sorbate and calcium sorbate.

The process developed by *J.S. Mehring, R.J. Sayen, R.E. Schara, C.T. Stocker and J.G. Rodriguez; U.S. Patent 4,298,624; November 3, 1981; assigned to General Foods Corporation* is directed to the prevention of growth and reproduction of mites in intermediate moisture foods containing a water content by weight of between 15 to 50%. It has been found that inclusion of specific fatty acids and derivatives thereof is effective in limiting the survival of mites in an intermediate moisture that is shelf stable due to the preservation system. Additionally, if mites should invade the food environment, the use of fatty acids in given concentrations prohibits their tendency to reproduce as well as to survive in the intermediate moisture food environment.

The sources of fatty acids providing protection against mites contain carbon atoms from C_{4-10}. Additionally, fatty acid derivatives chosen from amides, esters and salts are desirable when the chain length is from C_{3-10}. Some fatty acid salts may contain sodium. Small concentrations of these fatty acid, fatty acid derivatives and combinations have been found effective in providing the requisite protection.

A desirable range of fatty acid or fatty acid derivative for mite control is about 0.5 to 3.0% by weight. The exact minimum percentage that will obtain optimum results against mites is dependent upon the numerous variables including the specific fatty acid, derivative or mixture, the type of food and its ingredients, the total moisture content and the type of source materials incorporated into the food for their preservation against bacteria as well as yeast and mold.

Additionally, it has been found that another fatty acid may act as an extender in conjunction with the named fatty acids and derivatives so that these primary materials affording mite protection may be reduced in their content and concentration. This fatty acid, propionic acid in conjunction with the primary mite inhibitor, permits the total weight concentration of these materials in conjunction with one another to be within the same range as though the primary fatty acid component were employed alone.

For example, with the desirable range of primary fatty acid and/or fatty acid derivative between 0.5 to 3.0% by weight, the total weight of the fatty acid and/or fatty acid derivative with the added weight of propionic acid will be within the same weight range of 0.5 to 3.0%. Thus, it is considered that the use of propionic acid may directly substitute on a weight basis for the primary fatty acid so long as the propionic acid is used as a supplemental or additive component to the class of fatty acids and derivatives disclosed. While propionic acid per se affords some protection against mites, it is the purpose of using propionic acid in this disclosure that it only serves as an extender.

Example: A cat food was prepared by chopping and grinding the meats into small pieces which are then added to a jacketed cooker along with the emulsifiers, animal fat, and fishmeal. These ingredients are brought to a boil and then the remaining slurry ingredients are added and the total slurry was cooked about one hour, thereby effecting pasteurization and producing a liquefied slurry composition. The slurry was then finely ground through an emulsifier into a more or less pulpy, pumpable, flowable purée consistency.

The dry ingredients were mixed together and added to the hot slurry in a jacketed double sigma-bladed mixer. The total mix was heated for 15 minutes until it reached 190°F.

The cooked dough was extruded hot into one-quarter inch pellets which were immediately cooled by passing them over a cooling screen, transmitting dry cool air, to an ambient temperature of 80°F. The cooled pellets were then packaged in cellophane.

The cat food had a moisture content of 45% and a fat content of 1%. It also contained 0.2% potassium sorbate, 1.5% phosphoric acid and the mite control agents as given below.

Into each cellophane container which contains the food, 10 female mites *(Tyrophogus putrescentiae)* are introduced and the cellophane resealed against the atmosphere. These samples are stored at about 70°F and 50% relative humidity for a six-week period of time.

In a control sample with no mite control additives 22 mites were found on examination. No mites were found in samples which contained mixtures of (1) 0.5% capric and 0.5% caproic acids, (2) 0.5% capric and 0.5% propionic acids, (3) 0.5% caproic and 0.5% propionic acids, and (4) 0.5% capric, 0.25% caproic and 0.25% propionic acids.

Lactic Acid Formation Using Selective Cultures of *Pediococcus pentosaceus*

A method has been described by *M. Raccach; U.S. Patent 4,303,679; December 1, 1981; assigned to Microlife Technics, Inc.* for fermenting meat using selected cultures of *Pediococcus pentosaceus* having rapid fermentation characteristics at meat temperatures of between 15.6°C (60°F) and 26.7°C (80°F) especially in the presence of a stimulatory metal ion preferably a manganese ion.

This process relates to a meat fermentation method including the steps of providing lactic acid producing bacteria in the meat with an assimilable carbohydrate and with meat spoilage and rancidity inhibiting preservatives and then fermenting the meat with the bacteria so that lactic acid is produced from the carbohydrate over a period of time in the fermented meat.

The improvement comprises: providing in admixture in meat a culture of a selected *Pediococcus pentosaceus* at a concentration of between about 10^5 and 10^9 of the *Pediococcus pentosaceus* per gram of meat with an assimilable carbohydrate, with meat spoilage and rancidity inhibiting amounts of preservatives including a hydroxyaryl antioxidant which preservatives substantially inhibit the *Pediococcus pentosaceus* at meat temperatures between 15.6°C (60°F) and 26.7°C (80°F) and with a stimulatory, food grade metal salt, preferably a manganese salt, in an amount sufficient to reduce the inhibition of the *Pediococcus pentosaceus* by the preservatives, wherein the selected *Pediococcus pentosaceus* culture is characterized by an ability to rapidly ferment in the meat admixture at meat temperatures between about 15.6°C (60°F) and 26.7°C (80°F) to produce a pH less than about 5; and fermenting the meat admixture at smokehouse temperatures between about 15.6°C (60°F) and 48.9°C (120°F) with the *Pediococcus pentosaceus* so that lactic acid is produced in the fermented meat product.

The preferred *Pediococcus pentosaceus* strain of this process has been deposited at the Northern Regional Research Laboratory of the USDA, and has been designated as NRRL-B-11,465. NRRL-B-11,465 or a strain of the same species which has substantially the same low meat temperature fermentation characteristics is used in this process, such as those produced by genetic manipulation including mutation. For comparative purposes *Pediococcus pentosaceus* American Type Culture Collection No. 25744 is also described which ferments in meat in a manner similar to *Pediococcus cerevisiae.*

Pediococcus pentosaceus NRRL-B-11,465 was derived from a culture originally deposited at the American Type Culture Collection as ATCC 10791. This original culture was described as being most active at 26°C in a standard culture broth; however, when selected members of this strain were used in a sausage fermentation, the fastest fermentation temperature was found to be 45°C (113°F). It is uncertain as to whether NRRL-B-11,465: (1) is a mutant; or (2) is a selected single strain variant having anomalous low temperature fermentation characteristics.

The selected *Pediococcus pentosaceus* cells can be used as a concentrate containing at least about 1×10^7 cells per ml, usually between about 1×10^9 and 5×10^{11} cells per ml mixed with the metal salt, preferably manganese salts.

Example 1: *Pediococcus pentosaceus* NRRL-B-11,465 was grown in a growth medium containing a carbohydrate (glucose or other assimilable sugar), a nitrogen source (yeast extract or other source of amino acids) and traces of essential minerals or inorganic substances. The medium particularly included a manganese salt which was manganese sulfate monohydrate in the amount of 0.01% by weight. The pH of the medium was initially adjusted to between 6.5 to 6.7 and the fermentor was set to maintain a pH of 6.0 during the growth by the addition of ammonia. NRRL-B-11,465 was grown at 26°C and at 32°C for 11 hours to determine the effect of temperature on the growth conditions. For comparative purposes *Pediococcus pentosaceus* ATCC 25744 (which optimally grows at about 36°C) was grown at 35°C in the same medium.

The resulting cell concentrates were checked for cell count by growth on APT agar. For *Pediococcus pentosaceus* NRRL-B-11,465 at 26°C, about 1.7×10^9 cells per ml were produced and at 32°C, 1.3×10^9 cells per ml were produced. Thus the lower temperature appeared to be optimum in this growth medium for NRRL-B-11,465; however, growth temperature produced no essential difference in acid producing ability in sausage meat mix incubated at 25°C (77°F). For *Pediococcus pentosaceus* ATCC 25744 about 17×10^9 cells per ml were produced at 35°C and it exhibited good growth at this temperature which was better than NRRL-B-11,465 at the lower temperatures. ATCC 25744 does not grow well at low temperatures (23°C).

Example 2: *Pediococcus pentosaceus* NRRL-B-11,465 and ATCC 25744 cells were used as the broth cultures of Example 1 without further concentration for making semi-dry sausage, without added manganese salt or any antioxidants, but with sodium chloride and sodium nitrite as preservatives. A commercially available culture of *Pediococcus cerevisiae* NRRL-B-5627 (known as *Pediococcus acidilactici*) grown in the same medium as Example 1 without further concentration was also tested for comparative purposes.

A semi-dry sausage (typical summer sausage) was prepared by standard methods except that three separate lots of the sausage each contained one of the three cultures mentioned above. Test data on the three lots of sausage showed that *Pediococcus pentosaceus* NRRL-B-11,465 is much better than both *Pediococcus pentosaceus* ATCC 25744 and *Pediococcus cerevisiae* NRRL-B-5627 in lowering the pH at low meat temperatures. A pH of 5.0 was attainable after 12 to 14 hours at the 25°C (77°F) dry bulb and a 23°C (73.4°F) wet bulb temperature which was very unexpected. By comparison *Pediococcus cerevisiae* NRRL-B-5627, which is widely used commercially, was similar to ATCC 25744 at low meat temperatures in this time period and produced a pH of only about 5.46 in 16 hours at 26.7°C (80°F). Eventually, the pH will slowly begin to drop with these bacteria.

Propylene Glycol and/or 1,3-Butanediol plus Acid

Food storage and stability of the stored product are well-known problems in the food art. Generally a high moisture content food is the most palatable, but the least stable. On the other hand, a low moisture content food is the most stable but the least palatable. This distinction applies to foods in general. A discussion directed to pet food, but applicable to all food exemplifies these problems.

It is the object of the disclosure by *M.P. Burkwall, Jr. and P.L. Gould; U.S. Patent 4,191,783; March 4, 1980; assigned to The Quaker Oats Company* to provide a high moisture, shelf stable food having improved softness.

A food having high moisture and high water activity in the range of 0.91 to 0.95 is predictably stabilized by a stabilizing mixture comprising 1,3-butanediol, propylene glycol, or mixtures thereof in combination with a sufficient amount of edible acid to provide a pH of 4.0 to 6.8. An antimycotic may be added to the food if necessary. This procedure is also applicable to a pet food.

Figure 2.1 depicts a graph of a stable food containing sufficient amount of antimycotic to prevent mold and shows the pH required for stability when the moisture content is within a certain range as plotted on the Y axis of the graph and the percent of the propylene glycol and 1,3-butanediol is within an appropriate range as plotted on the X axis of the graph.

Figure 2.1: Propylene Glycol and/or 1,3-Butanediol plus Acid

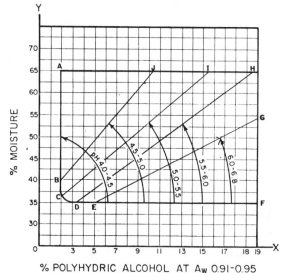

% POLYHYDRIC ALCOHOL AT A_w 0.91-0.95

Source: U.S. Patent 4,191,783

As depicted in Figure 2.1, the stabilizing mixture within a 35 to 65% moisture and a 2 to 19% range of propylene glycol, 1,3-butanediol, or mixture thereof, the food is stable at a certain pH. On the Y axis of the depicted graph, is the percent of moisture by weight of the food. On the X axis is the percentage of propylene glycol or 1,3-butanediol by weight of the food. The maximum limit of percent moisture is set at 65%, because, above this range the product becomes too difficult to process and the texture becomes too moist, soft and sticky. The percentage of glycol is set at the maximum of 19%, because, above this range the food becomes too bitter in palatability to be acceptable to the consumer. As is clearly indicated in the figure, when the food has the percentage of moisture and the percentage of glycol falling within the boundaries or area depicted, it is stable if the pH is adjusted.

A sufficient amount of at least one acidic compound is added to adjust the pH of the food to a range of 4.0 to 6.8. More preferably, the pH range is 4.0 to 6.0. To achieve these pH ranges, it is customary to use from a trace to about 5% of a food grade acidic compound such as an organic or inorganic acid or salt. Typical food grade organic acids include acetic, lactic, glucono delta lactone, adipic, succinic acids, and typical inorganic acids such as phosphoric, sulfuric, and hydrochloric acid or mixtures thereof can be used. Furthermore, the typical foodgrade acid salts may include monocalcium phosphate, monosodium phosphate, monopotassium phosphate, aluminum sulfate, sodium aluminum sulfate, sodium acid pyrophosphate, potassium acid tartrate, and mixtures thereof. Also, the salts and the acids can be mixed to achieve the desired pH.

By combining the acid, propylene glycol, 1,3-butanediol and mixtures thereof in the prescribed ratio, stability is achieved for a food having a high water activity, while avoiding or minimizing the problem caused by each of these components alone in an excessively high amount. If desired an antimycotic can be used in an amount sufficient to help prevent mold. More preferably, if used the antimycotic is present in a trace amount up to about 1% by weight. Even more preferably, if the antimycotic is used, from a trace to about 0.5% may be used. Potassium sorbate or sorbic acid is the preferred antimycotic due to availability and effectiveness. Other suitable antimycotics include the benzoates, the Parabens, the propionates, the acetates, or mixtures thereof.

Example: The following ingredients, in percent by weight, are assembled and formulated into a bacon flavored, rich protein, nutritious snack food suitable for human consumption: 25% soy protein isolate; 4% wheat flour; 1% bacon flavoring, vitamins, minerals, and coloring; 60% water; 9% propylene glycol; and 1% phosphoric acid.

After processing this snack-type product has a moisture content of 55%, a water activity of 0.95 and a pH 5.0 and is microbiologically stable.

OTHER ACIDS IN NONMEAT FOODS

Complex Salts of Monocarboxylic Acids

Saturated and unsaturated aliphatic carboxylic acids have only been sparingly used in industry and agriculture. The obnoxious odor of the free acids has made the handling of these acids unpleasant for the operatives. The obvious expedient of using the esters or neutral salts has been unsatisfactory since the acids on esterification or neutralization lose a considerable amount of their activity.

These problems have been overcome by the process used by *J.J. Huitson; U.S. Patent 4,179,522; December 18, 1979; assigned to BP Chemicals Limited, England.*

It was found that by adding a base to an acid in aqueous solution in an amount which is less than the chemical equivalent required for full neutralization, such compositions minimize to a substantial extent the odor and corrosivity of the acids without significant loss of activity of the free acid. In addition, it has unexpectedly been found that the base and acid combine under these conditions to form complexes which are stable in aqueous solutions. Such complexes also have the added advantage that they exhibit negligible vapor loss relative to the free acids and hence retain the preservative activity on the substrate for a longer period of time.

Accordingly, this process uses a liquid composition comprising ammonium ions and/or ions of a metal selected from Group I and Group II of the Periodic Table due to Mendeleef, a C_{3-8} carboxylic acid and water the ratio of acid to ammonium and/or metal ions being in the range of 2:1 and 4:1 on a chemical equivalent basis.

The compositions contain one or more carboxylic acids selected from saturated and unsaturated, aliphatic monocarboxylic acids having from 3 to 8 carbon atoms, preferably containing 3 or 4 carbon atoms. Propionic, n-butyric, iso-butyric, n-valeric, 2-methylbutyric, levulinic, acrylic and methacrylic acids are the most preferred.

The Group I and Group II metals of the Periodic Table due to Mendeleef are preferably selected from sodium, potassium, calcium and magnesium. Although metal ions such as copper, strontium and beryllium may also be used, it will be clear that such compositions can only be used for certain special applications, e.g., involving pesticidal or fungicidal activity, due to the known toxic nature of the cation. The chemical equivalent ratio of acid to cation is between 2:1 and 4:1. The amount of each component would naturally vary within these ranges depending upon the nature of the cation and the intended use of the composition.

These compositions may contain one or more complex acid salts. For example, when ammonia is added to aqueous propionic acid the resulting composition may contain ammonium dipropionate as the complex acid salt. Similarly by suitable choice of cations and acid any number of complexes such as ammonium diisobutyrate, sodium dipropionate, calcium tetrapropionate, magnesium tetrapropionate, etc., may be present.

The complex acid salt may be prepared by mixing a carboxylic acid with a calculated amount of a base of the desired cation in an aqueous medium. For example, in preparing compositions containing the ammonium ion the acid may be mixed with a concentrated aqueous ammonia solution. On the other hand, for preparing compositions containing the calcium ion, a full calcium salt of the acid may be dissolved in an appropriate amount of the free acid or the free acid may be partially neutralized by lime or reacted with limestone.

The composition may be prepared prior to use or the acid and base components forming the composition may be added separately but simultaneously at the point of application.

The compositions with a suitable cation may be used as a preservative for animal feedstuffs and agricultural crops to prevent growth of mold, bacteria and fungi. This may be achieved by applying the composition to the desired substrate.

By the term "substrate" is meant agricultural crops and/or compounded animal feedstuffs and materials used in preparation thereof such as barley, wheat, oats, rye, maize, rice, hay, straw, silage, dried grass, tick beans, soy bean, bagasse, sunflower seed, sugar cane, rape seed, groundnuts, fishmeal and bone meal, buckwheat chaff and wood shavings.

The preservative compositions contain one or more complex acid salts or a mixture of the free acid and the acid salt which may be formed "in situ" during the preparation of the composition. The composition may also contain other conventional additives, in particular those with fungicidal or bacteriocidal properties, such as formalin, formic acid, acetic acid, sorbic acid, dehydroacetic acid and bisulfites.

The amount of composition used for the preservation of a substrate would depend not only on the substrate to be preserved but also on the acidic and cationic ingredients thereof. For example, copper which is nutritionally valuable and is a known growth promoter in animal feed would be used in low concentrations. On the other hand, compositions containing ammonium ions can be used within a wide range of concentrations without any deleterious effect. Thus, these liquid compositions when applied as a preservative to a substrate suitably contain between 0.1 and 5% of the inorganic complex acid salts based on the weight of the substrate treated. It is preferably between 0.1 and 2.5% by weight of the substrate treated.

Examples were conducted using ammonium dipropionate, ammonium diisobutyrate, ammonium di-n-butyrate on barley (30% moisture content). When used at treatment levels of 0.75 to 1.0%, samples remained mold-free for over 365 days.

Sulfur Dioxide Addition to Must

It is recognized in wine making procedures that the introduction of sulfur dioxide into must at an early stage will reduce oxidation and inhibit the growth of natural wild yeast residing in the must. Oxidation of must will result in discoloration of the must juice while the presence of wild yeast may contribute in an undesirable fashion to the organoleptic properties of the wine.

Heretofore, rather rough empirical and sometime arbitrary procedures were deployed in introducing sulfur dioxide. Even manual applications of sulfur dioxide have been employed, and only upon trial and error could the wine maker determine whether or not enough or too much sulfur dioxide was used.

Proper addition of sulfur dioxide to must is an important facet of making quality wine. It has been recognized that accurate measurement, quick application and even distribution of sulfur dioxide in must are requirements of a satisfactory system.

A system and process for relatively quickly applying and evenly distributing accurate amounts of sulfur dioxide to must have been developed by *J.W. Lunt; U.S. Patent 4,302,476; November 24, 1981; assigned to Paul Masson, Inc.*

The system comprises a sulfur dioxide containment vessel and means for pressurizing the sulfur dioxide in the vessel with an inert gas. Conduit means delivers the sulfur dioxide to a must flow conduit. An inert gas under relatively low pressure is introduced into the conduit for the sulfur dioxide conduit means for ultimate delivery to the must flow conduit. Metering means delivers the precise proportions of sulfur dioxide to the must flow conduit.

The process enables the introduction of precise proportions of the sulfur dioxide to the must flow conduit and the introduction of an inert gas under relatively low pressure along with the sulfur dioxide into the must. In addition to the accurate delivery of sulfur dioxide to the must, the back flow of juice from the must is prevented when the supply of sulfur dioxide is stopped as a result of the continuous flow of the relatively low pressure inert gas into the must flow conduit.

Sulfur Dioxide plus Drying of Coconut

In the manufacture of desiccated coconut as an edible food product, greater care is given to the possibility that microbial contamination will make the product unsuitable for human consumption. The presence of salmonellae in desiccated coconut has been linked to human salmonellosis and extensive studies in 1962 have revealed that raw umprocessed coconut supports the growth of salmonellae as well as that of other enteric bacteria.

In the past it has been found possible to increase the shelf life of desiccated coconut without substantial yellowing by prior treatment of the wet coconut with sulfur dioxide (SO_2) bearing chemical solutions.

The resulting desiccated coconut product which is somewhat protected from yellowing has a residual free sulfur dioxide content not greater than about 500 ppm and a calcium content of at least about 150 ppm. Desiccated coconut with a residual SO_2 content in excess of 250 ppm is considered unacceptable because of off-flavor. Products for human consumption preferably have a free SO_2 content well below 100 ppm (e.g., 40 to 50 ppm).

A method of preparing dried coconut meat which retains its white appearance during extended storage while having increased flavor and odor imparting constituents has been described by *C.A. Escudero and C.P. Schaffner; U.S. Patent 4,307,120; December 22, 1981.*

According to this method for the manufacture of edible desiccated coconut, pieces or fragments of raw coconut meat are immersed in a sulfur dioxide containing water solution, after which the fragments are subdivided and the subdivided material treated to effect drying to a moisture content of the order of 2 to 4%, and to effect bacterial decontamination by subjecting the material to microwave irradiation. Treatment with sulfur dioxide containing solution serves to prevent yellowing during drying, and also ensures the production of a high quality white product which is not subject to yellowing over long storage periods at ambient temperature, and which has enhanced flavor.

Example 1: Harvested coconuts are shelled and the meat pared. The resulting white meat is cut into pieces approximately one inch square. These pieces are introduced into a water solution containing bisulfite and calcium at an ambient temperature. Calcium is provided by adding calcium chloride, and sulfur dioxide is provided by adding sodium bisulfite. Calcium metabisulfite solution is provided in which the ratio of calcium to metasulfite is about 1.75:1. The pieces of coconut are treated in the solution for a period of about one minute. The pieces are then removed from the solution and subdivided to form granules ranging in size from $\frac{1}{16}$ to $\frac{1}{32}$ inch in diameter. The resulting granular material has a moisture content of about 50%. This is fed to a microwave dryer of the type manufactured by Cober Electronics Inc., operating at a microwave frequency of 2,450 MHz. The dryer is provided with a conveyor for moving material continuously through the microwave energy fields. The moist coconut material is supplied to the conveyor belt as an even layer of about 2 inches thick and about 8 inches in width. The kW capacity of the oven would be such that the coconut is dried to a moisture content of about 4% in about 4 minutes, which suffices to kill all of the contaminating organisms.

Example 2: The coconut meat was prepared and contacted with a sulfur dioxide containing solution as described in Example 1. Upon removal from the solution the pieces are rinsed with wash water, and then subdivided to form granules ranging in size from about $\frac{1}{16}$ to $\frac{1}{32}$ inch diameter. In a typical instance the granular material has a moisture content of about 50%. This wet material is then fed continuously to a combination microwave hot air dryer. The microwave units are located at the entrance and exit ends of the hot air dryer housing. Both microwave units may be of the type as described above operating at a microwave frequency of 2,450 MHz. The combination dryer is provided with a conveyor for moving material continuously through the microwave energy fields.

The moist coconut material is supplied to a conveyor belt as an even layer of about 2 inches thick and about 8 inches in width. The kW capacity of the first oven is such that the coconut is rapidly heated to a temperature level of about 100°C. The total time period within the first microwave unit may be about 1 to 2 minutes. The moisture content of the material as it leaves the first microwave unit and enters the hot air dryer may be about 40 to 50%. As the material continues through the hot air dryer, it is contacted with hot air at a temperature of about 110° to 140°C, and the moisture content of the material as it exits from the air dryer may be about 20%. As the material passes through the second microwave unit it is rapidly heated to a temperature level of the order of 100°C within a period of the order of 1 to 2 minutes. The power input to the second microwave unit is such that within a period of time of the order of 5 to 10 minutes the material is dried to a final moisture content of about 2 to 3%.

Rice Vinegar in Breads

Rice vinegar has been used by *I. Kikuhara; U.S. Patent 4,165,386; August 21, 1979* as an antimold agent. The rice vinegar also improves the cellular structure, flavor and nutritive value of the bread.

Specifically, this process is intended to provide bread quite free of conventional antimold agent and yet good in preservation capability as well as in cellular structure, flavor, nutritive value, etc., and a method of producing the same. This process provides bread comprising rice vinegar and at least one kind of bread-making powder selected from the group of flour and rye powder, the amount of the rice vinegar ranging from 180 to 450 ml relative to 25 kg of the bread-making powder.

Incidentally, it is known that glacial acetic acid is added to flour in an amount of about 0.15% based on the weight of the flour in order to prevent rope bacteria generation on the product bread. But, glacial acetic acid is known to be detrimental to the dough and the bread quality or enhance the sourness of bread. Moreover, it is said that a satisfactory antimold effect cannot be expected of glacial acetic acid.

The rice vinegar used in this process is prepared as follows. In the first step, unhulled unpolished rice is polished with a rice-polishing machine. It is preferred that 0.1 to about 15% by weight of the rice skin be removed in the rice-polishing treatment.

The polished rice is water-washed and immersed in water for about 20 hours, followed by draining of the fully-water-absorbed rice and subsequent steaming of the rice so as to convert the starch of the rice into α-type. The steamed rice is gradually cooled down to 35° to 38°C and, then put in a koji-producing chamber.

Seed of koji mold such as *Aspergillus oryzae* is added to the steamed rice for multiplication of the koji mold so as to prepare koji. On the other hand, "Shubo", namely, the substance from which sake is prepared, is separately prepared by adding lactic acid or lactobacilli and yeast to a mixture of steamed rice, koji and water, followed by fermentation. Lactic acid is used because the optimum pH for the multiplication of yeast is slightly acidic. The resultant Shubo, which is obtained after about 20 days of fermentation, contains about 300 million yeast fungi per cubic centimeter of Shubo.

Steamed rice, koji and water are newly added to the Shubo, which is generally called "an initial addition". The day after the initial addition, "an intermediate addition" is carried out as in the initial addition. Likewise, "a binal addition" is conducted the day after the intermediate addition. In short, steamed rice, koji and water are added to the Shubo in three steps. This method is very effective for producing a large amount of sake with a small amount of yeast.

After the binal addition, the resultant mixture is subjected to alcoholic fermentation while being saccharified. It is necessary in this step to keep the fermentation temperature higher than for the case of producing a refined sake in order to permit a sufficient decomposition of the raw material. Unrefined sake having an alcohol content of about 18% is produced after about 25 days of fermentation. Vinegar is added to the unrefined sake, followed by squeezing the mixture so as to separate the mixture into liquor and lees. The separated liquor is subjected to aging for about one month. The aged liquor is diluted with water to an extent permitting acetic acid fermentation and then vinegar seed and acetic acid bacteria are added to the diluted liquid, thereby performing acetic acid fermentation and producing the desired product of rice vinegar.

The resultant rice vinegar has an amino acid content of 250 mg/ℓ or more and the extract content is 2.5% or more. It is quite reasonable to think that various nutrients such as vitamins B_1, B_2, B_6, B_{12}, calcium pantothenate, nicotinic acid amide, folic acid, peptides, inositol, choline and biotin are formed by the action of microorganisms added in the manufacturing step of the rice vinegar such as koji mold, lactic acid bacteria, yeast, and acetic acid bacteria. In this process, these nutrients are supposed to play vital roles in the production of bread.

Example: Various kinds of bread were produced by sponge and dough method using 100 kg of a high-strength flour. Specifically, various additives including rice vinegar, synthetic vinegar and antimold agent were added in the dough preparation step together with water in the six test preparations.

In adding rice vinegar, etc., the mixture was slightly kneaded, followed by hard kneading for preparation of the dough. Then the dough was subjected to fermentation and baking by the usual method so as to obtain 170 kg of bread for each case.

In the following test results, the amounts of the additions are reported as the amounts added to 25 kg of flour.

Tests showed that the product bread was kept free of mold growth for at least three days for the cases of Sample 1 (230 ml concentrated rice vinegar), Sample 3 (230 ml ordinary rice vinegar), Sample 4 (50 g calcium propionate) and Sample 5 (230 ml synthetic vinegar). Particularly, Sample 1 prepared by adding concentrated rice vinegar, was kept free of mold growth for 3.5 days or more. In contrast, mold was seen to grow on Sample 6 (no additive case) only one day after the manufacture. Incidentally, Sample 2 in which 150 ml concentrated rice vinegar was added, an amount below the range specified in this process, was kept free of mold growth for 2.5 days.

For every case of using the additive, the mold growth on the product bread was found to consist essentially of *Aspergillus niger* which is an acid-fast bacteria. It was also found that the pH of bread bearing mold had been lowered to about 4.0 to 5.0.

Acetic Acid and Salts in Soy Sauce

A method for preventing the putrefaction of soy sauce has been described by *K. Shibata, G. Yamaguchi, K. Takeda and H. Masai; U.S. Patent 4,241,095; December 23, 1980; assigned to Nakano Vinegar Co., Ltd., Japan.* This method comprises mixing soy sauce with acetic acid and a salt of acetic acid, propionic acid, butyric acid, malic acid, tartaric acid, citric acid or lactic acid. The amount of the acid salt is in excess of the amount of acetic acid. The method provides a sufficient antiseptic effect without producing a sour taste and without greatly lowering the pH.

This method for preventing foods from putrefying, may be applied effectively to many kinds of foods such as kamaboko (boiled fish paste), chikuwa, boiled noodles, pickles, miso, soy, season soy for grilling, packed rice cake, bread, cakes, hamburgers, tofu (bean curd), ham, sausage, household dishes, bottled nameko, salad, etc.

Acetic acid materials to be used in this process include fermented vinegar, synthetic vinegar and aqueous acetic acid, and acetic acid salts, for example, sodium acetate, potassium acetate, magnesium acetate, calcium acetate.

Other organic acid salts such as sodium propionate, sodium butyrate, sodium malate, sodium tartrate, sodium citrate, potassium citrate, sodium lactate and potassium lactate can also be used.

A solution to be used in this process may be prepared by adding one or more materials selected from the group consisting of egg shell, sodium bicarbonate and sodium hydroxide (hereinafter referred to as egg shell etc.) to a vinegar in such an amount as 0.29 to 1.10 parts, more specifically in such amount as 0.6 to 1.10 parts for the case of egg shell or sodium bicarbonate each added singly, or in such amounts as 0.29 to 0.53 part for the case of sodium hydroxide added singly, each by weight based on the weight of acetic acid (as 1 part) in vinegar, so that the weight percent of acetates (Ca, K and other salts) may be higher than that of acetic acid. The solution prepared may be added to raw and uncooked soy either before or after sterilization of the soy sauce.

Example 1: To 1 ℓ of fermented vinegar having an acetic acid concentration of 10%, 30 g egg shell, 14.0 g sodium bicarbonate and 6.67 g sodium hydroxide were added and they were mixed. The solution thus prepared has an acetic acid concentration of 5% and pH value of 4.8.

To 100 parts by volume of fresh uncooked strong soy which was brewed and squeezed in a conventional manner, the solution prepared above was added and mixed respectively in amounts of 1, 2, 3, 4 and 5% each by volume.

Up to 500 ml of each of the mixed solutions so obtained was charged into a 900 ml glass bottle and then sterilized in a water bath at 75°C for 10 minutes. After sterilization, the solution in each glass bottle was sealed and served as the final products for testing. Fresh and uncooked soy before sterilization contained total nitrogen 1.65%, NaCl 18.0% and ethyl alcohol 0.95%.

Final products prepared as described above were placed in an incubator at 30°C and examined for the growth of film yeasts therein for a certain period. When 3, 4 and 5% of the vinegar solution was added to the soy sauce no trace of yeast growth was found after 30 days. For 2% addition scanty growth was observed on the 15th day, for 1% addition scanty growth occurred on the 10th day. In the control with no addition, yeast growth was observed on the 5th day.

Example 2: Fresh uncooked weak soy which was brewed and squeezed in a conventional manner containing total nitrogen 1.30%, NaCl 17.8%, ethyl alcohol 0.76% was sterilized at 75°C for 15 minutes to obtain a weak soy product.

Up to 500 ml of the soy product was added to 900 ml glass bottles, and then 15 ml of the respective solutions prepared in Example 1 added to each of the bottles in an amount of 3% by weight and each of the bottles was then closed to serve as the final products. The weak soy product showed no growth of film yeasts even after 30 days storage in an incubator at 30°C.

In the sensory evaluation test (paired difference tests), the final products with the additive were not distinguishable at all from the products without the additive (15 ml of tap water was added in place of the solution).

Propionic Acid on a Particulate Carrier

It has long been known that propionic acid either alone, or on a carrier material, may be used as a preservative for foodstuffs. The use of propionic acid on a carrier material will generally give better results because of the better distribution of the propionic acid that may be obtained.

Even though much research has been conducted on the use of propionic acid and other compounds as preservatives, the problem of spoilage and mycotoxin formation resulting from fungus (mold) activity is still a big problem in the industry.

It is thus an object of the disclosure by *B.J. Bland; U.S. Patent 4,199,606; April 22, 1980* to provide an improved particulate composition comprising propionic acid on a carrier material which is useful as a preservative.

This particulate composition comprises a particulate carrier material having absorbed thereon at least about 0.1 part by weight of propionic acid in the liquid form per part by weight of the particulate carrier material, the particulate carrier material being one which at ambient temperatures within the range of about 0° to 60°C catalyzes and effects the formation of propionic acid in the monomeric form in the propionic acid vapors which evaporate from time to time from the propionic acid absorbed on the particulate material.

Carrier materials which will provide the desired catalytic effect include those aluminum silicates containing, in addition to the aluminum and silicon, iron and one or more metals selected from the group consisting of the alkaline earth metals and the alkali metals.

The preferred aluminum silicate is a vermiculite, especially hydrobiotite which is a form of vermiculite. Vermiculite is the name applied to a group of hydrated magnesium-iron-aluminum silicates of the mica group with chemical composition varying according to the locality from which it is obtained. Vermiculite, including the hydrobiotite form, may be thermally exfoliated (expanded) to result in a product having relatively large pores or capillaries formed between groups of platelets, and this expanded product is preferred in this process. The thermal expansion results in a void volume/surface area relationship which would allow greater amounts of propionic acid to be carried. Hydrobiotite which has been thermally expanded is known as verxite and may be readily obtained commercially. Verxite has been approved for use in animal feeds and in the U.S. Food and Drug Administration Regulations. Verxite is a known carrier for some types of nutrients for animal feeds although its use for carrying propionic acid is not known.

This process may generally be applied to preservation of any raw or processed agricultural crop product, or by-product or derivative thereof, which is subject to microbiological degradation and which is low in sugars and high in one or more of cellulose, starch or lignin. The treatment of fruits is not included and the application of this process to fruit preservation is not recommended.

Typical of the products of the type to which this process may be applied are hay (baled or pelletized), silage, crop residue such as corn stubble, milo stubble and wheat stubble, spent brewers grain, fishmeal, peanut meal, spent tea leaves, spent coffee grounds, pea shell forage, soybeans, sugar beet pulp, cottonseed hull and meal, sugar cane pulp (bagasse) and cassava root (tapioca). Cereals are also one of the major types of products that may be preserved according to this process. The term "cereals" includes not only the raw cereal grains themselves but cereal grains which have been processed by chopping, grinding or the like to produce products such as mash, meal and flour. The cereal grains include corn, wheat, rice, barley, sorghum, milo and rye.

Example: An experiment was performed in order to determine the effectiveness of propionic acid absorbed onto expanded perlite as a fungicidal composition. The experiment was conducted on cracked corn having 23% by weight of moisture. To one sample of 200 g of the cracked corn there was added 0.3 g of a mixture of propionic acid absorbed on expanded perlite in a 1/1 weight ratio. To a second sample of 200 g of the cracked corn was added 0.15 g of liquid propionic acid. A third sample of 200 g of the cracked corn was not treated with any preservative and served as a control. All three samples were then stored at room temperature (about 25°C) and observed daily for an indication of fungal activity.

The first sample containing the propionic acid-perlite composition did not show any noticeable fungal activity until about 22 days. The second sample treated with the liquid propionic acid (neat) showed noticeable fungal activity after only 10 days, and the untreated third sample showed noticeable fungal activity after only 5 days.

Sugar-Acid Solution for Preserving Cherries

A method has been used by *L.K. Anderson and H.B. Allen; U.S. Patent 4,298,623; November 3, 1981* for processing fresh cherries which renders them storable. The cherries are immersed in an aqueous storage solution containing an edible bacteriostatic acid and dissolved sugar. Sweetened cherries may be produced by draining stored cherries of this storage solution, and reimmersing them in a syrup prepared by increasing the sugar concentration of the storage solution.

The usual cherry, such as a Bing, which is sold on the market for consumption as a fresh cherry, contains substantial amounts of sugar (fructose). In producing a maraschino cherry, the cherries are stored in an aqueous solution of sulfur dioxide, which functions to extract this natural sugar, and this extracted sugar is discarded when the sulfite brine is discarded. Following this process, the natural cherry sugar is retained to a large extent during the storage process within the cherry. The solution which is used to immerse the cherry for storage itself may be used as the base for the solution used to sweeten the cherry to produce a garnish cherry.

In general terms, the present storage of cherries is performed by immersing clean, fresh, uncooked cherries in a dilute, aqueous, bacteriostatic, edible acid solution, which also contains controlled amounts of dissolved sugar which inhibits shrinking, splitting and other disfigurement of the cherries by reason of osmotic conditions in the solution.

The acid used is what is termed an edible acid, in that such can be included in food products without harmful effects on health. Exemplifying such edible acids are citric and phosphoric acids. The acid is a bacteriostat, in that the acid inhibits bacterial growth on the immersed cherries. As a general rule, it has been noted that inhibition of bacterial growth occurs at pH levels below about 4.5, and preferably, according to this process, the pH of the immersing or storage solution is below about 3.5. In general terms, and considering a solution made up of 100 parts water (parts herein refer to parts by weight), up to about 3 parts acid may be included without imparting to the cherries, which have been immersed in the solution, an unpleasantly acidic taste, and there being sufficient acid present to inhibit bacterial growth.

It will be understood that the pH of an unbuffered solution is subject to change over a period of time. By way of example, a storage solution containing 100 parts water and 2.66 parts commercial food grade, white, 75% orthophosphoric acid, after preparation exhibited a pH of 1.4. After a year of immersion in this storage solution of pitted Bing cherries, the pH of the solution rose to 3.

The presence of the acid in the storage solution has the additional effect of acting as a color stabilizer.

Example 1: A storage solution was prepared by mixing 100 pbw water, 33.3 pbw cane sugar, 2.67 pbw 75% food grade orthophosphoric acid, 0.21 pbw sodium benzoate and 0.021 pbw U.S.P. dicalcium phosphate (2 molecules attached water).

Orchard run fresh Bing cherries were culled, stemmed, pitted, and half-sliced. Two hundred pounds of a storage solution prepared as above was sufficient to cover 250 pounds of the half-sliced cherries with the cherries deposited in a suitable vessel. After four months of storage at ambient exterior temperatures (above freezing), the cherries were unspoiled. Skin appearance, color, flavor and texture were good.

Unpitted cherries with stems, were immersed in the same storage solution. In this instance, however, twice as much storage solution was required to cover the cherries by reason of the larger volume taken up by the whole cherries and the attached stems. As in the first run described, after four months of storage, the cherries were unspoiled with good flavor, texture and color.

Example 2: A storage solution was prepared by mixing 100 pbw water, 33.3 pbw cane sugar, 2 pbw 75% orthophosphoric acid and 0.1 pbw sodium benzoate.

One hundred twenty pounds of unstemmed and unpitted Bing cherries in a fresh and cleaned condition were placed in a vessel and covered by introducing into the vessel approximately 200 pounds of the storage solution. After four months of storage, at ambient exterior temperatures, the storage solution was drained from the cherries and the cherries were then stemmed and pitted. The storage solution had a pH of about 1.5 initially and rose to pH 2.2 after a month of storage.

The acidity of the storage solution was then reduced, to a pH 3.5, by the addition of powdered calcium hydroxide in small increments adding such slowly and with agitation of the liquid. Cane sugar was then added to the solution to raise the percentage of sugar to 35%. The pitted cherries were then reimmersed in this solution for one day.

The solution was then drained from the cherries, and by the evaporation of water, the sugar concentration of the solution was raised to 38%. The cherries were then reimmersed in the syrup so produced for another day. This process was repeated for two additional days to raise the sugar concentration of the solution to 47%.

It was observed that after such processing the pulps and skins of the cherries had undergone substantially no shriveling or wrinkling, the treated cherries were colorful and had a flavor and texture similar to a fresh cherry. The additional sugar content of the cherries was sufficient to give them a pleasing sweetness rendering them suitable as a garnish cherry.

The syrup produced as described above had a dark cherry color and a distinctly fruity cherry flavor, rendering the syrup suitable for food use.

Sodium Diacetate for Cereal Grains

Field corn brought into the farm headquarters and kept on hand for feeding purposes, despite some drying in the field, usually has a natural moisture content of at least 20% by weight and is readily attacked by mold. In fact, in order to prevent mold from growing during storage of the corn, the moisture content must be reduced to 12.0%. Corn having a moisture content above this level supports mold growth which then grows increasingly profusely as the moisture content of the corn is increased above 12.0%. Thus, corn having a 20.0% moisture content will grow mold much more rapidly and easily than corn at 14.0% moisture.

The mold grows any place on the surface of the corn kernel. It has its greatest hold in the cracks or interstices at the germ end of the kernel. This makes it particularly difficult to inhibit, since growth can easily get started in these folds in the seed coat. Any inhibitor which is used must, therefore, penetrate into these folds.

Molds growing on corn are generally the common penicillium and rhizopus varieties, although many others are also commonly found. Some molds produce significant quantities of mycotoxins. These are particularly virulent for fowl, since they show a high degree of mortality if fed on feeds made from corn which has become moldy during storage.

A process for inhibiting mold growth in a cereal grain having a seed coat has been developed by *E.F. Glabe, P.W. Anderson and S. Laftsidis; U.S. Patent 4,299,854; November 10, 1981; assigned to Food Technology Products.*

In accordance with this process, mold growth on cereal grains having a seed coat is inhibited by contacting such grains which have not been artificially dried and contain their natural moisture with sodium diacetate in such amounts and under such conditions that the sodium diacetate penetrates the seed coat and inhibits growth of mold in the grains.

It has been discovered that approximately 0.5 to 0.8% of sodium diacetate applied uniformly to the surface of whole corn kernels at 21.0 to 23.0% moisture content renders these corn kernels impervious to mold growth when stored under typical farm storage conditions. Smaller quantities of sodium diacetate provide protection for shorter periods of time.

Example 1: Whole kernel corn having an initial natural moisture content of 29.5% as it came from the field without artificial drying was divided into five equal portions and four portions were mixed with different amounts of sodium diacetate, namely, 0.25, 0.50, 1.00 and 1.25%, both micronized and unmicronized, the fifth portion, which was untreated, being a control.

Two types of tests were made, one in open pails and another in closed bags, at room temperatures around 70° to 75°F. After five weeks of storage at room temperatures it was observed that mold was effectively inhibited at a level of 1.25% in both types of tests. The micronized sodium diacetate was effective at a 1.00% level which may have been due to better distribution. In all other cases, including the controls, mold developed in one to two weeks.

In these tests, mold developed in the controls while the moisture content dropped in the corn in open pails from 29.5 to 16.1% over a three-week period and rose in the closed bags to 35.2%. In tests at a level of 0.5% sodium diacetate the moisture content after three weeks dropped to 15.7% in the open pails and rose to 39.5% in the closed bags. At a level of 1% sodium diacetate the moisture content after three weeks dropped to 22.9% in the open pails and rose to 40.4% in the closed bags. At a level of 1.25% sodium diacetate the moisture content after five weeks dropped to 10.8% after rising to 39.6% by the end of the second week in open pails and in closed bags the moisture content rose to 40.4% in a week and remained at this level.

This example demonstrates that the amount of sodium diacetate, to be effective in controlling mold on whole corn kernels containing their natural moisture, varies with the initial moisture content at the time of treatment and that the treatment is effective regardless of whether the moisture content rises or falls thereafter.

Example 2: Samples of the corn treated with sodium diacetate were fed to beef cattle and to swine in comparison with untreated corn. Observations were carefully made. It was very obvious that both types of animals preferred the corn treated with sodium diacetate to the untreated corn. When given a free choice of selection by being confronted with quantities of treated and untreated corn on the floor of the feeding pen, the animals chose the treated corn and consumed it first before going to the untreated grain.

OTHER ADDITIVES

Ascorbic Acid Acetals and Ketals in Meat

Acetal and ketal derivatives of ascorbic acid having utility in controlling the formation of undesirable nitrosamines in cooked, nitrite cured meat products are disclosed by K.R. Bharucha, C.K. Cross and L.J. Rubin; U.S. Patent 4,146,651; March 27, 1979; assigned to Canada Packers Limited, Canada.

The preferred ascorbic acid derivatives are selected from the group consisting of the acetals of tetradecanal, hexadecanal, octadecanal and octadec-9-en-1-al with ascorbic acid and the sodium salts of these compounds.

The compounds may be introduced into nitrite cured meat in any convenient manner so as to provide an amount in the cured meat which is effective to reduce or eliminate the nitrosamine content of the meat treated by the compounds and cooked at frying temperatures. In general, the compounds are applied in the brine solution which is used to cure the meat or are applied after cure in solution in a suitable solvent to the surface of the cured meat. Combinations of both methods may also be used.

The optimum amount of the particular compound to be incorporated in the meat varies to some extent from one compound of the general formula to another. In general, the useful range is from about 100 to 1,000 ppm on a weight basis of active compound to cured meat, with a preferred range being from about 500 to 1,000 ppm. Although the quantity of the active compound in the treated meat is extremely low, it nonetheless is effective in materially reducing or eliminating the known nitrosamines of cooked meat products.

Example 1: The acetal of dodecanal and ascorbic acid — Dodecanal (7.36 g, 0.04 mol) and L-ascorbic acid (17.6 g, 0.1 mol) were stirred together with toluene-p-sulfonic acid (2.0 g, 0.01 mol) in N,N-dimethylacetamide (40 ml) at 60°C for five hours. The reaction mixture was poured into water and extracted with diethyl ether (3 x 50 ml). The precipitate which formed on addition to water dissolved in ether. The ether layer was washed with water, dried over anhydrous sodium sulfate and evaporated to dryness on a rotary evaporator. A white solid (12.0 g) was obtained. In two crops from dichloromethane (50 ml) and methanol (3 ml) 9.0 g of white powder was obtained. The first crop (7.5 g) melted at 118°–121.5°C with sintering at 115°C. Crystallization from dichloromethane (300 ml) gave 6.4 g of white crystalline solid. The infrared spectrum in KBr showed two strong acetal bands. The PMR spectrum showed that the compound was in fact the acetal and not the isomeric enol ether. This compound was found to have a melting point of 122° to 124.5°C with sintering at 120°C.

Example 2: *Preparation of the sodium salt of the acetal of dodecanal* — The acetal of dodecanal and L-ascorbic acid prepared as in Example 1 (approximately 700 mg) was placed in a beaker with water (3 ml). A solution of sodium hydroxide (300 mg) in water (3 ml) was added dropwise until the solid just dissolved. The pH was 8.1. Water was removed under vacuum. Methanol was added as the evaporation proceeded until a pale yellow powder was obtained. An infrared spectrum of the powder showed that the lactone band and the acetal bands were still present.

Example 3: *The inhibition of nitrosamine formation in bacon* — The bacon used in the following experiments was obtained from the production line. The acetal was applied to the bacon as a slurry in soybean oil (2.7 ml/lb). The sodium salt was applied as a solution in water. After application of the additive, the bacon was fried without delay and the cook-out fat analyzed for volatile nitrosamines.

Use of 500 or 1,000 ppm of acetal additive gave 85 to 96% reduction of nitrosamine production. Use of 100 and 250 ppm of acetal reduced the nitrosamine production only 57 and 78% respectively.

Use of Sterile Casein in Processed Cheese

Processed cheeses are the products obtained by melting a cheese or a mixture of cheeses optionally containing other derivatives of milk at around 85° to 90°C in the presence of emulsifying salts known as "melting salts". This melting process is accompanied by fairly intense pasteurization so that, providing there is no secondary contamination and providing they are packaged under heat, the processed cheeses obtained no longer contain vegetative germs and the enzymes present in them are largely inactivated. Nevertheless, these processed cheeses cannot be considered as sterile cheeses because they contain sporulated germs which are not destroyed by pasteurization, such as, for example, *Clostridium sporogenes* and *Clostridium tyrobutyricum,* which can be brought to germination whenever favorable conditions prevail and all the more easily, the higher the water content of the cheese.

It is well known that sterilization destroys the texture of the processed cheese. Thus, a processed cheese which has been thermally sterilized is immediately recognized by its glutinous texture and differs from a normal nonsterile processed cheese which is characterized by a creamy texture.

R. Invernizzi and G. Prella; U.S. Patent 4,329,374; May 11, 1982; assigned to Societe d'Assistance Technique pour Produits Nestle SA, Switzerland provide a sterilized processed cheese having the texture which is normally typical of a thermally nonsterilized processed cheese, i.e., the creamy texture mentioned above, irrespective of its water content.

This process for producing a sterile processed cheese of this type, comprises preparing a mass to be melted, sterilizing this mass by melting at a temperature above 120°C, reducing the sterilized mass to a temperature of from 60° to 100°C and developing at that temperature the normal texture of a processed cheese in the mass by dispersing in it a texturizing agent consisting at least partly of a casein which is sterile in its native state.

Accordingly, it is necessary to prepare a first sample of sterile texturizing agent which, of course, cannot be obtained by thermal sterilization. To this end, a sample of nonsterile texturizing agent (casein, processed cheese, etc.) is initially prepared and then added to a thermally sterilized mass, i.e., a mass of undesirable texture. The mass then reassumes a suitable texture, as described above, but contains a distinctly reduced concentration of germs, particularly sporulated germs, by comparison with the concentration of nonsterile texturizing agent initially used. Thereafter it is in turn used as texturizing agent and so on. Accordingly, the procedure adopted is based on successive dilutions and "sub-cultures". It has been found that, for a dilution level of 10% (1 part by weight of texturizing agent to 9 parts by weight of thermally sterilized mass), 5 successive operations of this type are sufficient to obtain a degree of sterility which, although not absolute, is at least satisfactory for practical purposes.

The quantity of texturizing agent to be dispersed in the mass, of which the temperature has been reduced to from 60° to 100°C, for example to around 85°C, may be selected within a fairly wide range, advantageously from 0.5 to 20%. The mass obtained is preferably stirred in order to accelerate the creaming process.

Example 1: A mixture containing 41.5% by weight of finely divided Cheddar, 13.8% butter, 7% whey powder, 1.2% melting salts (sodium polyphosphate and/or citrate) and 36.5% water is prepared. The mixture contains approximately 45.6% solids, including 25.1% fats, and has a pH of 5.75. This mixture is then sterilized under pressure at 145°C (20 sec) in a steam injector. By expansion on leaving the injector, the sterilized molten mass is cooled to a temperature of 86°C. 4% of a sample of sterile processed cheese from a preceding production batch are then added to it, followed by stirring for 3 to 6 minutes at that temperature in a Stephan apparatus. The product obtained is then packaged in heat-sealed containers.

Analysis of the product reveals a sporulated germ content of less than 1 per 100 g, the lower limit of detection of the analysis method.

This sterile processed cheese is kept for 3 to 4 months at ambient temperature. When the containers are opened, they reveal a pleasant-tasting processed cheese of mellow texture showing no signs of microbial degradation or browning. This cheese separates without difficulty from aluminum foil.

Example 2: A traditional processed cheese is prepared from the starting mixture of Example 1 by pasteurization at 85°C in a scraped-surface heat exchanger. Although the processed cheese obtained has the desired unctuous texture, it does not keep for more than 3 to 4 weeks, depending upon the sporulated germ content of the starting materials.

Example 3: The procedure described in Example 1 is repeated omitting the step in which the sample of sterile processed cheese from the preceding production batch is added. The sterile processed cheese obtained in this case has an unpleasant, glutinous texture. This texture does not develop during storage; it remains glutinous. The cheese can no longer be separated from aluminum foil.

Maltol in Meat Products

D.L. Jones and F.R. Conant; U.S. Patent 4,279,936; July 21, 1981; assigned to William Underwood Company have devised a system for preserving pink meat color in canned, cooked, cured meats without the use of nitrites.

To effect this, this process employs a selected quantity of gamma pyrones at one or more of the various stages of the process, and in some instances, also in connection with the addition of iron salts.

Gamma pyrones are commercially available under the common name maltol (3-hydroxy-2-methyl-γ-pyrone) from Pfizer Chemical Company and others. It is postulated that maltol mixed with comminuted red meat complexes the iron atom of the myoglobin and prevents the degradation of myoglobin to an iron-free porphyrin-globin complex which tends to break down to a greenish porphyrin moiety. The presence of maltol will insure that such degradation does not occur.

The gamma pyrones, referred to generically as maltols, useful in this process include 3-hydroxy-2-methyl-γ-pyrone; 3-hydroxy-2-ethyl-γ-pyrone; 5-hydroxy-2-methyl-γ-pyrone and 5-hydroxy-2-ethyl-γ-pyrone. The iron salts optionally used with maltol are preferably initially ferrous salts of organic acids such as gluconates, citrates and the like, selected such that when ionized the organic acid will not materially shift the pH of the meat mixture.

The pickle brine used in the process is important primarily as a matter of taste and texture to the consumer but also is desirably compatible with maltol in terms of taste and chemistry as well. To this end, the pickle preferably is a nitrate-nitrite free pickling solution formed of common salt (NaCl) and a sweetener (e.g., cane sugar, dextrose, fructose, corn syrup and the like) in water in concentrations selected largely at the manufacturer's choice.

This process particularly useful with fresh or frozen pork comprises the addition of maltol in sufficient quantity so that in the finished product, the gamma pyrone is present in its free form in a range of about 0.02 to 0.2% by weight. The inclusion of maltol can be done at one of several stages, depending upon the procedure used to cure and cook the pork. Maltol can be added to a pickling solution free of nitrates and nitrites, such solution typically being composed of common salt and a sweetener. For example, a ham is artery or stitch pumped with the pumping pickle and then allowed to cure in a maltol-free cover pickling bath. The ham can then be precooked to green weight, processed to coarse ground form, hot water cooked, reground with spices, blended and canned. Additional maltol and iron salts can be added just prior to canning.

Instead of injecting the pumping pickle into the fresh or frozen ham, the latter can be processed through coarse grinding, mixed with the pumping pickle and allowed to cure prior to precooking.

It is desirable that the free maltol content in the final product should be in the range of 0.02 to 0.2% by weight. A smaller amount of maltol will often not provide the desired coloration and a larger amount of maltol does not improve the color and may impart an undesirable taste. In the cases where iron salts are added during the process, the latter should be limited to not greater than 0.04% by weight of the final product. The valence state of the iron is not generally important in that it is expected to be converted to the ferrous state inasmuch as the meat is a reducing matrix. Excessive amounts of the ferrous ion should not be used because the complex with maltol tends to become brown at high iron levels.

Example: A fresh or frozen ham is processed to remove the skin and bones, and the fat is separated from the lean meat. Both the lean and fat are ground (½" holes), the ground lean meat then being mixed with a pumping pickle in weight proportion five parts lean meat to one part pickle. The brine pickle is typically composed by weight of about 3% sweetener such as sugar, 20% common salts, 0.7% maltol, and the balance is water.

The ground meat is thoroughly mixed with the pickle and maintained at 40°F for three days, then being precooked in boiling water to green weight. The precooked pickled lean is mixed with the ground fat and with any desired spice mix in weight proportion about 1 to 2% spice, balance mixed meat. The mixture of fat into the precooked lean meat thus avoids rendering the fat. This latter mixture is ground, mixed and reground (⅛" holes) and then heated to about 160°F for 40 minutes to deactivate any enzymes present.

If desired, at this stage of the process a dry mixture of maltol 0.1% and ferrous gluconate 0.04% (by weight of the meat mixture) may optimally be added before filling cans with the mixture. The maltol-ferrous salt addition, if made, should preferably not bring the total free maltol concentration above 0.2% by weight of the final meat product.

Analysis of the meat mixture placed in the cans, where the total maltol is derived from the pickle, indicates that the free maltol is present in a range of 0.03 to 0.05 weight percent of the meat mixture, some of the original maltol having been leached in the boiling water precook stage.

Finally, the cans are filled with the processed meat, sealed and heat sterilized, for example, for 211 x 101.5 cans, at about 240°F for 40 minutes and then water-cooled prior to storage. When the stored can is later opened, the meat is found to have a desirable pink color which has not faded with aging and without evidence of any greenish surface color which would occur if the meat had been processed in the same manner but with the maltol omitted.

Fat-Based Coating Materials

A preservation coating composition containing lard, tallow and lecithin in specific ratios has been disclosed by *J.J. D'Atri, R. Swidler, J.J. Colwell and T.R. Parks; U.S. Patent 4,207,347; June 10, 1980; assigned to Eterna-Pak.* The mixture is heated and applied in molten condition to the chilled food to be preserved. The coated food is cold stored.

This process uses a coating comprising a portion of a fat more saturated than lard, an equal or smaller portion of lard and a minor amount of a wetting agent. Lard is an edible animal fat which may be defined as the fat rendered from certain fresh, clean, sound fatty tissues of hogs.

Commercial lard may consist of lard and hydrogenated, or hardened, lard. Commerical lard may also contain minor amounts of certain FDA-approved additives such as propyl gallate, BHA, BHT or citric acid. In some of the examples, commercial lard which had been obtained from a local supermarket was used. It is expected that most variations or additives in the lard would not be of such proportion as to materially affect the coating composition within the parameters of this process. The portion of lard in the preferred coating composition will be between approximately 30 and 50% of the total composition.

The second fat may be selected from a group consisting of tallow or of partially hydrogenated lard. In the preferred embodiment the second fat is tallow.

The edible tallow used in the examples below was obtained from a local supplier. The portion of tallow or other second fat may vary in the preferred coating mixture from between 70 and 50% by weight of the total composition.

A further addition to the composition is a minor amount of a wetting agent. A suitable wetting agent will be present in the mixture generally not in excess of 1% by weight. A preferred wetting agent is lecithin.

Example 1: A mixture of tallow, lard and lecithin in a 60:40:1 ratio was melted and adjusted to approximately 85°F (29.4°C). Meat samples at approximately 35°F (1.7°C) were dipped into the coating mixture. Among the meat samples dipped were portions of top round steak approximately 2 x 2 x ¾ inches (5 x 5 x 1.9 cm) as well as slices from a corned beef. It was not known how long the meat had already been stored before purchase but it is expected that it would have been 2 to 3 weeks old at that time. The samples were immersed in the coating mixture until the coating was at least ⅛ inch (0.32 cm) thick. The samples were refrigerated in a standard refrigerator for 15 to 20 minutes to harden the coating and were then stored in the refrigerator for three weeks. At the end of that time the odor, color and taste of the meat samples were very acceptable and there were no indications of developing rancidity. There was also no drip loss or weight loss.

Example 2: Ten-ounce prechilled New York strip steaks were dipped in a 60:40:1 (lard:tallow:lecithin) formulation which was held at about 80°F. The meat was maintained at 50° to 55°F prior to coating and the meat after coating was allowed to harden in an environment of 50° to 55°F. One-third of the samples were coated with the composition of this process, one-third were coated, allowed to harden and enclosed in a standard Bivac wrap, and one-third were wrapped in Bivac only, without being coated. All steaks were then stored at a temperature of 30°F for 5.5 weeks at which time the meat samples that had been wrapped in Bivac alone were brown and smelled and tasted sour and were inedible even with cooking. In contrast the odor and color of the samples coated with the above formulation, including both those wrapped with Bivac and those coated only, were good and the samples tasted fresh, even cooked to a medium rare stage. The perimeter fat was tasty and flavorful and had no hint of rancidity. There was no loss of meat drip.

Example 3: A number of oranges were also dipped in tallow:lard:lecithin (60:40:1) formulation and maintained in a refrigerator for 3.5 weeks. At the end of that time the coated oranges still had a "just-picked", fresh appearance and flavor. Several oranges that had been stored in excess of six months under home refrigerator conditions were still fresh and juicy when peeled at that time and had no signs of a long storage life. Their flavor was good, there was no rind breakdown and general appearance was excellent.

Similarly, carrots, lemons, potatoes, cherries, bananas and grapes have been coated according to this process and their storage life has been unexpectedly extended by several months. Extended shelf lives have also been obtained with cold cuts, wieners and corned beef when treated according to this process.

Cottage Cheese Containing *Streptococcus diacetilactis*

In the process developed by *E.L. Sing; U.S. Patent 4,318,928; March 9, 1982* cottage cheese can be made with the maximum possible shelf life and yet have virtually any amount of flavor that may be desired.

This process uses a blend of a first type of *Streptococcus diacetilactis* containing both a normal acetaldehyde- and diacetyl-producing strain and a second mutant type which retains its ability to inhibit food spoilage bacteria but does not have an appreciable amount of diacetyl or acetaldehyde production when grown in a milk substrate. The blend is added to achieve in the final cheese a cell count of at least one million cells per gram of the finished cheese.

In the preferred embodiment, a normal and a mutant strain are used. Many of the normal strains of *Streptococcus diacetilactis* are known. The preferred strain is ATCC No. 15346.

Mutant strains which are suitable for use with this process include strains 818 and 819, or their equivalent. Strain 818 has been assigned the accession number of NRRL B-12070. Strain 819 has been assigned the accession number NRRL B-12071.

After selected a normal and a mutant strain, the *Streptococcus diacetilactis* are separately grown in conventional fashion and thereafter the cells are separated from the growth media in the form of a paste. From 5 to 95 parts of the cell paste normal strain are added to from 5 to 95 parts of the cell paste containing the mutant strain of 818 or 819 (or the equivalent). To this blend are added additional parts of a suitable carrier for maintenance of viability but not in sufficient amounts to dilute the total *Streptococcus diacetilactis* cell count to as low as 3×10^9 cells per gram.

The resultant bacteria-containing composition is then placed in small containers. These containers are sealed and cooled to below 0°C, typically below –20°C, more preferably below –30°C. With the preferred carrier, the contents become frozen at this low temperature.

When it is desired to make cottage cheese, a sealed container of the bacteria-containing composition is warmed to above 0°C and preferably added to and mixed with a cottage cheese creaming mixture. The creaming mixture is then added to cottage cheese curds and blended in the conventional manner. The bacteria-containing composition is added in an amount to achieve at least 1.0×10^6 cells of *Streptococcus diacetilactis* per gram of cottage cheese.

Example: Fifty liters of citrate-containing heat treated milk substrate medium is cooled to about 30°C and then inoculated with an active culture of *Streptococcus diacetilactis* ATCC No. 15346 in sufficient amount to provide a luxurious growth after about 12 to 16 hours. After the luxurious growth is obtained, the culture is then centrifuged to obtain a cell-containing paste which is separated from the supernatant. The harvested paste is diluted with a phosphate buffered diluent containing 2% monosodium glutamate to obtained an optimum pH of 6.6 to 6.8 for maintenance of viability and to standardize the precipitation of this first strain to a known cell concentration of 1.5×10^{11} cells per gram.

While the preparation of the first strain is progressing, 450 liters of heat treated milk substrate medium is cooled to about 30°C and then inoculated with an active culture of *Streptococcus diacetilactis* strain 818 in sufficient amount to provide a luxurious growth after about 12 to 16 hours. After the luxurious growth is obtained, the culture is then centrifuged to obtain a cell-containing paste which is separated from the supernatant. The harvested paste is diluted with a phosphate buffered diluent containing 2% monosodium glutamate to obtain an optimum pH of 6.6 to 6.8 for maintenance of viability and to standardize the preparation of this second strain to a known concentration of 1.5×10^{11} cells per gram.

Ninety parts of the standardized preparation of the second strain are added to and mixed with 10 parts of the standardized preparation of the first strain. This mixture is then placed in 60, 240 and 400 gram containers which are sealed and cooled to -40°C, more preferably to -50°C, until the contents are frozen. These frozen containers are then shipped to dairy plants and stored at temperatures below -20°C, more preferably to -30°C until needed.

At the dairy plants, cottage cheese curd is prepared in conventional fashion. Cottage cheese creaming mixture is also prepared in conventional fashion except that the frozen mixture is thawed, removed from its container and added to and mixed with the creaming mixture. The amount of frozen mixture used is sufficient to achieve a cell count of about 10 million cells per gram in the finished cottage cheese. (Roughly 30 grams of the mixture per 1,000 pounds of the cottage cheese.) The creaming mixture is then blended with the curd in conventional fashion. The finished cottage cheese is kept cool to prevent significant growth of bacteria.

The cottage cheese produced has an excellent mild flavor and excellent shelf life.

Whole Crab Cooked in Sugar Solutions

An improved quick-freezing process for whole blue crabs to be subsequently served whole as steamed crabs has been provided by *K.B. Ross and C.R. Jones; U.S. Patent 4,336,274; June 22, 1982.* The process includes the steps of quick cooking whole blue crabs while maintaining maximum water content using a sugar-liquid bath, chilling while maintaining maximum water content using a sugar-liquid bath and quick freezing to at least -15°F without cracking, and storage at a uniform temperature of at least -15°F.

Sugar content of the sugar-liquid bath may be one-eighth pint dry measure per gallon of water, but may vary to ±50%. The solution in batch processing, should be replenished at the rate of at least 1% per minute to prevent contamination and "boiling-up" during processing.

The sugar acts as a flavor enhancer as well as a preservative and, surprisingly, reinforces natural fresh flavor of the crab rather than change it.

Salt not only does not help the flavor but also any taste-detectable amount tends to degrade the product and is to be avoided entirely according to this process.

Example 1: Blue crabs taken commercially from South Carolina waters in early December, measuring 5½ to 6½ inches across, and weighing accordingly, purposely selected to include typical proportions of "mixed quality", some poor, some medium and some heavy, were promptly, upon receipt in good condition, immobilized by chilling, stacked in layers 6 inches high (2 or 3 crabs thick) in an openwork plastic basket, boiled by immersion for 10 to 17 minutes at substantially 212°F in a solution of water containing one-eighth pint dry measure of sugar per gallon, prechilled by immersion for 5 to 10 minutes in a sugar-water bath of the same constituency maintained at 36° to 42°F, drained for 1 to 3 minutes, and then frozen by immersion in a bath of "Freon 12" liquid gas held at -23°F until boiling stopped, promptly removed, hermetically sealed by plastic wrapping in small lots in closely packed cartons, and stored at a uniform temperature below -20°F.

During the two months following freezing, periodic taste, odor and texture comparisons of the blue crabs so-processed and freshly thawed were made by experts in seafood preparation, using as standards of comparison both freshly thawed crab meat treated similarly but without sugar and liquid-content maintenance, and freshly cooked fresh blue crab meat. In every case the conventional frozen product was found distinctly inferior in taste, odor and texture, and the product of this process was found equal to freshly cooked fresh crab meat. Substantially more sugar than noted was found to impart a sweet taste; substantially less failed to give the desired effect; tasting was the principal criterion.

Example 2: Later similar tests extending over periods in excess of six months showed substantially no deterioration in the product and gave some indication that the flavor might actually improve.

The tests were also made using the same technique described but with "Mexican" commercial blue crabs of the same season taken commercially along the Gulf of Mexico coast. These measured an average of 7½ to 8 inches across and were proportionally heavier, grade for grade. Cooking time was toward the upper end of the range noted. Results were the same, fresh blue crab taste, odor and texture being uniformly preserved by this process over the two-month period.

It is extrapolatable that the uniformly good results obtained using the improvement of this process will continue over indefinitely prolonged periods of storage because of absence of noticeable degradation over the test periods.

Bacteria plate counts of about 250 maximum have been found in whole crabs processed according to this process in sharp contrast with picked crab meat plate counts of several hundred times that much (100,000 or more in freeze preserved picked crab meat). It is pointed out that bacteria plate counts of up to 250,000 are allowed.

Some variations in this process may be practiced, but will cause some degree of reduction of product quality. For example, the sugar in one of the cooking and chilling solutions may be omitted, but the end product is not as desirable in that the cooking solution apparently causes the meat to be permeated, and the use of sugar solution in the chill bath prevents the sugar from the cooking solution from being diluted. Further, as an optional step, at the second draining stage seasoning may be applied to the shell exteriors, and may be prefrozen thereon, if desired, as by a liquid nitrogen jet. However, this is to a degree contrary to the quick, nearly simultaneous freezing of the whole crab.

It is pointed out that the prechill sugar-water solution may have the equivalent "sugar content" by the use of sugar syrup, honey or similar substances rather than adding conventional sugar thereto.

VEGETABLE GUMS

PECTIN

Low D.E. Pectin

J.R. Mitchell, K. Buckley and I.E. Burrows; U.S. Patent 4,143,172; March 6, 1979; assigned to Mars Limited, England have used crude pectinaceous material of which the degree of esterification is or has been reduced to less than 20%, and preferably less than 10%, as the thickening or gelling agent in food products.

It was found that if a naturally occurring crude pectinaceous material, in which the degree of esterification (D.E.) of the pectin content is or has been reduced to below 20%, is employed as gelling agent levels of thickening or gel strengths can be achieved which are substantially higher than those obtained by the use of an equivalent amount (calculated on a galacturonic acid basis) of extracted or purified pectin. The D.E. of the crude material employed is preferably 10% or less.

Crude pectinaceous material means natural sources of pectin which have not passed through purification procedures to separate the pectin from its cellulosic matrix. Consequently these materials on a dry basis contain only some 5-45% of pectic acid (expressed as galacturonic acid), typically 25-30%, the remainder being composed of cellulosic materials, soluble sugars and mineral salts.

Useful crude pectinaceous materials include a wide variety of abundant and cheap materials such as apple, citrus or sugar beet residues, which have been subjected where necessary to deesterification, e.g., by alkali or enzymes, to reduce their D.E. to below 20%. Usually, natural pectic substances will have to be deliberately deesterified, for example by alkaline hydrolysis or by enzyme such as pectinesterase. However, deliberate deesterification may not be necessary if a natural protopectin source is used having a D.E. below 20%, for example because the source contains pectinesterase. Moreover, if the source contains such an enzyme the mere act of macerating the source material or adding it to a substantially neutral medium prior to pasteurization or sterilization may cause a reduction in the D.E. sufficient for the purposes of this process.

90

Thus if orange peel is ground to a small particle size, preferably neutralized to pH 7-8 with an alkali such as sodium carbonate, for example over a period of 30 minutes, washed to remove sugars and objectionable soluble compounds and dried, the ground product may be employed as a thickening and gelation agent in canned foods. Since the majority of the orange aromatic constituents and color lies in the surface layer of the peel (the flavedo) it is desirable when using the treated peel to thicken or gel products such as meat or fish that the flavedo be removed prior to processing, thus leaving the less characteristically flavored albedo. Shaving techniques for removing the orange flavedo from the crushed orange halves are well reported in the literature.

The reaction mechanisms by which the protopectin in the peel is converted into a gelling agent is not well understood. One possible explanation is that enzymes liberated during the grinding of the orange peel at least partially demethoxylate the pectin molecule, which is thought to be linked to hemicelluloses and other materials in the protopectin configuration, and that on subsequent treatment in a substantially neutral medium further demethoxylation and solubilization of the pectin may occur. It is believed that the resulting pectate then reacts with alkaline earth ions in the food product to form a thickened or gelled system.

Example 1: 10 kg of South African navel oranges were halved, squeezed to remove the juice and then passed through a mincer fitted with a $^3/_{16}$-inch plate. The ground peel was washed with tap water, pressed, slurried in water and sufficient anhydrous sodium carbonate was added with mixing to raise the pH to 9.0. The slurry was allowed to stand for 18 hr and was then pressed, washed, pressed and roller dried. The roller dried flakes were ground into a fine powder. The yield was 520 g.

Example 2: A comparison was made of the strengths of a gel containing 1% treated peel (30% galacturonic acid) with that of a gel containing 0.46% sodium polypectate (65% galacturonic acid) i.e. gels containing the same pectin content. This was carried out as follows:

882 g of distilled water was heated to 90°C. 3.3 g of sodium tripolyphosphate was added with stirring, followed by 10 g treated peel (prepared as described in Example 1) and mixed using a Silverson mixer for 5 minutes at 90°C. 5 g of dicalcium phosphate was added and mixed for one minute. 8 drops of an antifoam agent was added and the solution was evacuated in a vacuum chamber to ensure that no air was present in the solution. This solution was transferred to an open-necked 3-liter flask and a solution of 15 g glucono-delta-lactone in 85 g water was stirred in for 15 seconds.

This solution was poured into square Perspex jelly boxes and covered with polythene cover slips.

The preparation of the sodium polypectate gels was carried out in an identical manner to that described above except that the treated peel was replaced by 4.6 g sodium polypectate (Sigma Chemical Co.).

After standing for 5 hr at room temperature, two of the treated peel gels and two of the sodium polypectate gels were tested using the FIRA jelly tester. The force in grams required to turn the gel tester paddle through 90°C, i.e. to rupture the gel was recorded.

The treated peel gels (containing the same pectin level) were 25% higher in break strength value (214 g) than the purified sodium polypectate gels (171 g). Subjectively the treated peel gels were significantly tougher and firmer than the sodium polypectate gels. The pH of both sets of gels was 4.1.

Buffered Pectin Preparation

Partially esterified pectins, i.e., pectins with a degree of esterification less than 50% have gelling or thickening properties which are similar to those of agar or gelatin. Partially esterified pectins are used in the food industry, for example, for the production of low solids jellies, milk puddings and sugar-free jellies.

The gelling property of partially esterified pectins depends upon the amount of pectin, the solid content, the pH value and the buffer salt content and on the amount of calcium ion present. Moreover, the calcium ion concentration plays a significant role in which the optimum amount of calcium ion changes as a function of the degree of esterification. The optimum calcium content, expressed in mg Ca/g of pectin, which forms a solid gel, is a defined quantity for a particular pectin. If this calcium concentration is exceeded, a brittle gel results with a strong tendency towards syneresis. At the same time, the dependence on the pH value and the solids content plays only a subordinate role. For the technological application of these pectins, the following factors are, however, unfavorable:

(1) the lower the degree of esterification of a pectin, the less is its solubility in water;

(2) for maintaining an optimum calcium ion concentration, an addition of calcium is usually necessary.

G. Fox; U.S. Patent 4,136,209; January 23, 1979; assigned to Pektin-Fabrik Hermann Herbstreith KG, Germany has discovered a method for making a pectin preparation on the basis of partially esterified pectin with an amount of calcium that is optimum for its degree of esterification in a manner that a sufficient amount of buffer salt is simultaneously present in order to keep this pectin in aqueous solution.

This process comprises the following steps:

(a) a pectin is used as a starting material, which has a metal binding power of 30 to 140 g of metal salt, calculated as $CaCl_2$, per kg of pectin with a degree of esterification of 30 to 40%, preferably 35 to 38%,

(b) at a temperature between 10° and 90°C, a solution is prepared from this pectin, water and a phosphate selected from the group consisting of orthophosphates, pyrophosphates, and polyphosphates, and preferably sodium pyrophosphate, so that a pH value between 4.4 and 4.8 is obtained,

(c) a certain amount of soluble calcium, magnesium, aluminum or iron salts or mixtures of these salts is added in dry or dissolved form to the solution with stirring, the amount corresponding stoichiometrically to an amount of 30 to 140 g or $CaCl_2$/kg of pectin, and

(d) the pectin preparation thus formed is precipitated from the solution and dried.

A partially esterified pectin, with a degree of esterification of 30 to 40%, especially of 35 to 38%, is used as the starting pectin in the examples.

Example 1: In order to determine the necessary calcium chloride requirement, cooks are prepared with increasing amounts of calcium chloride. The following formulation used as basic cook contains:

> 500 g strawberry pulp,
>
> 500 g partially eserified pectin solution, 2.5% concentration (= 12.5 pectin),
>
> 650 g fructose syrup, 70% concentration, and
>
> ca. 3 ml of 50% citric acid for adjusting the pH value to 2.9.

By cooking to a solid content of 50%, a 1 kg cook is obtained.

Increasing amounts of a 2% $CaCl_2$ solution are added to 100 g samples of this book. The calcium chloride solution is stirred into this cook immediately after it is poured out and while it is still very hot. These samples are set aside to cool and gel.

As a rule, an addition of ca 2 to 7 ml of $CaCl_2$ solution is adequate. A 2% solution of anhydrous $CaCl_2$ is used. The consistency of the cook was used as a measure of the amount of $CaCl_2$ required. In the series of increasing strength, the sample with the best gel strength, immediately preceding the sample which begins to be gritty, was evaluated in the calculation. In so doing, an addition of each 1 ml $CaCl_2$ solution (2%) to 100 g of the cook containing 1.2% of partially esterified pectin, corresponds to a later dosage of 16.67 g of $CaCl_2$ (anhydrous)/kg of pectin. A calcium dosage of 6 ml to 100 g of cook corresponds to a later weighed-in quantity of 100 g of $CaCl_2$ for each 1 kg of partially esterified pectin.

Example 2: A receptacle was filled with 1,200 ℓ of water at a temperature of 50° to 60°C. Into this water, 30 kg of the above starting pectin were weighed in while stirring rapidly. Subsequently, 16.0 kg of tetrasodium pyrophosphate decahydrate ($Na_4P_2O_7 \cdot 10H_2O$) in solid form were added in order to obtain a pH value of 4.4 to 4.8 in the solution. After the addition of pyrophosphate, stirring was continued for about a further 15 minutes at a reduced speed, until the pectin was completely dissolved.

Subsequently, the calculated amount of food-grade quality $CaCl_2$ with a degree of purity of 96% is added in the form of a 10% solution, preferably through an injector, while stirring rapidly. Intensive stirring is necessary in order to achieve a rapid and good distribution of $CaCl_2$. After addition of the $CaCl_2$, rapid stirring must be continued for a period (generally about 5 minutes), in order to ensure an optimum reaction.

The $CaCl_2$ addition should be carried out in such a manner, that pectinate formation is avoided. In place of calcium chloride, other soluble calcium salts and/or magnesium salts, aluminum or iron salts or mixtures of these may also be added. On the whole, salts of multivalent metal ions may be added which meet the requirements for food-grade additives.

A further amount of pyrophosphate is added subsequently. This amount is determined by testing a sample in the laboratory for its pH value and its viscosity. The pH value should be adjusted to 5.0±0.2. The viscosity is, however, a better criterion and viscosities between 30 to 35 cp at 20°C (measured at 2,770 sec^{-1}) are acceptable. Higher viscosities should not be accepted. Viscosity measurements were carried out on a Rotovisco at n = 512 rpm with decreasing frequency of rotation, 100 S calibration with NV measuring equipment. Stirring is also rapid during this addition of pyrophosphate, which is in solid form.

After this amount of pyrophosphate has been dissolved, the pH value and viscosity of the solution are checked once again.

If the laboratory finding for the solution of the pectin preparation is positive, the solution is precipitated by the well-known procedure in alcohol.

After the precipitation is completed, the precipitation is squeezed out using an extrusion press. The squeezed-out pectin preparation is dried in a stream of air heated to not more than 60°C. Subsequently, the preparation is ground in the usual manner.

In Composite Fruit Gel

A flavored composite ice confection is peach Melba in which a portion of ice cream sits in a peach segment or half. Mass production of such a product has always seemed an almost hopeless task. This is because soft-fleshed fruits such as peaches, cannot be handled individually on a large scale.

Use of molded gelled fruit puree or pulp based on calcium alginate or calcium low-methoxy pectate, avoids some of these problems but cannot readily be used to prepare a satisfactory composite product such as peach Melba. This is because techniques for molding a product with a depression are either excessively complicated or reintroduce problems of handling. For example producing the depression by cutting out a portion of the gel is difficult to engineer on a large scale.

H. Göringer, T.R. Kelly, D. Ries and H. Silberzahn; U.S. Patent 4,190,676; February 26, 1980; assigned to Thomas J. Lipton have devised a method of overcoming these difficulties.

Incipiently gelling calcium alginate or calcium low-methoxy pectate is dosed into a mold and a portion of an extrudable foodstuff is then dosed onto the still incipiently gelling calcium alginate or calcium low-methoxy pectate to form a depression in which the portion of the extrudable foodstuff sits.

The mold is conveniently the container, e.g., a beaker, cup or tub, in which the product is finally packaged; the gel is less likely to move about for example during packing and transport. It should also be noted that the product has a further related advantage: the portion of the extrudable foodstuff fits the depression very closely and so is less likely to move during packing and transport than a product in which the portion of the extrudable foodstuff is dosed into a preformed depression let alone than a product in which it simply sits on the flat surface of the gel. The depression formed in the process is uneven and so gives, with the close fit of the extrudable foodstuff, firm anchorage to the portion of the extrudable foodstuff.

Although the method is illustrated below in terms of peach Melba, it will be appreciated that the process can be used to produce such products whatever the type of flavoring or ingredient used in the gel and whatever the extrudable foodstuff.

Example: Two streams of the following compositions were mixed rapidly and then immediately dosed in 60 g portions into plastic goblet-shaped beakers.

Stream 1 comprises 0.85 part sodium alginate, 0.15 part calcium hydrogen phosphate, 19 parts sucrose and 50 parts soft water. Stream 2 comprises 0.1 part sodium citrate, 0.7 part malic acid, 14 parts sucrose and 35 parts pureed canned peaches (23% solids, 12% syrup) and 0.2 part flavor and color.

Conventional ice cream was delivered in 35 g portions straight from the whipperfreezer to the surface of the incipiently gelling peach calcium alginate within 2 seconds of mixing of the two streams. The ice cream displaced the incipiently gelling peach calcium alginate to form a 4 mm shell in which the ice cream sat.

A suitable Melba sauce was then dosed onto the surface of the ice cream, a lid was placed on the top of the beaker and the product was then passed to a conventional hardening tunnel where the ice cream was hardened.

XANTHAN GUM

Isolation of Xanthan Gum

Several methods have been known for the isolation of the microbial polysaccharide (acidic water-soluble gum) from fermentation broth. One requires dilution of the broth with a large amount of a water-miscible solvent such as 2-propanol, whereupon the polysaccharide precipitates and may be removed by filtration or an equivalent procedure. Others are the precipitation of an insoluble calcium salt of the polysaccharide, followed by acidification, and precipitation by the use of a long chain amine or quaternary ammonium salt. All of these methods are either cumbersome (e.g., because of the large amount of diluting solvent required), expensive (e.g., because of the relatively high cost of lime, quaternary ammonium salts, etc.), or both.

An improved method of isolation of such water-soluble gums has been provided by *C.W. Schroeck; U.S. Patent 4,254,257; March 3, 1981; assigned to The Lubrizol Corporation.*

Acidic polysaccharides (e.g., xanthan gum) form amine salts with aliphatic or alicyclic polyamines having at least three amino nitrogen atoms and a molecular weight of at least 150. The amine salts may be used for isolation of the microbial polysaccharide from its fermentation broth by the steps of (A) acidifying, (B) forming the amine salt by adding the amine or a salt thereof, and (C) reducing the inorganic salt concentration as necessary (e.g., by dilution).

The amines from which the amine salts of this process are prepared are aliphatic or alicyclic polyamines having a molecular weight of at least 150 and containing at least three amino (i.e., basic) nitrogen atoms. Many amines of this type are known, including the followng: alkylene polyamines, including the ethylene, propylene, butylene and pentylene polyamines.

Specific examples of such polyamines are di(heptamethylene)triamine, tripropylene tetramine, tetraethylene pentamine and pentaethylene hexamine, as well as similar compounds in which the various alkylene groups are of differing chain lengths.

Step (A) of the method is the reduction of the pH of the mixture comprising the fermentation broth to a final value within the range of about 2.5-5.5. (By "final value" is meant the value during or just prior to precipitation of the microbial polysaccharide, that is, after steps (A), (B) and (C) have been completed.) The preferred range is about 2.8-4.2, and the optimum range is 3.7-3.8. The pH adjustment is normally done by adding an acidic material to the mixture comprising the fermentation broth. The acid used may be an inorganic or an organic acid but is preferably inorganic.

Step (B) is the addition to the mixture comprising the fermentation broth of at least one amine as described above, or a salt of the amine (preferably a salt of a mineral acid such as hydrochloric acid).

In step (C), the inorganic salt concentration of the mixture comprising the fermentation broth is reduced. The reduction may be accomplished by such methods as ion exchange, but it is usually convenient merely to dilute the mixture with water, about 0.5-5.0 and most often about 1.0-5.0 parts by weight of water generally being employed per part of fermentation broth.

The amine salts are easily dispersed in neutral or slightly acidic aqueous solutions, yielding low viscosity dispersions which do not exhibit the agglomeration or "clumping" phenomenon usually encountered when xanthan gum and similar acidic microbial polysaccharides are contacted with water. Upon addition to the aqueous dispersion of strong bases (e.g., sodium or potassium hydroxide), salts (e.g., sodium chloride, potassium nitrate) or an aqueous formaldehyde solution, a thickened aqueous solution is obtained by the action of the freed xanthan gum or similar polysaccharide.

If recovery of the amine from the amine salt is desired, the latter may be treated as a solid with a solution of strong base in a nonsolvent for the free acidic polysaccharide (e.g., aqueous methanol), whereupon the amine is released in an interchange reaction.

Example 1: A seed culture is prepared from sterile solutions comprising 860 g of water, 19.4 g of glucose, 86 g of an aqueous solution comprising 2% dipotassium hydrogen phosphate and 0.45% ammonium nitrate, 86 g of an aqueous solution comprising 0.1% magnesium sulfate heptahydrate, and 2.6 g of soy peptone. One hundred grams of the resulting solution is inoculated with a fresh culture of *Xanthomonas campestris* NRRL B-1459 and shaken in the dark at 29°C for 25 hr. A 70 g portion of the resulting broth is combined with the remainder of the aqueous solution and shaken at 29°C for 54½ hr. The pH of the solution is periodically measured and adjusted to 6.8-7.2 by the addition of a sterile 10% aqueous solution of a commercial ethylene polyamine mixture approximately corresponding in molecular weight to pentaethylene hexamine (and referred to as such hereinafter); a total of 8 ml of the pentaethylene hexamine solution is added. The resulting broth is used as a seed culture in later fermentations.

Example 2: A sterile system comprising a resin flask, stirring means, liquid and gas addition means, temperature measuring means and reflux condensing means is charged with sterile solutions comprising 270 g of glucose, 12 g of dipotassium hydrogen phosphate, 7.2 g of ammonium nitrate, 1.2 g of magnesium sulfate heptahydrate, 3.6 g of soy peptone and 12 ℓ of water. To the mixture is added 750 g of the seed culture of Example 1, and the solution is purged with air and stirred at 28°C in the dark for about 49½ hr. Periodic pH measurements are made and the pH is adjusted to 6.8-7.2 by the addition of a sterile 10% aqueous solution of pentaethylene hexamine. The glucose content of the solution is also checked periodically by means of Clinistix.

After 49½ hr, the broth tests negative for glucose and 800 ml of a 2.6% aqueous phosphoric acid solution is added to reduce the pH to 3.5. Water, 18 ℓ, is added slowly followed by 50 g of the pentaethylene hexamine solution. The total pentaethylene hexamine charged to the system by this time is 22.8 g. The desired xanthan gum precipitates as a fine precipitate which is separated by centrifuging; the supernatant liquid is cloudy, indicating the presence therein of substantial quantities of microbial cells. The xanthan gum amine salt is washed with a methanolic solution of sodium hydroxide and then with methanol, and is dried in a vacuum oven. The yield is 150 g.

In Peanut Butter Table Syrup

Methods of preparing a stable homogeneous peanut butter containing table syrup have been disclosed by *B.B. Deretchin; U.S. Patent 4,152,466; May 1, 1979.*

This method comprises forming a slurry of xanthan gum and peanut oil, hydrating the xanthan gum in the slurry by mixing the slurry with water while subjecting the slurry and water mixture to agitation, mixing a sugar syrup with the peanut oil, xanthan gum and water slurry and heating the mixture, dispersing peanut butter in the heated mixture, adding additional sugar syrup to adjust the solids content to the desired level with additional syrup, while maintaining the temperature and then cooling the syrup.

The xanthan gum may be employed at levels of about 0.07-0.13% by weight and preferably at about 0.1% by weight.

The peanut oil may be employed at a level of about 0.25-0.6% by weight and preferably at about 0.4% by weight. The peanut butter is employed at a level of about 9.5-11.0% by weight and preferably at about 10.5% by weight.

In carrying out the process the sequence of steps employed is of the most importance in obtaining the desired composition. The xanthan gum and peanut oil are combined and formed into a slurry with agitation. The slurry of xanthan gum and peanut oil is then added to the water at ambient temperature with agitation and the agitation continued until the xanthan gum is hydrated, generally 5-10 minutes.

One-half of the syrup is added to the xanthan gum, peanut oil and water admixture and the combination is heated with agitation to about 150°F. When the temperature reaches about 150°F, the peanut butter is added to the combination with sufficient agitation to disperse the peanut butter. Although somewhat higher or lower temperatures may be employed, care must be exercised at higher temperatures to avoid a break-down of the peanut butter–syrup system before the xanthan gum stabilizes the same.

When the peanut butter has been completely dispersed, the remaining syrup and salt are added and the total combination is heated to about 190°F to provide a bacteriologically stable product. Preferably the Brix of the syrup is adjusted to 67°Brix and the product is then packaged, preferably at about 190°F.

If desired potassium sorbate at about 0.05% by weight dissolved in just sufficient water to dissolve the same may be added just prior to packaging as a preservative.

In addition to the above, it has been unexpectedly found that the addition of about 0.4% citric acid to the final composition substantially enhances the sweetness thereof.

Example 1: The formula (parts are percent by weight) for a typical peanut butter table syrup product contains 77 parts of high fructose corn syrup (42% fructose, 71% solids), 10.7 parts of water, 10.5 parts of creamy peanut butter, 1.30 parts of salt, 0.4 part of peanut oil, and 0.10 part of xanthan gum.

Example 2: The procedure used for making the peanut butter table syrup was as follows: The peanut oil and xanthan gum were charged to a steam-jacketed vessel and then agitated with the stirrer blades rotating at about 200-400 rpm for about 5 to 10 minutes until a slurry was formed.

The slurry was then combined with water at ambient temperature and the agitation continued with the stirrer blades rotating at about 200-400 rpm for about 5 to 10 minutes until the xanthan gum hydrated.

One-half of the high fructose corn syrup to be used was charged to the vessel containing the xanthan gum, peanut oil, water mixture and the temperature increased to 150°F while maintaining agitation.

When the temperature reaches 150°F, the peanut butter was added and agitation continued until the peanut butter was dispersed. The remaining syrup and the salt was then added and blended in with agitation. The mixture was then heated to 190°F while maintaining agitation at which time the Brix was adjusted to 67°F and the product packaged.

In Low Calorie Starch Substitutes

Cellulosic flour substitutes currently available suffer from the major disadvantage that they can only be used up to a replacement level of about 20%, which leads to a caloric reduction in the final baked goods of only about 10%. When the available cellulosic flour substitutes are used at replacement levels greater than about 20%, the baked goods obtained are of unsatisfactory quality from the standpoint of taste and texture.

A. Torres; U.S. Patent 4,219,580; August 26, 1980; assigned to Pfizer Inc. has developed improved flour substitutes, which produce highly satisfactory baked goods at use levels which can be as high as 70% replacement of the flour component. Baked goods produced using 70% replacement of the flour by an equal weight of a flour substitute of this process have about 30 to 35% fewer calories per unit weight than conventional baked goods.

In this process, flour substitutes such as purified plant cellulose and modified starches can be improved by adding 1-3.5% by weight of xanthan gum, and from 2-7% by weight of an emulsifier, such as lecithin.

A wide variety of cellulosic components can be used in preparing these compositions, since all cellulose is essentially nondigestible. However, a preferred form of cellulose for use in the compositions is the purified, crystalline, alpha-cellulose obtained from plant fiber (Solka-Floc).

The preferred xanthan gum for this process is the colloid obtained by fermentation of *X. campestri* using glucose as the substrate. It is a heteropolysaccharide made up of building blocks of D-glucose, D-mannose and potassium D-glucuronate. The potassium can be replaced by several other alkali and alkaline earth metal cations.

The preferred emulsifiers for use in this process are lecithin, monodiglyceride mixture, sodium stearoyl-2-lactylate and triglycerol monostearate.

In preparing the flour substitutes, the components can be combined in any conventional manner, usung any method which will effectively admix the components. However, it is a preferred method of preparing the flour substitutes to first heat the emulsifier to a temperature of about 60°C, then thoroughly admix this with the xanthan gum. The cellulose or nondigestible, acid-treated starch derivative, or mixture thereof, is then added and the resultant mixture is blended until a homogeneous, free-flowing powder is obtained.

Example 1: Citrated starch, Avicel PH-101 and Solka-Floc B-200 (95.4 parts each) were each coformulated with 3.5 parts lecithin and 1.1 parts food grade xanthan gum, to obtain flour substitutes, I, II and III.

The lecithin in each case was heated to 60°C and thoroughly mixed with the xanthan gum. Subsequently the citrated starch in Formula I, or Avicel or Solka-Floc in Formulas II and III, was added and blended to obtain a free-flowing powder of homogeneous appearance.

The resultant powders were found to have a mild corn meal-like flavor and were insoluble in water. However they absorbed water and behaved in a manner analogous to wheat flour when incorporated in the preparation of sweetened leavened baked goods. In addition they allowed the reduction of shortening from 25-50% depending on the type of baked product and thus provided a product with significantly reduced calories.

Example 2: A typical cake (Cake No. 1) was prepared, using 31.73 g of emulsified shortening, 56.10 g of sugar (sucrose), 3.30 g of nonfat milk solids, 23.10 g of whole eggs (beaten), 30.04 g of water, 56.20 g of cake flour, 1.05 g of sodium bicarbonate, 2.20 g of glucono-delta-lactone and 0.28 g of vanilla extract.

In a 400 ml stainless steel beaker, the nonfat milk solids, sugar and shortening were creamed for 3 minutes. The eggs were added and the mixture was beaten for 2 minutes. The water and the vanilla extract were combined and added to the above, and then the resulting mixture was mixed for 2-3 minutes until a homogeneous creamy liquid was obtained. Meanwhile, the cake flour, sodium bicarbonate and glucono-delta-lactone were premixed and added to the other hydrated ingredients. The resulting mixture was mixed 3-5 minutes until a dough of smooth consistency was obtained. A portion of this dough (120 g) was placed into a lightly-greased, tared, 250 ml rectangular pan, and then baked at 162°C for 32 minutes.

Three further cakes (Cakes Nos. 2, 3 and 4, respectively) were prepared, in which 50% of the cake flour was replaced by an equal weight of flour substitutes I, II and III, respectively, of Example 1, and only 16.04 parts of shortening were used.

The procedure for making Cakes 2, 3 and 4 was the same as Cake 1, except that the flour substitute was added to the hydrated ingredients, and creamed for 3 minutes, before the premixed flour, sodium bicarbonate and glucono-delta-lactone were added.

Cakes 2, 3 and 4 were comparable in displacement volume, color and texture to Cake 1, but had only 2.82-2.84 calories per gram compared to 4.11 calories per gram for Cake 7.

In Modified Gluten Product

N.S. Singer and D.W. Murray; U.S. Patent 4,198,438; April 15, 1980; assigned to John Labatt Limited, Canada have provided a modified gluten product which when hydrated has improved maleability and moisture-imbibing characteristics.

It has been found that xanthan gum, under selected conditions, will interact with normal vital wheat gluten and the properties of the modified vital gluten are thereby very favorably enhanced.

The actual composition of the product is not known for certain. However, there are indications that the gluten is modified by at least a significant proportion of its free gliadin content being complexed with the xanthan gum and it is believed that the presence of the gliadin/xanthan gum complex imparts the desired properties to the product. For example, when xanthan gum in aqueous solution is added to an aqueous solution of isolated gliadin, a complex precipitates. The gliadin: xanthan gum ratio in the precipitate (at a set pH) can be increased until a maximum is reached indicating that all available free gliadin has then been complexed; hence, no advantage would be gained by exceeding that limit. Moreover, it has also been found that it is not necessary to complex all the free gliadin, a satisfactory product being produced if a significant proportion, at least 40%, of the free gliadin in the gluten is in complexed form. Also, the gliadin:xanthan gum ratio is dependent on the pH.

The modified gluten product of the process is prepared by mixing vital gluten with xanthan gum under conditions selected to ensure that the protein/gum complex is formed. For example, the gluten may be dispersed in aqueous solution and the xanthan gum, either dry or in, preferably aqueous, solution, added thereto and the mixture stirred until complexing occurs. Obviously, the gluten tends to ball up but, surprisingly, the xanthan gum is able to react under those conditions. The reaction mixture is generally dried by normal procedures to obtain the desired product as a dry powder.

However, the product formed by reacting xanthan gum with previously undried wet vital gluten direct from the dough washing process (containing about 30 to 40% solids) has been found more effective as a baking additive giving better loaf volumes, etc., and for that reason is the preferred method. This modified gluten product may be formed by extensively mixing wet normal vital gluten with xanthan gum, preferably in the presence of a buffering and/or metal ion sequestering agent for a predetermined period of time. The amount of mixing is extremely important since below a predetermined minimum, it is believed the desired complex is not formed, and the resulting baked products are poor in volume; and above a predetermined maximum, although adequate product volumes may be obtained, the cell-structure is very coarse and unacceptable, possibly because of adverse effects on the earlier formed complex.

Example: Vital gluten was prepared by the "batter-process" from hard Manitoba spring wheat, first clears flour. It was "washed" to the extent that the "solid" contained about 80% protein. The total solids content was determined to be 33.3%; the ash content was found to be 0.26%. Sodium citrate (a metal ion sequestering agent) and dry xanthan gum (Keltrol) were added to the wet gluten mass. The mixture obtained comprised 96.4% wet gluten, 0.4% sodium citrate, 2.4% xanthan gum, and 0.8% water.

The components were extensively mixed in a Brabender 600, sigma-blade mixing bowl at 250 rpm. Upon addition of the dry xanthan gum, the wet gluten mass, already mixed with the sodium citrate, disintegrated into relatively large lumps. Mixing was continued for 11 minutes during which time, the gluten lumps decrease in size until, at the expiry of that period, they had been converted into a relatively uniform highly maleable extensible mass. The mass was also somewhat softer than before the addition of the xanthan gum. (The torque, which is proportional to the apparent viscosity, had decreased from about 0.3 m-kg to 0.2 m-kg.)

Continued mixing resulted in the mass becoming softer and more uniform until, after the lapse of the short period of only two minutes, the mass attained a soft "maleable" state or condition which persisted essentially unchanged for 30 minutes. All uniform products obtained upon arresting such mixing during that period had the desired properties and constitute products of this process. The products may be utilized immediately in the wet condition or may be stored in a frozen or dried condition. Any conventional drying method may be used provided the conditions utilized prevent overheating and denaturing of the protein in the product. Continued mixing beyond the 30 minute mark produced an undesirable pale rubbery inextensible mass.

The product of the example was found to be easily reconstituted by simple, and brief, stirring with water thereby producing a soft extensible mass as obtained originally. Moreover, the full amount of water extracted during drying was taken up again by the product.

Regular dry vital gluten when rehydrated in the same manner, reverted to the normal firm rubbery mass typical of gluten but about 3-5% of the water extracted during drying was not taken up by the gluten mass.

In Agitated Milk Products

A process for increasing the viscosity of cold milk under agitation is provided by *R.S. Igoe; U.S. Patent 4,219,583; August 26, 1980; assigned to Merck & Co., Inc.*

It has been found that TSPP (tetrasodium polyphosphate) and xanthan gum, or other gums such as guar, and carboxymethylcellulose, when added to cold milk under agitation, significantly increase the viscosity of the cold milk system. This viscosity lasts long enough to be able to use this process to prepare a milk-shake type preparation, and serve it to a customer for consumption within a reasonable time, and have it remain in its thickened state during consumption and extended preparation.

The system to which this process is particularly suited is a continuously stirred or circulating drink dispenser, such as is commonly seen to serve fruit juice drinks or the like. When milk, sugar, flavoring, and the gum blend are employed in such a machine, the resulting cold milk mix is milk-shake-like in viscosity.

The cold milk used is at normal refrigeration temperatures, for example, between 35° and 45°F.

Although TSPP is the best phosphate used, and is preferred in this process, other phosphates, such as sodium hexametaphosphate, sodium tripolyphosphate and others, can be used.

The greatest viscosity increase comes from a premixed blend of xanthan gum and TSPP. This blend can be in dry form, or in an aqueous concentrate, which can be diluted up to about five or more times with milk to get the final use range.

The TSPP and the xanthan gum are employed in approximately equal amounts in the mix. Operably from 40:60 to 60:40 of each can be used (weight basis). However, up to two times as much TSPP can be used as xanthan gum, if desired. The final level of both in the milk is between about 0.05 to about 1% (weight/volume basis). A more preferred range is about 0.075% to about 0.6%, and even more preferred, from about 0.09 to about 0.3%.

Surprisingly, the increase in viscosity was only observed in the xanthan gum-TSPP where used in milk systems. In water, no significant viscosity increase is noted.

In a milk system, the following general procedure was used: A dry blend of xanthan gum or other gums, i.e., guar, carboxymethylcellulose, and TSPP mixture was dispersed with five times its weight with sugar and added to cold mix under a constant mixing speed (about 400 rpm) for 5 minutes. This equipment used was a T-line laboratory stirrer although any suitable mechanical stirrer would function. Viscosity was measured on a Brookfield viscometer, Model LVT. The results show a significant viscosity increase is obtained with the inclusion of TSPP. The gum concentration was 0.6% and TSPP was added at 0.8%.

When the mixture of TSPP and xanthan gum was used the viscosity was 1500 cp. When TSPP or xanthan gum was used alone, viscosities of only 70 cp and 510 cp respectively, were obtained. The mixture of guar gum and TSPP provided a viscosity of only 840 cp.

Further testing showed that the maximum viscosity is found at about a 1:2-2:1 ratio of the xanthan gum and TSPP.

OTHER GUMS

Guar Gum in Gelatin Desserts

One of the most important uses of gelatin in the food industry today is in the manufacture of gelatin desserts, which are prepared from gelatin dessert powders typically composed of gelatin, sugar, an edible organic acid, flavor and color. In the standard preparation of such desserts, the gelatin powder is dissolved in hot water and the mixture is then cooled by adding cold water and allowing it to stand in a refrigerator, whereupon a translucent gel is formed. Much of the use of gelatin desserts today is in institutional programs such as school lunches and hospital meals. Because of the uncertainty in the production and supply of institutional foods, the meals prepared therefrom are usually stored in a freezer. Gelatin desserts, however, cannot tolerate the freeze-thaw cycles which are undergone when the gelatin is thawed before use or when the storage temperature fluctuates, and consequently break down through syneresis, losing their acceptable texture, set and mouth-feel. The gelatin desserts must therefore be prepared and stored separately from the main meal.

Guar gum has been known for its use as a thickener and stabilizer for many foods such as ice cream, cheeses and salad dressing; however, it is unexpected that guar gum will impart low-temperature stability to a gelatin dessert while not producing an undesirable cloudy product or reacting with the gelatin, as do many other known stabilizers such as food starches and various other gums.

B.H. Nappen; U.S. Patent 4,272,557; June 9, 1981; assigned to National Starch and Chemical Corporation has developed a gelatin dessert powder which, when incorporated into a gelatin dessert, provides the dessert with low-temperature stability.

According to this process, guar gum is blended with the other ingredients of the dessert powder in any order, and in an amount which may range from about 0.5 to 5% by weight of the powder, depending on the setting properties desired in the gelatin dessert and on whether fruit particles are to be dispersed, in which a larger amount is added. The preferred amount of guar gum employed is 1-3% by weight.

In the preparation of the gelatin dessert, the standard technique is followed whereby initially the gelatin or complete dessert powder is dissolved in warm or boiling water, accompanied by stirring to achieve dissolution. The remainder of the ingredients of the powder are added at this point, if not already added, followed by the addition of cold water. The amount of total water added depends on the type of gelatin powder employed but ordinarily is such as to be four to ten times the weight of the dessert powder. The resulting mixture is allowed to set at refrigeration temperature until a gel is formed. Fruit pieces derived from fresh or canned fruit (such as fruit cocktail) may be optionally added to the gelatin mixture prior to the setting of the dessert.

The gelatin dessert obtained using the powder of this process successfully withstands break-down caused by syneresis from freeze-thaw cycles, while essentially retaining the taste, mouthfeel, texture, color and clarity of a gelatin dessert which does not contain any stabilizer. In addition, the dessert remains as a gel at room temperature overnight, whereas standard gelatin desserts break down at room temperature in a few hours.

Example: Five gelatin dessert powders were prepared using 7.2 g of unflavored powdered gelatin, 56.5 g of sugar, 1.3 g of a drum-dried mixture of 30 parts lemon and 70 parts starch and one of the gums listed below.

Samples A and B contained 2.0 g and 2.5 g of guar gum respectively. Comparative samples C, D and E contained other gums, e.g., C contained 3.0 g xanthan, D contained 2.5 g locust bean gum and E contained a mixture of 1.25 g xanthan gum and 1.25 g locust bean gum.

In the preparation of the gelatin dessert, 338 g of water at ambient temperature were placed in a saucepan and the gelatin was added thereto. The water and gelatin were stirred at low heat until the gelatin dissolved. The mixture was taken off the heat, and the remainder of the ingredients was added with stirring until the sugar dissolved. The mixture was thereafter chilled for 24 hr.

Samples A and B gelled with a slight cloudiness as compared to the control, but with a heavier set. The drum-dried fruit mixture remained dispersed throughout Samples A and B with little or no settling. When Samples A, B and D were prepared in a separate experiment without chilling thereof, Samples A and B remained stable at room temperature without breaking down (with no visible separation of the liquid phase), whereas D broke down in 2 hr. When Samples A, B and D were frozen for 12 to 24 hr and subsequently thawed, the gel structures of Samples A and B were completely intact, giving excellent low-temperature stability, but comparative Sample D had only moderate low-temperature stability.

Comparative Sample C did not gel or set up at all and thus was not evaluated for low-temperature stability. Comparative Sample E gelled, but the fruit particles settled to the bottom. When frozen and thawed under the same conditions as those employed for the samples above, Sample E retained its gel structure but was cloudy and had only moderate low-temperature stability.

Locust Bean Gum in Milk Beverages

J. Takahata; U.S. Patent 4,212,893; July 15, 1980; assigned to Honey Bee Corp., Japan has provided an acidified whole milk beverage having an acidic pH and containing locust bean gum as a stabilizer.

It has been discovered that use of a very small amount of locust bean gum as a stabilizing agent enables whole milk to maintain a stable emulsified state without causing the curd to be segregated from the whole milk even when mixed with fruit juice or organic acid.

In a preferred embodiment, the whole milk-containing acidified beverage can retain a more stable emulsified state by using pectin or agar as an auxiliary stabilizing agent together with the locust bean gum used as the main stabilizing agent. The acidified beverage of whole milk is prepared through the following steps:

(1) locust bean gum is dissolved in whole milk;

(2) the whole milk emulsion which is formed is mixed at a temperature of 35° to 60°C with at least one acidifying agent selected from the group consisting of fruit juice and organic acid; and

(3) the acidified whole milk emulsion is homogenized.

In the first step, the locust bean gum may be added in the form of powder, but it is more convenient to apply the gum in the form of an aqueous solution. Part or all of the usable additives can be dissolved in the aqueous solution of locust bean gum. It is preferred to add sweetening agents and table salt to the solution, followed by thermal sterilization. The locust bean gum is readily soluble in water at a higher temperature than about 80°C. The locust bean gum should be added in an amount of 0.1 to 1.0% by weight or preferably 0.2 to 0.3% by weight based on the total weight of the finished product.

In the second step, an emulsion of whole milk containing the locust bean gum is mixed with fruit juice, organic acid or a mixture thereof, followed by thorough stirring. The fruit juice or organic acid should be added after the locust bean gum is fully dissolved in whole milk. The reaon is that if a whole milk free from locust bean gum is added to fruit juice or organic acid, then fat and casein will be segregated in the form of curd.

The emulsion of whole milk containing the locust bean gum should be mixed with fruit juice or organic acid at a temperature of 35° to 60°C, or preferably 40° to 50°C. The mixed whole milk should preferably be stirred about 10 to 30 minutes at this temperature or while the solution is allowed to cool, in order to stabilize the emulsified acidified whole milk.

In the third step, the acidified emulsion of whole milk prepared in the second step is homogenized for greater stability. This homogenizing step can be effected under the customary condition using a homogenizer, for example, the Gaulin homogenizer generally used in the homogenized milk-producing industry.

Example 1: 5 kg of cane sugar, 1.5 kg of grape sugar, 20 g of table salt and 200 g of locust bean gum were thoroughly dissolved with stirring in 20 ℓ of water at 80°C. The mixture was boiled 20 minutes at 90°C for sterilization. After cooling, the solution was mixed with 15 ℓ of commercially available whole milk and a proper amount of orange flavor, followed by full stirring. 7.5 ℓ of orange juice and 150 g of citric acid were dissolved in 10 ℓ of warm water. This solution was added to the abovementioned solution containing the milk and locust bean gum at a temperature of about 50°C in about 5 minutes, followed by stirring. Later, stirring was continued about 20 minutes at the same temperature. At this time, citric acid was added in a sufficient amount to set the pH value of the entire solution at 3.5. Last, homogenization was carried out by the Gaulin homogenizer, providing an acidified beverage of whole milk maintained in a stable emulsified condition free from segregated curds. The product had a saccharinity of 11° as measured by the Abbe's saccharimeter.

After being sterilized at 80°C, the acidified beverage of whole milk was placed into a metal container, followed by quenching to 15°C. After being stored one year, the beverage did not contain any sediment, but maintained a stable emulsified condition.

Example 2: Acidified beverage of whole milk was produced substantially in the same manner as in Example 1, except that not only 200 g of locust bean gum but also 40 g of agar were jointly used as a stabilizing agent. An acidified beverage of whole milk thus prepared maintained a stable emulsified state free from segregated curds.

Example 3: 100 g of cane sugar, and 3 g of locust bean gum were dissolved with stirring in 200 cc of water at 85°C. 400 cc of whole milk was added to the solution. Later, 1 g of citric acid, and 4 cc of lactic acid were added at a temperature of about 50°C in about 5 minutes. 0.3 cc of lemon flavor was added to the mixture. The whole mass was stirred about 20 minutes at a temperature of about 50°C. Further, citric acid was added in a sufficient amount to set the pH value of the milk beverage thus prepared at 3.5. Finally, the mixed mass was homogenized by the Gaulin homogenizer, providing an acidified beverage of whole milk lacking fruit juice. This product also maintained a stable emulsified state free from segregated curds.

Kappa-Carrageenan in Dessert Gels

It is known that carrageenan extract from *Chondrus crispus* forms a gel in a water solution; that this carrageenan contains two carrageenan components, kappa and lambda-carrageenans; and that only the kappa-carrageenan exhibits gel-forming characteristics in a water solution. Also recognized is that the strength of the water gel that is formed is greatly affected by the cations which are present, for example, with the presence of sodium cations imparting little gel strength, while the gel-forming properties of the kappa-carrageenan are greatly enhanced in the presence of potassium cations. Thus, kappa-carrageenan is often referred to as "potassium-sensitive carrageenan."

Notwithstanding its desirable gel-forming properties, kappa-carrageenan is not without shortcomings which have limited its use in the preparation of water dessert gels.

A.L. Moirano; U.S. Patent 4,276,320; June 30, 1981; assigned to FMC Corp. provides an improved aqueous gel-forming composition consisting of kappa-carrageenan, a sodium salt of a sequestering agent, and potassium in an ionizable form. The sequestering agent and the potassium are present in the composition in amounts sufficient, respectively, to sequester substantially all of the polyvalent cations when the composition is dissolved in an aqueous medium, and to provide this aqueous medium with a potassium ion content of from about 200 to 800 ppm. For the sake of simplicity, it will be understood that all ppm (parts per million) concentrations set forth throughout the description and claims are based upon the total weight of the aqueous phase which includes the water and materials dissolved therein.

In this process, the carrageenan composition is merely solubilized in water which need be at only such temperature as to dissolve the carrageenan composition, after which the sol is allowed to set. Depending upon the potassium cation concentration present, but within the range specified, gelling occurs at a temperature of about 25°C or less and the resulting gel is thermally reversible, remelting at about 35°C. The temperature of the water required for preparation of the sol generally need not exceed about 45°C.

The carrageenan composition is well adapted for use in preparing acidified water dessert gels, neutral nondairy puddings, and aerated desserts, which are referred to herein as "dessert gel," "water gel," or "gel." Dessert gels prepared with this carrageenan composition possess satisfactory elasticity and gel strength and are subject to negligible syneresis. Yet more significant is that the resulting water gel softens or melts readily within the mouth of the consumer, providing for

excellent flavor release, a pleasant mouth feel, and other desirable eating qualities.

In the carrageenan composition, the sequestering agent renders inactive substantially all of the polyvalent cations which are present during gel formation and,
therefore, the resulting gel exhibits good elasticity, gel strength and resistance to
syneresis. As carrageenans are normally extracted from sea plants in the presence
of lime, it is not uncommon for conventional kappa-carrageenan to contain from
3 to 4% of calcium cations, based upon the weight of carrageenan. When such carrageenan is used at a 1% concentration in water, the concentration of calcium
cations in the aqueous solution will range from 300 to 400 ppm. Moreover, as
the carrageenan composition is intended to be placed in solution with ordinary
tap water, both calcium and magnesium cations are likely to be present during gel
preparation. Of course the hardness of the water used will vary with location, and
a very hard water can well introduce perhaps some 200 ppm of calcium cations.
Accordingly, to satisfy its intended function, the sequestering agent is present in
amounts ranging from 0.1 to 0.3%, based upon the weight of the water employed.

A variety of nontoxic sequestering agents, used alone or in combination, are useful in the composition of this process. With a carrageenan solution having a low
pH, as for example a pH of from about 3.0 to 5.0 as in preparing an acidified water dessert gel, polyphosphates, such as, sodium hexametaphosphate, are most
effective. At a higher pH, sodium salts of orthophosphates, carbonates, pyrophosphates, and organic carboxylic acids, as well as polyphosphates, are satisfactory.

The presence of potassium cations in the carrageenan, of course, enhances gelation during the water gel preparation. Significant, however, is that this carrageenan
composition assures the necessary gel formation, yet limits the potassium cation
concentration to from 200 to 800 ppm. The necessary potassium cations may be
supplied by employing potassium carrageenate in the composition formulation,
or by one or more ionizable potassium salts, as when sodium kappa-carrageenan is
used, or by a combination of such potassium carrageenate and potassium salts.
Nontoxic salts which are suitable include, for example, potassium chloride, potassium phosphate, potassium citrate, and potassium tartrate.

To attain the objects of this process, it is essential that the described carrageenan
composition, upon solubilization in water, provide a potassium cation concentration which does not exceed 800 ppm and is not less than 200 ppm. At the upper level of potassium cation concentration of 800 ppm, gelling will occur at approximately 30°C and remelting of the formed gel at about 41°C, while at a potassium concentration of less than 200 ppm, the gel strength of the resulting product
is much too weak.

The carrageenan composition may be used with iota-carrageenan or locust bean
gum if still further improvement in the elasticity, gel strength and syneresis of the
resulting gel is desired.

Example 1:   An acidified water dessert gel was prepared from 5 g potassium
sensitive (kappa) carrageenan, 2 g sodium hexametaphosphate, 2.5 g adipic acid,
1.0 g sodium citrate, 85.0 g sugar, and flavor and color to suit.

Two (2) cups (1 pint) of deionized water at 45°C were poured over a mixture of
the dry ingredients in a bowl and spoon stirred until dissolved (about 2 minutes).

The solution was poured into molds and refrigerated. The gel formed at 21°C. It melted readily in the mouth with fine flavor release. Melting temperature was 34°C.

This system contained 450 ppm of K^+ and 400 ppm of Ca^{2+}. In this case the cations all came from the carrageenan.

Example 2: A nondairy custard pudding was prepared from 2.50 g potassium sensitive (kappa) carrageenan sodium form, 0.75 g sodium hexametaphosphate, 30.00 g Whip-treme 3296 (Beatrice Foods Co.), 50.00 g sugar, 6.00 g sodium caseinate, and 0.18 g potassium bitartrate.

The dry mix was added to two cups (475 ml) of hard water (290 ppm Ca^{2+}) at 45°C and spoon stirred until dissolved (about 2 minutes). The solution was refrigerated until set.

In this case, the system contained a total of 100 ppm K^+ (77 ppm from the potassium bitartrate and 23 ppm from the carrageenan) and 300 ppm Ca^{2+} (294 ppm from the hard water and 6 ppm from the carrageenan).

Beta-1,4-Glucan–Polymer Composition

There are numerous instances in manufacturing and in home operations, notably in food processing but also in operations entirely unrelated to food, where it is desirable to obtain a uniform stabilized dispersion of a powdered or granular form of an oleaginous or other difficult to disperse material in a cold aqueous system. For example, in making bread it would be desirable to have a dry, free-flowing mono- and diglyceride powder which could be added directly to bread dough, the bread dough being a cold, aqueous system. Mono- and diglycerides are widely used in making bread but normally a predispersed hydrate form is necessary to achieve dispersion in the bread dough cold aqueous system.

E.J. McGinley and J.M. Zuban; U.S. Patent 4,231,802; November 4, 1980; assigned to FMC Corporation have provided a mechanism whereby oleaginous or other difficult to disperse material can be quickly and conveniently formed into microdispersions in cold water.

According to this process a matrix of beta-1,4 glucan having a water-soluble polymer intimately associated therewith, is distributed throughout the difficult to disperse material and subsequently is formed into a powdered or granular mass. The individual particles of such mass are impregnated with the difficult to disperse material. The matrix may comprise by weight from about 70 to 99 parts of beta-1,4 glucan and from about 1 to about 30 parts of the water-soluble polymer.

In the preferred form, the matrix comprises by weight from about 85 to about 95 parts of disintegrated beta-1,4 glucan and from about 5 to about 15 parts of sodium carboxymethylcellulose having a degree of substitution of 0.75±0.15. This preferred form of matrix and means of making it are described and claimed in U.S. Patent 3,539,365. As disclosed this matrix material is water-insoluble but is readily dispersible in water to form a thixotropic gel. Thus, this process is not directed toward forming a dispersion of the water-insoluble beta-1,4 glucan-containing material but involves using such material as a means or vehicle for attaing a dispersion of other water-insoluble materials as well as difficult to disperse water-soluble materials.

This is accomplished by incorporating the difficult to disperse material into a paste consisting of the matrix component and water. The difficult to disperse material can be mixed with the matrix component prior to adding the water to form a paste or may be added to the paste. A suitable paste is generally attained when the amount of water is between about 10% and about 85% of the total weight of the paste. This limited amount of water causes the individual particles of the matrix to swell thereby opening up the interior of the particles and greatly enlarging the surface area thereof. The difficult to disperse material is then thoroughly stirred into the paste and during the course of this stirring the difficult to disperse material is spread throughout the pasty mass.

Example 1: Twenty grams of a liquid mono- and diglyceride was poured into a five quart Hobart mixing bowl. Eightly grams of the preferred matrix substance in dry powder form was added to the bowl. The bowl was then put on a Hobart mixer fitted with a paddle attachment and mixed on slow speed No. 1 for 2-3 minutes to form a crumb. At this point 200 cc water was added and mixing was continued on faster speed No. 2 to form a pasty mass. While heat was applied to the bowl, mixing continued on speed No. 2 until substantially all the added water was evaporated. At this point, application of heat was discontinued while continuing to operate the mixer until the product cooled. The resulting product was a fine free-flowing granular mass. These granules consisted of the aforesaid matrix substance impregnated with the mono- and diglyceride. Five grams of this powder was poured into 95 g of distilled water at 23°C and two minutes of mild hand stirring resulted in a smooth, viscous dispersion. Examination of the dispersion under a microscope showed a fine distribution of emulsifier globules.

Example 2: Ten grams of oleoresin paprika was placed in a 5 quart Hobart mixing bowl. Ninety grams of the preferred matrix substance in dry powder form was added to the bowl. The bowl was then put on a Hobart mixer fitted with a paddle attachment and mixed on speed No. 1 for 2-3 minutes to form a crumb. At this point 200 cc water was added and mixing was continued on speed No. 2 to form a pasty mass. With heat applied to the bowl, mixing was continued until substantially all the added water was evaporated. At this point heat was discontinued while continuing to operate the mixer until the product cooled. The resulting product had the same characteristics for dispersing in cold water as in Example 1.

Example 3: Fifteen grams of a solid mono- and diglyceride was melted in a heated five quart Hobart mixing bowl. Eighty-five grams of the preferred matrix substance in dry powder form was added to the bowl. The bowl was then put on a Hobart mixer fitted with a paddle attachment and mixed on speed No. 1 for 2-3 minutes to form a crumb. At this point 200 cc water was added and mixing was continued on speed No. 2 to form a pasty mass. With heat applied to the bowl, mixing was continued until substantially all the added water was evaporated. At this point heating was discontinued while continuing to operate the mixer until the product cooled. The resulting product had the same characteristics for dispersing in cold water as in Example 1.

Cyclodextrin in Citrus Foods

Citrus fruit generally contains a large variety of flavonoid compounds and, among these compounds, hesperidin and naringin tend to detract from the quality of the citrus fruit products as is described below. Thus, hesperidin occurs as solubilized in fresh citrus fruit, but when the fruit is processed into canned fruit packs, bottled beverages, etc. and stored, this compound begins to emigrate from the first tissues and is precipitated as white insoluble crystals. These crystals cause the so-called "clouding" in citrus food which, in turn, significantly decreases its sales value.

On the other hand, naringin is known to be a bitter material in citrus fruits such as grapefruit, summer oranges, buntan oranges, etc. and although its bitter taste is sometimes preferred, such citrus fruits rich in this substance are not suitable for ingestion or are of limited value because of their intense bitterness.

Furthermore, limonin is also known as one of the bitter constituents of citrus fruits such as navel oranges, Valencia oranges, iyo oranges, etc. These citrus fruits are not used extensively for canning or for the manufacture of juice products because of the bitter taste which develops in the juice and the segments when the fruits are allowed to stand exposed to air, or when pasteurized.

It is an object of the disclosure by *M. Miyawaki and A. Konno; U.S. Patent 4,332,825; June 1, 1982; assigned to Takeda Chemical Industries, Ltd., Japan* to provide an improved citrus food that has little or no undesirable taste or flavors and does not cloud. This has been achieved by treating the citrus products with cyclodextrin.

Many species of cyclodextrin are known, such as α-cyclodextrin consisting of 6 pyranoglucose units linked by α-1,4-glucosidic bonds, β-cyclodextrin consisting of 7 linked pyranoglucose units, γ-cyclodextrin consisting of 8 units and so forth. Among these species, β-cyclodextrin is most successfully employed in this process. Mixtures of these cyclodextrins also can be employed as well.

In accordance with this process, cyclodextrin is added to natural citrus materials in a proportion of 0.005 to 1.0 weight percent, preferably 0.01 to 0.7 weight percent, and more preferably 0.05 to 0.5 weight percent, based on the weight of the citrus material. It is undesirable to add cyclodextrin in an amount exceeding 1 weight percent because a white sediment is often formed in the citrus food. A natural citrus material means a fresh fruit as harvested, and even when it is used as peeled, concentrated or diluted, cyclodextrin is added in the above proportional range based on the amount converted to natural citrus material.

For example, in preparing a fruit juice from 500 ml of a ½ concentrate of Unshiu mandarin juice, twice the volume, i.e 1 ℓ, is considered to be the amount of natural citrus material present and the level cyclodextrin added is based on 1 ℓ. The amount of cyclodextrin to be added to any specific citrus food is determined with reference to the variety of citrus food, the time of harvest, and the type of citrus food, e.g. whether the desired food is a canned fruit or a marmalade etc. When the bitterness is to be removed, it may sometimes be desirable to leave some of the bitter-causing material in the food for improved delicacy, and, therefore, the level of addition of cyclodextrin is selected taking this into consideration.

Example 1: Peeled Unshiu mandarin oranges were pretreated with acid and alkali in the conventional manner and 250 g of the segment was packed into a No. 4 (Japan Agricultural Standard, 74.1 mm in diameter and 113 mm in height) can together with 200 g of a syrup made of 20% sucrose and 20% glucose, fructose and liquid sugar in which 0.75% of β-cyclodextrin had been dissolved. The can was sealed, heated at 82°C for 20 minutes and cooled to give a canned mandarin orange pack. As controls, a canned product without any anticlouding agent, a product with 0.002% (based on syrup) of methylcellulose and a product with 0.05 unit/ml (based on syrup) of hesperidinase were also prepared.

These mandarin orange packs were stored at room temperature for a month and each of the cans was opened and examined for the clarity of syrup. The clarity measurement was carried out in the following manner. A white sheet of paper was marked with a black dot (5 mm in diameter) and a transparent flat-bottomed glass column 25 mm in diameter was placed on the sheet in concentric relation with the black dot. Each syrup was poured into the glass column with the black dot being watched and the height (mm) of the body of liquid at which the black dot had just ceased to be visible was recorded and regarded as representing the clarity of the syrup. The clarity values thus found were 50 mm in the case of the blank control (no additive), 110 mm for the methylcellulose group and 150 mm for the hesperidinase group. The value for the β-cyclodextrin group was more than 200 mm, evidencing the excellent cloud prevention effects of this method.

Example 2: One kg of summer oranges was sorted into peel and segments and the peel was rinsed to remove any foreign matter and chipped. The segments were separately processed into fruit juice.

To the peel and juice were added 500 g of water and 5 g of β-cyclodextrin, the mixture was heated under stirring, 1 kg of sucrose was added, and the mixture was concentrated by heating to give a marmalade.

The marmalade with added β-cyclodextrin was markedly less bitter than the control marmalade.

Example 3: A mixture of 60 g of Valencia orange juice, 60 g of summer orange juice, 15 g of orange juice sacs, 100 g of sucrose, 1.5 g of citric acid and 0.5 g of β-cyclodextrin was made up with water to 1000 ml and pasteurized at 95°C. The pasteurized mixture was filled into a 200 ml bottle and stored at room temperature for 3 months. Then, the contents were examined for sediments.

Whereas the cyclodextrin-free control product showed a white sediment, the product with added β-cyclodextrin did not show any white sediment. Moreover, the product containing β-cyclodextrin was not bitter to the taste.

LOCUST BEAN – XANTHAN GUM MIXTURES

Direct Acidified Yogurt

The preparation of yogurt traditionally is accomplished by bacterial fermentation. Such a process is inherently time-consuming and requires careful control for commercial success.

R.S. Igoe; U.S. Patent 4,169,854; October 2, 1979; assigned to Merck & Co., Inc. has developed a process for preparing yogurt which does not require bacterial fermentation.

It has been found that a direct acidified yogurt can be obtained from milk, acidulent, and a thickener blend of specific ingredients in specific ratios. Sugar may be added, if desired, for taste considerations.

The direct acidified yogurt may be prepared using whole milk, reconstituted nonfat dry milk solids, or nonfat dry milk solids and water.

The acidulent may be any common food acid such as, for example, citric, tartaric, acetic, malic, lactic, fumaric, ascorbic, adipic, succinic, oxalic, and phosphoric acids.

The optional carbohydrate sweetener may be sucrose, fructose, lactose, maltose, invert sugar, glucose, corn syrup solids having a dextrose equivalent of at least about 28, and the like.

In parts by weight, the direct acidified yogurt of this process comprises 80-98 parts of milk, 0-18 parts of carbohydrate sweetener, 0.5-1.2 parts of acidulent, and 1.8-4.5 parts of thickener blend.

The thickener blend contains starch, CMC, xanthan gum, and locust bean gum. The starch may be any starch which is stable at a low pH. Examples of such starches are waxy maize starch and tapioca starch. The CMC preferably has a degree of substitution of from about 0.65 to about 0.95, and most preferably from about 0.7 to about 0.9. The xanthan gum is food grade xanthan gum and the locust bean gum is food grade locust bean gum. The thickener blend (in parts by weight) contains 40-80 parts of starch, 20-45 parts of CMC, 3.0-8.0 parts of xanthan gum, and 0.7-3.0 parts of locust bean gum.

The direct acidified yogurt is prepared by adding the thickener blend to the milk, or if a carbohydrate sweetener is used, adding a mix of the thickener blend and carbohydrate sweetener to the milk, with moderate agitation. The resulting liquid is then subjected to pasteurization treatment after which acidulent (and flavor) if any are added with vigorous agitation, followed by shearing treatment, typically single-stage homogenization at from about 5 to about 45 atmospheres. The resulting product is then packaged and cooled.

Example: A thickener blend is prepared by blending 18 g of waxy maize starch, 8 g of CMC, 1.6 g of xanthan gum, and 0.4 g of locust bean gum.

The foregoing blend is then mixed with 74 g sugar and added to 890 g milk. The resulting liquid is heated at 160°F (71°C) for 30 minutes. It is then removed from heat and 8.0 g of citric acid added with mechanical stirring. It is then passed through a single-stage homogenizer at 350 psi, packaged and cooled. The resulting product has a tender delicate gel with a body and consistency resembling that of a cultured yogurt.

For Stabilized Icings

It is an object of the disclosure by *H. Cheng; U.S. Patent 4,135,005; January 16, 1979; assigned to Merck & Co., Inc.* to provide an improved icing composition.

It has been found that an improved freeze-thaw and heat-stable icing is obtained from icing compositions containing a high-melting-point fat and a gelling composition comprising from about 25 to about 75 parts by weight xanthan gum and from about 25 to about 75 parts by weight of locust bean gum, the gelling composition being present in an amount effective to impart freeze-thaw stability and heat stability. Any edible fat having a high melting point, e.g. from about 32° to about 55°C may be used, preferably from about 45° to about 55°C. Typically, an icing composition of the process consisting essentially of in % by weight from about 70 to about 85% sugar, from about 15 to about 30% water, from about 0.5 to about 4% fat, from about 0.1 to about 0.5% emulsifier, from about 0.1 to about 0.5% salt, and from about 0.001 to about 0.1% of a gelling system consisting essentially of from about 25 to about 75% xanthan gum and from about 75% to about 25% locust bean gum. Other hydrophilic colloids may optionally be present to modify the properties of the icing, e.g. propylene glycol alginate to increase viscosity and enhance sheen, or agar to form a gel.

In the following examples, the parts cited are parts by weight.

Example: Part I of the icing recipe contains 16.6 parts of water, 3.0 parts of granulated sugar, 0.025 part by weight of xanthan gum, and 0.025 part by weight of locust bean gum. Part II contains 16.5 parts of granular sugar, 1.5 parts by weight of fat, MP 51°-54°C, 0.2 part of emulsifier (high melting point mono- and diglyceride mixture, and 0.3 part by weight of salt. Part III contains 61.85 parts of 10 X sugar which contains 3% starch. Coloring and flavoring may be incorporated as desired in Part III.

The dry ingredients of Part I are well blended and added to the water while mixing. The resulting solution is heated to boiling and boiled for 2 minutes. The ingredients of Part II are added to the boiled solution of Part I ingredients and heated to boiling. The resulting solution is added slowly to a small Hobart mixer equipped with paddle and a steam-heated water bath to which the Part III ingredients have previously been added while mixing at No. 1 speed and heating to maintain temperature above 57°C until smooth.

The resulting icing is applied at 57°±1.1°C to fresh cooled donuts (37°-43°C). The icing is coated on the donuts using donut coating equipment with excellent adhesion with no icing rundown on the sides of donuts. The coating marks (from the donut coating equipment) heat immediately and skin forms within 20 seconds. The icing hardens and dries in less than 7 minutes. The icing not only has excellent sheen and opacity but also has excellent eating qualities. After standing at room temperature for a total of 10 minutes, the donuts are packed in a sealed paper carton and frozen in a freezer. When frozen, the donuts are thawed at room temperature for 2 hr. This is considered as one freeze-thaw cycle. Visual inspection of the icing after four freeze-thaw cycles reveals no weeping or cracking and no change in sheen, icing structure, adherence or eating qualities. The frozen donuts are also thawed by heating at 160°C for 3-5 minutes in an oven. The icing does not crack, melt, change structure or lose adhesion upon heating. The icing is still stable after four freeze-thaw cycles and upon heating.

In Jelly Confectionery Products

*H. Cheng; U.S. Patent 4,219,582; August 26, 1980; assigned to Merck & Co., Inc.*has found that from about 0.08 to about 0.50% (weight basis) of a 1:3 to 3:1 gum blend of xanthan gum-locust bean gum can be employed in formulating starch jellies which decreases the setting time. Preferably about 0.2-0.4% of a 1:1 blend is employed.

This particular xanthan gum-locust bean gum combination is especially valuable in formulating starch jelly candies. As these candies are produced by cooling a gellable syrup, which requires a prolonged gelling time in the mold, a food gelling agent which can reduce setting time will be commercially valuable.

The starch jelly candies most benefited by the addition of the gum blend are blends of cooked starch, water, sugars, and flavoring agents. The starches used vary but are generally thin-boiling corn, sorghum, and wheat starches, modified to the 30-90 fluidity level, or high amylose starches; these usually contain amylose at levels up to about 70%.

Example 1:   A gum confection was prepared from 560 g of 42 DE corn syrup, 380 g of sugar, 120 g of Eclipse G starch (thin boiling starch, A.E. Staley Mfg. Co.), 3 g of Keltrol F (xanthan gum, Kelco Co.), 3 g of locust bean gum, and 9.34 g of water.

Add water in a steam-heated kettle. Dry blend all the dry ingredients well and add to water with good agitation. Add corn syrup, cook and boil the mixture until 228 °F is reached and the solids content is 78-79%. The hot mixture is then deposited into starch molds and allowed to set at room temperature. Within 30 minutes after depositing, the gum confection has attained sufficient gel structure to permit removal and sugaring without further holding time. Without adding the xanthan gum and locust bean gum, the regular thin boiling starch confections require 42-48 hr in the hot conditioning room. These two types of confections are similar in body, texture and mouth-feel.

Example 2:   Confections were prepared from 48 g of 42 DE corn syrup, 32 g of sugar, 7 g of Mira-Quick C starch (high amylose starch, A.E. Staley Mfg. Co.), 0.2 g of Keltrol F (xanthan gum, Kelco Co.), 0.2 g of locust bean gum, and 12.6 g of water.

Water is added to a steam-heated kettle. Keltrol F, locust bean gum and 3 lb sugar are dry-blended well, added to the water with good agitation and heated to 180°F to hydrate the gums. The other ingredients are added and the mixture preheated to 160°F and cooked through a steam injection cooker at 340°F. Refractometer readings on the mixture showed actual solids content in the range of 82-83%. The mixture was then deposited into starch molds and allowed to set at room temperature. Within 30 minutes after depositing, the gum confection had attained sufficient gel structure to permit removal and sugaring without further holding time. Without adding the xanthan gum and locust bean gum, the regular high amylose starch confections require 8-12 hr in the hot conditioning room to gain the required gel structure for demolding.

In Cheese Products

An acid-stable, heat-reversible, cheese-like gel is prepared by the process cited by *R.H. Whelan and F.R. Conant; U.S. Patent 4,143,175; March 6, 1979; assigned to CPC International Inc.*

The process comprises mixing natural cheese with an acidulant and a locust bean gum/xanthan gum stabilizing system.

The cheese food product of the process is prepared by shredding mozzarella and provolone cheeses and adding these and grated romano cheese to a mixer cooker. By mozzarella cheese it is understood that any of the varieties including whole milk mozzarella, low moisture whole milk mozzarella, part skim mozzarella, and low moisture part skim mozzarella may be used alone or in combination with other members of this group. Typical of the mixer cookers used are steam-injected high-shear mixers (Waring type), or screw mixers (Reitz or Damrow type), as well as surface heating screw mixers (Reitz type) or a sigma-style mixer (J.H. Day).

To this mixture is added an aqueous dispersion of citric acid, sodium citrate, sucrose, potassium chloride, locust bean gum and potassium sorbate. This dispersion is prepared in a high shear mixture until a homogenous slurry is produced. Typical high shear mixers include Waring blender and the Lanco mixer. If a direct steam injection cooker is used, reductions in the slurry moisture must be made to compensate for the addition of water in such processing.

This locust bean gum slurry is added directly to the cheese mixture followed by the addition of a xanthan gum/olive oil slurry. Agitation is begun with cooking at from about 80°C to about 93°C. In a screw-type cooker, agitator rpm may vary from 20 to 100 rpm, mixing and heating time may vary from 2 to 30 minutes. The final composition of the cheese food product generally has about 60 to 64% moisture content, but the moisture content may be within the range of about 55 to about 70% depending upon required final structure. Fat content will be from about 12 to about 16%, pH from about 4.1 to about 4.4, protein from about 14 to about 17%.

This cheese food product may be filled into any standard size block or slab as desired (e.g., 5, 10, 20 or 40 pound) and cooled. After cooling, to about −1.1° to +4.4°C, the cheese is cut and shredded as desired for use in the hot sauce product formulation.

The cheese food product may also be prepared utilizing more sophisticated methods such as a master batch premix which is cooked in segments and then extruded onto a continuous chill-belt to form variously sized slices or slabs. These slices or slabs can then be further cut or shredded for use in the hot sauce as indicated above.

Example 1: In a one-gallon Waring blender 2.31 g citric acid, 1.67 g xanthan gum (Keltrol Ktl 27580, Kelco Co.), 1.67 g locust bean gum, 1.49 g sucrose and 1.59 g sodium citrate were mixed together and added to 41.8 g of water. This mixture was mixed and heated at the same time until well blended. Then 72.66 g of whole milk mozzarella cheese and 11.58 g provolone cheese were added gradually and continuously blended in the mixer. The temperature was raised up to about 74°C. The cheese was mixed until homogenized and then poured. While the cheese flavor was not strong enough, the cheese product did hold up when stirred vigorously into a 93°C acid sauce. The product grated well and appeared to be slightly too moist.

Example 2: 231.99 g of whole milk mozzarella cheese and 22.23 g provolone cheese were melted in the top of a double boiler. 7.58 g of citric acid, 5.27 g xanthan gum, 5.27 g locust bean gum, 5.04 g sodium citrate and 4.72 g of olive oil were added to a blender and were mixed at stir speed with 15 g of olive oil until the mixture was well blended. In turn this gum mixture was heated resulting in a relatively smooth mixture. This smooth gum and mixture was stirred into the melted cheese while still being heated in the double boiler. When all these ingredients were well blended, having the consistency of a roux, 135.8 g of water were added to the mixture. This was stirred in a blender and whipped at high speed until a homogeneous mixture was obtained. The resultant cheese product had a fairly homogeneous texture and a slight sandy taste. The flavor was very definitely that of cheese.

OTHER GUM MIXTURES

In Freezable Gel Confections

A. Braverman; U.S. Patents 4,140,807; February 20, 1979; 4,216,242; August 5, 1980; and 4,264,637; April 28, 1981 has prepared freeze-it-yourself pops by sealing an edible, room-temperature-storable, storage-stable, aqueous, opaque, flavored confection in a plastic film pak. The confection composition is in the form of an aqueous gel having a pH within the range of from about 3.0 to about 5.0; it has a pudding consistency when maintained at room temperature, but has a chewy consistency when frozen. Either low-methoxyl pectin (LMP) alone, carrageenan (CGN) alone, a combination of low-methoxyl pectin and carboxymethylcellulose (CMC) or a combination of carrageenan and carboxymethylcellulose provides the composition with homogeneity, stability and the noted consistency in both frozen and unfrozen states.

All the noted characteristics are achieved through use of 1) specific gum stabilizers, such as CGN, LMP, CMC and combinations thereof and 2) opacity-, milkiness- or turbidity-producing agents, such as chocolate or cocoa, milk products, nondairy whitener, and natural or artificial color, e.g. titanium dioxide. These are used individually or in any combination. Milk or dairy products include skim milk, nonfat milk powder (NFDM), evaporated milk, condensed milk, whole milk powder, whey, cream, casein and derivatives. Among nondairy products known as coffee whiteners, emulsified vegetable fat is the principal ingredient.

Edible acid is incorporated in the composition according to taste preferences for the particular flavor employed. Acidity (pH 4.6 and below) also restricts the growth of pathogenic microorganisms.

Specified means to suspend and stabilize solid particles are combined in an opaque, flavored confection composition to provide a substantially homogeneous stable edible product which is stirrable (but will hold its shape when maintained in an undisturbed condition) at room temperature and can be frozen to a chewy condition. When maintained at room temperature, the composition is in gel form and contains preservative, sweetener, flavor-imparting ingredient means, sufficient water to provide a desired density and acid means to adjust the pH. The confection composition is ordinarily sealed in a plastic film pak in which it is sold. (It is optionally placed, e.g., in a sealed plastic cup or other suitable container for subsequent use, for example as pie filling or dessert topping.) By placing the plastic film pak or sealed plastic cup and enclosed flavored-confection composition in a home freezer, a freeze-it-yourself pop is produced.

In weight percentages the compositions contain from 0.08 to 0.2% of preservative, from 0.10 to 0.8% of stabilizer, from 20 to 38% of sweetener, sufficient food acid to obtain a desired (based on taste) pH, natural and/or artificial flavor to taste and/or from 1 to 4% of chocolate liquor or cocoa, from 1 to 3% of NFMS and/or from 0.01 to 0.05% of artificial color. A chocolate-flavored composition is optionally prepared with chocolate flavor, an opacifier and artificial color, i.e. without chocolate or cocoa. The compositions are aqueous compositions containing sufficient water to provide a total-solid range between 20° and 40° Brix. Natural or artificial colors are incorporated in the compositions as desired, and salt is optional.

The stabilizer serves to suspend solid particles, to thicken the composition to a point of gelation, to impart storage stability and shelf life to the composition and to provide the composition with a chewy or fudge-like consistency when it is frozen. The compositions have from 0.01 to 0.05% by weight of artificial color or from 1 to 4% by weight of some other ingredient to render it opaque; such ingredient is, e.g., chocolate liquor or cocoa when either of these is present, NFMS, an artificial water-dispersible color or pigment or a milk or other dairy product in other compositions. NFMS is an optional component in those compositions which contain chocolate liquor and/or cocoa.

The stabilizer is a key ingredient of these compositions, which need not contain any xanthan gum. A suitable single stabilizing agent is LMP or CGN. References to "pectin" or to "pectins" in connection with confection compositions are uniformly references to HEP (high-ester pectins) which is materially different and which will not gel a milk-containing product. Compositions (which are free from NFMS or other dairy product wherein the stabilizer comprises LMP) must contain at least traces of a polyvalent ion. Divalent metal ions, such as those of a calcium salt, e.g. from 0.05 to 0.2% by weight of calcium chloride hydrate, are suitably employed for this purpose.

Irrespective of the opaqueness-imparting ingredient, the subject compositions advantageously contain from 0.3 to 0.4% by weight of LMP, preferably in pectinic-acid-amide form, and, optionally, from 0.1 to 0.45% by weight of CMC. This particular stabilizer suspends and stabilizes solid particles present in the composition by imparting to the admixture a thickness or texture (at room temperature) which is on the verge or at the point of being semisolid; it is referred to herein as a pudding-like texture.

A similar texture is imparted to the composition when the stabilizer is in the form of CGN alone or a combination of CGN and CMC. When this stabilizer is employed, the individual components are preferably in the range of 0.15 to 0.4% by weight of CGN to from 0.1 to 0.4% by weight of CMC. Accordingly, the preferred percent by weight of this stabilizer in the subject compositions is from 0.25 to 0.8%. Unlike LMP or a combination of LMP with CMC, no polyvalent-metal ions are required in compositions wherein the stabilizer is CGN alone or a combination of CGN and CMC. All of the indicated stabilizers yield frozen confections which have a chewy or fudge-like character. At lower pHs, i.e. at or about pH 3.0, adverse effects with regard to gelation are particularly well avoided by using CGN in combination with CMC.

Naturally, the specified stabilizer, i.e. LMP, a combination of LMP and CMC, CGN or a combination of CGN and CMC, is suitable when employed as the sole stabilizer of the confection compositions. Each, however, can also be employed in combination with any other and/or in combination with one or more further stabilizers. A critical limitation of the stabilizer is that it yields a thickened confection composition having storage stability and in which syneresis (unsightly separation of water from the liquid mass) does not occur when the composition is maintained at room temperature. A further critical limitation of the stabilizer is that the composition (when frozen and in ready-to-eat form) has a chewy and "fudgy" texture.

The combination of LMP and CMC or the combination of CGN and CMC protects a composition from syneresis and from separation of insoluble ingredients even when the composition is kept hot for several hours in a vat prior to being filled [at filling temperature (160° to 180°F)] into a suitable receptacle, container or ultimate package. When LMP (with the required polyvalent metal ions) or CGN is used as the sole stabilizer, the product (at room temperature) tends to have an appearance of not being completely homogeneous. The noted combined stabilizers (with CMC) are thus preferred also from an esthetic viewpoint.

There are a number of critical features in the process. The compositions are opaque; the opacity thereof is brought about by any of many dairy products, e.g., milk, cream and caseinates, by nondairy creamers, e.g. Cremora Powder, by artificial color and/or by the component which imparts a chocolate or cocoa flavor to the confection. The flavor of the confection is that of a fruit, chocolate, cocoa, butterscotch, caramel or other nonfruity confection. The composition has a pudding-like texture and total soluble solids reflected by a range of from 20° to 40° Brix, corresponding to a density from approximately 1.083 to about 1.1787. When frozen, the composition has a chewy or fudge-like character. The pH of the composition can be as high as pH 5.0 or even higher, but is conveniently at most pH 4.6 in view of federal regulations. The lower end of the range is about pH 3.0.

Soft Serve and Hard Frozen Yogurts

Most known gums and blends have been found to react with the milk protein during the processing procedures resulting in yogurts with whey-off and coarse bodied, grainy yogurt. By the term, "whey-off" is meant the separation of fluid material from solid material.

The disclosure by *R.S. Igoe; U.S. Patent 4,178,390; December 11, 1979; assigned to Merck & Co., Inc.* is directed to compositions which act as a stabilizer for soft serve and hard frozen yogurt. The stabilizer is a combination of ingredients having the following composition.

> Propylene glycol alginate – 45-60%
> Sodium alginate – 15-35%
> Guar – 10-20%
> Carrageenan – 2-10%

The stabilizer composition can in turn be utilized in combination with an emulsifier and is referred to as a blend. The stabilizer blend, therefore, is a mixture of the stabilizer as described above with an emulsifier. The stabilizer may constitute a range of about 55-75% by weight of the stabilizer blend and the emulsifier may constitute about 45-25% by weight of the stabilizer blend. The amount of emulsifier may vary, however, depending on which particular one is chosen. The stabilizer and stabilizer blend components are mixed by conventional methods including dry blending techniques which are known to those skilled in this art.

The stabilizer and/or blend composition can be added to the milk before or after processing, i.e., pasteurization, homogenization or incubation.

The stabilizer can be utilized in the soft serve or hard frozen yogurt at a range of about 0.1-0.4% by weight. A preferred range for stabilizer would be about 0.2% by weight.

The stabilizer blend can be utilized in the soft serve or hard frozen yogurt at a range of about 0.15-0.60% by weight. A preferred range for stabilizer blend would be about 0.3% by weight.

Typical emulsifiers that may be utilized include monoglycerides, diglycerides, lecithin, polysorbate 80 and polysorbate 65. These emulsifiers may be used alone or in combination with each other. For example, the polysorbates may be utilized with the monoglycerides or diglycerides and the like.

Example: The parts of ingredients in the following yogurt recipe are percentage by weight. The soft yogurt was prepared from 54 parts of 3.5% fat milk, 6.0 parts of milk solids nonfat, 11.0 parts of sugar, 8.0 parts of 36 dextrose equiv.-corn syrup solids, 0.27 part of stabilizer blend, 0.40 part of flavor concentrate, trace of color and 20.33 parts of water.

The 0.27 part of stabilizer used in the above recipe contained 0.096 part of propylene glycol alginate, 0.046 part of sodium alginate, 0.028 part of guar, 0.010 part of carrageenan, 0.072 part of mono/diglyceride, and 0.018 part of polysorbate 80.

Under medium to high speed agitation, the stabilizer/blend and milk solids nonfat (MSNF) are added to the milk. The milk solution is pasteurized at 180°-185°F for 30 minutes and then homogenized at 2000 pounds per square inch (psi).

After cooling the milk to 90°F, the yogurt culture is then added. The inoculated milk is incubated at 90°F for 15 hr to obtain a yogurt of pH 4.0 which is then stirred to a smooth consistency.

With mild stirring of the yogurt, the sugar, flavor concentrate, color and water are added. The yogurt is then homogenized at zero pressure to reduce the viscosity. The yogurt is then cooled over cooling coils at 40°F and then frozen on a Sweden Single Barrel Freezer. The product is drawn at 19°-20°F from the freezer to obtain a desired 50% overrun.

A formulation is run repeating the above ingredients and amounts, except that the emulsifier is omitted. The product obtained is of the same texture, appearance and smoothness as the product in which the emulsifier is utilized.

Gelling Composition for Yogurt

Yogurt is a product having a certain gelatinized texture obtained by adding to milk, which is usually pasteurized, certain ferments which acidify the milk through a coagulation of the casein. The transformation is made at a temperature close to 40°C.

If the yogurt is preserved at an ambient temperature, the ferments continue their action, the acidity continues to develop, the product loses its qualities and its preservation is thereby limited.

This preservation is appreciably improved on destroying the ferments by heat. But as this operation usually modifies the texture of the product, it is advisable to add a gelling agent to palliate this disadvantage. This addition is usually made to milk before pasteurization or to yogurt after fermentation.

This heating, called thermization, is nevertheless limited to approximately 70°C with common gels. Beyond this temperature, the protecting effect shades off and the casein undergoes a dehydration process leading to a heterogeneous texture which is characteristically "sandy" and very disagreeable which can lead to a total separation if the temperature rises too much.

An improved gelling agent which eliminates this disadvantage has been provided by *G. Brigand, H. Kragen and R. Rizzotti; U.S. Patent 4,200,661; April 29, 1980; assigned to Société CECA SA, France.*

This agent comprises a first gelling agent which is a galactomannan and a second gelling agent which is an agar and/or a xanthan, wherein the galactomannan (locust bean gum) has undergone a depolymerization treatment so that its solution in water at 1% has a viscosity in the range of 10 and 1,000 cp at 25°C.

With use of this gelling agent, yogurt can be treated at very much higher temperatures while preserving the product's complete homogeneity. This thermization at a higher temperature, i.e., higher than 70°C allows a longer preservation period on assuring a better destruction of the ferments.

Thermization can be carried out either directly on the product which is then conditioned in the most aseptic way possible, or on the product which has been previously put into hermetically sealed containers.

The product may be yogurt by itself or yogurt mixed with other dairy products: fresh cheese or cottage cheese; or nondairy products: fruit, jams, preserved fruit, vegetables, seaoning herbs and spices, as well as with sugar, salt, flavorings, colorings, etc.

It was found that a gel of very agreeable consistency and having none of these disadvantages can be obtained by using a depolymerized galactomannan, i.e., whose depolymerization rate is reduced by physical, chemical or biochemical means. An easy method of achieving this depolymerization consists in the action in solution of hydrogen peroxide or of an acid. More precisely, the recommended depolymerization rate is such that the solution of the galactomannan in water at 1% has a viscosity in the range of 10 and 1,000 cp at 25°C.

In addition, the mouth feel of the gel is again improved by using agar and the galactomannan in the form of a homogenized mixture obtained by coprecipitation by a solvent from the mixture of the agar extract and the galactomannan extract.

Example 1: 30 kg locust bean meal are introduced by stirring into 2 m^3 water at 80°C. Once the gum has been well solubilized, it is heated to 90°C and 0.6 ℓ hydrogen peroxide at 110 volumes is added to effect the depolymerization.

The viscosity is measured at regular intervals. After 4 hours, the viscosity is 200 cp. A concentration check indicates that there is about 1% of soluble locust bean gum. This is then filtered through a press filter with 13 kg/m^3 of a filtering agent. The clear solution thus obtained is cooled to 40°C.

Under the same conditions 33 kg of locust bean splits may be used instead of 30 kg of locust bean meal to obtain the same results.

A similar treatment may be carried out by using sulfuric acid instead of hydrogen peroxide until a pH of 4 is obtained. At the end of 5 hours, a viscosity close to 200 cp is obtained; the degradation may then be stopped by neutralization and the treatment ended.

Example 2: Solutions of more or less depolymerized locust bean gum were coagulated with an agar solution to obtain a final mixture containing 25% agar and 75% locust bean gum. The depolymerization of the locust bean gum was measured by taking a sample of its solution and coagulating it; after drying, the viscosity at 1% of this locust bean extract was measured. The complexes obtained were used at 1% to make water gels by dissolution on boiling and cooling of the solution.

Example 3: The yogurt is prepared in a standard manner. 3 kg of milk powder are added to 100 ℓ milk; the milk is heated for several minutes at 90°C or sterilized by passing through a UHT sterilizer, then homogenized. The enriched milk may also be prepared by light concentration after thermic treatment of the intial milk on vacuum cooling.

The milk thus obtained is cooled at 40° to 45°C. The ferments are added and are left to act until the desired acidity is obtained. The maintenance time is variable according to the chosen ferments and temperatures.

The obtained yogurt may be cooled or not to prolong its shelf-life. The ferments stop acting when the temperature is lower than 10°C.

The gelling agent (0.4% of depolymerized locust bean gum/agar mixture) is added by vigorously stirring it into the 99.6 parts of yogurt. The mixture is heated in a vat at 95°C and maintained for a period ranging from several minutes to 1 hour. During all these heating operations, it must be ensured that the difference in temperature between the product and the heating fluid is as small as possible.

The mixture is homogenized or not and put into pots at high temperature to maintain the sterility. The gelatinization, giving the product its traditional yogurt look, takes place during cooling.

Three-Component Beta-1,4-Glucan Stabilizer

A known cellulosic powder capable of forming a stable, thixotropic gel comprises beta-1,4-glucan intimately associated with a small amount of a specific form of sodium carboxymethylcellulose (CMC). This powder has been used extensively in a variety of products for a variety of purposes, including use as a stabilizing agent in ice cream, ice milk, etc. However, in some countries, notably Japan and a number of western European countries, the use of CMC in food products is objectionable.

It is also known that by associating with the beta-1,4-glucan particles a single additive of various gums other than CMC a good body can be imparted to a frozen dairy type food but the texture is coarse, icy and totally lacking in heat shock resistance. Microscopic analysis of the beta-1,4-glucan/hydrocolloid gum dispersion in the frozen product revealed an incompleted dispersion of beta-1,4-glucan microcrystals in every case and in many cases destabilization of the dairy emulsion was apparent. The hydrocolloid gums studied included guar gum, locust bean gum, gum arabic, sodium alginate, propylene glycol alginate, carrageenan, gum karaya and xanthan.

In addition to associating a single gum additive with the beta-1,4-glucan, single sweeteners were associated with the beta-1,4-glucan, particularly carbohydrate sweeteners such as sucrose, dextrose and hydrolyzed cereal solids (HCS). When the foregoing beta-1,4-glucan/carbohydrate sweetener compositions were employed in a typical ice milk formulation it was found that dispersion of the beta-1,4-glucan microcrystals was much improved over the beta-1,4-glucan/hydrocolloid gum compositions. The hydrolyzed cereal solids (HCS) were found to produce the best results. However, even though these formulations resulted in improved dispersion of the beta-1,4-glucan particles and also in improved texture of the frozen dairy type product, the body of the product was not appreciably improved.

A water-dispersible powder useful as a stabilizing agent and for enhancing the body and texture of frozen dairy type foods, is disclosed by *E.J. McGinley; U.S. Patent 4,263,334; April 21, 1981; assigned to FMC Corporation* as having individual particles comprising beta-1,4-glucan codried with a carbohydrate sweetener and a hydrocolloid gum.

Even though no one additive other than CMC, either a gum or a carbohydrate sweetener, gave fully satisfactory results when used as a component of a frozen dairy type product, it has been found that certain three-component systems give excellent results. The key to the three-component system is the carbohydrate sweetener additive which allows dispersibility of the beta-1,4-glucan microcrystals in conjunction with a variety of hydrocolloid gums. Thus, the beta-1,4-glucan functionality can be utilized in combination with various diverse properties of the gums.

Certain three-component beta-1,4-glucan, carbohydrate sweetener and hydrocolloid gum compositions have been found to be very effective as a stabilizing agent and in improving both body and texture of frozen dairy type foods.

Effective compositions include those containing 60% beta-1,4-glucan, 30% HCS and 10% of a gum such as guar gum, sodium alginate, carrageenan, locust bean gum, gum arabic, karaya gum, or propylene glycol alginate.

Other carbohydrate sweeteners useful in carrying out the process include sucrose, fructose, lactose, maltose, invert sugar, molasses, corn syrup solids, dextrins, maltodextrins and galactose.

Other compositions may include 30-80% beta-1,4-glucan, 10-65% of a sweetener, and 5-20% of the various gums.

The single most effective composition appears to be 60% beta-1,4-glucan, 30% HCS and 10% xanthan gum.

The two-component additive, namely the carbohydrate sweetener and the hydrocolloid gum may be introduced to and associated with the beta-1,4-glucan in several ways but always before the newly released and attrited beta-1,4-glucan is fully dried. The sweetener and gum may be added as a dispersion to the aforementioned filter cake of beta-1,4-glucan and the three components then attrited and bulk-dried followed by grinding the bulk-dried mass into powder form. On the other hand beta-1,4-glucan filter cake may be attrited after the addition of sufficient water and the sweetener and gum added to the attrited dispersion, after which the dispersion is spray-dried to produce a powder the individual particles of which consist of beta-1,4-glucan having intimately associated therewith the carbohydrate sweetener and the hydrocolloid gum. Intimate mixing of the three components in the wet state followed by codrying is necessary to produce a powder which upon reconstitution in an aqueous system provides a colloidal dispersion of beta-1,4-glucan particles which functions as a stabilizer in frozen dairy type foods and enhances both body and texture.

Seamoss and Gum Arabic in Milk Drinks

Attempts have also been made to incorporate seamoss, a nutritive ingredient, into foods for human consumption. Typically, the addition of seamoss serves merely to stabilize compositions such as chocolate milk and the like. These additions are limited for stability and are not included to serve as the basis for a nutritive composition for human consumption.

An edible seamoss composition containing water, gum arabic, dry linseed, seamoss, milk sweeteners and flavoring has been provided by *B. Lauredan; U.S. Patent 4,180,595; December 25, 1979.*

In the method of this process, water, gum arabic, dry linseed and seamoss are first combined and heated to boiling until a slightly brown texture is obtained. Thereafter, the blend is filtered to remove undissolved solids. Milk and the sweetening agent as well as the flavoring agent may be then included to produce the nutritive seamoss composition. The composition may be prepared and canned, bottled or processed as an anhydrous instant product by conventional means.

Water which serves as a basis for formulating the composition is added in an amount of 100 parts by weight.

For each 100 parts by weight of water, arabic gum is added in an amount from about 0.5 to about 0.8 part by weight and desirably from about 0.6 to about 0.7 part by weight. For each 100 parts by weight of water, dry linseed or flaxseed may be also added in an amount from about 0.5 to about 0.8 part by weight and desirably from about 0.6 to about 0.7 part by weight. For each 100 parts by weight of water, seamoss, desirably as Irish moss, may be added in an amount from about 1.0 part to about 1.6 parts by weight and desirably from about 1.2 parts to about 1.4 parts by weight.

Although seamoss is generally referred to in the present context, it will be appreciated that Irish moss and similar marine vegetable matter may be used herein such as agar, kelp, algins, alginates, and the like.

Milk may also be included in this composition in the form of condensed milk, sweetened or unsweetened, nonfat dry milk, or various commercial forms of milk as desired. Milk is added in an amount of about 10 to about 40 parts by weight per 100 parts of water when condensed, or equivalent amount thereof when powdered or whole milk. Preferably, an amount of about 15 to about 30 parts by weight of condensed milk is included per 100 parts by weight of water.

Sweetening agents may be included to suit taste preferences. Typically, sugar is the normally used sweetening agent herein although sugar substitutes which are commercially available may be included if desired.

The amount of sweetening agent included on the basis of water in this formulation when sugar is used varies from about 5 to about 20 parts by weight and preferably from about 8 to about 15 parts by weight. Sugar substitutes may be included proportionally and consistent with taste preferences.

Flavoring agents may be added in amount to suit preferences of taste. Useful flavoring agents include salt, nutmeg, cinnamon, vanilla, rum and the like. For each 100 parts by weight of water, the flavoring agents may be included in amounts of about 1.5 to about 1.6 parts by weight and desirably about 1.2 to about 1.4 parts by weight. The flavoring agents may be added either separately or preferably in combination.

Salt, for example, as are the other ingredients used herein, may be added or salt derivatives such as disodium phosphate or the like can also be used if desired.

Artificial color may be added as desired using colors certified for human consumption.

Example: Water, taken in an amount of 100 parts by weight to serve as a basis for additions of the remaining ingredients, was combined with 0.65 part by weight gum arabic, 0.65 part by weight dry linseed, and 1.3 parts by weight Irish moss. The blended ingredients were heated to boiling until a slightly brown, textured liquid resulted. The textured liquid was filtered to remove nondissolved solids which were found to be typically linseed particles and Irish moss. This blend was found to be slightly tasteless. Thereafter, sugar in an amount of 10 parts by weight was combined with 20 parts by weight of milk. This blend was mixed into the filtered ingredients and a product resulted which lacked stability. The blend could only sustain eight days refrigeration or 24 hr at room temperature. The product was noted to lack taste.

Following the milk-sugar addition, a further flavoring agent addition was made comprising salt 0.2 part by weight; nutmeg 0.1 part by weight; cinnamon 0.05 part by weight; and vanilla 0.4 part by weight. Thereafter, 10 parts stout combined with 0.4 part by weight rum were added and the product blended to uniform consistency.

The product was found to be a delicious natural drink stabilized over 3-4 weeks when refrigerated while being a highly nutritive seamoss composition.

Milk Shake Stabilizer

A blend of gums is disclosed by *R.S. Igoe; U.S. Patent 4,242,367; December 30, 1980; assigned to Merck & Co., Inc.* which is useful as a milk shake stabilizer. The blend comprises guar, xanthan gum, carrageenan, and, optionally, locust bean gum.

At a usage level of 0.14-0.25%, this blend of gums is particularly useful for stabilizing mild compositions containing milk solids, fat, sugar, and water such as milk shakes, soft serve frozen desserts, and like frozen confections.

These gums are all available in food grade quality with variations depending on source of supply and processing techniques. All of the commercially available products are useable in this process. It has been found, however, that coarse mesh gums (i.e., those passing through a 20 mesh screen but retained on a 200 mesh screen) are easier to disperse in an aqueous mixture although harder to dissolve whereas finer mesh gums are harder to disperse but easier to dissolve. Where a mix utilizes another ingredient such as sugar which acts to aid dispersion, the finer mesh gums are useable. Where the mix is low in sugar, coarse mesh gums are preferred.

The gum blend of this process is used to stabilize milk products, specifically those containing fat (either animal or vegetable), sugar and milk, either as whole milk, cream or milk solids to which water is later added. These products can be used to prepare milk shakes, soft serve frozen desserts, and like frozen confections.

In the past, the gums of this process have been used in various ratios to stabilize milk products. However, it has been found that guar and xanthan gum alone in the ratio 70/30 when used at the usage levels of this process do not prevent whey-off in, for example, a milk shake mix for prolonged periods of time. Such a mix sould be stable for about 10 days in order to ensure usability at retail distribution sites. This allows time for preparation at the factory, distribution to such sites, and storage at the retail site prior to sale to the public. The inclusion of carrageenan, and optionally locust bean gum to the 70/30 guar/xanthan blend imparts such prolonged stability to a milk shake mix, the length of stability depending on the amount of carrageenan added.

The gums of this process comprise 53-68% of guar, 20-35% of xanthan, 9-13% of carrageenan and 0-5% of locust bean. The preferred composition contains 62.63% of guar, 25.95% xanthan, 10.52% of carrageenan and 0.9% of locust bean gums.

In varying these ratios, it is preferred to keep the guar/xanthan/carrageenan in approximately the ratio 2.41:1:0.40. Where such a blend is used in a milk shake mix at a level of 0.14-0.25% (preferably 0.17 to 0.20), the milk shake mix is stabilized for 10 days, which is necessary and sufficient for the commercial distribution of such a mix. Where lesser amounts of carrageenan are used in such a blend, the length of stabilization is decreased.

Example: Gum blend, milk solids not fat (MSNF), and sugar are added to cream and water in a vat at room temperature and mixed for 10 minutes. The mix is pasteurized via HTST (high temperature, short time) at 79.4°C for 25 seconds, homogenized at 1800/500 psi and then cooled to 4.4°C. This mix is observed for whey-off and viscosity. The mix is frozen in a Taylor freezer, spindled on a multimixer for 45-60 seconds, and evaluated for firmness, overrun, iciness, and body. During spindling, chocolate or other flavor can optionally be added.

The milk shake formulation contains 9-14% milk solids, 2-4% fat, 7-10% sucrose and 72-82% water. When 0.2% of a 70/30 guar/xanthan stabilizer was added, the viscosity obtained was 135 cp and the whey-off at 7 days was very slight. When the added stabilizer was 0.2% of the 70/30 guar/xanthan gum plus 0.015%

carrageenan, the viscosity obtained was 210 cp and no whey-off was observed at 7 days.

Clouding Agents for Beverage Mix

J.M. Serafino, S. Yadlowsky and J.S. Witzeman; U.S. Patent 4,187,326; Feb. 5, 1980; assigned to General Foods Corporation have prepared a clouding agent by drying an aqueous dispersion comprised of, on a dry basis, a major amount of malto dextrin (in solution) and minor amounts of xanthan gum and titanium dioxide. The clouding agent is added to a dry beverage mix in amounts effective to produce the desired opacity and remains suspended for a time sufficient to consume a beverage prepared from the dry mix.

Various malto dextrins (hydrolyzed cereal solids) are those starch hydrolyzates produced by converting refined corn starch into nutritive saccharides through the use of acids or specific enzymes. The carbohydrate composition of malto dextrin is arranged to yield a DE (dextrose equivalent) of less than 20. They are typically bland in flavor and without appreciable sweetness. Preferably, the malto dextrin as used in this process will have a dextrose equivalent of between 10 and 20. Fro-Dex 15 has been found most preferable. It is a white free-flowing powder extremely bland in taste with little or no sweetness. Additionally, it has a quality of contributing a slight opacity to the clouding composition.

The xanthan gum as used in this process is a high molecular weight natural carbohydrate or more specifically polysaccharide. Xanthan gum defines the exocellular biopolysaccharide which is produced in a pure culture fermentation process by the microorganism *Xanthomonas campestris*.

An important functional quality of the xanthan gum is its ability to control aqueous fluid rheology. Water solutions of xanthan gum are extremely pseudoplastic; when shear stress is applied, viscosity is reduced in proportion to the amount of shear once the yield point has been exceeded. Upon release of the shear, total viscosity recovery occurs almost instantaneously. Another attribute of xanthan gum that makes it useful as a suspension stabilizer is that it not only has a yield value but also a viscosity which is almost independent of temperature and pH. A particular xanthan gum found most useful in this process is Keltrol F (Merck & Company).

Titanium dioxide as used herein is preferably a purified inorganic white, named by the 1971 Color Index, pigment white 6, CI 77891. It is available from H. Kohnstamm & Co.

The cloud composition of this process is preferably added to constituent ingredients necessary to fomulate a dry beverage mix. Therefore, unless otherwise indicated all percentages herein referred to will be in relation to the total dry mix (as in the instant example) composition. Thus, the malto dextrin may be present in the range of 0.5 to 5.0% by weight of the total dry mix composition. Preferably the range will be 1.5 to about 2.0%. The xanthan gum will be present in the range between 0.01 and about 0.1% by weight of the dry mix composition. Preferably the xanthan will be present in the amount of 0.015 to about 0.035% by weight of the dry weight mix composition. The titanium dioxide will be present in amounts between 0.01 and 0.1% by weight of the dry mix composition, and preferably between 0.025 and 0.075%. Preferably the titanium dioxide will be finely divided, 0.1 micron to about 0.7 micron in particle size.

A number of unexpected results arise from the combination of the three components of the instant clouding composition. Most importantly it has been found that it is critical that the titanium dioxide be added to an aqueous solution of malto dextrin and xanthan gum and the resultant suspension be dried concurrently or codried. The combination of each component by mixing in its dry form does not have utility. The titanium dioxide immediately begins to precipitate when the mix is put in a beverage. Likewise, codrying titanium dioxide with either malto dextrin or xanthan gum separately does not work. The titanium dioxide again immediately precipitates when used in a beverage. Surprisingly the codried combination permits a cloud which is useful in a beverage for periods of at least 24 hr and preferably at least 48 hr when the beverage is stored at 50°F.

Example 1: To a high shear planetary-type mixture is added a preblended dry mix of 958.9 pounds of malto dextrin (Fro-Dex 15, American Maize-Products Co.) and 13.7 pounds of xanthan gum (Keltrol F, Merck and Company) and 290 gallons of water with constant mixing. To the resulting solution is added 27.4 pounds of food grade titanium dioxide.

During the mixing the temperature of the solution is elevated to about 150°F. Next the mixer speed is increased to maximum speed and shear and is maintained for a 15 minute period. The product is then transferred to a separate holding tank preparatory to spray drying. The mixture enters the spray drying tower by means of passage through a two stage Manton Gaulin positive displacement pump. The first stage and second stage are operated at 2000 psi pressure and 3000 psi pressure, respectively. The solution enters the upper portion of the drying tower through a series of spray nozzles. The inlet air temperature of the drying tower is about 400°F. The exit air temperature in the lower portion of the tower is about 190°F. The product exiting the bottom of the tower is a dry flowable powder having a moisture content below about 5%.

Example 2: A standard fruit-flavored beverage mix is prepared employing 1.825 parts of the cloud prepared as in Example 1.

For comparison, a fruit-flavored beverage mix is prepared containing 1.574 parts of a cloud containing 76.44% modified starch, 19.6% hydrogenated coconut oil, 1.96% titanium dioxide and 2.0% tricalcium phosphate.

Thirty-three grams of Beverage Mix B is reconstituted in 8 ounces of water and produced a beverage with an opacity reading of 60 on a Bausch & Lomb Spec 20 opacity meter.

Thirty-three grams of the above Beverage Mix A is reconstituted in 8 ounces of water. The opacity is the same as that of the Beveraage Mix B. Additionally the body and mouth feel of the beverage of A and that of B are judged equivalent by an expert panel.

The product of Beverage Mix B after being reconstituted in 50°F water and held at that temperature for 24 hr displayed precipitation of the cloud system. Titanium dioxide came out of solution and settled at the bottom of the sample flask with an attendant reduction in opacity. The product of Beverage Mix A has no apparent precipitation or change in opacity after 24 hr at 50°F. Additionally the product of Beverage Mix A is readily dispersible in cold water and does not display the clumping or streaking problems associated with beverage mixes which have oil.

Spoonable Deep Freeze Ice Cream

A problem with conventional ice cream is that at deep freeze temperatures, e.g., $-20°C$, they cannot be served or eaten as readily as when they are at normal eating temperatures, e.g., $-10°C$.

For an ice cream to be spoonable at $-20°C$ its log C (C is its penetrometer value) should preferably be less than 2.8, particularly preferably less than 2.5; a correlation exists between spoonability and log C.

I.C.M. Dea and D.J. Finney; U.S. Patent 4,145,454; March 20, 1979; assigned to Thomas J. Lipton, Inc. have prepared ice creams scoopable at deep-freeze temperatures by incorporating stabilizer mixtures comprising (a) locust bean gum and/or tara gum and (b) kappa-carrageenan and/or xanthan gum and/or agar-agar.

Although as (a) a mixture of locust bean gum and tara gum can be used, locust bean gum is preferred.

Replacement of up to 50%, by weight, of (a) by guar gum can occur provided the total remaining (a) is greater than 0.7% by weight of the ice cream mix. Such replacement gives a creamier mouth feel and improves the resistance of the hardened ice cream to temperature cycling. The amount of (a) should generally be greater than 0.07% by weight of the ice cream to ensure preferred melt-down and stand-up characteristics.

Substances such as low-DE malto dextrins and modified starches can be added to improve mouth feel, although preferably not in amounts such that the viscosity of the ice cream mix is more than 4 poise reciprocal seconds at $5°C$.

It has further been found that starches, in particular modified starches such as the modified starch "Instant Cleargel" (Laing National) and believed to be adipate/acetate modified waxy maize starch, can usefully be incorporated to improve the mouth feel. The amount of the modified starch should preferably be, by weight of the ice cream, in the range 0.05 to 0.4%, most preferably in the range 0.05 to 0.2%.

At least for simplicity simple combinations of (a) and (b) are preferred. For this reason use of agar-agar as (b) is a preferred form of the process. When (b) is agar-agar, the amount of agar-agar should, in an ice cream, preferably be from 0.05 to 0.15% by weight; the amount of (a) should preferably be from 0.1 to 0.20% and the weight ratio of agar-agar to (a) should preferably be in the range 1:1 to 1:3. The total amount of (b) and (a) and the ratio of (b) to (a) depend to some extent on other ingredients present but the above is a useful general rule when (b) is agar-agar. The total amount of (a) and (b) is preferably, when (b) is agar-agar, in the range 0.15 to 1.0%, particularly preferably 0.25 to 0.5%.

Because of the particularly good properties, e.g., stability, which are obtained, (b) is preferably xanthan gum. When (b) is xanthan gum the weight ratio of (a) to (b) is preferably 7:1 to 1:7 particularly preferably 4:1 to 1:2, when (a) is locust bean gum, and preferably 7:1 to 1:2 particularly preferably 4:1 to 1:1, when (a) is tara gum. With a low excess of (a) to (b), i.e., weight ratio 4:1 to 1:1, aeration of the ice cream and the texture of the ice cream are particularly good. Total (a) plus (b) is preferably in the range 0.15 to 1% particularly preferably

0.25 to 0.5%. Preferably the amount of (a) is at least 0.1% and is less than 0.4%. Preferably the amount of xanthan gum is at least 0.05%.

Use of kappa-carrageenan as (b) also is a preferred form of the process; improved properties, in particular stability, are obtained. When (b) is kappa-carrageenan, the weight ratio of (a) to (b) is preferably in the range 1:1 to 7:1. The total of (a) and (b) is preferably in the range 0.15 to 1%, particularly preferably 0.15 to 0.5%. The ratio of (a) to (b) is particularly preferably in the range 3:1 to 7:1. The amount of (a) is preferably greater than 0.1%.

The following example gives an ice cream mix (in % by wt) that with conventional processing provides an excellent ice cream scoopable at –20°C.

The stabilizer used in this recipe contained locust bean gum and kappa-carrageenan in a ratio of 4:1. The recipe contained 27% of made-up skimmed milk (32.5% solids), 13% sucrose, 2% glucose syrup, 9.5% liquid oil blend, 0.45% monoglyceride emulsifier, 0.03% color and flavor, 0.05% salt, 3.0% glycerol, 0.25% stabilizers (LBG and kappa-carrageenan in ratio 4:1), and water to 100%.

In Fatty Substances

Fatty substances having a quasi-vitreous structure have been produced by *C.M. Cousin and P.G. Cavroy; U.S. Patent 4,145,452; March 20, 1979.*

It has been discovered that through intimate dispersion of a vegetable gum in the dry state in predetermined proportions in fatty sbustances which are solid at ambient temperature and which may or may not contain dry fillers, it is possible to obtain these fatty substances in a quasi-vitreous state or a state in which these fatty substances no longer have the character of a crystallized solid at determined solidification temperatures, but behave like glasses whose viscosities increase continuously when the temperature is lowered, until solidification occurs. This quasi-vitreous structural state is likewise manifested by a widening of the working range or temperature range within which the substance is sufficiently plastic to be shaped. In this quasi-vitreous state, contraction on crystallization no longer occurs and the reduction of volume accompanying cooling becomes progressive, so that it is possible to control contraction in the course of the solidification. In this state, moreover, the tendency of the oils to separate from the mixture is considerably reduced.

This technique of dispersion of a gum appears to depart fundamentally from previous techniques of utilizing gums, particularly because of the feature comprising their incorporation in the dry state in the fatty substance to be treated, since in this case the presence of water can be considered to be detrimental. If the products contain a certain amount of water, a corresponding amount of gum will be associated with this water, and the unassociated gum will then be the dry gum in the sense of this description.

The process accordingly provides a fatty substance containing fatty constituents which are crystallizable at ambient temperature and also oils, characterized by a quasi-vitreous structure which is determined by filling with dry gum in a mechanically dispersed state.

The filling is advantageously composed of a dry gum in an amount between 0.01

and 1.5% by weight of the fatty substance, the latter being held divided into small zones bounded by elongated chains of the gum.

The gum is selected from vegetable gums comprising the guaranates, the carraghenates, alginates, carob gum, the pectins, gum arabic, gum tragacanth, and resin and is added to the fatty substance in the pulverulent, dry state.

Furthermore, the dry gum preferably constitutes between 0.02 and 0.8% by weight of the fatty substance.

As a variant, a fatty substance of quasi-vitreous structure is obtained by mixing two starting batches, one of the two batches being a fatty substance of the kind described above. As a variant, the fatty substance also contains a dry filler substance.

In a process for the preparation of a fatty substance of the kind described above, the fatty substance is heated beyond its melting point, the dry gum is added, the gum is dispersed mechanically in the fatty substance, and the resulting dispersion is allowed to cool, with kneading until the consistency is firm.

Example 1: 100 kg of 42°/44°C hydrogenated palm oil and 50 kg of 38°/40°C hydrogenated palm oil are placed in a conching vessel. The mass is heated to 75°C and 150 g of guaranate and 30 g of carraghenate suspended in about 1 kg of hydrogenated palm oil are added. Kneading is started, temperature is lowered to 34°C, kneading is continued at full power for half an hour, and the treated fat is poured into plastic containers which are then packed in cartons.

Example 2: 22.2 kg of fat treated the day before as in Example 1 are placed in a roller mixer, 46 kg of sugar, 22.5 kg of skimmed milk, 3.2 kg of cacao powder containing from 10 to 12% of cacao butter, and 35 g of lecithin are added. These ingredients are mixed and are tempered in the conventional manner for making chocolate. After tempering the mixture is ground in a cylinder grinder and then placed in a conching vessel with an addition of 3.8 kg of fat treated as in Example 1 and 265 g of lecithin, and is worked for 6 hours. The mixture is cooled to about 40°C to form a relatively thick paste and is then poured into bars.

Example 3: 200 kg of cocoa butter were placed in a double-walled stainless steel vessel and heated to 70°C. Then, 120 g of pulverulent guaranate and 80 g of pulverulent carraghenate were added to the melted cocoa butter, and the mass was stirred with a Rayneri mixer provided with a deflocculator blade, at 1500 rpm for 10 minutes.

The mix was then cooled to 12°C; the stirring was continued during cooling. When a noticeable crystallization of cocoa butter occurred on the inner wall of the vessel, the mass was reheated before further cooling. The cycle was repeated until no noticeable crystallization appeared (usually, four cycles were needed). Then the mix was poured into containers.

Thickener from Taro Corms

Aroids (family Aracae) possess a corm, or a rounded thick modified underground stem base bearing membranous or scaly leaves and buds and acting as a vegetative reproductive structure in certain monocotyledonous plants, which may be prepared for human consumption. The most noted of the Aroids is taro, an important food crop in tropical and subtropical regions. To prepare taro corms for eating complicated cooking procedures are required because the corms contain acrid matter that produces irritation of the mucous membranes in the mouth and throat. The cooked product from taro called poi is very sticky and viscous and has a high moisture content. Consequently, poi is difficult to handle, dry, and store, thus, limiting its usefulness as a food.

J.C.C. Tu; U.S. Patent 4,246,289; January 20, 1981; assigned to U.S. Secretary of Agriculture has devised a process which overcomes these difficulties. In addition a thickener is obtained from the waste product.

In this process taro corms are peeled, washed and cut into pieces of any geometrical shape such as cubes, squares, rectangules, etc. Typically, the pieces have the following dimensions: cubes, about 10 mm; squares, about 20 x 20 x 2 mm; rectangles, 20 x 10 x 2 mm.

The taro pieces are next extracted with an aqueous alkaline agent, such as sodium hydroxide, potassium hydroxide, calcium hydroxide, and the like, which is nonpoisonous. Usually, the concentration of alkaline agent is about 2 to 5%, based on the weight of taro corms. The alkaline mixture is allowed to stand at room temperature (about 20°C) for a period of about 10-24 hours or a period of time sufficient to remove the acrid principles from the taro. It is within the scope of the process to conduct the alkaline treatment within the temperature range of about 20°-30°C.

Following this alkaline treatment the taro pieces are separated from the alkaline extract and washed with water to remove the alkaline agent. The pH of the taro pieces should be about 9-10.

The so-extracted pieces are mixed with water in the ratio of 2-4 parts of water per part of taro. Then, sufficient food grade hydrogen peroxide is added such that the final concentration of hydrogen peroxide in the water is about 0.7-10% based on the weight of taro. Alternatively, the hydrogen peroxide can be mixed with water prior to combination with the taro pieces. In general, the concentration of hydrogen peroxide should be sufficient to decolorize the taro. For practical purposes the treatment with hydrogen peroxide is conducted at ambient (room) temperature although temperatures in the range of 20°-30°C can be employed.

The treatment with hydrogen peroxide is carried out for a period sufficient to decolorize the taro. Usually, this occurs after a period of about 10-30 hours. Whether the decolorizing procedure has been effective can be determined by placing a piece of treated taro into boiling water. Absence of darkening on the surface of the taro piece indicates that the decolorization is completed.

After treatment with hydrogen peroxide, the pieces are separated and washed with water until the pH of the taro is neutral, i.e., about 6.5-7.5. Alternatively, the separated pieces can be washed and then allowed to stand in water until neutral pH is realized. The neutralized pieces are separated from the water and dried to a moisture content of about 5-10%. The pieces may be dried in air or by the application of heat as, for example, in an oven.

The new taro products may be stored and/or transported without refrigeration.

To reconstitute (hydrate) them for consumption they may be placed in about 5-10 parts of water per part of dry taro. The reconstituted pieces are easily cooked requiring about one hour in boiling water and may be included in a number of recipes.

The alkaline extract, after removal of the taro pieces, can be acidified with dilute aqueous food grade acid such as hydrochloric acid, sulfuric acid, phosphoric acid, and the like. Acid should be added to attain a pH of about 3.5-4.5. The precipitate that forms is separated by conventional means such as by centrifugation, filtration, etc., and then dried as described above. The dried material swells in water and becomes highly hydrated to form a gum that can be used as an emulsifying, thickening, and smoothing agent in food preparations.

Example 1:   Ten grams of uncooked taro corms were peeled and cut into small cubes, 1 cc, and then were mixed with 20 ml of 0.25 N sodium hydroxide in a 150 ml jar. The jar was stoppered and allowed to stand at room temperature for 24 hours. The cubes were removed from the jar and washed with water several times. Then, the cubes (pH 9) were transferred to another jar containing 20 ml of water. To the jar was added 0.17 g of hydrogen peroxide (100%). The jar was stoppered and the contents were allowed to stand at room temperature for 10 hours or more. The cubes were removed from the jar and washed several times with water. The cubes were then suspended in water with stirring and held for a period of 2 hours, this process being repeated until the pH of the cubes was about 7.0.

The above procedure was repeated except that a 1.0 N sodium hydroxide solution was used in place of the 0.25 N solution.

The so-prepared cubes were separated and dried in air to a moisture content of 10%. The dried cubes were rehydrated by suspension in water (50 ml) and then cooked. The cooked taro pieces did not discolor and did not irritate mucous membranes in the mouth and throat when eaten.

Example 2:   The sodium hydroxide extract from Example 1, after removal of the taro cubes, was acidified to pH 4.0 by addition of hydrochloric acid. The precipitate that formed was separated by centrifugation and then dried to a moisture content of 10%.

The precipitate was hydrated in water (20 ml) to yield a gum-like substance, which may be used as a thickening agent in food preparations.

EMULSIFIERS

IN DAIRY TYPE PRODUCTS

In Low Calorie Imitation Dairy Products

C. Gilmore, D.E. Miller and R.J. Zielinksi; U.S. Patent 4,199,608; April 22, 1980; assigned to SCM Corporation provide low-calorie imitation dairy products that are free of added fat, and thus have substantially reduced caloric contents. Specifically, this process resides in the discovery that the caloric content can be significantly reduced by replacement of the normal fat content with a partial glycerol ester emulsifier, the major constituent of which is a diglyceride, present in an amount of about 38 to 48%. The triglyceride content is less than the mono- and diglyceride content combined but may be about equal to the diglyceride content, the balance being essentially monoglyceride.

For purposes of this process, percentages and ratios are on a weight basis.

A small amount of an additional hydroxy-containing emulsifier may be employed to obtain hydrophilicity if necessary.

The emulsifier is usable in the amount of preferably 3 to 12% for a whippable topping composition, on a wet basis (i.e., containing the amount of water to obtain a fluid composition), which is substantially lower than the fat content normally required for imitation dairy products. At the same time, this process provides compositions having all of the desirable attributes required of conventional imitation dairy products. For instance, in the case of a whippable topping composition, the product has good flavor and eating qualities, good foam stability, good overrun, defined as the ability to incorporate air up to 200 to 300% of the composition initial volume; and good whipping time, defined as the ability to whip to the desired consistency or density, using a household type mixer, in 5 to 10 minutes. Similar results have been obtained with compositions formulated to simulate other dairy products, for instance sour cream, coffee whitener, mellorine, chip dip and cream cheese.

In the case of dry mixes, the compositions are readily reconstituted by admixture with water or milk.

Preferably the plastic partial glycerol ester emulsifier has in addition to about 38 to 48% diglyceride, a diglyceride to monoglyceride ratio of about 5:1 to about 1.5:1, the balance being essentially triglyceride (with small amounts of free glycerin and free fatty acids, e.g., less than 1%).

In a preferred embodiment, the emulsifier is prepared by blending together three partial glycerol ester fractions, a monodiglyceride having a low monoglyceride content, which shall be referred to as the (low mono) monodiglyceride fraction; a soft monodiglyceride; and a smaller amount of a hard monodiglyceride. Up to 10% hard monodiglyceride (based on the total lipid content) can be used although the hard monodiglyceride preferably is employed in amounts as low as about 1 to 2% to obtain the properties desired.

In one embodiment, there was prepared a blend containing three lipid fractions, about 74.2% of a (low mono) monodiglyceride made from 70 I.V. soybean oil having a monoglyceride content of only about 13%, a diglyceride content of about 43% and a triglyceride content of about 43%; about 24.7% of a soft mono-diglyceride having a monoglyceride content of about 40 to 48%, a diglyceride content of 40 to 48% and a triglyceride content of 8 to 12%; and about 1.1% of a hard monodiglyceride having a monoglyceride content of about 40 to 48%, a diglyceride content of 40 to 48% and a triglyceride content of 8 to 12%. (On a dry basis, total formulation, the lipid blend proportions were 23.44, 7.82 and 0.35, respectively.)

The soft and hard monodiglycerides from SCM Corporation are known as Dur-em 114 (made from a 75 to 85 I.V. soybean oil, and has a capillary melting point (CMP) of 110° to 125°F) and Dur-em 117 (made from 5 maximum I.V. soybean oil, and has a CMP of 145° to 150°F). These emulsifier fractions in the proportions stated gave combined mono-, di- and triglyceride contents of about 22, 43 and 35%, respectively. The lipid blend may contain up to about 1% free glycerin and free fatty aicds. In this particular example, the lipid blend had a CMP of about 109°F and an I.V. of about 72.

Example: A spray-dried whippable topping base was prepared by the following process wherein the parts of ingredients represent the percentage of the composition. First 2.9 parts of Avicel RC 581 (a mix of 89% cellulose gel and 11% cellulose gum, FMC Corp.) was added to water sufficient to disperse the Avicel. A dry mix of 7.71 parts of sodium caseinate, 0.19 part of Gelcarin MMR (Marine Colloids Inc. carrageenan), 19.28 parts sugar, and 38.31 parts 36 DE corn syrup solids was then added to the above Avicel mixture. This was heated to 120°F and a premelted mixture of 23.44 parts of monodiglyceride (low mono made from 70 I.V. soybean oil), 7.82 parts of Dur-em 114 and 0.35 part of Dur-em 117 was added.

The entire mixture was pasteurized at 165°F for 30 minutes, homogenized at 3,000 to 500 psi through a two-stage homogenizer, and then was spray-dried to remove water. The resultant powder was tempered at about 40°F for 24 hours. Then 42.53 g of this dried topping base was mixed with 4.0 g sugar, 0.15 g salt, 0.3 g cellulose gel and cellulose gum and 0.06 g imitation flavor.

The above 47.04 g of topping mix was readily mixed with 113.4 g water and 1.70 g vanilla extract, and when mixed gave a composition that was readily whippable in 5 to 10 minutes to the desired density; gave good overrun, up to 200 to 300%; and provided good foam stability. The total lipid content on a wet basis was about 8%.

In Low Fat Liquid Spreads

It is an object of this process by *P.M. Bosco and W.L. Sledzieski; U.S. Patent 4,292,333; September 29, 1981; assigned to Standard Brands Incorporated* to provide an improved low-fat, butter-flavored, liquid spread and a process for preparing it.

The products prepared by this process simulate good-quality liquid margarine in flavor, texture, apperance, mouthfeel and stability, yet have caloric densities of less than 50 calories per 14 g serving compared to about 100 for margarine and butter. The spreads are liquid at refrigerator temperature of about 40°F, and preferably down to about 32°F. They are based on oil-in-water emulsions which remain stable even after standing at room temperature.

The ingredients which are essential to the formation of the product are: (1) fat which is suitably selected to have a solids profile which enables the formation of a liquid product at 40°F; (2) a water-soluble emulsion stabilizer; and (3) an effective emulsifier system comprising both lipophilic and hydrophilic emulsfiers.

The fat will preferably be present in amounts within the range of from about 5 to 40% based on the weight of the spread, and most preferably in an amount of from 10 to 30% of the weight of the spread.

The use of water-soluble emulsion stabilizers is essential to provide the necessary stability. These will preferably be a hydrophilic colloid, and can be selected from the group consisting of microcrystalline cellulose, carrageenan, guar gum, alginate, xanthan gum, soy protein isolate, methylcellulose, carboxymethylcellulose, ethylcellulose, hydroxypropylmethylcellulose, dextrins, starch, gelatin, locust bean gum, pectin, and the like and mixtures of these.

Commercial stabilizers known as Frimulsion Q8 and Frimulsion 10 (Polak's Frutal Works, Inc.) have been found effective, especially when used combined. The Q8 product is a blend of modified food starch, locust bean gum, guar gum, gelatin and pectin, and is preferably employed at a weight ratio within the range of from about 1:1 to 3:1 to the Frimulsion 10 which is a blend of locust bean gum and guar gum.

The relative and total amounts of the emulsfiers are selected to be effective to provide a stable emulsion and a product which is liquid at 40°F. Typical of effective levels will be levels of from 0.1 to 4.0%, based on the total weight of the spread, of the total emulsifier system which employs each of the hydrophilic and lipophilic emulsifiers at levels of at least 0.025%, on the same basis. The lipophilic emulsifier will typically have an HLB (hydropile-lipophile balance) of less than 7, and the hydrophilic emulsifier will typically have an HLB of from 10 to 20, preferably from 11 to 17.

The emulsifier system is preferably present at a level of from 0.15 to 2.0%, and the lipohilic and hydrophilic emulsifiers are preferably each present at levels of at least 0.05%, all percentages based on the total weight of the spread.

The hydrophilic emulsifier will preferably comprise a member selected from the group consisting of polyoxyethylene (20) sorbitan monostearate, polyoxyethylene (20) sorbitan monooleate, and mixtures of these.

These emulsifiers, commonly known as polysorbate 60, and polysorbate 80, repectively, are preferred. However, it is believed that other hydrophilic emulsifiers with an HLB of between 13 and 16, will be operable.

The lipophilic emulsifier will preferably comprise a member selected from the group consisting of mixed fatty acid monoglycerides; mixed fatty acid diglycerides; mixture of fatty acid mono- and diglycerides; lipophilic polyglycerol esters; glycerol esters, etc.

In addition to emulsion stability, the products are preferably stable against microbiological and oxidative deterioration. To control mold and yeast growth, the products desirably contain one or more preservatives such as benzoic acid, sorbic acid, phosphoric acid, lactic acid and the soluble salts of these and other like materials.

The pH of the aqueous phase is desirably maintained at a value below 6.0, and preferably within the range of 5.0 to 5.9, to provide effective microbial control and good flavor with the lowest necessary levels of preservatives.

Example: This example describes the preparation of a preferred liquid spread according to this process. The spread is made from 19.4 parts of corn oil (SFI's 19-23 at 50°F, 11-14 at 70°F and <3 at 92°F), 0.5 part Dur-em 114, 0.1 part polysorbate 60 (Durfax 60), 0.005 part beta-carotene (30% in oil) and vitamins, 1.65 parts salt, 0.02 part butter flavor, 1.0 part Frimulsion Q8, 0.4 part Frimulsion 10, 76.6743 parts water, 0.13 part potassium sorbate, 0.1 part sodium benzoate, 0.015 part phosphoric acid and 0.0057 part calcium disodium EDTA.

An aqueous phase is prepared by heating the water to 190°F and adding the Frimulsion Q8 and 10 stabilizers predispersed in a portion of the fat phase, and the other dry ingredients to it with agitation. Mixing is continued until the stabilizers are uniformly dispersed and hydrated.

A separate fat phase is prepared by melting the fat and the emulsifiers at a temperature of about 150°F. The color and flavor are then admixed with the melt to obtain a uniform blend.

The aqueous and fat phases are then blended at about 160°F to provide an emulsion. The emulsion is then homogenized in a Gaulin Laboratory Homogenizer Model 15M set at 250 atmospheres.

The emulsion is then cooled to 50°F in 10 minutes by slowly agitating with a Hobart Model N50 mixer fitted with a wire whip and a jacketed (water/alcohol coolant at approximately -20°F) 5-quart mixing bowl. After cooling, the viscosity is determined by shaking for 10 seconds and then measuring at 40°F using a Brookfield RVT viscometer fitted with a number 3 spindle, rotated at 10 rpm. The Brookfield viscosity is found to be 5,850 cp.

Low fat spreads with other flavors have also been disclosed by *P.M. Bosco and W.L. Sledzieski; U.S. Patent 4,273,790; June 16, 1981; assigned to Standard Brands Incorporated.* The same combinations of lipophilic and hydrophilic emulsifiers are used. Flavors described in this patent include garlic, blue cheese, sour cream, bacon and yogurt flavored spreads. The process used is the same as that described in U.S. Patent 4,292,333 except that other flavoring ingredients replace the butter flavor.

Encapsulated Enzyme Emulsions in Cheese Making

A process for microencapsulation of substrates by way of a stable multiple-phase emulsion has been developed by *N.F. Olson and E.L. Magee, Jr.; U.S. Patent 4,310,554; January 12, 1982; assigned to Wisconsin Alumni Research Foundation.* This process will be described with reference to the microencapsulation of enzymes for enhancing the ripening of cheese and the development of desired aroma and flavor.

Microencapsulation of curing agents in milk fat capsules has been found to be effective for concentrating the curing agent in an environment to accelerate reaction and to release such curing agents at a time and place where their effectiveness can be optimized.

Cheese ripening consists of complex enzymatic and chemical modifications of casein and milk fat and, to a limited extent, residual sugars in cheese. The reactions occur in sequential steps to produce the body, texture and balanced flavor of mature Cheddar cheese. Many investigators have attempted to accelerate or characterize the ripening process by incorporating into cheese various catalysts, such as selected enzymes, microbial cell free extracts, viable pregrown microbes and miscellaneous curing agents. Generally, these modifications have resulted in poor quality cheese presumably from an inability to control or coordinate the activity of the added catalyst in the normal events of cheese ripening. The inability of additives to accelerate satisfactory flavor development may have resulted, in part, from an improper spatial arrangement between the enzymes, microbes, and substrates.

The process will be illustrated by the microencapsulation of a cell free extract of *Streptococcus lactis, subsp. diacetilactis* and substrates with a view towards producing diacetyl plus acetoin in cheese.

Example: Cell free extracts and/or substrates (hereinafter referred to as carrier) were encapsulated into milk fat capsules prepared by the following method. Frozen, unsalted butter was melted in free-flowing steam and 200 g of milk fat was decanted into a beaker. The milk fat was maintained at 62°C and 3 g of Glycomul TS (Glyco) was added and held for 10 minutes until melted. Three g of Span 60 (Ruger Chemical Co.) was similarly added and melted.

The milk fat emulsifier mixture was transferred to a water bath at 37°C and stirred at 250 rpm with a three-bladed propeller, until the milk fat emulsifier mixture was also 37°C. The stirring rate was increased to 500 rpm while slowly adding 40 ml of carrier consisting of a specified mixture of cell free extract and/or substrates. After 5 to 10 minutes of stirring, the emulsion was ejected into acidifed skim milk at 15°C to form the capsules. A Wagner ST350, airless sprayer with nozzle extension and 0.4 mm orifice was used to form the capsules.

The above intact capsules were added to skim milk and made into cheese by a direct acidification procedure to avoid interference of the lactic starter culture in production of diacetyl and acetoin in cheese. Also, a low temperature was used to cook the curd to maintain integrity of the capsules. This resulted in a soft bodied, high moisture, low fat cheese.

Cheeses were also made with broken capsules (heat-treated to disrupt the capsules) as well as control cheeses with unencapsulated enzymes.

At specified times of aging, the cheese was analyzed for diacetyl and total diacetyl-acetoin.

A broad comparison of total aroma development indicated cheese with the intact capsules produced approximately six times more total diacetyl-acetoin and diacetyl than the various control cheeses. In cheese containing intact capsules, there was a gradual rise in the concentration of total diacetyl-acetoin produced during the first days of aging which peaked between four to six days. This indicated diacetyl-acetoin production occurred within the capsules in the cheese rather than a result of formation from an instantaneous enzyme reaction when the enzyme was mixed with the substrate prior to encapsulation.

Aroma development in the control cheeses was almost nil. Diacetyl concentrations were below levels detectable by the assay procedure; total diacetyl-acetoin was measurable but extremely low. Cheese containing broken capsules contained approximately 0.9 to 1.3 μg total diacetyl-acetoin/g cheese and the diacetyl was stable over the aging period. Correcting this value to account for presumed 40% loss of enzyme activity would increase the diacetyl-acetoin concentration to 3.2 μg/g cheese which is still four times lower than concentrations in cheese with intact capsules.

Cheese made with the unencapsulated cell free extract and substrates developed 1.3 μg diacetyl-acetoin/g cheese and undetectable levels of diacetyl. This value was six times less than the value of 8.3 μg diacetyl-acetoin/g cheese which developed in cheese with intact capsules which also developed 1.52 μg diacetyl/g cheese.

Substrate was required in the capsules for diacetyl-acetoin production since very little diacetyl-acetoin was produced in intact capsules with encapsulated cell free extract but not substrate. Substrate encapsulation without a cell free extract produced very little diacetyl-acetoin indicating no contamination occurred in these trials.

Lecithin in Cheese Production

R.R. Bily; U.S. Patent 4,277,503; July 7, 1981 has disclosed a method to increase the yield of cheese from milk and thereby reduce the amount of whey waste while not adversely affecting the taste and edibility of the cheese.

This method comprises the addition of lecithin to the milk used in making the cheese, preferably in quantities of 0.001 to 0.066% by weight of the milk used in making the cheese. The lecithin is added to the prepared milk prior to coagulation or precipitation of the curd. Lecithin is a natural mixture of phospholipids or phosphatides derived from vegetable or animal sources and, as normally used in commerce today, is the phospholipid complex derived from soybeans.

Example: In the preparation of part skim low moisture mozzarella cheese, vat one was the control vat and vat two was the vat containing lecithin. Milk for both vats was drawn from the same holding tank. The milk was pasteurized at 161.6°F for 16 seconds, cooled to 90°F and pumped into identical Damrow Double O vats. Samples of milk were tested for the percentage of fat, percentage of protein and percentage of total solids. In this particular test, milk in vat one was found to contain 1.77% fat, 3.15% protein and 10.54% total solids. The milk in vat two was found to contain 1.78% fat, 3.10% protein and 10.44% total solids.

Since this milk had not been homogenized, slight variations in lab analysis are to be expected in different samples taken from the same holding tank, although these milks are considered similar for production purposes.

Vat one was filled with 4,860 gallons of pasteurized milk and at one gallon of milk to equal 8.6 pounds, this equates to 41,796 pounds of milk. After 15 minutes, 140 gallons of bulk starter were added to vat one. Forty minutes later, the milk was ready for the addition of a milk coagulating catalyst. In this instance 5 gallons of a diluted microbial enzyme, *Mucor miehei,* was mixed in a ratio of 68 ounces of microbial enzyme to 5 gallons of water. The milk was agitated for 5 minutes and the vat covered.

After 30 minutes the curd in vat one was cut and allowed to remain undisturbed in the whey for approximately 15 minutes with only periodic gentle agitation. The curd was then cooked to 110°F in 35 minutes while using slow mechanical agitation through much of the cooking period. The curds and whey were pumped onto an automatic Damrow DMC belt where the curd was drained, matted and cheddared prior to being mechanically stretched in a cooker/molder (Stainless Steel Fabricating Co.).

The cheese curd was carefully segregated throughout these processes to guarantee that the curd from the control vat was isolated at all times. When the curd blocks were properly acid-ripened, they were milled (cheddared) and mechanically stretched in heated water at a temperature of 170°F. The cheese was then removed from the molder in 5-pound loaves and placed in stainless steel forms to coalesce. The forms were then placed in cold water until the body was firm and the cheese was then removed from these forms and placed in brine tanks for proper salting.

After the curd from vat one had all been removed from the DMC belt, the cooker-molder machines were broken down to remove all remaining curd left inside this equipment. This cheese was then processed as before and added to the rest of the cheese from vat one. All cheese from vat one was carefully segregated in its own brine tank and allowed to cure for the specified amount of time. This cheese was then packaged, boxed and weighed and the weight of the finished cheese from control vat one was 3,406 pounds. The cheese was then placed under refrigeration at 40°F for later testing.

Vat two was started filling while vat one was filling. Seventeen and one-half pounds of pure powdered lecithin were mixed slowly into a small body of milk that was being run through a reciprocal pump under strong pressure until the lecithin was thoroughly incorporated into the milk. The mixture was then pumped into vat two for further mixing with the main body of milk. Vat two continued to fill to 4,860 gallons or 41,796 pounds of milk. The percentage of lecithin in the milk equaled approximately 0.042%. After 15 minutes 140 gallons of starter were added and 35 minutes later 5 gallons of the diluted milk coagulating enzyme, *Mucor miehei* was added. The batch was cut after 30 minutes and all procedures described for the control vat one were followed in an identical manner. A higher yield of 3,528 pounds representing an increased yield of 3.58% over that obtained from vat one was obtained.

Samples of cheese from vats one and two were later tested for percentage of fat, moisture, total solids, pH and percentage of protein. The cheeses from vats one and two were similar in percentage of moisture, total solids, pH and percentage of protein but there was a consistent dissimilarity in percentages of fat between vats one and two. The fat content of the cheese from vat one (without lecithin) was 16.47% while fat content of the cheese from vat two (with lecithin) was 15.39%.

No explanation was reached for the lower percentage of fat in the samples made with lecithin. Since milk coagulation involves the precipitation of milk protein, colloidal chemistry is obviously involved. Cheese yield is dependent upon the recovery of fat and casein by the curd during cheesemaking and by the composition of the milk and the moisture content of the final cheese. Since lecithin is a valuable colloid, emulsifier and water binder, it is understandable that it can affect cheese yield.

Lecithin in Dry Milk Products

An improved lecithin-based wetting agent has been described by *W.P. Jackson and M.R. Warseck; U.S. Patent 4,164,594; August 14, 1979; assigned to Carnation Company.* When used in the production of dry fat-containing products the wetting agent of this process provides products which are spontaneously wettable in aqueous liquids even in ice water. This wetting agent comprises oil free, granular phosphatides containing at least 95% acetone-insoluble matter dissolved in an oil carrier which has a bland taste, a maximum iodine value of 2.0 and which is a liquid at temperatures as low as about 0°C.

Such vegetable oil carriers which may be used are glyceryl esters of capric acid and caprylic acid and mixtures thereof. The oil-free phosphatide is combined with the carrier in an amount sufficient to provide the wetting agent with a level of acetone-insoluble matter in the range of from about 25 to 65%. If desired, an antioxidant may also be included in the wetting agent.

The wetting agent may be applied to the fat-containing powder by conventional techniques such as spraying the wetting agent as an aerosol onto the powder, dry blending the wetting agent with the powder and the like.

The resulting product, which has a phosphatide content of from about 0.5 to 0.35% by weight, will wet and disperse in 6°C water in less than 10 seconds and will remain stable with respect to wetting for at least one year at room temperature. The wetting agent imparts no adverse taste or flavor to the reconstituted beverage.

The wetting agent of this process utilizes, as one component, granular lecithin which is substantially oil free and which contains 95% or more acetone-insoluble matter or phosphatides. Such oil-free lecithin may be produced by contacting commercial lecithin, such as obtained by the solvent extraction of soybeans, with acetone to remove the glyceride oils, fatty acids, sterols, traces of bitter principles, etc., from the phosphatides. The acetone-washed phosphatides are then freed from residual acetone by agitation and distillation under vacuum. The oil-free phosphatides have a very bland flavor and are free from the undesirable off flavors inherent in commercial lecithin. Such granular oil-free phosphatides are available commercially, such as granular lecithin known as Centrolex R (Central Soya Co.).

An important component of the wetting agent is an oil carrier which remains liquid at low temperatures, that is, temperatures as low as about 0°C, has a maximum iodine value of about 2.0, and has a bland flavor. It has been found that oils having very low unsaturation and a high content of short chain acids have this desired combination of properties. Such oils include glyceryl esters of capric acid and caprylic acid and mixtures thereof.

A preferred oil for use as the carrier in the wetting agent is a mixture of 70% glyceryl esters of capric acid and 30% glyceryl esters of caprylic acid known as PVO 1400 (Pacific Vegetable Oil Co.). This oil has a very bland flavor and a settling point below −5°C, so that the oil remains a liquid at reconstitution temperatures as low as 0°C and does not interfere with the wetting function of the phosphatides at such low temperatures. Also, PVO 1400 oil has a maximum iodine value of 2.0 so that it is highly stable against oxidation, and has a very low viscosity which enables the oil to form a very fine aerosol and evenly distributed over a fat-containing powder. Moreover, PVO 1400 oil exhibits a defoaming capacity in reconstituted products.

The wetting agent is prepared by dissolving the oil-free phosphatides in the oil carrier in an amount sufficient to provide the wetting agent with an acetone-insoluble phosphatide content of from 25 to 65%, preferably from about 30 to 40% phosphatide. Within these ranges, the wetting agent has a very bland flavor and provides good wetting even at temperatures as low as 0°C. At a phosphatide level above about 65%, good wetting is obtained but flavor problems are encountered, while at a phosphatide level below about 25%, flavor is acceptable but wetting ability is impaired.

Example 1: A wetting agent is prepared by dissolving 36.8% by weight granular, oil-free lecithin which contains 95% acetone-insoluble whole phosphatides into 63.1% by weight of an oil consisting of 70% glyceryl esters of capric acid and 30% glyceryl esters of caprylic acid, and 0.1% by weight of an antioxidant which consists of 52% propylene glycol, 40% butylated hydroxyanisole and 8% citric acid. In preparing the wetting agent, the antioxidant and oil are combined and heated to a temperature of about 140°F and the granular oil-free lecithin added with stirring until dissolved. The wetting agent thus formed contains 35% acetone-insoluble matter and is free from foreign or off odors and flavors. It is liquid at room temperatures and has a maximum Brookfield viscosity of 100 cp at 25°C (No. 4 spindle).

Example 2: An instant, low butterfat milk is prepared by adding about 1.9 pounds of powdered lactose, and 0.008 pound of a dry blend of Vitamin A and D to 97.9 pounds of dry, low fat milk powder which has a butterfat content of about 5% (dry basis), a moisture content of about 3.5% and a particle size such that 1% is +12 mesh and 10% is −200 mesh (U.S. Standard Sieve). The dry ingredients are blended for 3 to 5 minutes. About 0.2 pound of the wetting agent of Example 1 is heated to a temperature of 130° to 140°F and sprayed, at that temperature, onto the dry ingredients mix. After spraying, the mix is blended in a ribbon blender for 15 to 30 minutes to insure thorough distribution of the phosphatides. The resulting mix, which has a moisture content of 2.5 to 3.5%, is then subjected to an instantizing step wherein the individual particles of the dry mix are moistened and bonded together to form porous agglomerates which are subsequently dried to a moisture content of 1.5 to 3.5%. The resulting product will wet and disperse in 6°C water in less than 10 seconds, and has no off flavors or odors.

Purified Phosphatides for Margarines

Conventionally, phosphatides are obtained from beans, in particular soybeans and other phosphatide-containing materials. In the processing of beans, the phosphatides are, for instance, obtained by pressing or solvent extraction of the beans and by separating the phosphatides from the resulting crude oils by a treatment with water or aqueous solutions.

The aqueous sludge obtained contains phospholipids, a certain proportion of oil, fatty acids, carbohydrates, proteins, mineral salts, sterils, some remainders of the bean shells and occasionally other materials. The sludge may be dried to obtain a yellow to black mass with a wax-like consistency. The phosphatides may be subjected to various treatments such as removal of the oil, replacement of the oil by another oil, hydroxylation and hydrolysis, either by enzymatic action or by acidic or alkaline hydrolysis.

It has been found by *F.G. Sietz; U.S. Patent 4,166,823; September 4, 1979; assigned to Thomas J. Lipton, Inc.* that a purified transparent phosphatide product is obtained when phosphatides are subjected to the combined action of a hydrophobic liquid and water. The two liquids are separated and the purified phosphatide product is recovered from the hydrophobic liquid. According to the process a transparent phosphatide product is obtained.

The purified phosphatide product obtained is superior to the untreated one with respect to odor, taste and oil solubility. The product can especially usefully be applied in food preparations like margarine. Still another advantage of the purified phosphatide product is its good oil solubility as compared with conventional products.

The advantages of the process are especially pronounced when the process is applied to phosphatides which have been hydrolyzed by pancreatin.

The hydrophobic liquid to be used may be any hydrophobic liquid in which phosphatides are soluble. Examples are aliphatic, preferably saturated hydrocarbons, preferably alkanes such as heptane, hexane and pentane. Cyclic alkanes such as cyclohexane are also suitable. Aromatic compounds such as benzene may also be used. Hexane is preferred.

The presence of substantial amounts of polar organic liquids tend to make the subsequent separation of hydrophobic liquid and water more difficult. Moreover, it has been found that polar organic liquids cause the recovery of the phosphatide product from the hydrophobic liquid to be more troublesome and affect the quality of the product adversely.

The process is preferably carried out while the amount (by volume) of hydrophobic liquid is about 1 to about 3 times, preferably about twice, as large as the amount (by volume) of phosphatides plus water. The sequence in which the two liquids are added to the phosphatides is immaterial. In a preferred embodiment, the hydrophobic liquid is added to the aqueous sludge described above containing the phosphatides, optionally after some other treatment like, for instance, hydrolysis of the phospholipids. This procedure has the advantage that the need for an intermediate drying step is obviated.

The manner in which the two liquids are separated is not essential. Centrifuging is preferred.

Recovery of the purified phosphatide product from the hydrophobic liquid is preferably by evaporation.

The temperature and pressure at which the process is carried out can vary within wide limits. Atmospheric pressure at ambient temperature is preferred.

Example 1: 10 kg of crude soybean phosphatides were homogenized with 10 ℓ water. The sludge was then homogenized with 40 ℓ cyclohexane and the mixture was centrifuged. The white solvent phase was evaporated. The phosphatides thus obtained were transparent. Analysis showed that the sugar content decreased to about 40% and the iron content to about 55% of the starting value. The P-content had risen by 0.04%. Color, acetone-insoluble matter and acid value had remained substantially constant.

Example 2: 10 kg crude soybean phosphatides were dissolved in 40 ℓ hexane. Subsequently 10 ℓ water was stirred into the solution. After 10 minutes stirring the solution was allowed to settle. After a resting period of 5 hours and separation of the phases the hexane solution was evaporated. The phosphatides thus obtained were transparent. The analytical data corresponded to those of the preceding example.

Example 3: 10 kg crude phosphatide sludge, which has been hydrolyzed enzymatically according to U.S. Patent 3,652,397, were homogenized with 20 ℓ hexane and centrifuged. The hydrolyzed phosphatides isolated from the hexane phase were transparent. The sugar and iron contents had been reduced to about half the starting values.

The P-value had increased by 0.05%. The other values (color, acetone-insoluble matter, acid value) had remained constant.

Phosphatide plus Metal Oxide in Nonspattering Margarines

According to a process described in Netherlands Patent 7,314,936, a margarine having a reduced tendency to spattering is produced by incorporating in the oil phase a finely divided metal or metalloid oxide which was pretreated with alcohols, fatty acids, aldehydes or lecithins in order to render the oxide hydrophobic. Specific examples mentioned in this patent are methylated silicon dioxide and lecithinated silicon dioxide. A drawback associated with the use of this method is that the antispattering effect achieved, diminishes on storage.

J.H.M. Rek and P.M.J. Holemans; U.S. Patent 4,325,980; April 20, 1982; assigned to Lever Brothers Company have found a process which alleviates the above disadvantage to a great extent.

The improved process for producing a water-in-oil emulsion of the margarine type, which displays a reduced tendency to spattering consists of adding separately (a) a phosphatide and (b) a hydrophilic finely divided metal and/or metalloid oxide, into an oil phase essentially consisting of a fat or fat blend which is appropriate for producing a nonliquid margarine.

It is an essential feature of this process that the phosphatide and the oxide components are added separately to the oil phase. By added "separately" is meant that both components are physically separated before and during their incorporation in the oil phase and that a possible interaction between the components can consequently only occur in situ in the oil phase.

Suitable phosphatides which are preferred are those obtained by treating vegetable oils in which phosphatides are present, such as soy oil, groundnut oil, sunflower oil and rapeseed oil, cottonseed oil, and the like, with steam or water and separating out the phosphatides, which are also less accurately referred to as lecithin. Naturally, phosphatides of other origin, such as egg yolk, or synthetically prepared phosphatides can also be used for the process.

The phosphatides are usually added in an amount of about 0.01 to 5% by weight, preferably between about 0.1 and 1% by weight, based on the weight of the margarine.

The finely divided metal and/or metalloid oxide suitable for use in this process are well-known substances. They are normally prepared by thermal decomposition (vapor phase hydrolysis) of the corresponding volatile metal and metalloid halogenides at very high temperatures (about 1100°C), the particle size and the surface area of the oxides obtained being regulated by means of the concentration of the halogenides, the temperature and the reaction time. The bulk density is regulated by the degree of compression of the oxides obtained.

It was found that useful hydrophilic oxides should preferably have (a) a specific surface area of at least 130 m^2/g and preferably a specific surface area ranging from 200 to 400 m^2/g and (b) a particle size ranging from 1 to 100 mμ.

The oxide preferably used is highly voluminous, light silicon dioxide having the physical characteristics specified above. Mixtures of highly voluminous metal and metalloid oxides, for example, a mixture of silicon dioxide and aluminum oxide, can also be used.

This process is particularly concerned with nonliquid margarines or plastic margarines. By this term is meant margarines having an oil phase essentially consisting of a fat or fat blend which has a slip melting point ranging from 30° to 40°C.

The proportion of oil phase in the margarine lies normally within 75 to 90% by weight and the proportion of the aqueous phase lies within the range of 10 to 25% by weight.

It has been further found that the particularly good antispattering effect could be achieved by ensuring that the pH of the aqueous phase ranges from 4.0 to 6.5.

Example: With the aid of a Votator a margarine was prepared in the usual manner starting from 83.786% by weight of a fatty phase consisting of 20 parts coconut oil, 15 parts palm kernel oil, 15 parts palm oil, 50 parts soy oil, hydrogenated to a melting point of 32°C (D_0 = 800, D_{15} = 550, D_{20} = 450, D_{25} = 250, D_{30} = 125, D_{35} = <25) and 16% of an aqueous phase consisting of 50% of buttermilk and 50% of water (pH = 4.6).

Before the fatty phase was emulsified with the aqueous phase, the following were added separately to the fatty phase: 0.168% by weight of raw soy phosphatide (that consisted of about one-third soy oil) and 0.168% by weight of Aerosil 200 (finely divided silicon dioxide having a surface area of 200 m^2/g and an average primary particle size of 16 mμ).

This margarine was subjected after 3 and 10 days storage to the following spattering test: in an enameled pan with a smooth bottom surface, 50 g of the margarine was heated each time to 175°C (regulated by a thermocouple). The spattering fat was caught, at a distance of 21 cm above the pan, on a piece of paper that had been weighed beforehand. After the test the piece of paper was weighed again.

The degree of spattering was determined on the basis of the increase in the weight of the paper and this was converted as follows into an assessment according to points: 10 = very good antispattering behavior = less than 10 mg; 6 = moderate antispattering behavior = 50 to 100 mg; and 2 = very bad antispattering behavior = ⩾500 mg.

The points in between express a corresponding spattering behavior. The results were as follows: after 3 days storage, 10; and after 20 days storage, 10.

When the test was repeated under identical conditions, but with, as the only difference, the oxide first having been mixed with the phosphatide before it was added to the fatty phase, the results of the spattering test were: after 3 days storage, 9.5; and after 10 days storage, 6.

Protein-Free Coffee Whiteners

Improved fluid nondairy creamers commonly referred to as coffee whiteners have been developed by *C.E. Rule; U.S. Patent 4,341,811; July 27, 1982; assigned to SCM Corporation.*

The nondairy coffee whitener, adapted to be added to an acidic, hot environment, can be prepared by pasteurizing and homogenizing a water-rich lipoidal emulsion consisting essentially of about 6 to 15% edible fat, about 0.6 to 2% mixed lipoidal emulsifier, and water, the fat having a Wiley melting point below about 120°F, the emulsifier constituting about 0.3 to 1% low HLB mono- and diglyceride or propylene glycol partial ester of fat-forming acids and about 0.3 to 1% high HLB polyglycerol ester having an HLB value above about 10, a hydroxyl value of about 400 to 600, a saponification number of about 60 to 100 and acid values of less than about 10.

Preferred polyglycerol esters are octaglycerol monooleate and octaglycerol monostearate.

It is a totally surprising discovery that this coffee whitener formulation, being free of protein, can be added to an acidic, hot environment, such as hot coffee, without coalescence of the fat globules. It is believed that this is due to a unique combination of factors; namely, concentrations, wet basis, of the lipid ingredients, the combination of specific lipid ingredients stated, and homogenization preferably carried out to obtain an average particle size of less than about 2 to 3 microns for the fat globules.

Preferred low HLB emulsifiers are partial fatty acid esters of glycerol and propylene glycol having an HLB value not substantially greater than about 5 and a capillary melting point sufficiently high to have a normally hard consistency at room temperature of about 70°F. Emulsifiers falling within this category are mono- and diglycerides which are in normally solid form. One such mono- and diglyceride is Dur-em 117 having an HLB value of about 2.8, made from 5 maximum I.V. hydrogenated soybean oil, having a capillary melting point of about

145° to 150°F and 40% minimum alpha-monoglyceride content. This emulsifier is marketed with an amount of citric acid to help protect flavor.

A preferred high HLB polyglycerol ester emulsifier component is selected from the group consisting of octaglycerol monooleate known as Santone 8-1-O and octaglycerol monostearate known as Santone 8-1-S (SCM Corp.). Santone 8-1-O is described as a normally liquid polyglycerol ester of fatty acids having an I.V. of about 25 to 35, an HLB of about 13 and a sponification value of about 77 to 88. It contains about 60 to 68% oleic acid and has a hydroxyl number of about 500 to 570 and an acid number of under 4.1. Santone 8-1-S is a similar such ester, but normally solid, having an I.V. of about 3 maximum, a Mettler dropping point of 52° to 57°C, an HLB of about 13, and a saponification value of about 77 to 88. By normally solid or liquid, is meant that phase which exists at room temperature (about 70°F).

One fat which can be used in this process is a nonlauric fat (that is, one having a low lauric acid content, C_{12}) which is hydrogenated and elaidinized to provide a desired hardness or high solids content sufficient to maintain a substantially plastic consistency throughout a wide temperature range, and at the same time a rapid melting at elevated temperatures above about 110°F without retention of a waxy mouth feel.

One suitable such elaidinized fat is a partially hydrogenated vegetable oil (cottonseed or soybean) known as Duromel (SMC Corp.) having an I.V. of 60 to 65, a Wiley melting point of 101° to 105°F, and a solid fat index of: 56 at 50°F; 43 at 70°F; 36 at 80°F; 16 at 92°F; and 4 maximum at 100°F. Duromel has a free fatty acid content of 0.1 maximum.

Example: A whitener was produced from 10% partially hydrogenated vegetable fat (Duromel); 0.55% polyglycerol ester (Santone 8-1-S); 0.5% mono- and diglyceride (Dur-em 117); 89.5% water; and optional amounts of sweetener, flavor and color.

The coffee whitener was prepared by heating water to 130°F and adding the lipids (Duromel, polyglycerol ester and mono/diglyceride) to it. The mix then was heated to 160°F and held for 15 minutes, following which it was homogenized at 2,500/500 psi through a two-stage homogenizer. Homogenization follows conventional dairy technology. The mix was rapidly cooled to 40°F, packaged and refrigerated at 40°F.

In the above formulation, 36 DE corn syrup can be used at a level of 12%, with excellent results.

The whiteners of this example were found to resist oil separation and rancidity for long periods under refrigeration, up to two months' time. Only very slight oil droplets on top of the emulsion were noted. The amount was considered acceptable. The whitener had a pH of about 6 to 6.2.

Homogenization is critical to produce a desired size of the fat globules, less than about 2 to 3 microns (average particle size), preferably less than about 1 micron, for the purposes of emulsion stability and whitening power. Whereas homogenization pressures of 2,500/500 in a two-stage homogenizer produces excellent results, good results can also be obtained at other pressures ranging from 1,000 to 6,000, using either a single or two-stage homogenizer.

In the above example, pasteurization was carried out at 160°F and held for 15 minutes. It is an aspect of this process that the coffee whitener formulation can also be subjected to ultra-high pasteurization at temperatures of 280° to 300°F for 2 to 6 seconds, without adverse effects. Protein-containing coffee whiteners subjected to such temperatures are likely to suffer from protein breakdown. Ultra-high pasteurization offers the advantage of extended shelf life.

FOR MEAT AND FISH PRODUCTS

Low Calorie Emulsifier for Meat Analog

A stable emulsion system that can be used to deliver flavor, color or lipids in a cooked food product and/or to reduce caloric density of food products has been described by *N.B. Howard; U.S. Patent 4,226,890; October 7, 1980; assigned to The Procter & Gamble Company.*

This stable hydrated emulsifier composition comprises: (a) an emulsifier selected from the group consisting of polyglycerol monoesters of fatty acids; monoacyl-glycerol esters of dicarboxylic acids; sucrose monoesters of fatty acids; polyol monoesters of fatty acids; phospholipids; and mixtures thereof; (b) water; and (c) a food additive, such that the emulsifier is in a liquid crystalline state which is thermally stable at temperatures of from about 98° to 200°F.

The key to the formation of a thermally and gravitationally stable emulsifier system to which both polar and nonpolar food additives can be added is in the selection of an emulsifier system which, when hydrated, forms a stable, liquid crystalline state in the temperature region of 98° to 200°F.

The term "liquid crystalline" is synonymous with a "mesomorphic state," e.g., a lamellar or neat phase. It refers to a fluid state between the perfectly ordered structure found in solid crystals (emulsifier) and a disordered state in an amorphous structure, which shows a birefringence under polarized light.

By "food additive" is meant flavoring agents, coloring agents, fats, sugars, and other ingredients which are added to food products to enhance the color, flavor, or nutritive value of the product. Flavor and color precursors are included in this term. The use of fat (triglycerides) as a food additive in this system allows the use of less fat than would normally be present in the food or beverage.

Polyglycerol monoesters suitable for use in this process have an average of from 2 to 10 glycerol units and an average of one fatty acid acyl group per glycerol moiety. Preferred polyglycerol esters have an average of 2 or 3 glycerol units and one fatty acyl group having from 14 to 18 carbons per polyglycerol moiety.

The polyglycerol essentially is a polymer which is formed by the dehdyration of glycerol. For each unit of glycerol that is added to the polymer chain there is an increase of one hydroxyl group. In this process, from about 1 to 4 of these hydroxy groups of the polyglycerol molecule is esterified by reaction with fatty acids. This esterification is similar to that of glycerol or other polyols.

To prepare the stable hydrated emulsifier composition, an emulsifier selected from the group consisting of polyglycerol monoesters of fatty acids, monoacyl-glycerol esters of dicarboxylic acids, sucrose monoesters of fatty acids, polyol monoesters of fatty acids, phospholipids, and mixtures thereof, is heated above the melting point to liquefy the material.

This melt is dispersed with high shear and cooling, the shearing and cooling being done either simultaneously or in sequence, into an aqueous medium to form a dispersion, the ratio of emulsifier to water being from 10:1 to 1:10, preferably from 8:1 to 1:8.

The water-soluble food additives, i.e., flavors, flavor precursors, coloring dyes, color precursors, are predissolved in the water.

Similarly, the oils and oil-soluble food additives are dissolved or dispersed in the emulsifier. The additional ingredients which are added to the emulsifiers, for example, fat or oil are preferably added at a temperature above the melting point of the emulsifier crystals and cooled with high shear mixing.

In forming the emulsion, it is required that the emulsifiers be in the form of a melt and then subjected to high shear prior to or during cooling.

The hydrated emulsifier compositions as prepared above can be incorporated within the food or beverage product or laminated on the surface of the food product.

The food additive composition is mixed with the texturized protein, binder, and water used to prepare a meat analog. No criticality exists with regard to the choice of texturized protein or binder material.

Example 1: Pork flavor volatiles (0.46 g) are blended with 461.8 g of polyglycerol ester of palmitic acid which has been warmed to about 100°F. Water (1,078.7 g) is added to this solution and the mixture heated to 140°F using microwave energy. The mixture is then placed in a mechanical blender and blended for three minutes at ambient temperature, and then for 10 minutes at ice bath temperatures. The sides and bottom of the container are scraped periodically during both mixing periods.

A flavored emulsion which is in a liquid crystalline or mesophase state is produced. The emulsion is stable up to 190°F.

When whey solids (77.1 g) are mixed with the mesophase prepared above, the mesophase system turns brown on heating.

Example 2: Lecithin (12.7 g), hardstock (38.0 g) and pork triglyceride (707.6 g) are heated to 75°C to solubilize the components. The pork flavor volatiles (0.76 g) are then added to this fat solution, the mixture cooled to 45°C, and the whey solids are added using a blender at low speed.

The stable hydrated emulsifier system prepared in Example 1 is then added to the melted fat phase using a blender at low speed over a three-minute period. The dispersion is further mixed as follows: two minutes at low speed followed by six minutes at a medium speed. The sides of the containers are scraped periodically during the mixing.

The stable hydrated emulsifier system and oil dispersion prepared by the above method is birefringent under polarized light and is heat stable up to about 190°F. The product has a mild pork-like taste.

The product can be used to replace the fat (triglyceride) in a pork analog product. The fat level of such a product is about 50% lower than the real pork product.

Colloidal Systems for Fish and Meat Products

The process disclosed by *S. Inamine, T. Matsuda and T. Shimomura; U.S. Patents 4,252,834; February 24, 1981 and 4,168,323; September 18, 1979; both assigned to KK Ueno Seiyaku Oyo Kenkyujo, Japan* relates to a method for preparing a food additive composition comprising a colloidal solid prepared from a sugar alcohol, a sugar, a food-grade surface-active agent and an edible oil or fat.

This process is based on the discovery that a food-grade surface-active agent, either alone or together with an oil or fat, forms a colloidal solid in a dispersing medium consisting of a sugar alcohol and/or a sugar. This makes it possible to render slightly water-soluble or water-insoluble substances readily dispersible or soluble in water, convert liquid surfactants and liquid oils or fats into solid powders, and thus to greatly broaden the area of utilization of surfactants for food production.

The sugar alcohol may, for example, be sorbitol, mannitol, and maltitol. Examples of the sugar are sucrose, glucose, fructose, lactose, and maltose. Of these, sorbitol and sucrose are especially preferred.

Useful surface-active agents include glycerin fatty acid esters, propylene glycol fatty acid esters, sorbitan fatty acid esters, sucrose fatty acid esters, and lecithin. Preferred fatty acids for preparing these surfactants are, for example, stearic acid, palmitic acid, myristic acid, lauric acid, capric acid, caprylic acid, oleic acid, linoleic acid, and linolenic acid. Mixed fatty acids such as cottonseed oil, soybean oil, colza oil, corn oil, safflower oil, palm oil, beef fats, fish oils, and hardened products of these oils or fats, can also be used.

This composition contains 60 to 99%, preferably 70 to 98%, of the sugar alcohol and/or the sugar, 30 to 0.5%, preferably 15 to 1%, of the surface-active agent, and 30 to 0%, preferably 20 to 0%, of the edible oil or fat.

The composition must form a colloidal solid in which the surfactant, either alone or with the oil or fat, is dissolved or dispersed in the sugar alcohol and/or the sugar as a dispersing medium. If the proportion of the surfactant or oil is larger than the above-specified range, phase inversion occurs, and the surfactant or oil becomes a dispersing medium. As a result, the composition will not be readily soluble in water, and the desired effect cannot be achieved. When the surfactant is hydrophilic, it tends to dissolve in the dispersing medium, but does not impair the good solubility of the dispersing medium.

This composition can be produced, for example, by (1) a process which comprises heat-melting the sugar alcohol and/or sugar, adding the surfactant either alone or together with the oil or fat to the melt, stirring the mixture to dissolve or disperse it, cooling the resulting solution or dispersion to solidify it to a colloidal solid, stabilizing it by crystallization, and then pulverizing the stabilized colloidal solid to a powder with a size <20 mesh; (2) a process which comprises adding the surfactant either alone or together with the oil or fat to an aqueous solution of the sugar alcohol and/or the sugar, stirring the mixture to form a solution or dispersion, evaporating it by heating it under reduced pressure while the sugar alcohol and/or the sugar is maintained in the dissolved or molten state, cooling it to solidify it to a colloidal solid, and then treating it in the same way as in the method (1) above to form a powder having a size smaller than 20 mesh. In these methods, a suitable amount, for example, 3 to 30%, of a seed crystal such as a sugar alcohol and/or a sugar can be added to promote crystallization, before the mixture is solidified. When the sugar alcohol and/or sugar is used as the seed crystal, the amount of the starting sugar alcohol and/or sugar needs to be decreased correspondingly.

In one preferred embodiment of this process, the surfactant either alone or together with the oil or fat is added to an aqueous solution (preferably in a concentration of at least 50%) of the sugar alcohol and/or the sugar and stirred to form a solution or dispersion, after which a powder of sugar alcohol and/or sugar is added as a seed crystal and mixed with the solution or dispersion, and then the mixture is powdered and dried (or the mixture is dried and then powdered) to form a powder having a size smaller than 20 mesh. If the aqueous solution of sugar alcohol and/or sugar has a high concentration, it may be heated. The drying can be performed by usual methods such as vacuum drying or hot air drying.

The colloidal solid obtained by this process can also be used as a seed crystal.

For the production of a good-quality colloidal solid, an organic solvent such as glycerin, propylene glycol or ethanol may be added before or after the addition of the surfactant.

A composition which has the same constituents and proportions as this composition but is obtained by merely mixing the constitutents cannot be a colloidal solid. Even when such a mixture is put into water, the unique effects of the colloidal solid of this process cannot be obtained.

This composition can be utilized in a very wide range of applications by varying the type or proportion of the surfactant according to the type of food to which the composition is to be added. For example, by using stearic acid monoglyceride, the composition can be used as a softener for bread. When propylene glycol stearate or sucrose monostearate is used, the composition of this process can be used as a foamer for sponge cakes. If containing sorbitan trioleate, the composition can be used as an effective defoamer. The use of a monoglyceride derived from a cottonseed oil, or a mixture of propylene glycol oleate and a normally liquid oil such as cottonseed oil or soybean oil can afford a composition which has an effect of increasing the whiteness of frozen fish mince and preventing spoilage during refrigeration.

Capric acid, caprylic acid, lauric acid, oleic acid, linoleic acid and linolenic acid have an antimicrobial activity, and glycerin fatty acid esters, propylene glycol fatty acid esters and sucrose fatty acid esters derived from these fatty acids also have an antimicrobial effect. Compositions prepared by using them as surfactants can be used in applications which require antimicrobial activity.

Example 1: Sorbitol (850 g) was heated to 95°C and 100 g of linoleic acid monoglyceride (purified by distillation) was added. They were stirred to form a uniform mixture. As a seed crystal, 50 g of sorbitol powder was added. The mixture was cooled, solidified, crystallized and then pulverized to form a powder having a size of 35 mesh or under.

Example 2: In this example, the composition obtained in Example 1 was added to sausage, and its effect was confirmed. Pork (1,500 g), 1,000 g of beef, 1,000 g of mutton, 750 g of lard and 750 g of ice water were used as main raw materials, and mixed with 12.5 g of white pepper, 3.5 g of nutmeg, 2.5 g of cinnamon and 12.5 g of sodium glutamate as secondary materials. Furthermore, 60 g of δ-glucono-lactone as a pH adjuster and 0.2% sorbic acid were added. They were mixed for 10 minutes by a silent cutter. In each run, the mixture was filled into twenty entrails of sheep, dried at 75°C for 40 minutes, boiled in water at 75°C for 20 minutes, and allowed to cool to afford 20 packs of sausage.

Two packs of suasage were placed in each of ten sterilized Petri dishes, and pre-served at 25°C. The state of spoilage was examined.

No spoilage occurred until the 6th day and not until the 20th day were all packs spoiled. For comparison, when the sorbic acid and the sorbitol and emulsifier of Example 1 were added without formation of the colloidal solid, 3 packs were spoiled by the 4th day and all 20 packs were spoiled by the 8th day. With no additives, 5 packs were spoiled on the 2nd day and all packs were spoiled on the 4th day.

Example 3: A 70% aqueous solution of sorbitol (3,572 g) was heated to 70°C, and 120 g of a monoglyceride of cottonseed oil and 80 g of sorbitan mono-stearate were added. They were stirred by a homomixer to form a uniform dis-persion. The dispersion was heated to 95°C and dehydrated under reduced pres-sure. As a seed crystal, 1,500 g of sorbitol powder was added. The mixture was cooled, solidified, crystallized, and then pulverized to form a powder having a size of 35 mesh or under.

Example 4: Sixty kilograms of minced flesh of fresh Canadian pollack was divided into three lots each weighing 20 kg. A first lot was mixed with 4.2% (4% as sorbitol) of the composition obtained in Example 3, 4% of sugar, 0.1% of sodium pyrophosphate and 0.1% of sodium polyphosphate for 15 minutes. As a comparison, a second lot was mixed for 15 minutes with 4% of sorbitol, 0.12% of a monoglyceride of cottonseed oil, 0.08% of sorbitan monoleate, 4% of sugar, 0.1% of sodium pyrophosphate, and 0.1% of sodium polyphosphate. In the second lot, the types and amounts of the additive components were the same as in the first lot, but they were added as a mere mixture and not as a solid colloid. A third lot was mixed for 15 minutes with 0.1% of sodium pyrophosphate, 0.1% of sodium polyphosphate, 4% of sorbitol, and 4% of sucrose as a comparison.

The mixtures were each frozen at -22°C, and after one month, melted. Then, the whiteness of each lot was examined.

Packed kamaboko was produced using the frozen mince as a material, and its quality and the antiseptic effect of the additive composition were examined.

Packed kamaboko was produced by mixing the minced flesh with 3% of common salt, 0.6% of δ-gluconolactone (pH adjuster) and 0.2% of potassium sorbate (aux-iliary antiseptic) for 15 minutes by a silent cutter. The mixture was packed into plastic tubes, heated for 45 minutes in hot water at 90°C, and rapidly cooled with cold water.

For the sake of reference, packed kamaboko was also prepared from the third lot without adding potassium sorbate.

Preservation tests were run at 30°C on 20 packs of kamaboko for each lot. In lot 1, one pack was spoiled on day 4 and all 20 were spoiled by day 10. In lot 2, 8 packs were spoiled on day 4 and all 20 packs were spoiled by day 8. For lot 3, all 20 packs were spoiled by day 6 and a control sample with no additives had complete spoilage on day 4.

Lecithin for Foam Frying

A method described by *J.M. Rispoli, M.A. Rogers, R.J. Sims and R.H. Waitman, Sr.; U.S. Patent 4,188,410; February 12, 1980; assigned to General Foods Corporation* comprises frying a comestible in a medium containing an oil or fat and an amount of an emulsifier effective to foam the oil or fat during frying.

Appropriate emulsifiers include citric acid esters of mono- and diglycerides, phosphated mono- and diglycerides, sodium stearyl fumarate and sodium sulfo-acetate derivatives of mono- and diglycerides. However, the preferred emulsifier is lecithin (a phospholipid). Lecithin is preferred due to its ability to sustain a desired level of foam for relatively long periods of time while preventing sticking and burning, as well as foaming without substantial splattering.

The fat or oil and emulsifiers need be present in an amount effective to provide a level of foam sufficient to cook the comestible in the foaming frying medium. Due to the foaming effect an artifically high level of heat transfer medium is created thus enabling substantially reduced levels of fat or oil to be employed to cook or fry the comestible.

Preferably the amount of fat or oil in the frying medium is sufficient to create a height of fat or oil on the frying surface (without foaming) within the range of about 0.4 to 15 mm. This translates into about one-eighth cup (30 ml) to 2 cups (474 ml) of oil or fat needed in a 12-inch pan to foam-fry a comestible. When foam-frying, to be included in the amount of fat or oil in the drying medium is the fat or oil added directly to the frying surface as well as any fat or oil contributed from the comestible or the coating on the comestible. In a comparison of the levels of oil or fat needed in foam-frying versus ordinary frying, when foam-frying with one-quarter cup (59 ml) of oil and 7.0% of lecithin by weight of the oil, to create the same level of heat transfer medium in ordinary pan-frying generally at least 2 cups (474 ml) of oil would be needed.

In order to create the foaming action, moisture need be present in some form, i.e., either in the food, on the surface of the food or added extraneously to the frying medium, as the foaming action is caused by the release of moisture in the form of a foam. The comestible itself will generally provide sufficient moisture to induce the desired foaming action.

The emulsifier can be incorporated by any convenient method into the fat or oil of the frying medium in order to obtain the foaming effect during frying. These methods include adding the emulsifier in solid or liquid form to the fat or oil; in-coporating or blending (dry or wet) the emulsifier with the fat or oil in the processing of the fat or oil; incorporating the emulsifier into a fat or oil to produce a solid form (e.g., stick, bar, flake, granule); and adding the emulsifier as part of a dry coating mix for coating the comestible prior to frying. Conceivably the emulsifiers can even be added to the comestible itself, so long as during frying sufficient emulsifier is released into the frying medium to obtain the foaming effect during frying.

The emulsifier is preferably added as part of a dry coating mix. The dry coating mix, aside from the emulsifier, contains ingredients which are common to the coating mix art such as fat or oil (e.g., hydrogenated vegetable oil), protein (e.g., egg solids, soy, milk), flour, bulking agents (e.g., bread crumbs, fines) and seasonings (e.g., salt, paprika, monosodium glutamate, pepper, etc.). The dry coating mix preferably contains lecithin at a level of about 3 to 35% (preferably 4 to 10%), fat or oil at a level of 0 to 75% (preferably 4 to 30%), protein at a level of 0 to 30% (preferably 4 to 18%), flour at a level of 0 to 80% (preferably 10 to 55%) and seasonings at a level of 4 to 40% (preferably 12 to 32%), all being by weight of the dry coating mix. The fat or oil is not a necessary ingredient of the dry coating mix as long as a sufficient source of fat or oil for the frying medium is added directly to the frying surface or is available from the comestible itself.

Generally, the comestible will be foam fried at a temperature within the range of about 260° to 400°F, preferaby 320° to 370°F, with care being taken not to burn the comestible or emulsifier or cause excessive splattering. Appropriate comestibles include such categories as meat, fish, poultry and vegetables. For example, chicken pieces, chicken cutlets, pork chops, beef steaks, fish fillets, or vegetable strips are all suitable.

Example 1: A dry coating mix was prepared by blending the following ingredients: 41.7 parts of all purpose flour, 20 parts hydrogenated vegetable oil, 12.2 parts egg white solids, 7.4 parts of dry lecithin, 2 parts silica gel and 16.7 parts of seasoning (salt, paprika, monosodium glutamate, pepper).

A 2½ pound (1.14 kg) cut-up chicken was wetted with water and each chicken piece was coated (about 85 g of coating mix to coat all the chicken pieces). The coated chicken pieces were placed skin side down in a preheated (350°F), 12-inch skillet with ¼ cup (59 ml) of oil added thereto. The pieces were foam fried for 10 to 15 minutes on each side until golden brown; then the heat was reduced to about 320°F, the chicken pieces were turned, and the frying was continued for about 25 to 30 minutes longer to fully cook the chicken, followed by draining the chicken pieces. During cooking the foam level was generally at a height of about 50 to 70% (within the range of 35 to 95%) of the height of the chicken pieces. There was minimal splattering observed and no stocking or burning of the chicken onto the pan surface. The level of lecithin was about 7.0% by weight of the fat or oil in the frying medium.

The resultant foam-fried chicken pieces had a golden brown appearance, with a crisp and crunchy coating and had the taste, texture and appearance of ordinary pan-fried, coated chicken pieces. The chicken was not greasy and had tender and moist texture and taste.

Example 2: The dry coating mix of Example 1 was added at a level of 100 g to ½ cup (119 ml) of water and blended until a smooth batter-like consistency was obtained. Pork chops (five chops, about 1.3 cm thick) were completely covered with the batter, letting the excess run off. The batter-coated chops were placed in a preheated (350°F), 12-inch skillet with a ¼ cup (59 ml) of oil added thereto. The chops were foam fried for 10 minutes on each side until golden brown, then the heat was reduced to 325°F and the chops were fried for another 25 minutes to fully cook the chops, followed by draining. During foam frying the level of foam was generally at about 80 to 95% (within the range of 50 to 95%) of the height of the pork chops. The level of lecithin in the frying medium was about

5.6% by weight of the fat or oil in the frying medium. Minimal splattering and no sticking or burning was observed. The resultant foam-fried, batter-coated pork chops had a flaky, golden brown coating and had the taste, texture and appearance of ordinary pan-fried, batter-coated pork chops. The meat was very moist and tender and the pork chops did not shrink to the extent normally observed in ordinary pan-frying.

IN BAKERY PRODUCTS

Sodium Calcium Alginate in Yeast Dough Products

Commercial wheats vary widely in properties depending on the variety and area where grown. They may be loosely grouped into two general headings as hard and soft.

The hard wheats are the types most desirable for bread production. They mill well and yield good quantities of flour that is high in good qualities of protein, from which strong, elastic doughs can be made. These doughs have excellent gas-holding properties and will yield bread with good volume, grain, and texture under a wide range of conditions. The hard wheat doughs have high water-absorptive capacities and have excellent waterholding properties when properly matured.

The soft wheats are used primarily in the production of flour for cake, pastry, cookies, and so forth. They are characterizied, for the most part, as being lower in protein and they yield flours which have low water-absorption capacities and poor tolerance to mixing and fermentation. They handle poorly in bread baking equipment.

In the milling of both hard and soft wheat there is a by-product known as "clear flour" which although containing a high level of protein, is unsuitable for use at high levels in yeast-raised dough compositions.

Alginates have been used in prior work to improve the properties of baked goods. Sodium or ammonium alginate have imparted to breads the characteristics of high gluten breads when the breads were prepared from low gluten flours.

Alginate has been disclosed as one of several emulsifying agents which have been found to be very desirable in retaining the fatty material of the dry mix in dispersion during the baking operation, and especially in retaining air and gas in the mix.

L.G. Fischer, P. Kovacs, A.W. Russell and J.E. Vey; U.S. Patent 4,244,980; January 13, 1981; assigned to Merck & Co., Inc. and DCA Food Industries, Inc. have developed a method of using alkali metal calcium alginates in preparing yeast dough products from soft wheats.

This process is based on the discovery that the addition of from about 0.20 to about 1.00 part of an alkali metal calcium alginate per 100 parts of soft flour and/or clear flour, the pH of which has been reduced to about 6.0 or below, imparts to the resulting yeast-raised dough product gas retention and structure-forming properties comparable to yeast-raised dough compositions made entirely with high-quality hard wheat flour.

It is a requirement of this process that both the soft wheat flour and/or the clear flour be treated to reduce its pH to about 6.0 or lower. A preferred method of treating the flour to reduce its pH to about 6.0 or lower is to bleach the flour by means of chlorine gas. Preferably, the flour is treated to reduce the pH to from about 4.3 to about 6.0, most preferably to a pH of from about 4.5 to about 5.8.

It is another requirement that the soft wheat flour and the clear flour each have a minumum of 6.0 weight percent wheat protein, that is to say, gluten.

The alkali metal calcium alginate is preferably sodium calcium alginate but other salts such as potassium calcium alginate and ammonium calcium alginate may also be used. The mixed alkali metal calcium salts may be prepared, for example, by reacting alginic acid with an excess of a sodium salt and thereafter reacting the resulting sodium alginate with a calcium salt, or by reacting alginic acid with a calcium salt in the presence of a lower alkanol and reacting the resulting calcium alginate with a sodium salt. Alternatively, the alginate compositions may comprise finely ground calcium alginate in admixture with finely ground alkali metal alginate. The sodium calcium alginate which contains from about 5.5 to about 7.5 weight percent sodium and from about 2.5 to about 3.5 weight percent calcium is preferably used in a readily soluble form.

This process can be applied in preparing yeast-raised doughnuts. A typical doughnut recipe comprises about 100 parts by weight flour containing 0 to 60 weight percent hard flour, from 0 to 100 weight percent bleached soft wheat flour, and from 0 to 70 weight percent bleached clear flour and, relative to the flour, from about 5 to 8 parts by weight sugar, from about 5 to 6 parts by weight of a triglyceride, from about 1.5 to 2.0 parts by weight milk solids, from about 1.0 to 1.5 parts by weight monoglycerides or diglycerides or a mixture thereof, from about 1.25 to 1.50 parts by weight salt, from about 1.25 to 1.50 parts by weight chemical leavening agent, from about 0.25 to 1.0 part by weight egg yolk, from about 0.15 to 0.35 part by weight bromate blend, and per 100 parts of bleached soft wheat flour and/or bleached clear flour, from about 0.20 to 1.00 part of alkali metal calcium alginate.

Fluent Emulsifiers for Shortenings Used in Breads

Fluid shortenings are useful in the preparation of baked goods and bread-making processes. The function of fluid shortenings is similar to plastic shortenings in baking processes, but fluid shortenings are much preferred for use in commercial baking processes due to their ease in handling, pumping, and metering.

An emulsifier concentrate which is mechanically dispersible into a comestible mixture has been developed by *I. Gawrilow; U.S. Patents 4,137,338; January 30, 1979 and 4,234,606; November 18, 1980; both assigned to SCM Corporation.* This process is particularly applicable to the preparation of an emulsifier concentrate which is mechanically dispersible into a fluent edible oil to produce a liquid shortening.

This fluent emulsifier concentrate comprises a normally liquid partial glycerol ester food emulsifier vehicle having stably dispersed therein ethoxylated fatty acid esters and solid phase fine crystalline food emulsifier components, the concentrate being mechanically dispersible into a comestible mixture for the emulsification thereof.

By "stably dispersed," is meant that the solid phase emulsifier components remain at least in part in suspension and in solid phase in the partial glycerol ester vehicle, whereas the ethoxylated fatty acid esters may or may not be in suspension. That is, they may be in whole or in part dissolved in the emulsifier vehicle.

In a particular embodiment, the fluid concentrate is mechanically dispersible into an edible oil, such as a vegetable or animal derived oil, in the preparation of a fluid shortening suitable for bread and dough mixes providing improved anti-staling and dough-conditioning properties.

Usable liquid partial glycerol esters include both pure monoesters of glycerin and mixtures of monoesters and diesters of glycerin. They are generally a mixture of unsaturated and saturated glycidyl esters of fatty acids typically derived from hydrogenated and nonhydrogenated vegetable oils such as soybean oil, corn oil, olive oil, peanut oil, safflower oil, cottonseed oil, palm oil and like vegetable oils, and animal fats such as tallow and lard.

The ratio of monoglycerides to diglycerides typically is about 40 to 60 weight percent monoglyceride to about 35 to 45 weight percent diglyceride and minor amounts of 5 to 14% triglycerides. They have an iodine value in the broad range of about 40 to 150 although the mono- and diglycerides derived from vegetable oils preferably have an iodine number between about 65 to 150. The preferred iodine number range of the mono- and diglycerides is between about 40 and 85. They are fluid in consistency. The acid number of the mono- and diglycerides is less than 2 and the peroxide value should be less than 1 in accordance with conventional specifications of mono- and diglycerides commercially available.

The class of ethoxylated fatty acid esters useful in the concentrate are the ethoxylated fatty acid esters of glycerol, hexitol, hexitan and isohexide, as well as the fatty acid esters of ethoxylated glycerol, hexitol, hexitan and isohexide. A preferred class of compounds are the ethoxylated mono- and diglycerides, which are the polyethoxylated fatty acid esters of glycerol, and may be conventionally described as a mixture of stearate, palmitate, and lesser amounts of myristate partial esters of glycerin condensed with about 18 to 22 mols, preferably about 20 mols, of ethylene oxide per mol of alpha-monoglyceride reaction mixture. The fatty acid radicals of ethoxylated monoglycerides preferably are higher fatty acid chains having about 12 to 18 carbons.

Representative fatty acid mono- and diesters of glycerin from which the ethoxylated monoglycerides are derived are glycerol monostearate, glycerol distearate, glycerol monopalmitate, glycerol dipalmitate, glycerol monooleate, glycerol dioleate, and others.

The solid phase crystalline food emulsifier component can be any beta-forming emulsifying ingredient which remains in whole or in part suspended and in solid phase form in the glycerol monooleate, in the presence of the ethoxylated fatty acid esters, in the proportions as defined by this process. The term "beta-forming" describes the predominant crystal form adopted by the solid phase component on rapid chilling of the concentrate system from a melt form at an elevated temperature, followed by agitation for a prolonged period of time to develop and maintain a homogeneous dispersion of the desired beta-crystals in the fluid concentrate.

Examples of functional solid phase fine crystalline food emulsifier components, which are beta-forming, are an alkali or alkaline earth metal salt of an acyl lactylate, or alternatively a succinylated mono- and diglyceride.

The described partial glycerol ester, ethoxylated fatty acid esters and solid phase ester components are contained in the fluid concentrate in a wide range of weight ratios, preferably varying from about 10 to 70% glycerol ester to about 90 to 30% of other emulsifier components including the solid phase ester emulsifiers.

A preferred formulation comprises about 10 to 70 weight percent liquid partial glycerol ester, about 15 to 72 weight percent ethoxylated fatty acid ester, the remainder being essentially normally crystalline food emulsifier, the ratio of solid phase ester emulsifier to ethoxylated fatty acid ester not exceeding about 1.5, the proportions being such as to produce a fluent cloudy suspension. By "fluent" it is meant flowable or pumpable. By "cloudy" it is meant other than clear. In the case of the use of an acyl lactylate salt, the ratio of such salt to ethoxylated fatty acid ester should not exceed about one. A preferred formulation comprises about 40% glycerol monooleate vehicle in which is suspended and dissolved 30% sodium stearoyl-2-lactylate and about 30% ethoxylated mono- and diglyceride, to provide a cloudy, viscous but flowable or pumpable concentrate.

Another suitable formulation comprises about 10% glycerol monooleate vehicle in which is suspended or dissolved about 45% ethoxylated mono- and diglyceride and about 45% succinylated mono- and diglyceride.

Preferably, the emulsifier concentrate is processed by physically, uniformly blending the respective emulsifier components at about 130° to 150°F, sufficient to liquefy all of the components, using a high-speed mixer. The concentrate is then subjected to rapid indirect chilling or nucleating to initiate beta-crystal formation within the concentrate, and then is mildy agitated and mixed following known procedures or techniques to permit the substantially complete development of the desired beta-crystal form of the beta-phase-forming emulsifier component.

In a preferred procedure, the concentrate is subjected to quick chilling in a swept-surface heat exchanger such as a Votator "A" type unit chilling machine. The primary function of the Votator "A" unit is to quick-chill the concentrate to initiate the formation of beta-crystals. The chilled blend is then passed to a Votator "B" unit wherein the blend is mildly agitated or worked to produce a product of desired consistency. The primary function of the "B" unit is to allow the beta-crystallization to further develop. The exit temperature of the "B" unit should be maintained between about 70° and 100°F. As the "B" unit has no cooling capability, the "A" unit exit temperature is maintained in whatever range is necessary to achieve the desired "B" unit exit temperature.

Following votation, the resulting stable fluent concentrate is subjected to a prolonged mild agitation for an extended period to achieve substantially complete conversion to or development of the beta-crystal form, on the order of at least 90% conversion, and preferably 95 to 100% conversion. This agitation may be carried out for a period of from about 18 to 48 hours, and is carried out at a temperature in the range of about 70° to 100°F, preferably ambient temperature or about 75° to 85°F. The agitation may, if desired, be carried out in a jacketed kettle through which a coolant (e.g., tap water) is circulated to effect the purpose of abstracting heat generated by the mechanical agitation and crystallization of the mass. One suitable jacketed vessel is known as a stehling tank.

Example: Two food emulsifier concentrates were produced in accordance with this process and were evaluated in continuous mix bread by adding the concentrate to liquid soybean oil and then from that blend preparing a fluid bread shortening. The two emulsifier concentrates were: (A) 90% glycerol monooleate, 5% sodium stearoyl-2-lactylate and 5% ethoxylated mono- and diglyceride; and (B) 40% glycerol monooleate, 30% sodium stearoyl-2-lactylate and 30% ethoxylated mono- and diglyceride.

These two concentrates were compared against liquid soybean oil without any emulsifying agents as a control. The concentrate systems were utilized at a level of 0.5% based upon the weight of the flour in the bread, the total lipid content being about 3.0% based on the weight of the flour. The control used 3.0% oil only.

Both of the concentrates (A) and (B) exhibited higher specific volume, less percent shock loss, higher bread score and smaller Instron readings (softness) after 3 and 7 days than the control. Concentrates (A) and (B) were substantially equivalent in Instron readings, but concentrate (B) exhibited significantly less shock loss than concentrate (A), and had a slightly higher specific volume. Concentrate (B) also had a better bread score. On the basis of shock loss and bread score, concentrate (B) is the preferred form of this process.

Soybean stearine is added to formulation of the emulsifier concentrate disclosed in U.S. Patent 4,137,338 in the modification disclosed by *I. Gawrilow; U.S. Patent 4,226,894; October 7, 1980; assigned to SCM Corporation.*

More specifically, the fluid shortening of this process comprises on a weight basis about 40 to 70 parts of a liquid vegetable oil and about 8 to 29 parts of an emulsifier fraction or concentrate consisting essentially of about 4 to 10 parts normally liquid or soft partial glycerol ester food emulsifier vehicle into which is stably dispersed about 2 to 8 parts ethoxylated fatty acid ester, about 2 to 8 parts solid phase fine crystalline food emulsifier component, and about 0 to 3 parts soybean stearine, blended with water in the proportion of about 25 to 55% shortening to about 75 to 45% water.

Preferably the solid phase ester emulsifier is a succinylated mono- and diglyceride or alkali or alkaline earth metal salt of an acyl lactylate. Preferred glycerol ester vehicles are a fluent monoglyceride such as glycerol monooleate, or a soft mono- and diglyceride.

The advantage of this process is that the foregoing emulsifiers can be mixed together to make a stable concentrate and then shipped to a remote location for blending with an edible oil and water to make the shortening. For this purpose, the concentrate preferably contains about 10 to 70% glycerol ester and about 90 to 30% of other emulsifier components including the solid phase ester emulsifiers. A preferred formulation comprises about 10 to 70 weight percent liquid partial glycerol ester, about 15 to 72 weight percent ethoxylated fatty acid ester, and about 4 to 54% normally crystalline food emulsifier. The ratio of solid phase ester emulsifier to ethoxylated fatty acid ester preferably does not exceed about 1.5, the proportions being such as to produce a fluent cloudy suspension.

Example: A fluid shortening was produced having the following components: 81.45% refined soybean oil; 6.50% mono- and diglycerides (50% minimum α-monodiglyceride, Dur-em 204); 4.90% ethoxylated mono- and diglycerides (Durfax EOM); 4.90% sodium stearoyl-2-lactylate (Emplex); and 2.25% soybean stearine.

The emulsifier ingredients (minus the soybean oil) were charged into a holding tank and heated to a temperature in the range of about 130° to 150°F and agitated at this temperature until a molten mixture was obtained. The ingredients were then subjected to quick chilling in a Votator "A" unit having an exit temperature in the range of about 80° to 90°F, the chilling being at a rate sufficient to initiate beta-crystallization. The mixture was then subjected to mild agitation in a Votator "B" unit, having an exit temperature of 70° to 100°F, to promote further beta-crystallization. Following this, the ingredients were subjected to prolonged agitation (stehling) at a relatively constant temperature, between about 70° to 100°F, until beta-crystallization was substantially complete. The stehling time was 18 to 48 hours.

Subsequent to manufacture of the concentrate, the ingredients were again charged into a holding tank, mixed with the soybean oil, and heated to a temperature of about 140° to 145°F and agitated at this temperature until a molten mixture was obtained. To this mixture was added water in the proportion of about 60% water to about 40% shortening (above). Mixing was conducted using a high-speed mixer, after which the emulsified mixture was then pumped to a first Votator unit "A" wherein the mixture was rapidly cooled to a chilled blend. The exit temperature of the Votator unit "A" was 65° to 75°F. The mixture was then pumped into a Votator "B" unit where it was maintained at 70° to 80°F with mild agitation to produce a stabilized uniform dispersion. Holding time of the chilled blend within the Votator "B" unit was about one minute.

The hydrated fluid shortening had a stable viscosity profile with regard to time and temperature, providing good handling properties.

Alternatively, the water, emulsifier and oil may be stably blended by physically mixing the ingredients using a conventional high-speed mixer.

Mixed Glycerides for Pastas

J.L. Suggs and D.F. Buck; U.S. Patent 4,229,488; October 21, 1980; assigned to Eastman Kodak Company provide a pasta or potato conditioner comprising a wettable powder composed of both saturated and unsaturated vegetable oil. This additive, referred to as a starch complexing agent, is readily dispersible in pasta products and potatoes, overcoming problems of stickiness, pastiness and sliminess when no such additives are used. The starch complexing agent prevents these undesirable properties by inhibiting the starch from swelling and absorbing water.

The conditioning additives of this process comprises from about 25 to 95% distilled monoglycerides prepared from saturated vegetable oil and from about 75 to 5% distilled monoglycerides prepared from unsaturated vegetable oil. The resulting conditioning additive has an iodine value of between about 5 and 35, a melting point of above about 50°C and is capable of forming a dry, free-flowing powder which is sufficiently wettable to form a particulate dispersion in water at 25°C. Preferably, the iodine value is between about 20 and 35, and the wettability value is between 0 and 40. The percentages are by weight, based on the total weight of the saturated and unsaturated monoglycerides.

At least about 5% of unsaturated monoglyceride must be present to obtain adequate dispersibility, i.e., a wettability value of between 0° and 80°C. Up to about 75% of unsaturated monoglyceride may be present without any significant decrease in conditioning properties (stickiness, pastiness, or sliminess).

The preferred saturated monoglyceride is prepared from saturated soybean oil, but palm oil, cottonseed oil, peanut oil, sesame oil and the like, which have been substantially fully hydrogenated, may be used. Commercially available saturated monoesters include Myverol 18-00, 18-04, 18-06 and 18-07 distilled monoglycerides (Eastman Chemical Products, Inc.).

A preferred unsaturated monoester is a distilled monoester product made from cottonseed oil. Other useful unsaturated monoesters include the distilled monoester made from palm oil, soybean oil, peanut oil, corn oil, sesame seed oil and the like. Commercially available unsaturated monoglycerides include Myverol 18-30, 18-35, 18-40, 18-50K, 18-85 and 18-98 distilled monoglycerides.

The powder may be prepared by melt-blending the ingredients until a homogeneous mass is obtained and then forming a powder from the mass. Melt blending may be accomplished by individually maintaining or raising the temperatures of the compounds to a point above their respective melting temperatures so each is a molten mass and then thoroughly blending, or by mixing the ingredients at room temperature and then raising the temperature of the mixture at least to the melting point of the highest-to-melt ingredient followed by thoroughly blending to form a homogeneous mass. Preferably, melt blending is accomplished at a temperature of between about 80° and 120°C. Powdering may be accomplished by conventional means, e.g., spray chilling, freezing and pulverizing, etc.

Laboratory experiments indicate that the quality of product produced on small scale by powdering in a blender using dry ice is quite satisfactory. Such powdering is accomplished by first heating a mixture of the selected ingredients until a molten or liquefied mass is formed, and then rapidly stirring until the mass is homogenized. For example, 100 g of molten mixture in a 250 ml beaker may be stirred until the mixture is found to be homogeneous. The mixture may then be poured out and cooled until solidified, typically for about 3 to 4 hours at room temperature. The solid may then be powdered in a high-speed stirring device such as a Waring Blendor using dry ice. The dry ice is subsequently evaporated and the powder residue sieved to an approximate size of 50 to 300 microns.

The total monoglycerides concentration in the alimentary pasta or potato product is in a range from about 0.3 to 2% by weight of the flour content (considered in the as-obtained or as-added condition).

The pasta products are prepared by contacting the flour or potato component with the monoglyceride composition at any stage following the formation of the flour up to and including the point of formation of the alimentary pasta product. The monoglyceride composition can be added as a dry powder to the flour or potato or at the blender before the kneading operation.

Example: Fifteen pounds of monoglyceride powder prepared from fully hydrogenated palm oil and monoglyceride prepared from refined palm oil in a 1:1 ratio, having an iodine value of 21, MP of 65°C and wettability value of 30 is mixed with 1,000 pounds flour and dry mixed for 5 minutes. It is transferred to an extruder and mixed with water to a moisture content of about 28%, making a stiff dough. The dough is extruded at about 2,000 psi to form spaghetti and dried.

The spaghetti is overcooked (cooked 20 minutes in boiling water) and examined. Subjective tests indicate an absence of stickiness, absence of surface starch, very little water absorption, and in general, excellent eating characteristics. This product was then compared with spaghetti to which no starch complexing agent had been added and also cooked 20 minutes, and the subjective tests mentioned above indicated a very significant advantage with use of the monoglyceride starch complexing agent.

Mixed Glycerides for Bakery Products

J.L. Suggs and D.F. Buck; U.S. Patent 4,229,480; October 21, 1980; assigned to Eastman Kodak Company also provide dry, powdered emulsifiers which can be incorporated directly into dough used for bakery products such as bread, buns, pastries and other yeast-raised baked goods without first being dispersed in a carrier.

In accordance with this process, monoesters unsaturated to the extent of having an iodine value of between about 5 and 33 and a melting point of above 50°C, are found to form a dry, free-flowing powder and be sufficiently wettable to form a particulate dispersion in water at 25°C. Normally, to prepare the monoesters, a solid solution of (1) saturated monoesters, or mixture of saturated monoesters and succinylated monoglycerides and (2) unsaturated monoesters is prepared and used as a powdered emulsifier.

The unsaturated monoesters are prepared from glycerin and/or propylene glycol and fatty acids which are usually obtained from edible vegetable oil or animal fat containing unsaturated or partially hydrogenated fatty acids as further described herein. The saturated monoesters are prepared from glycerin and/or propylene glycol and fatty acids which are usually obtained from edible vegetable oil and/or animal fat containing saturated fatty acids. This solid solution is useful in enhancing the physical properties of bakery products, and is normally mixed in the dough or sponge.

The solid solution preferably contains at least about 80 to 90% of the above compounds. As a practical matter, however, diesters, triesters, unreacted material, etc., may be found in the solid solution. The solid solution is characterized, however, by the absence of a carrier.

A preferred saturated monoester is a distilled monoester product made from fully hydrogenated lard with an iodine value in the range from about 0.4 to 1. Other preferred saturated monoester products include the distilled monoester products made from fats such as tallow, palm oil, cottonseed oil, soybean oil, peanut oil, sesame oil and the like, which have been fully hydrogenated, the distilled monoester products made from such saturated fatty acids as palmitic acid, stearic acid and the like, and blends of distilled monoester products. Commercially available saturated monoesters include Myverol 18-00, 18-04, 18-06 and 18-07 distilled monoglycerides (Eastman Chemical Products, Inc.).

A preferred unsaturated monoester is a distilled monoester product made from lard. Other preferred unsaturated monoesters include the distilled monoester made from fats such as tallow, cottonseed oil, palm oil, soybean oil, peanut oil, corn oil, sesame seed oil and the like, the distilled monoester products made from such unsaturated fatty acids as oleic acid, linoleic acid and the like, and blends of these distilled monoesters. Commercially available unsaturated monoglycerides include Myverol 18-30, 18-35, 18-40, 18-50K, 18-85 and 18-98 distilled monoglycerides.

Succinylated monoglyceride which may be used in the solid solution is a half ester of succinic acid and of a monoacylated polyhydric alcohol, or its salt.

Stearic acid is the preferred fatty acid residue in the succinylated monoglyceride, but myristic, palmitic and behenic acids provide satisfactory residues. The fatty acid residue should be substantially completely hydrogenated. Also, the preferred polyhydric alcohol residue is a glycerol residue, thus providing a free hydroxyl group on the glycerol residue. However, a propylene glycol residue provides a compound having no free hydroxyl group and which is a satisfactory compound. Sorbitol or sorbitan provide compounds having two or more free hydroxyl groups and are satisfactory.

Succinylated monoglyceride known as Myverol SMG Type V (Eastman Chemical Products, Inc.) is commercially available.

The emulsifier mixtures of this process are also prepared by the method described in U.S. Patent 4,229,488 above.

The powder produced as described above is found to be especially useful in the production of bread. The powder is most conveniently added to the dough, but, of course, could be added to the sponge or brew if desired. The powder is preferably added in amounts of between about 3 and 12 ounces per 100 pounds of flour.

The sponge-dough method of bread production involves the mixing of part of the ingredients to form a sponge which is then fermented for approximately 4 hours. The sponge is then mixed with the remaining ingredients until the desired gluten structure is obtained. After the second mixing period and a suitable rest period, the dough is divided for further processing. The sponge-dough method produces bread which has better volume and flavor than bread produced with the continuous-mix method while the latter method is notable in that it produces bread with very fine grain at a lower manufacturing cost.

Example: A typical sponge bread was used for the following tests. The bread volume of a bread prepared using no emulsifier was 2,256 ml as determined by conventional methods using rapeseed displacement technique. However, when an approximate 1:1:1 powdered mixture of (1) Myverol 18-04 distilled monoglyceride, prepared from fully hydrogenated palm oil, (2) Myverol 18-35 distilled monoglyceride, prepared from partially hydrogenated refined palm oil (unsaturated), and (3) Myverol SMG Type V succinylated monoglyceride, prepared from fully hydrogenated vegetable oil was used the bread volume was increased to 2,587 ml.

Mixed Emulsifiers for Cakes

Emulsifiers useful in food products such as cakes, icings, cream fillings, whipped toppings, etc., are disclosed by *J.L. Suggs, D.F. Buck and H.K. Hobbs; U.S. Patent 4,310,557; January 12, 1982; assigned to Eastman Kodak Company.* The emulsifiers comprise a blend of monoglycerides, propylene glycol monoesters and an alkali metal salt of a fatty acid. The emulsifiers are essentially totally active and do not require the use of a carrier.

In detail this composition is in particulate form and the particles comprise a homogeneous blend of (a) about 20 to 40% by weight (preferably about 25 to

35%) of monoglycerides having an iodine value of about 2 to 15, (b) about 40 to 70% by weight (preferably about 50 to 60%) of propylene glycol monoesters, and (c) about 5 to 20% by weight (preferably about 12 to 18%) of an alkali metal salt of at least one fatty acid having 8 to 22 carbon atoms.

The monoglycerides have an iodine value of about 2 to 15 and are prepared by reacting glycerin with straight chain fatty acids such as those found in vegetable oils and animal fats having from 8 to 22 carbon atoms, and saturated to an extent to result in an iodine value of about 2 to 15. Such monoglycerides known as Myverol 18-06 are commercially available. The monoglycerides may be made by esterifying various fatty acids, and then blending to obtain the desired iodine number. On the other hand, acids with the proper degree of saturation may be selected to result in an iodine value of about 2 to 15.

The propylene glycol monoesters are prepared by reacting propylene glycol with straight chain fatty acids such as those found in vegetable oils and animal fats having from 8 to 22 carbon atoms. Preferred monoesters include Myverol P-06 distilled propylene glycol monoesters.

Alkali metal salts of fatty acids having 8 to 22 carbon atoms are well known and available commercially. Potassium and sodium palmitate, and potassium and sodium stearate are examples of suitable salts. Potassium stearate is preferred.

As a practical matter, the monoglycerides and propylene glycol monoesters referred to above may contain diesters, triesters, unreacted material, and the like. It is preferred, however, that they be of a purity of at least 60%. Such compositions conventionally contain such material even though they are commonly referred to in the art as monoglycerides and monoesters.

A preferred saturated monoester is a distilled monoester product made from fully hydrogenated lard with an iodine value in a range from 0.4 to 1. Preferred monoester products include the distilled monoester products made from fats and oils such as tallow, palm oil, cottonseed oil, soybean oil, peanut oil, sesame oil, and the like.

The emulsifier mixture in powder form is prepared as described in U.S. Patent 4,228,488.

The powder is white, free flowing, and the particles consist essentially of a homogeneous blend of the monoglycerides, propylene glycol monoesters and alkali metal salt of fatty acid.

The emulsifier is generally used in products such as cake batters at a level of about 1 to 5% based on the weight of flow and in products such as icings, cream fillings, whipped toppings at a level of about 1 to 4% based on the weight of shortening. The emulsifiers cause aeration of the food products as indicated by good volume, low density and desirable physical characteristics such as texture.

Similar emulsifier systems which also contain salts of fatty acids have also been disclosed by *J.L. Suggs, D.F. Buck and H.K. Hobbs; U.S. Patent 4,310,556; January 12, 1982; assigned to Eastman Kodak Company* .

This dry emulsifier composition comprises (a) about 15 to 40 parts by weight of succinylated monoglycerides, (b) about 25 to 45 parts by weight of monoglycerides having an iodine value of about 2 to 15, (c) about 15 to 40 parts by weight of propylene glycol monoesters, and (d) about 5 to 18 parts by weight of an alkali metal salt of at least one fatty acid having 8 to 22 carbon atoms.

A preferred saturated monoester is a distilled monoester product made from fully hydrogenated lard with an iodine value in a range from about 0.4 to 1. Other preferred saturated monoester products include the distilled monoester products made from fats such as tallow, palm oil, cottonseed oil, soybean oil, peanut oil, sesame oil and the like, which have been fully hydrogenated, the distilled monoester products made from such saturated fatty acids as palmitic acid, stearic acid and the like, and blends of distilled monoester products. Commercially available saturated monoesters include Myverol 18-00, 18-04, 18-06 and 18-07 distilled monoglycerides.

A preferred unsaturated monoester is a distilled monoester made from lard. Other preferred unsaturated monoesters include the distilled monoester made from fats such as tallow, cottonseed oil, palm oil, soybean oil, peanut oil, corn oil, sesame seed oil and the like, the distilled monoester products made from such unsaturated fatty acids as oleic acid, linoleic acid and the like, and blends of these distilled monoesters. Commercially available unsaturated monoglycerides include Myverol 18-30, 18-35, 18-40, 18-50K, 18-85 and 18-98 distilled monoglycerides.

Succinylated monoglyceride which may be used in the solid solution is half ester of succinic acid and of a monoacylated polyhydric alcohol, or its salt.

Stearic acid is the preferred fatty acid residue in the succinylated monoglyceride, but myristic, palmitic and behenic acids provide satisfactory residues. The fatty acid residue should be substantially completely hydrogenated. Also, the preferred polyhydric alcohol residue is a glycerol residue, thus providing a free hydroxyl group on the glycerol residue. However, a propylene glycol residue provides a compound having no free hydroxyl group and which is a satisfactory compound.

Succinylated monoglyceride known as Myverol SMG Type V is a commercially available product.

As an example, the emulsifier composition is made by blending the following dry ingredients: 40 parts by weight monoglyceride (iodine value of 2); 30 parts by weight succinylated monoglyceride; 30 parts by weight propylene glycol monoester; and 18 parts by weight potassium stearate.

The solid components are melted together, mixed as liquids, cooled to a solid mass then powdered. The emulsifier is used in cake batters, icings and cream fillings. The emulsifier is added over-the-side to the batter.

The emulsifier is generally used in cake batters at a level of about 1 to 5% based on the weight of flour and in icings, cream fillings, whipped toppings at a level of about 1 to 4% based on the weight of shortening.

Emulsifier Blends for Cake Mixes

A blend of commercial grade emulsifiers suitable for use in dry prepared cake mixes has been developed by *J.E. Morgan, A.J. Del Vecchio, B.L. Brooking and D.M. Laverty; U.S. Patent 4,242,366; December 30, 1980; assigned to The Pillsbury Company.*

The emulsifier blend of this process comprises a combination of food-grade emulsifiers wherein the functional emulsifying components consist essentially of three partial fatty acid esters of polyhydric alcohol in predetermined amounts, i.e., 45 to 68% propylene glycol monoester (PGME), 20 to 39% alpha-monoglyceride (MONO) and 9 to 22% glyceryl lacto ester (GLE).

The PGME component in the emulsifier blend is preferably supplied by a commercially available food-grade emulsifier containing from about 30 to 60% PGME, most preferably 50 to 60% PGME. These emulsifiers may be prepared by reacting dihydric alcohols with higher fatty acids, or fats containing fatty acids.

Particularly useful are the food-grade emulsifiers known as Durpro 107, Durpro 187 and EC-25 (Durkee Foods Division of SCM Corp.). Durpro 187 is a preferred emulsifier and comprises from 50 to 60% PGME and 10 to 15% MONO. It has a capillary melting point of 117° to 125°F and a hydrophilic-lipophilic balance (HLB) value of 2.5.

The MONO component of the emulsifier blend is preferably supplied by a commercially available food-grade, monodiglyceride emulsifier. This emulsifier may be prepared by known procedures involving the reaction of glycerin with certain fats, oils or fatty acids and is usually available as the crude reaction product which also comprises diglyceride, triglyceride, glycerin and free fatty acid.

As examples of food-grade emulsifiers which have been found to be particularly useful as the source of the MONO component in the emulsifier blend are the Dur-Em emulsifiers (Durkee Foods Division of SCM Corp.) and the Myverol emulsifiers (Eastman Kodak Co.). Dur-Em 104 is a preferred emulsifier and comprises at least 40% alpha-monoglyceride and no more than 1% free fatty acid and 1% free glycerin. This emulsifier has a capillary melting point of 115° to 121°F and an HLB value of 65 to 70.

The GLE component of the emulsifier blend is preferably supplied by a food-grade glyceryl lacto ester emulsifier which may be prepared by known procedures involving the reaction of a mono- and diglyceride concentrate with lactic acid. The lactopalmitates are preferred but the lactostearates are also satisfactory and may be used alone or in admixture with the lactopalmitates. A commercially available emulsifier which is particularly useful is Durlac 200, which has at least 24% water-insoluble combined lactic acid (WICLA) and no more than 7% alpha-monoglyceride and 10% free fatty acid. This emulsifier has a capillary melting point of 97° to 103°F, a saponification number of 290 to 305 and an iodine value no greater than 10.

This emulsifier blend provides significant improvements in the moistness and water-holding capability of cakes baked from dry mixes while providing a tender texture and fine even grain structure and excellent eating quality to cake. The emulsifier blend is preferably incorporated in the cake mix at 2 to 7% of the weight of the total dry ingredients in the mix. The improved results obtained with the emulsifier blend of this process are not achievable with any of the individual components alone or with two-component systems using the same or similar emulsifiers, and these results appear to be dependent upon a synergistic effect among the components. The components of the emulsifier blend may be incorporated directly into the cake as individual ingredients or the components may be combined prior to incorporation into a cake mix, or a shortening used therein. The effectiveness of the emulsifier blend is not dependent upon the type of process used to produce the cake mix, and advantageous results may be achieved by utilizing the blend in conventionally prepared cake mixes, those that utilize a spray-dried premix of flour, sugar and emulsifier, high-ratio cake mixes, etc.

Example: A yellow cake mix utilizing an emulsifier blend of this process was prepared from 41.8% flour, 41.7% sugar, 4.0% emulsifier blend, 3.4% leavening agents, 3.0% corn sugar, 2.0% shortening, 2.0% nonfat milk solids, 0.7% salt, 0.5% flavor and 0.5% gums.

The emulsifier blend was prepared by thoroughly mixing together a mixture comprising 43.75% of a food-grade propylene glycol emulsifier comprising about 50 to 60% propylene glycol monoester and about 10 to 15% alpha-monoglyceride (Durpro 187), 25.0% of another food-grade emulsifier comprising about 34 to 38% propylene glycol monoester and about 20 to 22% mono- and diglycerides (EC-25K), and a food-grade glyceryl lacto ester emulsifier having a WICLA value of at least 24%, and no more than 7% alpha-monoglyceride (Durlac 200). The resultant emulsifier blend comprised about 61% propylene glycol monoester (PGME), 24% alpha-monoglyceride (MONO), and 15% glyceryl lacto ester (GLE).

The emulsifier blend and shortening were melted together and held at 135°F, weighed and blended with the sugar at that temperature. The remaining dry ingredients were mixed thoroughly with the emulsifier/shortening mixture in a ribbon blender and then passed through a high-speed finishing machine to provide a homogeneous, free-flowing cake mix.

A cake batter was prepared from the cake mix by adding two eggs and 1⅓ cups of water to 15 ounces (425 g) of the dry mix. The mixture was mixed using a standard household mixer operated at low speed for about one minute and then at high speed for about two minutes to form a thick, fluffy, light-colored batter. The batter was subsequently placed in a standard 9" x 13" cake pan and baked for about 25 to 30 minutes at 350°F.

After cooling for 1½ to 2 hours at room temperature, three height measurements were taken along the longitudinal axis of the cake (one at the center and one on each side) while the cake was in the pan. The sum of the three measurements in millimeters was 79 and this value was considered to be a rough approximation of the cake volume. The resultant baked cake had a fine, even grain with the thin, delicate cell walls characteristic of a high-quality layer cake, a smooth crust surface free of wrinkles and water rings, and a moist, tender texture with a rapid rate of disappearance in the mouth.

Colloidal Solids in Breads and Sponge Cakes

S. Inamine, T. Matsuda and T. Shimomura; U.S. Patent 4,277,512; July 7, 1981; assigned to KK Ueno Seiyaku Oyo Kenkyujo, Japan have also used a colloidal solid prepared from a sugar alcohol, a sugar, a food-grade surface-active agent and an edible oil or fat in bakery products. Such colloidal solids are discussed in more detail in related U.S. Patents 4,168,323 and 4,253,834, where the colloidal solid emulsifiers were used in meat and fish. (See page 149.)

Example 1: Sorbitol (775 g) was heat-melted at 95°C. To the molten sorbitol were added 100 g of sucrose palmitate (HLB = 15) and 75 g of propylene glycol. They were then stirred by a homomixer to a uniform mixture. Then, 50 g of sorbitol powder was added as a seed crystal, and the mixture was cooled, solidified and crystallized. The crystallized product was pulverized to a powder having a size of 24 mesh or under.

Example 2: Sorbitol (850 g) was heat-melted at 95°C, and mixed with 100 g of stearic acid monoglyceride (purified by distillation) and 10 g of glycerin. They were stirred to form a uniform mixture. As a seed crystal, 40 g of sorbitol powder was added. They were mixed, cooled, solidified, crystallized, and pulverized to form a powder having a size of 35 mesh or under.

Example 3: This examples shows the effect of the composition of Example 1 in the production of sponge cakes.

Whole eggs (200 g), 200 g of sugar, 80 g of water, and 20 g of the product of Example 1 as a foamer were put into a mixer, and mixed for 30 seconds at a low to medium speed. Then, 200 g of wheat flour was added, and mixing was further performed at a low to medium speed. The mixture was whipped by performing the mixing at a high speed for 5 minutes. After the whipping, the density of the resulting dough was examined. The dough was placed in a former, and baked at 180°C for 40 minutes. The resulting cake was allowed to stand at room temperature for 1 hour, and then its volume was measured.

For comparison, the above procedure was repeated except that a mere mixture of sorbitol, sucrose palmitate and propylene glycol was used as a foamer instead of the product of Example 1, or that no foamer was used.

The density of the cake obtained using the product of Example 1 as foamer was 0.24 g/cm^3. When no foamer was used the cake density was 0.47 g/cm^3. When a mere mixture of the components used to prepare the foamer of Example 1 was used the density of the cake was 0.37 g/cm^3.

Example 4: In this example, the composition obtained in Example 2 was applied to bread, and the softening effect of the composition of this process was confirmed. Bread was produced under the condition of 700 g wheat flour standard by the sponge-dough method (AACC Method 10-11). The amount of the composition, calculated as the stearic acid monoglyceride, was 0.2% based on the wheat flour.

The bread produced was allowed to stand indoors, and when the temperature of the inside decreased to about 30°C, it was packed in a polyethylene bag. The polyethylene bag was sealed, and allowed to stand indoors at 20°C.

The staling-retarding effect was evaluated by measuring the load, which caused a 2.5 mm strain to a test sample (a 2-cm-thick slice), by means of a compressiometer (AACC Method 74-10).

Hardness on the third day of bread prepared using the colloidal solid of Example 2 was 14.2 g/cm^2. When only a mixture of ingredients was used, the hardness measured 18.2 g/cm^2 and when no additive was used, the hardness was 19.7 g/cm^2.

Powdered Hydrated Emulsifiers

Surface active agents (emulsifiers) are commonly utilized in the food and cosmetic industries for stabilizing and thereby enhancing the physical characteristics of various bakery products, cake icings, shortenings, whipped toppings, and cosmetics.

These emulsifiers are commercially available as powders, plastics or hydrates. As is to be expected, each of these three forms has certain advantages and disadvantages with regard to their handling characteristics and functionalities.

While powdered emulsifiers are most easily handled, they are most difficult to disperse in aqueous end use applications. Accordingly, while easier to handle, powders are the least functional. At the other end of the spectrum are the hydrated emulsifiers. The hydrated emulsifiers offer the best functionality in terms of being readily dispersed in aqueous systems, but are considered extremely inconvenient or messy to handle. The plastic emulsifiers might be termed the "happy medium" of the state of the art. Being semisolid the plastic emulsifiers are slightly more difficult to handle than the powders, but they do not exhibit the functionality of the hydrates.

A method for the preparation of powdered hydrated emulsifiers has been developed by *C.A. Jackson; U.S. Patent 4,159,952; July 3, 1979; assigned to Southland Corporation.*

The procedure for preparing the powdered hydrated emulsifiers is first to melt the chosen emulsifier or mixture of emulsifiers, which is usually in a temperature range of from 120° to 170°F. The melted emulsifier is then pumped through a conduit to a spray nozzle which communicates with a cooling chamber maintained at about 60°F. Just upstream of the spray nozzle water is metered into the flow of melted emulsifier so that the water/emulsifier mixture exits the spray nozzle into the cooling chamber. The water is maintained at a temperature above about 100°F. The flow of water is metered such that the resulting spray-chilled product contains from 2 to 15% by weight water of hydration. The resulting free-flowing powder comprises the powdered hydrated emulsifier composition.

It is believed that spray chilling of these minimal hydrates creates a matrix of water and emulsifier wherein the water is bound to the emulsifier by hydrogen bonding. It is believed that the spraying process first makes an intimate mixture of water and emulsifier and secondly locks the water of hydration into the system by rapid chilling. This theory is supported by the fact that as the particle size of the spray decreases, the limiting water content increases as does the functionality of the powdered hydrated emulsifier composition.

Virtually all of the common emulsifiers capable of hydrogen bonding are suitable for use for the purpose of forming the powdered hydrated emulsifier compositions. Accordingly, the emulsifiers may be defined as consisting of fatty acid partial esters of polyhydric alcohols, half esters of succinic acid of a monoacylated polyalcohol, alkoxylated condensates of monoglycerides, stearoyl 2-lactylate and calcium and sodium salts thereof, alkoxylated fatty acid partial esters of polyhydric alcohol, diacetyl tartaric acid esters of fatty acid partial esters of polyhydric alcohols, and mixtures of these emulsifiers.

Example: One mol of industrial distilled monoglyceride, made from fully hydrogenated fatty acid glycerol esters (principally C_{16} and C_{18}), containing 90% alphamonoglyceride, was reacted with one-half mol of succinic anhydride. The mixture was melted and heated with stirring until maximum formation of the succinic half ester was formed. One mol of water was added to the succinic half ester and the mixture was sprayed through a spray nozzle into a cooling chamber. The exit spray was cooled from about 170° to about 60°F.

The resulting cooled product, containing about 4 wt % water, was a free-flowing white powder. About 50% passed through a standard 20 mesh screen.

The self-emulsification properties of the hydrated powder were examined by placing about 40 g of the screened material into about 120 g of water maintained at about 72°F. The mixture was stirred just sufficiently to disperse the powder in the water. The mixture was then observed with no further mixing.

After about 10 minutes, noticeable swelling of the powder had absorbed virtually all free water, although spherical particles could still be observed. After one hour, self-emulsification had occurred so that all excess water had been absorbed and the spherical nature of the original powder particles could no longer be seen.

In the method for preparing the powdered hydrated emulsifier composition of this example it should be noted that the water was injected into the stream of melted succinic half ester at a point just upstream of the spray nozzle. Experimentation has revealed that if the water is injected into the emulsifier well upstream of the spray nozzle a gel will form which is not sprayable.

When used in bread making, the emulsifier of the above examples produced test breads with excellent grain, texture, volume and softness.

Pulverulent Free-Flowing Monoglycerides

In bread, the monoglycerides prevent or reduce staling and thereby increase softness and shelf life. The theory is that staling to a great extent is caused by the retrogradation of the gelatinized starch, particularly of the amylose fraction of the starch. Amylose forms a helix and a monoglyceride molecule fits into this helix. A monoglyceride molecule entrapped in the amylose helix represents a water-insoluble complex which will not take part in the progressing retrogradation of amylose. Thereby, the staling tendencies are reduced.

In spaghetti, macaroni, noodles and similar products the monoglycerides ensure a firm, nonsticky consistency when the product is cooked.

Monoglycerides are of a polymorphic nature. In the solid state two substantially different crystal forms exist, an unstable alpha crystal form and a stable and more compact beta crystal form. When suspended in a water phase and heated both crystal forms will hydrate and convert into an active state or form with affinity to starch and ability to form a complex with amylose. Monoglyceride powder in alpha crystal form gives the effects referred to above because the alpha form at room temperature easily converts into an active amylose complexing state whereas ordinary monoglyceride powder in the beta crystal form at room temperature is almost ineffective in this respect.

It is, however, very difficult or even impossible to prepare a composition of monoglycerides in the alpha crystal form retaining this crystal form for a reasonable period of time. This problem can be overcome in a rather satisfactory manner by ensuring that the stable beta crystal form is converted into an active form during the baking process for the bread or heating process for the other types of products mentioned. As this conversion is ordinarily limited by the time available and the temperatures prevailing during baking processes or the processes used for preparing macaroni, noodles etc., it is necessary that the beta crystals to convert are very small, i.e., that the monoglyceride product is very finely crystallized with a large surface exposed to the water phase. Furthermore, the monoglyceride must be extremely well distributed in the dough in order to achieve maximum effect thereof.

An improved dry, pulverulent monoglyceride product for use in preparing bread, extruded snack-products, macaroni, noodles and spaghetti has been prepared by *J.A.H. Gregersen; U.S. Patent 4,178,393; December 11, 1979; assigned to A/S Grindstedvaerket, Denmark.*

This can be achieved if the monoglycerides are incorporated into the dough for the products in the form of pulverulent mixture of the monoglyceride and a phosphatide (phospholipid) obtained by mixing the two components together in molten condition and spray-cooling the molten blend.

It has been found that the monoglyceride is present in the composition of this process in a stable physical beta-crystalline form able to convert into the active amylose complexing form during baking processes for making bread or similar products such as cakes, or during manufacturing processes used for making extruded snack products or farinaceous products such as noodles, macaroni, spaghetti. Furthermore, the product is in the form of a fine powder with excellent free-flowing properties which can very easily be admixed with flour and the other ingredients used when preparing the farinaceous product.

The monoglyceride used may be a commercial distilled monoglyceride containing for instance from about 80 or 85% up to about 98% of monoglyceride, for instance, about 90% by weight of monoglyceride (the balance being diglyceride, triglyceride and free glycerol). The fatty acid residues in the monoglycerides may be of any saturated fatty acid having from 12 to 24 carbon atoms. Various distilled monoglycerides on the basis of fully hardened, commercial fats are available and any such commercial product can be used. A very convenient monoglyceride product is a distilled monoglyceride containing about 90% monoglyceride and prepared on the basis of fully hardened lard. The fatty acid residues are about 62% stearic, about 32.5% palmitic, about 3.2% myristic and about 2.3% arachidic acid. As the predominant compound in this product is 1-monostearin, it is often referred to as GMS (glyceryl monostearate).

The phosphatide is preferably a lecithin and particularly preferred are various edible grades of soybean lecithin because such lecithins are readily available.

The monoglyceride normally will constitute the major portion of the combined monoglyceride/phosphatide, especially monoglyceride/lecithin product. Though the ratio is not very critical, it is preferred that the contents of phosphatide is 0.1 to 40%, calculated on the weight of the mixture. The higher contents especially apply when a comparatively crude grade of lecithin is employed.

This blend is prepared by melting a monoglyceride of saturated C_{12-24} fatty acids or a mixture of such monoglycerides, adding at least one substantially dry phosphatide (phospholipid) and spray-cooling the mixture without adding water. The particle size of a spray-cooled product can be influenced by the choice of spraying nozzles or the speed of a spinning disc. It is preferred to conduct the spray-cooling process in a manner so as to ensure a pulverulent product having an average particle size between 10 and 150 μm, especially between 20 and 100 μm and very convenient between 30 and 80 μm, for instance about 50 μm.

Example 1: 1,800 g of distilled monoglyceride, prepared from hardened beef tallow, were heated at 85°C. While stirring, 200 g of soybean lecithin were added to the melt. The lecithin was an alcohol-soluble fraction having the following analysis values: phosphorus, 0.84%; nitrogen, 0.75%; phosphatidylcholine, 27.4%; and phosphatidylethanolamine, 11.5%. The hot mixture of distilled monoglyceride and lecithin was sprayed on a spinning disc at 16,000 rpm in a spray tower (diameter 3 meters) under the simultaneous blowing of air at 15°C. The air under left the tower at a temperature of 30°C. The spray-cooled powder is discharged from the air by a cyclone.

Example 2: 1,640 g of distilled monoglyceride, prepared from hardened lard, was heated at 85°C. 270 g of fractionated soybean lecithin (same type as in Example 1) and 90 g of full fat soybean flour were added under stirring. The hot mixture was cooled as described in Example 1.

When the products of the above examples were used in breads, they produced products with greater volumes and less tendency to staling than control breads made without additive.

Tempered Powdered Emulsifiers

Improved powdered emulsifiers have been described by *T. Fukuda, H. Matsuura, Y. Koizumi and T. Yamaguchi; U.S. Patent 4,315,041; February 9, 1982; assigned to Riken Vitamine Oil Co., Ltd., Japan.* In another process, the same workers described a powdered emulsifier composition comprising a high purity distilled monoglyceride composed of 65 to 85% of saturated fatty acid monoglyceride and 35 to 15% of unsaturated fatty acid monoglyceride and having an iodine value of 10 to 40, and tempered for more than 30 minutes at a temperature of more than 45°C, which does not cause melting of the composition. In the composition, the number of carbon atoms for the saturated fatty acid is in the range of 12 to 22, and that for the unsaturated fatty acid is in the range of 16 to 22.

It was found that, by tempering this emulsifier composition for more than 30 minutes at a temperature which is within the range of from more than 45°C to a temperature which does not melt the composition, the properties of the composition are improved drastically and the preservation stability is also improved. If the tempering temperature is below 45°C, it takes longer to attain the required properties and the crystal sizes become coarse. Therefore, a lower temperature cannot satisfy both the production and expected property requirements.

When fatty acid monoglyceride is solidified by cooling, it forms the stable beta-crystalline form after undergoing successive crystal formation, from each of sub-alpha, alpha and beta' forms. But as the transition time becomes longer, the crystals thus obtained become coarser. Therefore, it is necessary to finish the transition as rapidly as possible in order to obtain minutely sized crystals. The most effective method is to use solvent such as water or alcohol.

The above emulsifier composition exerts prominent effects. But, since the product is made structurally of saturated fatty acid monoglyceride, the crystal surface of which is surrounded by a thin film of unsaturated fatty acid monoglceride, there sometimes arises problems such as caking or stickiness which develops on the surface of handling equipment under preservation and distribution environments at certain temperatures.

In order to overcome these shortcomings, this process solves these problems by including in the emulsifier composition described above, 5 to 20% of fat that has a melting point which is more than the tempering temperature of the monoglyceride composition.

Since the advantages of the properties of this emulsifier composition can be exerted principally by forming a complex with starch, its performance can be evaluated by Blue Value (BV). BV can be obtained by utilizing the reaction of iodine and starch. When the starch forms a complex with fatty acid monoglyceride, the color of starch is disturbed by iodine. Accordingly, the degree of complex formation, in other words, functional properties of the emulsifier composition can be evaluated by measurement of BV using spectrophotometry analysis. Therefore, the lower the BV, the more complex forming ability, indicating the excellence of the functional properties.

BV of less than 0.300 as indicated in the embodiments is adequate for actual application.

Example 1: A mixture of 70 parts of saturated fatty acid monoglyceride and 30 parts of an unsaturated fatty acid monoglyceride was melted and after spray cooling, a powder with a medium size of 60 mesh was obtained. This was tempered at 45°C for 4 days and the temperature lowered to room temperature. The first mix had an iodine value of 31.1 and a BV of 0.085.

Example 2: Fixed amounts of a fully hardened beef tallow (MP 61°C) was mix-melted with an amount of the first mix of Example 1 to provide a product having a mix ratio of fats of 10. The product was sprayed under a normal temperature and powdered (product temperature: 30° to 34°C). The temperature of products thus obtained rose to 47° to 53°C by the generated heat automatically under static condition. Following this, the products were tempered for 30 minutes at 50°C and cooled to normal temperature. This product had a BV value of 0.250.

Example 3: Bread was made from the product of Example 2 by the AACC sponge dough method. The addition of emulsifier composition of Example 2 was 0.4%. After 3 days, aging tests on the bread using the Texturometer gave a reading of 1.24 kg/10 mm for the bread prepared with the emulsifier of Example 2. Bread made with no added emulsifier gave a reading of 1.65 kg/10 mm while a bread using a commercially available emulsifier gave a reading of 1.27 kg/10 mm.

IN OTHER FOODS

Cleaning Agents for Foods

The surface of food and fodder of both plant and animal origin which are generally available are for the most part more or less seriously contaminated with both physiologically and toxicologically active noxious substances. These substances reach the food in the form of pesticides, fertilizers, preservatives, etc. Further, dirt and other noxious substances unavoidably reach the food and fodder, e.g., in the form of of industrial emissions, fuel residues, oil fumes, and road dust, etc.

The development by *E. Bossert and M. Aumann; U.S. Patent 4,140,649; February 20, 1979* relates to a method for cleaning the surfaces of foods and fodder of plant and animal origin by manual or mechanical washing in water.

It is generally known that cleaning agents have a stronger effect at higher temperature in the cleaning bath. However, in view of individual types of perishable foods, cleaning at low temperatures, or lukewarm temperature is preferred. The objects are achieved with compositions wherein at least one substance from each of the groups below is included.

(a) 1 to 10% by weight anionic tensides, selected from the group consisting of sulfates, ether sulfates, sulfonates, sulfoacetates, sulfosuccinates, sarcosinates and protein condensation products. Olefin sulfonates and lauryl sulfate are particularly suitable because they are nearly tasteless. In particular, the solubilizing effect of anionic tensides and the wetting effect of nonionic tensides favor the objects of this process.

(b) 0.1 to 4% by weight nonionic tensides, such as polyoxyethylene adducts. Additives of nonionic surfactants serve primarily to improve cold wettability and solubilization. When combined with anionic active ingredients, there is a pronounced improvement in the cleaning power when cold.

(c) 1 to 25% by weight sequestering agent. Complexing of noxious elements is of great importance for this process. Particularly advantageous is ethylene dinitrilotetraacetic acid as well as salts, since it is physiologically relatively harmless. Its metal complexes retain sufficient stability even in slightly acid solutions to dissolve metal oxides and other compounds, e.g., mercury, tin, nickel, cobalt and zinc, and to keep them in solution. In addition, the stability of ascorbic acid is increased by masking of iron and copper. Other suitable sequestering agents include hydroxycarboxylic acids such as tartaric acid, citric acid and luconic acid, as well as D-sorbitol and mannitol.

(d) 1 to 4% by weight of substances controlling the redox state.

(e) 50 to 30% by weight buffers. The wash water advantageously should have a pH value between 5 and 8, especially about 6. This may be achieved by adding hydroxycarboxylic acids or their alkali salts.

(f) 0.1 to 0.5% by weight preservatives.

(g) Balance up to 100% by weight composed of stabilizers, processing additives, extenders and diluents.

(h) Reductones: It is advantageous to have a reducing aqueous medium in order to protect the food for a short period of time against oxidative effects during cleaning, since it affects subsequent preparation. The color of the food is remarkably freshened thereby, and in the case of slightly damaged vegetables, discoloration by chemical and enzymatic oxidation is retarded.

(i) Disinfectants: It is not feasible to completely disinfect the surface of food because of the nature thereof. However, it is possible to have a nonspecific scouring of microorganisms under the influence of washing active substances, even in the case of massive contamination. Therefore, it is possible to obtain a decrease up to 95% in the total bacterial count on the surface of the food. Depending on the composition employed, the addition of a preservative may be necessary. Tartaric acid and lactic acid, slightly disinfectant compounds, can be added without objection. In addition, small amounts of other known food disinfectants and preservatives may be employed in nontoxic amounts.

The processing agent can be provided in dry form as powder, granules, tablets, or pieces, in a dissolved form as paste or gel, or as an aqueous concentrate. Anhydrous preparations can be stabilized by adding colloidal silicic acid (e.g., Aerosil 200). Potassium chloride and sodium chloride can be included for example as solid diluents.

According to the functions of the components described above for the individual groups within the overall effect of the processing agent, the following general formulation is recommended for the various possible solid, semisolid and liquid forms:

Formulation	Percent by Weight
Anionic tensides	1–10
Nonionic tensides	0.1–4
Ethylenediaminetetraacetic acid, disodium salt	1–5
Ascorbic acid	0.5–4
Sodium disulfite	0.5–2
Citric acid	1–12
Tripotassium citrate	4–20
Potassium carbonate	0–8
Polyalcohols	2–20
Preservatives	0–0.5
Inert fillers and additives, including water	5–90

As Solubilizers for Erythrosine

Red No. 3 which is also known as erythrosine, is a xanthene-type coal-tar-derived synthetic color. Chemically, it can be defined as the disodium salt of tetraiodofluorescein. It is an extremely bright pink/red color, is insoluble in acids and exhibits a very strong fluorescence and strong staining properties. It has been used in aqueous food products, but has heretofore been generally restricted to systems with a pH of above about 4.5. The insolubility is probably due to the fact that it contains a carboxylic group which is predominantly in the nonionized form below pH 4.5.

This development by *I.-Y. Maing, T.H. Parliment and R.J. Soukup; U.S. Patent 4,133,900; January 9, 1979; assigned to General Foods Corporation* relates to rendering erythrosine soluble and stable in aqueous food systems having a pH of less than 4.5.

According to the process, Red No. 3 is solubilized by forming a mixture of polysorbate, Red No. 3 color and an aqueous medium under conditions of shear effective to impart the degree of stability necessary for a commercially acceptable aqueous food product. Typically, this is accomplished by using a blender at speeds in excess of 20,000 rpm. The composition forms an optically clear solution when dispersed in water under acid conditions. It is believed that a colloidal suspension is formed. Further, it is believed the particle size of the Red No. 3 is reduced to less than 0.2 micron and the Red No. 3 is protected from the acid in low pH systems by an encapsulation effect of the polysorbate. In this description "solubilized" means that the colloidal suspension in the final product is discerned by the eye as a clear solution. Thus, the term "solubilized" in reference to Red No. 3 relates to colloidal dispersions.

The surfactants commonly known as polysorbate 60, polysorbate 65 and polysorbate 80 are the particular polysorbates employed in this process. While it is believed that surfactants with an HLB (hydrophile-lipophile balance) of between 10 to 17 and preferably 14 to 16 may be operable, factors such as off-flavor, off-color and general unsuitability for use in foods make the aforementioned three polysorbates the most desirable choices.

Another embodiment of this process relates to the use of solubilized Red No. 3 in a dry food system which may be subsequently reconstituted. This is accomplished by first preparing a solution as above, preferably using water to form the aqueous phase, although again a portion of the final aqueous food system may be used in place of water. It is preferable to use a minimum amount of water to form the solution since excess water will simply require longer drying time and the process will be energy intensive to produce a dry form of solubilized Red No. 3. Once the solution is obtained, it is next combined with a carrier and dried. The carriers employed may be any of those nontoxic edible systems or combination of substances known for use in foods of the type described herein. These include tricalcium phosphate, hydrophilic colloids such as dextrins and gum arabic, and the organic acids such as citric, adipic, fumaric, tartaric and malic acids. Preferably the carrier will be a dextrin.

Example 1: A mix consisting of one FD&C Red No. 3, 9 parts of polysorbate 80, 100 parts water and 25 parts Mor Rex dextrin having a DE of 15 is prepared in an Eppenbach mixer by operating the mixer at 30,000 rpm for 15 minutes. The solution obtained is spray-dried in a Niro laboratory spray dryer. The average inlet temperature of the dryer is 150°C and the average outlet temperature of the spray dryer is 90°C. The resulting products have a moisture content of less than 5% by weight of the total composition and have a bluish/red appearance.

Next, a dry beverage mix is formulated from the ingredients comprising 95.50% sugar, 1.85% citric acid, 1.55% monocalcium phosphate, 0.04% Vitamin C, 0.03% strawberry flavor and 0.3% of the above spray-dried product.

Extended storage stability tests of the dry mix, i.e., 90 days at a temperature of 100°F and 55% relative humidity, had no deleterious effects on final product color. The dry beverage mix, whether hydrated with water prior to storage or after 90 days storage, produces a clear beverage which is free from haze and/or precipitation for a period in excess of 48 hours when stored at a refrigerator temperature of 43°F.

Example 2: A spray-dried product is prepared according to Example 1, except that 7 parts polysorbate 80 is employed for each one part of FD&C Red No. 3. An unsweetened dry beverage mix is formulated consisting of 49.8% citric acid, 41.08% monocalcium phosphate, 1.18% Vitamin C and 0.75% cherry flavor.

Extended storage stability tests of the dry mix, had no deleterious effects on final product color. The dry beverage mix, when hydrated with water prior to storage and after 90 days storage, produces a clear beverage which is free from haze and/or precipitation for a period in excess of 48 hours when stored at a refrigerator temperature of 43°F.

Example 3: A mix consisting of one part FD&C Red No. 3, 9 parts polysorbate 60, 100 parts water and 25 parts Mor Rex dextrin having a DE of 15 is prepared in an Eppenbach mixer operated at 30,000 rpm for 15 minutes. The resulting solution is freeze-dried in a Thermovac vacuum chamber. The shelf temperature is maintained at 90°F and the pressure is maintained at 200 microns of mercury for 24 hours. The resulting product has a moisture content of less than 5% by weight of the total composition.

A dry beverage mix is formulated comprising 95.50% sugar, 1.85% citric acid, 1.55% monocalcium phosphate, 0.04% Vitamin C, 0.03% grape flavor and 0.3% of the above freeze-dried color.

Extended storage stability tests of the dry mix had no deleterious effects on final product color. The dry beverage mix, when hydrated with water either prior to or after 90 days storage, produces a clear beverage which is free from haze and/or precipitation for a period in excess of 48 hours when stored at a refrigerator temperature of 43°F.

As Antifoaming Agents for Food Colorants

Polymeric dyes have been proposed as satisfactory replacements for FD&C Red No. 2, which can no longer be used in foods, drugs and cosmetics in the United States. When polymeric colors are used in edibles, if the size of the molecules of polymeric color exceeds a certain limit, usually a molecular size of from about 1,000 to 2,000 daltons, and if the color compounds do not break down and thus maintain this size, the polymeric colors are not absorbed through the walls of the gastrointestinal tract. This means that when such materials are consumed, they essentially pass directly through the gastrointestinal tract. They are not taken into the body or its systemic circulation and thus any risk of possible systemic toxicity is eliminated. An excellent red polymeric colorant, which shows promise as an especially desirable replacement for existing red food colors such as FD&C Red No. 2, is an anthrapyridone known as Poly R-481.

The use of Poly R-481 in aqueous solutions, particularly in carbonated beverages, has unfortunately been limited by the excessive foaming which has been observed in such solutions. Foaming is considered a significant drawback to the use of this polymeric dye in carbonated beverages since it would result in filling problems during high-speed bottling operations. Foaming could result in excessive spillage and short-filled packaging in both bottling and canning operations.

T.E. Furia; U.S. Patent 4,185,122; January 22, 1980; assigned to Dynapol provides a method of preventing or eliminating the foaming of polymeric dye-containing aqueous solutions, particularly soft drink compositions, by use of glycerol monooleate, glycerol dioleate or a mixture thereof.

Glycerol monooleate (GMO) and glycerol mono- and dioleate (GMDO) are food-grade additives usually considered to be emulsifiers and solvents. They have been found to be surprisingly effective, at low concentrations, when used as defoamers for aqueous solutions of Poly R-481. This utility is completely unexpected and particularly surprising in view of the fact that other food-grade additives, including those generally considered to be food-grade defoamers, have been found to be ineffective or to actually increase foaming in aqueous solutions containing Poly R-481.

In the tests described below, foam height was measured as a function of time by filling a 250 ml graduated glass cylinder, fitted with a ground glass stopper, with 20 ml of test solution. The test solution and a control were simultaneously agitated manually for 25 vertical strokes after which the initial foam height and its decay as a function of time were measured in milliliters.

Example 1: To 10 ml of a 1,500 ppm solution of Poly R-481 containing 20 ppm GMDO and 20 ppm of ethanol as a cosolvent, 10 ml of carbonated beverage was added at room temperature in a 250 ml graduated cylinder. The final solution contained: 750 ppm Poly R-481, 10 ppm GMDO and 10 ppm ethanol. In another vessel, the carbonatd beverage was added to the dye solution containing only ethanol; this served as a control.

Immediately upon the addition of 10 ml of carbonated beverage to 10 ml of 1,500 ppm solution of Poly R-481, a thick copious foam with a height of 50 ml was produced. The foam was quite stable and after 10 minutes a 10 ml foam height persisted. With 10 ppm GMDO in the Poly R-481 solution, only 5 ml of foam was produced immediately and this decayed within 30 seconds to an insignificant amount (<0.5 ml). The use of ethanol (10 ppm) did not affect foaming.

These surprising results indicate that filling problems in a finished carbonated beverage due to foaming are greatly reduced or eliminated by the use of GMDO.

Example 2: GMDO was incorporated directly into a carbonated beverage at 20 ppm. The essential oils in the beverage were obviously sufficient to solubilize the small amount of GMDO.

To 10 ml of a 1,500 ppm solution of Poly R-481 in a graduated cylinder was added 10 ml of carbonated beverage. The GMDO concentration in the final test solution was 10 ppm. As a control, 10 ml of carbonated beverage without GMDO was added to 10 ml of 1,500 ppm Poly R-481 solution.

When carbonated beverage without GMDO was added to an aqueous stock solution of Poly R-25 481, foaming developed at an increasing volume during the addition. This foaming subsided slowly within 0.5 minute but considerable foam persisted throughout a 10 minute observation period. With 20 ppm GMDO in the carbonated beverage, foam developed only slightly during the addition and was <0.5 ml throughout the remainder of the observation period.

This was a particularly difficult test since the carbonated beverage was being dropped onto the Poly R-20 481 stock solution and there was a great tendency towards foaming from the liberation of CO_2. In production, the finished beverage containing dye would normally be prepared first and carbonation conducted as the final operation during filling.

Whey Colloidal Precipitate

This disclosure by *S.M.M. Shah and A.J. Luksas; U.S. Patent 4,143,174; March 6, 1979; assigned to Beatrice Foods Company* relates to food compositions which have been modified in their properties by the inclusion of a whey colloidal precipitate. The modified properties may include stabilization, emulsification, thickening, clouding, gelling and viscosity control.

Thus, broadly stated, this process provides food-grade compositions which comprise a mixture of a food-grade material or a food (the latter in at least a flavor producing amount) and whey colloidal precipitate and water. At least a modifying amount of the whey colloidal precipitate is contained in the composition, especially up to 30% by weight of the water therein.

The resulting compositions may be a liquid of controlled viscosity or in a gelled state, including thixotropic gels. Alternately, the compositions may be an emulsion with the whey colloidal precipitant functioning as the primary emulsifier. The emulsion may be a water-in-oil or oil-in-water type, or alternately, or in addition, include an air emulsion to provide a whipped and stabilized foodstuff. The composition prepared by this process may be placed in a dried form for later reconstitution by the addition of water. This modifier can pass through a drying state without substantial degradation of its properties.

This modifier is precipitated from the whey by various methods. Irrespective of the method, the precipitate obtained must be identifiable by reference to the following essential properties.

> (A) The modifier, in pure form, is a nonproteinaceous, complex precipitate of whey. The modifier forms a colloidal suspension in amounts up to about 30% by weight of the water at room temperature, i.e., 68°F.

> (B) The average particle size of the precipitate in a "suspension" may be in the range of less than 10 μ and more usually less than 5 μ. More often, the precipitate will have an average particle size of about 1 μ or less, particularly in the range of about 1 mμ to about 1 μ.

> (C) The precipitate will also be identifiable by virtue of its action on hydrocarbon liquid solvents. This precipitate not only will gel aqueous solutions and suspensions, but will gel hydrocarbon solvents, such as petroleum ether.

> (D) A further distinguishing characteristic of this precipitate is the essentially white color and substantially bland taste. The precipitate does not have the normal undesirable taste of whey.

> (E) The precipitate in essentially pure form is nonproteinaceous. "In essentially pure form" means that the precipitate as finally used in a food composition will consist essentially of the nonproteinaceous complex and the amount of protein associated with the precipitate is sufficiently low that a water suspension of the precipitate will not yield a substantial protein precipitate when treated with trichloroacetic acid.

> (F) Finally, this precipitate is capable of being dried to an essentially free-flowing powder; thus, it further distinguishes from other possible fractions of whey which are more usually hygroscopic in nature and cannot form dry, free-flowing powders.

This precipitate is obtained by causing a complex to form from the components in whey and causing the solubility of the complex to be exceeded such that precipitation thereof occurs. The precipitate must be substantially separated from the supernate (which contains the undesired lactose, etc.). Otherwise, these properties will be either greatly reduced or be lost altogether. One method of causing precipitation is that of heating whey to a temperature of at least 80°C. Another method is treating whey with a water-soluble, nontoxic base to raise the pH of the whey sufficiently to cause precipitation or growing a yeast (e.g., baker's yeast) under aerobic conditions to raise the pH. The whey precipitate may be obtained by removing the other whey components, e.g., lactose, lactate, riboflavin, etc., which will leave this complex in the whey liquid. If raw whey is used in the foregoing methods, the precipitate will contain substantial amounts of protein in addition to this whey colloidal precipitate.

Example 1: Twenty-five hundred gallons of raw acid whey from cottage cheese were fed to a Westfalia separator operated at a bowl speed of 1,600 rpm. The feed rate of the acid whey was 1,600 gallons per hour. The sludge from the separator, containing casein fines and other insolubles, was discarded and the clarified supernate (the clarified acid whey) was recovered. The pH of the clarified acid whey was about 4.4. The total amount of clarified acid whey recovered was about 2,450 gallons. To the recovered acid whey were added 45 pounds of calcium hydroxide, in 5 pound additions, until the pH stabilized at about 7.3.

This, essentially, neutralized whey was then fed to the Westfalia separator with a bowl speed of 1,600 rpm and at a feed rate of about 800 gallons per hour. About 315 gallons of wetted precipitate were recovered as the sediment from the separator. A portion of the wetted prcipitate was resuspended in water at 50% by weight concentration and immediately spray-dried (1,500 psi nozzle pressure, 290°F inlet, 190°F outlet) to an essentially white free-flowing powder. The remaining portion of the wetted precipitate was frozen for subsequent use.

Example 2: A solution of the precipitate of Example 1 was prepared with 20 parts of the precipitate and 100 parts of water. To 100 parts of this solution were added 30 parts of 5% vinegar and the mixture was stirred. Fifty parts of sucrose were added with stirring. Thereafter, 100 parts of liquid vegetable oil (soybean oil) were added and the mixture was homogenized. The emulsion which was prepared was stable and had the viscosity of a mayonnaise mixture.

Other examples show the use of the whey precipitate in orange- and grape-type drinks and in whipped toppings.

Natural Emulsifiers in Center-Filled Chewing Gum

The disclosure by *K. Ogawa, S. Tezuka, M. Terasawa and S. Iwata; U.S. Patent 4,292,329; September 29, 1981; assigned to Lotte Co., Ltd., Japan* relates to a method of improving the flavor-retaining capacity of a center-filled chewing gum.

It has been found that prevention of the flavored liquid fill from penetrating into the surrounding gum composition can be achieved by adding a natural emulsifier to the flavored liquid center fill. In other words, it has been found that the added natural emulsifier has not only its intrinsic emulsifying function but also an unexpected function of preventing the flavored liquid from penetration into the circumferential composition.

It has not clearly been known why the flavor of the liquid center fill can be prevented from penetration into the circumferential gum composition with use of the natural emulsifier. It is presumed the flavor component is coated with the emulsifier by its emulsifying and dispersing functions, thus lessening the chance of contact with the inner wall of the cavity formed in the chewing gum. As a result, undesired penetration of the flavored liquid fill into the gum composition may be avoided.

The flavored liquid fill used in the process may be any of the types conventionally used such as natural essential oils, flavoring preparations and the like.

The quantity of the flavored liquid center fill in the chewing gum may be varied widely by selection depending on type of liquid, formulation of the chewing gum base, etc. Further, the chewing gum base used may be any of synthetic resins such as polyvinyl acetate or natural resins such as chicle. The chewing gum base may contain any additives (such as microcrystalline wax, calcium carbonate, etc.), if desired.

The emulsifier suitable for use in the process may be any known natural emulsifier such as lecithin, a natural gum, alginic acid, gelatin, or the like or a mixture thereof. A preferred emulsifier is arabic gum. Natural gums having an emulsifying function and suitable for use in addition to arabic gum include arabinogalactan, ghatti gum, tragacanth gum, karaya gum, curdlan, carrageenan, agar, guar gum, tamarind, pectin, locust bean gum, xanthan gum, purée or the like or a mixture thereof. The quantity of the emulsifier to be added to the flavored liquid may be of 0.01 to 0.5% by weight, preferably 0.02 to 0.2% by weight of the flavored liquid center fill.

The object, that is to retain the flavor of the liquid center fill, may be achieved by adding only the emulsifier to the flavored liquid, but a more remarkable effect may be obtained by using the emulsifier in combination with a solvent. The solvent suitably used includes polyvalent alcohols such as propylene glycol, glycerol, sorbitol, mannitol, polyethylene glycol, maltitol, xylitol and the like as well as other solvents for emulsification. The quantity of the solvent to be added is determined so as not to deteriorate the flavor for which reason it is preferred to reside in the range of 0.01 to 0.1% by weight of flavored liquid center fill.

In the case of using the emulsifier in combination with the solvent, the quantity of the emulsifier to be added may generally be 0.01 to 0.5% by weight, and preferably 0.02 to 0.3% by weight of the flavored liquid center fill. Addition of the solvent per se has no direct influence on preservation of the flavoring effect. However, the solvent, when used in combination with the emulsifier, serves as a viscosity regulator for the flavored liquid and enhances the emulsifying (coating) and dispersing effects of the emulsifier, resulting in the indirect effect of preventing the flavoring components from penetrating into the gum composition.

In any case, the flavored liquid after the emulsifier has been added is preferably homogenized to increase the flavor-retaining capacity more remarkably, since the homogenization facilitates the emulsification (coating) to provide finely divided homogeneous flavor dispersions in size of not more than 0.1 μ. The apparatus for use in the homogenizing operation may be of any type known by those skilled in the art, and the homogenizing time is not critical, but preferably it may fall in the range of 10 to 40 minutes per 160 kg.

The flavor retention of the flavored liquid may be achieved either by adding the emulsifier alone, in an amount of 0.01 to 0.5%, preferably 0.02 to 0.2% by weight, or by adding the emulsifier in combination with the solvent, in an amount of 0.01 to 0.5%, preferably 0.02 to 0.3% by weight. In any case, the quantity over the lower or the upper limit may not be employed for practical use, because the lesser quantity results in insufficient coating of the flavor components with the emulsifier, whereas the larger quantity results in undesirably high viscosity of the flavored liquid and gives undesirable taste of the emulsifier per se.

Example: The flavored liquid center was prepared by adding 0.07% of arabic gum to 0.2% peppermint oil. The resulting solution was added to 99.73% of mixed syrup of saccharides consisting of 15 to 25% of sorbitol, 20 to 30% of invert sugar, 25 to 40% of malt honey, 4 to 10% of sucrose and 5 to 10% of water, and then the solution was either stirred alone or homogenized to give a uniform mixed flavored liquid.

The desired chewing gum base was prepared in the conventional manner by blending 0 to 20 parts natural resins; 15 to 25 parts natural waxes; 20 to 30 parts polyvinyl acetate resin; 15 to 25 parts ester gums; 5 to 20 parts synthetic rubbers; and 25 to 30 parts others.

The center-filled chewing gum was prepared according to the conventional manner, using the flavored liquid fill and the chewing gum base prepared by the manner as hereinbefore described. The ratio of the flavored liquid to the gum base was 12:88.

The penetration ratio (% after 30 days) of the flavor component was 5.0% in the chewing gum where homogenization took place. In another example where no homogenization was done, the penetration ratio was measured at 42.5% after 30 days.

Sugar Esters in Center-Filled Chewing Gum

K. Ogawa, S. Tezuka, M. Terasawa and S. Iwata; U.S. Patent 4,157,402; June 5, 1979; assigned to Lotte Co., Ltd., Japan have also used a series of polyol higher fatty acids as emulsifiers in center-filled gum.

The emulsifier suitable for use in this process may be any known natural or synthetic emulsifier such as sucrose-(higher) fatty acid esters, glycerol-(higher) fatty acid esters, propylene glycol-(higher) fatty acid esters, sorbitan-(higher) fatty acid esters and a mixture thereof; preferably it is selected from those having a HLB value in the range of 3 to 15. The quantity of the emulsifier to be added to the flavored liquid may be 0.01 to 0.5% by weight, preferably 0.02 to 0.2% by weight of the flavored liquid center fill.

Specific emulsifiers used in this process include: (1) sugar ester S-370, (HLB-3, from Ryoko Chemical Co.); (2) sugar ester S-1570 (HLB-15, Ryoko Chemical Co.); (3) Atmos 50 (a glycerol fatty acid ester, Kao-Atlas Co., HLB-3.2); and (4) Span 60 (a sorbitan fatty acid ester, Kao-Atlas Co., HLB-4.7).

When the above emulsifiers were used in amounts ranging from 0.005 to 0.5% in the flavored liquid center followed by homogenization, the penetration ratios after 30 days were from 5.0 to 8.6. With no emulsifier, the 30 day penetration ratio was 74.3.

Low Cholesterol Egg Yolk

Egg yolk has a large application in foods and as it is an efficient emulsifying agent it is an essential ingredient in mayonnaises, cake batters containing fats, cream puffs, bakery goods and candies.

Egg yolk contains a high level of cholesterol and saturated fats and is itself an emulsion comprising a dispersion of oil droplets in a continuous phase of aqueous components. It has a total solid content of approximately 50 to 52% composed of 15.5 to 16.5% protein, 31.5 to 34.5% lipid, 0.5 to 1.5% carbohydrate and 0.9 to 1.2% ash.

The egg yolk lipids comprise as their main components approximately 65% triglyceride, 29% phospholipid and 5% cholesterol. The high amount of self-emulsifying phospholipids, wherein the highest hydrophilic component is phosphatidylcholine representing 75% of the total, makes egg yolk a very stable emulsion in addition to being an emulsifying agent.

The main features of the organoleptic properties of egg yolks are the flavor and the color, the latter being due to naturally occurring pigments which are mainly alcohol-soluble xanthophylls.

A process of reducing the amount of cholesterol in egg yolk has been described by *U. Bracco and J.-L. Viret; U.S. Patent 4,333,959; June 8, 1982; assigned to Societe d'Assistance Technique pour Produits Nestle SA, Switzerland.* This process comprises reducing the pH of fresh wet egg yolk to destabilize the emulsion, treating the destabilized emulsion with an edible oil to form a fine dispersion and centrifuging the dispersion to separate the egg yolk phase from the oil phase.

The edible oil used may be any edible digestible oil substantially free of cholesterol and one that is normally liquid or partially liquid at room temperature. It may be a polyunsaturated oil, a partially hydrogenated oil or an oil in which the balance of polyunsaturated to saturated components has been adjusted according to the requirements. Conveniently the oil contains at least 50%, preferably at least 75% polyunsaturated component.

Examples of suitable polyunsaturated vegetable seed oils are corn oil, cottonseed oil, soybean oil, sesame seed oil, sunflower seed oil, safflower oil, rice bran oil, grapeseed oil, pumpkin oil or peanut oil. Oils with a high polyunsaturated content, in addition to removing cholesterol, produce an egg yolk with an increased polyunsaturated fat content. Conveniently, cholesterol-containing edible oil that has been separated from the egg yolk may be treated to remove cholesterol and other lipid material solubilized during the process for example, by molecular distillation and this treated oil may be reused to reduce the amount of cholesterol in egg yolk.

The destabilized emulsion is conveniently treated with a quantity of edible oil sufficient for the blend to have a free-flowing state but not in such quantity that the after-treatment of the oil is uneconomical. Preferably the destabilized emulsion is treated with from 2 to 10 times and especially from 3 to 5 times its own weight of edible oil.

The particle size of the fine dispersion of the destabilized egg yolk emulsion produced by treatment with the edible oil may be from 1 to 20 μ, and preferably from 5 to 10 μ. The fine dispersion of the destabilized egg yolk and edible oil may be formed by homogenization conveniently by using an agitator mill with microballs, a high-pressure homogenizer or intensive vibration as used in continuous disintegration devices. When a homogenizer is used the speed is conveniently from 5,000 to 10,000 rpm. The homogenization may conveniently be carried out for a period of 5 to 90 minutes, preferably from 10 to 20 minutes, and at a temperature from ambient to 65°C, preferably 50° to 60°C.

The fine dispersion of the egg yolk and edible oil is centrifuged conveniently in a batch-type centrifuge, preferably at a relatively low speed for instance from 1,000 to 5,000 rpm and especially from 2,000 to 4,000 rpm. The dispersion may conveniently be centrifuged for a period of from 10 minutes to an hour and optimally from 20 to 40 minutes. After centrifuging, the lower egg yolk phase may be separated from the upper oil phase.

The low cholesterol egg yolks obtained may be further treated to prolong their shelf life. For example, they may be pasteurized, frozen, freeze-dried and therefore stored before any technical utilization. The functional properties of low cholesterol egg yolk are unaffected by freezing and prolonged storage at -20°C.

Example 1: 2 kg egg yolk containing 1.24% cholesterol with an initial pH 6.68 were treated with 13 ml of 16% hydrochloric acid. The pH was lowered to 5.65 and the viscosity decreased from 280 to 175 cp at 20°C. The egg yolk was placed in a mixer (Homorex) with 8 kg peanut oil. The mixture was homogenized at 50° to 55°C for 1 hour at 7,000 rpm. After mixing, the oil/yolk mixture was centrifuged in a batch centrifuge for 30 minutes at 3,000 rpm. Two distinct phases, an upper oil phase representing 83.5% of the initial mass and a lower yolk phase representing 16.5% of the initial mass were collected. The amount of cholesterol was reduced to 0.25% which is a 79.84% decrease from the initial amount.

Example 2: The peanut oil used in Example 1 was substantially freed from the cholesterol by submitting it to a fall film molecular distillation on Leybold KOL-1 at 230°C and a pressure of 2.10^{-3} mm Hg.

4 kg of the distilled peanut oil were added to 1 kg of the egg yolk treated in Example 1 which had a pH of 5.63. The mixture was homogenized at 50° to 55°C for 1 hour at 7,000 rpm and batch centrifuged at 3,000 rpm to obtain a lower yolk phase representing 15.2% of the initial mass. The amount of cholesterol was reduced to 0.13% which is an 89.25% decrease from the amount present in the original untreated egg yolk.

Example 3: A process was carried out in a similar manner to that described in Example 1 and afterwards the separated peanut oil was added back to the treated egg yolk to reconstitute the fine dispersion. This dispersion was freeze-dried and then centrifuged. The amount of cholesterol present in the freeze-dried egg yolk before centrifugation was 2.74% and this was reduced to 0.55% after centrifugation, which is a decrease of 79.95%.

Isolated Soy Protein in Imitation Mayonnaise

Mayonnaise is defined under U.S. FDA Standards of Identity as an emulsified semisolid food prepared from edible vegetable oil, such as unhydrogenated soybean or safflower oil, acetic or citric acid, and egg yolk. Optional ingredients permitted include salt, natural sweeteners, spices, or spice oils, monosodium glutamate, and any suitable harmless flavor from natural sources. The oil level must be not less than 65% by weight of the mayonnaise and 77 to 82 is the usual weight range. Mayonnaise is an oil-in-water emulsion in which the egg yolk constitutes the major emulsifying component.

Salad dressing under the Standards resembles mayonnaise in that it is an emulsion of oil in vinegar using egg as an emulsifier. It differs from mayonnaise in that it also contains starch paste as a thickener.

In both products, the presence of egg as an emulsifier is, in some respects, a drawback as the product then contains some cholesterol and is also considered to be a "higher calorie" food.

It is known to substitute, for the egg emulsifier in a mayonnaise or salad dressing type of product, an emulsifier consisting mainly of an isolated soy protein (also known as soy isolate), e.g., Supro 620 or Promine. Such a substitution will provide an imitation mayonnaise or salad dressing product that is relatively low in calories, of high water content, and is cholesterol-free. The high water content, of between 20 to 40%, is necessary in order to dissolve the soy isolate. However, the resulting products have the following major drawbacks: unstable emulsion formation and high syneresis.

Improvements in formulation and preparation of imitation mayonnaise and salad dressings employing isolated soy protein (ISP) as the sole or dominant emulsifier has been disclosed by *P.U. DePaolis; U.S. Patent 4,163,808; August 7, 1979.*

The imitation mayonnaise or salad dressing of this process is an oil-in-water emulsion in which between 3 to 10% by weight of ISP is employed as an emulsifier. This amount of ISP used is substantially more than that conventionally employed and water in an amount of from about 20 to 60% is required to dissolve from 3 to 10% ISP, respectively, aside from the aqueous acetic acid (vinegar) added.

Vinegar in amounts of 0.1 to 2% by weight of the finished product is preferably added as 200 grain concentration (20% by weight) which is double the concentration conventionally employed.

The imitation mayonnaise or salad dressing is made in a three-step procedure. In the first step, water, ISP and spices, are added to a small fraction of the total salad oil to be utilized, and the ingredients are admixed for about 5 to 10 minutes, as compared to the more conventional one minute. In the second step, the vinegar is blended in. In the third step, the remaining major amount of oil is added and admixed in a blender having a shearing action such as a "Waring" blender.

The abovedescribed formulation and procedure, in combination, produces an imitation mayonnaise or salad dressing type product with the advantage of a stable emulsion and in which little, if any, syneresis takes place.

The salad oil source employed may be any of a variety of vegetable oils such as winterized cottonseed, unhydrogenated or hydrogenated and winterized soybean oil is normally utilized. The salad oil may range from about 40 to 80% by weight in the finished product and preferably is present at about 65% by weight. The ratio of water to ISP in the final formulation is at least 6:1 by weight.

The acid preferably used is 200 grain vinegar, it is present in amounts that range between 0.1 and 2.0% by weight, and preferably is present in about 0.5% by weight.

Spices and seasonings used may be mustard, paprika, garlic, cloves, onion, salt, sugar and the like. The spices and seasonings normally constitute 1% or less of the finished mayonnaise or salad dressing product.

Example: In step 1, 15 quarts water, 9.5 pounds ISP (Supro 620), 8 ounces mustard, 2 ounces garlic, 2 ounces onion, 2 ounces granulated lecithin and 4 quarts soybean oil are placed in a paddle-type motor drum mixer. The ingredients (about 49 pounds) are mixed for 5 minutes until the mixture becomes whitish in color.

In steps 2 and 3, the 49 pounds of admixture from step 1 is then placed in a Waring type blender in which the blades of the blender may be operated at very high speed resulting in great shearing forces. The shear forces aid the emulsification of the final product. 16 ounces of vinegar (200 grain) are added followed by the addition of 100 pounds (110 quarts) of oil. The oil is blended in very slowly to make 150 pounds of finished product.

It is to be noted that only about 4% of the total oil is added in step 1 for "seeding" purposes and in order to enable a stable emulsion to be readily prepared. The ratio of water to oil is about 11:1.

MODIFIED STARCH AND CELLULOSE

STARCHES—GENERAL

Dual Treated Starches in Puddings

R.W. Rubens; U.S. Patents 4,183,969; January 15, 1980 and 4,219,646; August 26, 1980; both assigned to National Starch and Chemical Corporation discloses a dual derivatizing process for the preparation of cold water swelling starches and foodstuffs thickened therewith.

According to this process the granular starch base is derivatized in a primary crosslinking step with sodium trimetaphosphate (STMP) at a pH of 10 to 12 and at a temperature sufficiently low to maintain the starch in an unswollen granular form. The slurry is then neutralized to a pH of 5 to 6.5, recovered and preferably washed in accordance with conventional known techniques. At this stage, however, rather than reslurrying and drum drying as is carried out in typical STMP reaction procedures, the starch is reslurried, adjusted to a pH of at least about 7.5 and additional STMP is added to the slurry. The resulting slurry is then fed onto a drum dryer and drum dried in accordance with conventional procedures to effect a secondary crosslinking reaction in situ.

The foodstuffs thickened with the starches resulting from this unique dual derivatizing reaction are characterized by superior viscosity-related properties as evidenced by their viscosity curves which, after peaking, maintain a high viscosity with little or no degradation or breakdown even after exposure to elevated temperatures for extended periods of time.

It is thought that this procedure is superior to conventional STMP crosslinking techniques in that the primary crosslinking reaction occurs only in the outer portions of the starch granule with limited penetration, while the subsequent secondary crosslinking is effected on a starch product which loses much of its granular form during the drum drying step so that the latter crosslinking reaction which occurs simultaneously during the drying takes place mainly on the dispersed starch and not on the granule.

The applicable starch bases which may be used in this process include any granular starch in raw or modified form. Useful starch bases include corn, waxy maize, grain, sorghum, wheat, rice, potato, sago, tapioca, sweet potato, high amylose corn, or the like. Also included are the conversion products derived from any of the starch bases including, for example, dextrins prepared by the hydrolytic action of acid and/or heat, oxidized starches prepared by treatment with oxidants such as sodium hypochlorite, and fluidity or thin boiling starches prepared by enzyme conversion or by mild acid hydrolysis. Particularly preferred starches, due to their end use applications, are corn, waxy maize and tapioca.

Example: This example illustrates the production of a crosslinked cold water swelling amioca starch. A water slurry of amioca at about 36% solids and 35° to 38°C was treated with 0.6% sodium chloride (based on starch solids). A water solution containing 3% by weight of sodium hydroxide was added until the total amount of sodium hydroxide was 0.6% of the weight of the starch. The pH at this stage was in the range of 11.1 to 11.4. To this system, 0.18% sodium trimetaphosphate was added and allowed to react for about 3 hours. After the reaction, the system was neutralized to a pH range of 5.0 to 5.5 with hydrochloric acid and the starch was recovered by filtration. The starch was then washed and dried to a powder containing about 12% moisture.

A portion of the resultant starch was reslurried in water, drum dried and milled according to prior art pregelatinization techniques and was set aside as a control. The remainder of the crosslinked amioca was then reslurried in water at a solids concentration of 38 to 42%. To this slurry was added 0.5% sodium chloride and 0.15% sodium trimetaphosphate based on the weight of the starch. The pH of the slurry was then adjusted to a range of 7.8 to 8.1 with sodium carbonate and drum dried with steam at about 100 psig pressure. The dried sheet was removed from the drum surface and milled to a powder, such that 85 to 90% by weight would pass through a 200 mesh U.S. standard sieve.

A viscosity curve obtained by Brabender analyses of the resulting dually crosslinked starch is shown in Figure 5.1 and compared with the curve obtained from the control distarch phosphate ester prepared above in accordance with prior art techniques.

The Brabender testing was performed on a simulated fruit pie filling mixture prepared by blending starch with sugar and water to produce a dispersion containing 4.6% starch solids and 23% sugar and adjusting the pH to 2.7 with acetic acid. The mixture was then premixed for 3 minutes, held at 30°C for 10 minutes, heated to 95°C and held at 95°C for an additional 10 minutes as indicated in the figure.

In Curve A (the control), the viscosity peak is achieved within 10 to 15 min after starting the heating cycle at a temperature of about 50°C. Thereafter, with continuous heating until a temperature of about 95°C is reached and maintained, there is an undesirable loss of viscosity or breakdown. In contrast, the improved properties obtained with the starch prepared using the dual crosslinking process are shown in Curve B. Here, the viscosity developed during the first 10 minutes, when the temperature is maintained at 30°C, is the result of pregelatinization/cold water swelling. After the start of heating and during a uniform increase in temperature from 30° to 95°C, the viscosity has peaked and stabilized, in fact increasing slightly from the start to the end of the heating cycle.

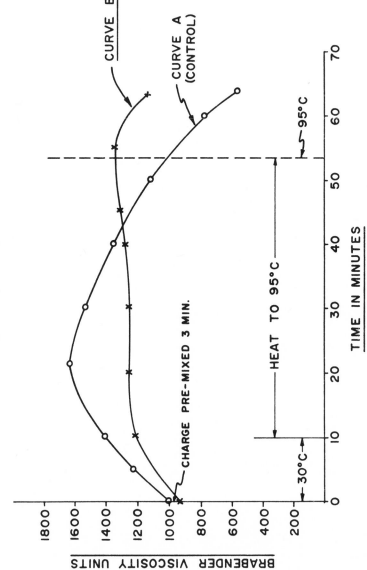

Figure 5.1: Viscosity Curve

Source: U.S. Patent 4,183,969

In addition to this significant resistance to viscosity breakdown, other disadvantages of the prior art materials are also practically eliminated. There is little or no loss in viscosity during heating, thus retaining thickening power, the texture is "short" without any tendency towards gumminess, and when used in fruit filling for tarts, the puncturing of the pastry shells is sharply reduced.

Other examples in the specification show this process on cornstarch and tapioca starch.

Starch Additives for Pie Crust Dough

Prepared pie crusts are available in frozen form in their own baking pans. However, the quality of such crusts is not very good. They are extremely fragile, require a significant amount of storage space in the household freezer, etc.

An alternative to the abovedescribed ready-to-use pie crusts was a refrigerated folded pie crust which was in sheeted form and the only steps necessary for use were unpackaging, unfolding, and placing the pie crust in the pie pan. However, this pie crust had excessive browning during the baking of the product which had been held for an extended shelf life (e.g., 45 days or more). Other major problems included off flavors and cracking upon unfolding the crust.

The method used by *M.J. Haasl, P.D. Pratt, R.W. Chen and H.M. Fett; U.S. Patent 4,297,378; October 27, 1981; assigned to The Pillsbury Company* overcomes this problem of excessive baked browning of a prefolded pie crust late in shelf life.

By controlling the composition of the formulation as described, a pie crust of acceptable quality can be made which exhibits good flakiness, nongreasiness and good browning characteristics after extended shelf life while substantially reducing cracking upon unfolding.

This dough broadly includes a high-starch, low-enzyme flour, shortening, water, salt, gluten (optional), a preservative system and food coloring.

The desirable flour has a high starch level, low enzyme level and minimal starch damage. Because such a flour is difficult to obtain as a standard product, it has been found desirable to modify a standard flour by adding additional starch to achieve a flour component having a high starch level and a low enzyme level (this will be referred to as the flour blend). Any suitable starch can be added, such as wheat flour starch, potato starch, cornstarch, tapioca starch, etc. In modifying typical standard flours, it is preferred that the flour blend composition consist of between about 30 and 70% by weight added starch on a dry starch/dry flour basis.

Thus, the flour blend contains between about 92 and 98% starch (dry starch basis), and the dry starch fraction of the dough is between 37 and 47% by weight. The flour blend is present in the dough in a range of between about 38 and 50% by weight of the dough, with this range being based on dry flour blend weight. Preferably, the flour blend contains less than about 0.2 absorption units of enzymes in the dry flour.

The protein level that is used to provide an adequate network to prevent excessive cracking depends on the way the dough is handled during the process. In the case of hand folding, protein should be present on a weight basis in the range of between about 1.5 to 6%. In the case of machine folding, it is preferred that protein be present in the dough in the range of between 2.2 to 6% by weight.

Shortening is necessary at relatively high levels to give the desired texture of baked pie crust. The preferred shortening level is in the range of between about 24 and 35% of the total weight of the dough.

The preferred shortening is prime steam lard, about 1 to 10% of which has been fully hydrogenated. It has a solid fat index (SFI) at 50°F of about 29 to 36, preferably about 30 to 35 and more preferably about 31 to 33 as measured by the standard AOCS dilatometry method CD10-57.

Also, it is preferred that the shortening be deodorized and have free fatty acids of less than or equal to about 0.2% by weight of total shortening and a peroxide value of less than or equal to about 3 milliequivalents per kg of total shortening. By having this level of free fatty acids and the proper level of peroxide value, rancidity can be eliminated.

Water is present in the dough. The water is introduced into the dough generally by two means; the flour blend contains a certain amount of water, and water is normally added to the dough as a separate ingredient. The total water in the dough is preferably in the range of between about 19 and 25% by weight of dough.

In the preferred formulation, an antimycotic preservative system is added to prevent molding. A preferred preservative system which has been found effective to prevent mold growth at 40°F for about 90 days is sodium propionate in the range of between about 0.05 to 0.6% by weight of the dough and potassium sorbate in the range of between about 0.05 and 0.15% by weight of total composition and citric acid in an amount up to about 0.07% by weight of dough.

Salt is added to the composition at a level of about 1.2 to 2% by weight of dough for flavor purposes and also to reduce water activity.

In the process of making dough, shortening, either in liquid form or solid form, is added to the chilled flour. The ingredients are mixed to disperse the shortening and flour to form a blend. Either during or after the shortening addition, water is added with additional mixing to disperse the water within the mixture and to hydrate and develop gluten. The other ingredients, such as salt, preservatives and color, are added with the water. In order to produce a long-flake crust, it is desirable to have small pellets of shortening up to about ⅜ inch in diameter in the dough.

To produce small pellets of the correct size, it is desired that the flour blend be chilled before mixing with the shortening.

The temperature of the flour, shortening and other components should be such that after the dough has been mixed, it is at a temperature in the range of between about 50° and 70°F.

It has been found that exposing the dough to vacuum above a certain level and for a certain time period, significant improvements in dough characteristics and product quality can be obtained. The dough becomes significantly stronger and more extensible. Also, the baked product is significantly flakier.

The application of vacuum should be such that the dough is exposed to a vacuum of at least about 27.5 inches of mercury. The time of exposure to the minimum vacuum level should exceed at least about 0.5 second, more preferably at least about 1 second. The dough should be allowed to expand during the application of the vacuum or the previously described characteristics will not occur.

The dough can be used and further processed as desired. For making pie crusts or the like, the dough is sheeted with any suitable sheeting apparatus. During the sheeting operation, it is preferred that the dough be at least about 55°F to prevent disruption of the continuous sheet, and the temperature should not exceed about 75°F, as the material may become difficult to process.

After sheeting, the dough is cut and folded. Preferably, a separating sheet is positioned on opposite sides of the pie crust. The pie crust is then packaged in a heat sealable pouch or the like.

In the final form the dough may be packaged in an atmosphere of fat and/or water-soluble gas. This aids in maintaining the highly flaky character of the dough which has been vacuum processed.

The gas should have a solubility at 20°C and atmosphere pressure of at least about 0.1 volume of gas per volume of water and/or at least 0.05 volume of gas per volume of shortening. Particularly suitable gases are CO_2 and N_2O.

Starch Coatings for Amino Acids

Many synthetic foods and special dietary products exist which contain, among other nutrients, amino nitrogen-containing (proteinaceous) compounds (e.g., one or more free amino acids or their salt derivatives, protein hydrolysates, intact whole proteins, or a combination of these) plus vitamins, including amino nitrogen containing vitamins, plus other, nonamino nitrogen-containing compounds, e.g., ammonium compounds such as ammonium sulfate, plus carbohydrates, including reducing sugars.

Over time, in the presence of moisture, and in even moderate heat (i.e., at temperatures above the freezing point of water), the Maillard/browning reaction results from the interaction of the nitrogen compounds with the aldehyde groups of the reducing sugars or other carbonyl compounds.

K. Mohammed; U.S. Patent 4,144,357; March 13, 1979; assigned to Johnson & Johnson has found that starch having a DE number up to about 24 can be used successfully to separate the nitrogen-containing components from the reducing sugars or other carbonyl group containing components that are present in such mixture, whereby the Maillard reaction is prevented or substantially delayed during storage.

More particularly, the process provides a synthetic dietary composition in solid form, typically a powder, for supplying nutritional requirements to a mammal, comprising: (1) a proteinaceous material (preferably including at least one free amino acid and optionally also including a nitrogen containing vitamin); (2) a starch; and (3) an aldehyde group-containing dietary compound (generally a reducing sugar) wherein the Maillard reaction between the proteinaceous material and the aldehyde compound is prevented or substantially retarded during storage by coating at least a major amount of at least one reactant selected from the proteinaceous material and the aldehyde compound with at least a portion of the starch, the starch used for the coating having a DE number of between 0 and about 24.

When the elemental diet product is intended for use in liquid form, it is preferred that the coating starch have a DE number in excess of 0 (but not in excess of about 24). A preferred range for the DE number is between about 5 and about 15. More particularly, it was found that coating the amino acids and other proteinaceous materials with hydrolyzed cereal solids having an approximate DE of 10 has eliminated the Maillard reaction in product aged at 30°C for 28 months, a significant improvement over the shelf life of products containing uncoated amino acids.

The source of the starch can be from corn, rice, casava, arrowroot, potato, taro, or any other vegetable from which starch can be made.

The starch may be hydrolyzed by known techniques, for example, by acid hydrolysis, with neutralization of excess acid with an alkali or by enzymatic hydrolysis, in which either the enzyme carbohydrase is introduced directly into a starch slurry or the starch slurry is passed through a column containing embedded carbohydrase.

Partially hydrolzyed starches having specified DE number or ranges suitable for use as the coating starch in this process are available commercially, for example, Maltrin 5 and 10 (DE 5 and 10) (Grain Processing Corp.), Frodex products (DE 15 and 24) (American Maize), Morex-1918 (DE 10) (Corn Products Corp.), and Clintose products (DE 15 and 24) (Clinton Corn Processing), and others. In each of the above cases the DE numbers indicated by the manufacturer represent an average, and the product actually has a range of DE values close to, but straddling, the average.

Example: A large batch of an amino acid mixture of a dietary formula was coated with a modified starch hydrolysate having a DE of about 10. The ratio of the amino acid mixture to starch hydrolysate in the coated product was 1:1. Both the coated product and uncoated amino acid control were blended with the vitamins, minerals, safflower oil, dextrose, corn syrup solids (DE 36) and flavoring materials of a representative dietary food powder. All samples had a moisture content of about 5%.

The dietary formula was packaged in 30 gram amounts. These materials were dispensed into packets which were then nitrogen flushed and heat sealed. The packaging material was a laminate of polyethylene, pin-hole-free one-mil aluminum foil and cellophane. The filling of the nitrogen flushed pouches with the dietary food powder was accomplished with a commercial filling and packaging machine.

Three flavor variations were packaged (unflavored broth base, flavored broth and orange flavored beverage). Identical samples were placed in aging at 30°C and 4.4°C, the latter temperature serving at the appropriate control.

Nutrient and organoleptic analyses after 12 months aging at 30°C showed no appreciable changes in potency or organoleptic properties when compared with the 4.4°C controls. After 28 months at 30°C no detectable signs of the Maillard reaction were found by an appraisal panel.

Pregelatinized Starch in Marshamallow Products

For many years, marshmallow flavored variegating syrups have been combined with ice cream, ice milk, mellorine and other frozen desserts for the purpose of adding flavor, variety, and eye appeal to these confections. These ingredients or confectionary additives of marshmallow and similar whipped variegating materials have often been purchased in a prewhipped or complete state, and have been added to the ice cream or frozen confection either by simultaneously pumping the variegate and the frozen confection through a common line or extrusion nozzle into the final container, or by placing the marshmallow or similar material on the top or bottom of the frozen confection by means of special filling attachments designated for this purpose.

A great difficulty is incurred in the incorporation of certain variegates with certain types of frozen confections with the variegate losing its character and blending improperly with the remainder of the frozen confection. Prewhipped marshmallow, for example, contains about 80% sugar solids in contrast to about 40% sugar solids for the ice cream or other frozen dessert with which it is blended. As a result there is a differential of freezing and melting points so that the consumer often ends up with the marshmallow melting and sinking to the bottom of the finished packages even when the confection is stored at conventional temperatures used for ice cream.

I.H. Rubenstein; U.S. Patent 4,189,502; February 19, 1980; assigned to Maryland Cup Corporation provides a formulation for marshmallow and other whipped variegates which are readily combinable with ice cream and other frozen desserts. This new product also maintains a sharp line of delineation or demarcation between the frozen dessert matrix and the variegate.

This process produces a marshmallow product and other variegates in which an emulsifier provides the whip, and wherein the stretch normally associated with marshmallow and other whipped marshmallow-like products or variegates is provided by certain materials, preferably pregelatinized starch. Additionally, a stabilizer such as a gum is generally preferred in such a mixture and such gums as carboxymethylcellulose, locust bean gum, guar gum, sodium alginate, gelatin or similar acceptable colloids are utilized to improve the body and stiffness of the finished marshmallow product. The emulsifiers are chosen with an HLB number (hydrophile-lipophile balance) in the range of 3 to 9. The range of the emulsifier is 0.2 to 0.8% by weight of the marshmallow or variegate mix while from about 1 to 3%, preferably 2 to 2.5% by weight of pregelatinized starch is employed in the formulations.

Materials other than starch can be employed to provide the stretch component characteristic of marshmallow in this process. For example, about 0.5 to 1%, preferably 0.7%, by weight of sodium alginate or about 0.2 to 0.4%, preferably 0.3%, by weight of carrageenan can be substituted for part or all of the starch. Other suitable substitutes for the starch include about 0.5 to 1% by weight of pectins or about 0.75 to 1.5%, preferably 1%, by weight of the product known as Avicel (micro-crystalline cellulose).

Where it is desired to make a whipped variegating syrup rather than a marshmallow variegate, that is, where the stretch of the marshmallow is not desired, the amount of pregelatinized starch or other stretch component is reduced and the level of gum or stabilizers such as guar gum, gelatin, locust bean gum, carrageenan or the like is increased.

The preferred starch should be in a pregelatinized state and can be of corn, tapioca or other vegetable origin. This is to avoid the necessity of cooking the starch, but if it is a matter of economy or convenience to use a conventional starch, this can be done if the mixture is heated to an appropriate temperature of at least 180°F and held at that temperature for 20 minutes.

The formulations of this process are designed for ice cream plants and can be handled in conventional ice cream freezer equipment which is available at such plants.

Example 1: A marshmallow product in which the quality of stretch is desired is prepared from 25% corn syrup (wet basis), 25% cane sugar syrup (dry basis), 0.8% emulsifier 6-2-S, 2.5% pregelatinized starch (instant Clearjel, National Starch Co.), 0.5% gelatin; and water to make 100%.

Example 2: A variegate syrup where stretch is not a desired characteristic of the formulation is prepared from 0.4% emulsifier (sodium stearyl-2-lactate), 25% corn syrup (wet basis), 25% cane sugar (dry basis), 1.0% pregelatinized starch (Nu:Col, A.E. Staley Co.), 0.5% gelatin, 0.3% citric acid, 48% strawberry (or other fruit flavor) plus water to make 100%.

Grinding Aids for High Fat Foodstuffs

B.H. Nappen; U.S. Patent 4,232,052; November 4, 1980; assigned to National Starch and Chemical Corporation has found that high fat containing foodstuffs may be readily powdered to produce free-flowing powders by adding a grinding agent. This agent comprises a food-grade film-forming material which has been spray dried from a solution in the presence of a latent gas and which is characterized by a bulk density within the range of 3 to 25 pounds per cubic foot. The specially prepared grinding agent is added to the foodstuff in an amount of 25 to 400% by weight based on the weight of the fat content of the foodstuff and, if the foodstuffs is in solid form, the mixture then ground using conventional methods.

The resulting food product is characterized by its free-flowing consistency with no paste formation or tendency to agglomerate. The product when desired, may be used directly in its powdered form or may be incorporated into other processed foodstuffs.

Generally, these film-formers fall into the three major classifications of starches, dextrins and gums, although a few of the useful materials fall outside these classes. Useful starches include food-grade products derived from corn, rice, potato, sago, tapioca, waxy maize, wheat, etc., as well as any food-grade modifications thereof, for example, the acetate, propionate and butyrate esters, as well as the hydroxyethyl, hydroxypropyl and carboxymethyl ethers.

Suitable dextrins for use include the enzymatic, chemical or heat degradation products of starch, such as the dextrins, maltodextrins, British gums, white dextrins, etc. Since it is necessary to employ the film-former in solution form and, since efficiency considerations dictate the use of as high a solids level as possible, it will be recognized that the choice of film-former will be governed in part by the solution viscosity thereof. Suitable gums for use include gum arabic, alginates, pectinates, low viscosity carrageenans, as well as low viscosity synthetic gums such as carboxymethylcellulose, hydroxypropyl cellulose, methylpropyl cellulose, cellulose xanthate, etc. Other useful film formers include gelatin, soybean protein, zein protein, etc.

The term "latent gas" as used herein refers to any material, whether solid, liquid or gaseous, which can be incorporated in the solution of the film-forming material and which can be converted into a gas, i.e., which produces a gas or is rendered gaseous at an elevated temperature, preferably a temperature at which the film-forming solution may be dried.

Particularly useful gases which may be employed include dissolved carbon dioxide and ammonia. Additionally, there are a large number of liquid and solid substances which are decomposable at elevated temperatures or react with other substances to produce gases and are known in the art as blowing agents. Any blowing agent, the residues of which are foodgrade materials, may be employed herein provided it can be incorporated in the solution of film-forming material.

The film-forming material must be solubilized (dispersed) so as to be substantially free of granules. It will be recognized that, depending upon the particular film-forming material, the material may be solubilized directly in water or an aqueous alcoholic solvent or it may be necessary to heat the mixture in order to effect solubilization.

After solubilization of the film-forming material and incorporation of the latent gas, the solution is spray dried in accordance with conventional techniques in order to produce a grinding agent of suitable bulk density. For use herein, agents having a bulk density of about 3 to 25, preferably 4 to 10, pounds per cubic foot are required.

The grinding agent produced as described above is merely blended with the foodstuff and then the mixture ground using conventional techniques.

Example 1: K-Dex 4484 (National Starch and Chemical Corp.), a tapioca based dextrin, in an amount of 2,200 grams was added to 1,500 grams tepid water and the slurry heated to 88°C and held at this temperature for 20 to 30 minutes to effect solubilization. Another 1,800 grams of water was then added to cool the solution to 38° to 48°C. A solution of 110 grams ammonium carbonate in 624 grams of water was then added to the cooled dextrin solution and the mixture was spray dried at 250° to 400°C using an Anhydro spray drier. The bulk density of the resulting grinding agent was measured as 4.6 to 6.55 pounds per cubic foot.

Example 2: This example illustrates the grinding of almonds to produce a free-flowing powder in accordance with this process. 67 grams of almonds (containing approximately 54% fat) were mixed with 33 grams of the modified dextrin grinding agent produced in Example 1. The mixture was then ground in a Moulinex grinder to form a dry, powdered almond foodstuff.

Example 3: In order to reduce the level of the modified dextrin employed, the procedure described in Example 2 was repeated using a mixture of 10 parts of the modified dextrin of Example 1, with 23 parts of unmodified dextrin.

A satisfactory powdered, free-flowing almond product was produced, however, some caking occurred during the grinding operation itself, rendering the procedure more difficult than when the modified dextrin was employed as in Example 2.

Example 4: In order to ascertain whether other conventional materials could be used as grinding agents herein, the procedure described in Example 2 was repeated using: (a) 80 parts almonds with 20 parts modified dextrin of Example 1; (b) 90 parts almonds with 10 parts of modified dextrin of Example 1; (c) 67 parts almonds with 33 parts unmodified dextrins; (d) 67 parts almonds with 33 parts cornstarch.

Only the almonds ground using 20 grams of the modified dextrin produced an acceptable product. Thus, the use of low levels of modified dextrin as well as higher levels of other additives did not facilitate grinding of the high fat containing almonds.

Starches Modified with Sodium or Calcium Stearoyl-2-Lactate in Puddings

A process for preparing a pregelatinized modified starch suitable for use in instant puddings has been developed by *W.A. Mitchell and W.C. Seidel; U.S. Patent 4,260,642; April 7, 1981; assigned to General Foods Corporation.* This process comprises forming an aqueous slurry containing an ungelatinized starch and an effective amount of sodium or calcium stearoyl-2-lactylate, heating the slurry to a temperature and for a period of time sufficient to gelatinize the starch, and recovering the starch. The resultant pregelatinized modified starch possesses a smoother, creamier mouthfeel and has a high sheen.

This process first involves forming an aqueous slurry containing an ungelatinized starch and sodium stearoyl-2-lactylate (SSL) or calcium stearoyl-2-lactylate (CSL). Generally, the solids level of the slurry employed when treating the starch is such as to facilitate processing and handling of the slurry. For example, when the means employed is drum drying, the solids levels in the slurry can be up to about 50% by weight, preferably from 20 to 40% by weight. When the means of heating the slurry is a swept surface heat exchanger, preferably the solids level in the slurry is at most 10%, in order to prevent significant granule breakage during processing and to facilitate pumping and handling of the slurry.

When drum drying the starch slurry preferably the level of SSL is within the range of about 0.04 to 0.75% by weight of the dry starch and the level of CSL is within the range of about 0.08 to 2.25% by weight of the dry starch. When the starch is to be heated in a swept surface heat exchanger, preferably the level of SSL is within the range of about 0.04 to 1% by weight of the dry starch and the level of CSL is within the range of about 0.08 to 3% by weight of the dry starch.

Various types of ungelatinized starch can be modified by the process including starches such as tapioca starch, potato starch, cornstarch or waxy maize starch, with the preferred starch being tapioca because of its clean taste and minimal flavor masking effects when used in food products, such as instant puddings.

While the pH of the slurry is not critical to the process, preferably the pH is within the range of about 4 to 8 to further aid in granule stability. However, extremes in pH are to be avoided as they can be expected to cause chemical reactions when heated.

After forming the slurry of starch and SSL or CSL, the slurry is heated to a temperature and for a period of time sufficient to gelatinize the starch, and then the starch slurry is dried. In one preferred embodiment of this process, the slurry is heated and dried simultaneously in a drum dryer. The drum drying shatters the starch granules and forms light porous agglomerates which when in solution swell providing a smoother, creamier mouthfeel, a higher gloss and sheen and a less grainy texture without a loss of viscosity or body than a starch processed without SSL or CSL.

In another preferred embodiment, the starch can be heated in a conventional heat exchanger such as a swept surface heat exchanger (e.g., a Thermutator or a swept surface jacketed kettle), followed by recovering the starch by drying, e.g., air drying, spray drying, vacuum drying, alcohol precipitation, etc. Through the use of SSL or CSL in the above process, significantly more whole starch granules are retained which have been uniformly swelled to the maximum resulting in a high quality modified starch which has a creamier, smoother mouthfeel, a higher sheen and a less mucid and grainy texture than a starch processed without SSL or CSL.

Example 1: An aqueous slurry at a temperature of 20° to 25°C was formed containing ungelatinized tapioca starch at a level of 20% by weight of the slurry, SSL at a level of 0.15% by weight of the dry starch and polysorbate 60 at a level of 0.40% by weight of the dry starch. The slurry was pumped to a double drum dryer where the starch was gelatinized and dried. The drum dryer was operated under the following conditions: steam pressure of 5.8 kg/cm^2, drum speed of 3 rpm and drum gap of 0.003 cm. The dried modified pregelatinized starch was then ground and screened through a 230 mesh U.S. Standard screen.

The modified starch was then added at a level of 23.5 grams to an instant pudding mix containing approximately 76 grams of sugar, 27 grams of a cocoa blend, 1.5 grams of tetrasodium pyrophosphate, 3 grams of disodium orthophosphate, flavor and color.

Upon reconstituting the pudding mix with 474 ml of milk, a pudding was obtained which had a smooth, creamy mouthfeel and a high gloss and sheen. The textural and flavor attributes were judged to be at parity with an instant pudding prepared using a tapioca starch which had been chemically modified with epichlorohydrin and propylene oxide.

Example 2: An aqueous slurry was formed containing ungelatinized potato starch at a level of 20% by weight of the slurry, SSL at a level of 0.75% by weight of the dry starch and polysorbate 60 at a level of 1.0% by weight of the dry starch. The slurry was drum dried on a double drum drier, then ground and screened under the conditions specified in Example 1. When evaluated in an instant pudding mix, the pudding was judged to have a smoother, creamier mouthfeel and a higher sheen without a loss of viscosity or body than a starch pregelatinized without the above modifiers. The resultant modified starch possessed properties which mimicked the properties of chemically modified starches.

Texturized Starch in Tapioca Pudding

Among the advantages which characterize the ultra-high temperature short time (UHTST) method in ready to eat puddings are better preservation of product texture, better flavor, and better overall appearance. The UHTST method avoids the long heat treatments which characterize typical retort systems, and thus the natural pudding color is maintained, while at the same time off-flavors caused by heat itself or by reaction of the metal of the cooking unit under high heat, are avoided. Thus, the UHTST method is one well adapted to a light-colored and bland-flavored pudding such as tapioca.

Retorted puddings are subjected to high temperature sterilization after the individual containers are filled, and the retort method typically requires longer heat treatment than does the UHTST method. Thus, in a typical retort system, the pudding ingredients are heated in a steam bath to about 175°F, after which the containers are filled with the heated ingredients and sealed. The sealed containers are then placed in a continuous retort, and heated at 250°F with agitation for 10 to 20 minutes or more until sterilization is complete. The time required will depend on the pudding ingredients, size of container, and heat penetration rate. In view of the harshness of pressure and temperature conditions typical of the retort method, a method such as the UHTST is preferred in preparing the tapioca-style pudding.

H. Cheng; U.S. Patent 4,192,900; March 11, 1980; assigned to Merck & Co., Inc. has prepared an improved retorted or aseptically packaged, tapioca-style pudding wherein the improvement comprises the use of uniform starch particles. These particles are obtained by:

(a) preparing a dry homogeneous mixture of one or more starches selected from tapioca, corn, waxy maize, potato, sago, arrowroot and cereal; and one or more gelling hydrocolloids selected from sodium alginate, sodium pectate, hydroxypropylcellulose, methylcellulose, methylhydroxypropylcellulose, methylethylcellulose, carrageenan, furcellaran, agar, gelatin, a mixture of xanthan gum and locust bean gum, and curdlan;

(b) adding water and mixing until smooth;

(c) cool extruding the mixture through a $\frac{1}{32}$ to $\frac{1}{4}$ inch hole die and cutting to produce starch particles;

(d) where the gelling hydrocolloid selected is sodium alginate or sodium pectate, soaking the starch particles in a solution of calcium, barium, or magnesium ions;

(e) drying the starch particles to a moisture content of from 1 to 12%; and

(f) screening the starch particles to pass No. 10 and be retained by No. 14, U.S. Standard Sieve Series.

The starch particles having heat and shear stability during retorting or processing for aseptic packaging, yet swelling without dissolving during the processing, produce a tapioca-style pudding of improved tapioca-like appearance and texture.

The starches employed are selected from tapioca, corn, waxy maize, potato, sago, arrowroot and cereal starches such as rice and wheat. Derivatives of these starches, such as crosslinked etherified and esterified starches, may also be employed. Blends of two or more of these starches together may be employed. However, it is preferred, overall, to use tapioca starch in preparing the improved tapioca-style pudding of this process.

The gelling hydrocolloids employed in the process are selected from sodium alginate, sodium pectate, hydroxypropylcellulose, methylcellulose, methylhydroxypropylcellulose, methylethylcellulose, carrageenan, furcellaran, agar, gelatin, a mixture of xanthan gum and locust bean gum, and curdlan. Two or more of these hydrocolloids may be employed together. Overall, however, it is preferred to employ sodium alginate.

Sodium alginate, the preferred hydrocolloid along with sodium pectate, is gelled or crosslinked by contacting it with calcium, barium, or magnesium ions, as is explained in more detail below. It is preferred to employ sodium alginate with a molecular weight range of from 75,000 to 100,000.

The following hydrocolloids employed in this process are cold-water-soluble but thermal gelling with hot water: hydroxypropylcellulose, methylcellulose, methylhydroxypropylcellulsoe, and methylethylcellulose. The following hydrocolloids employed in this process are hot-water-soluble but gel on cooling: carageenan, furcellaran, agar, gelatin, a mixture of xanthan gum and locust bean gum, and curdlan. Thus, all of these hydrocolloids are gelled during the manufacturing process for the tapioca-style pudding of this process.

Example: A pudding was prepared from 59.9% tapioca starch, 0.1% sodium alginate, 2.0% pregelatinized modified starch (Redisol 88, A.E. Staley Mfg. Co.) and 38.0% water. The pudding was prepared in accordance with the following procedure:

(1) Blend all dry ingredients well in a Hobart mixing bowl.

(2) Add water and mix until smooth.

(3) Cool extrude through a $\frac{1}{16}$ inch hole die on a Hobart meat grinder.

(4) Cut the extruded strands into $\frac{1}{16}$ inch lengths.

(5) Soak in 3% $CaCl_2$ solution for 30 minutes and rinse with water.

(6) Dry in an air-forced oven at 110° to 130°F to a moisture content of less than 12%.

(7) Screen through No. 10 and No. 14 mesh screens.

(8) Collect products on No. 14 mesh screen as samples.

The uncooked product does not dissolve in cold water on stirring. The cooked product is firmer in texture and body than a control of tapioca starch only, and has excellent heat and shear stability.

Stabilized Pregelatinized Starches

R.R. Leshik and J.H. Katcher; U.S. Patent 4,341,809; July 27, 1982; assigned to General Foods Corporation have prepared a stabilized starch composition which is resistant to acid under dry storage conditions by adding a buffer to a slurry of starch, followed by drying the starch-buffer slurry.

Various starches which are subject to acid degradation under dry acidic storage conditions can be stabilized by this process including chemically or physically modified starches, raw starches or pregelatinized starches, as well as starches from various sources, such as tapioca, corn, waxy maize, potato, etc. The preferred starch which is useful in products such as instant pudding mixes is a pregelatinized modified starch, such as a crosslinked and hydroxypropylated starch, a crosslinked and acetylated starch, a crosslinked starch, etc. The slurry is preferably an aqueous slurry and the solids level of the slurry is at a level convenient for subsequent handling and drying, generally less than 60% solids by weight.

The buffer which is added to the starch slurry prior to drying, by definition will provide and maintain a neutral to alkaline environment for the dry starch particles to protect the starch from acid degradation. That is, the buffer will neutralize the acid which comes into contact with the starch particles to maintain an environment for the starch particles wherein acid degradation is prevented. Generally, the environment of the starch particles to prevent acid degradation must be maintained between a pH of 5 to 12, preferably 6 to 9. Suitable buffers include hydroxides of various metals or alkaline salts, such as trisodium citrate, trisodium phosphate, sodium hydroxide, etc. The buffer, critically is added to the slurry of starch in an amount effective to stabilize the dry starch by maintaining a pH environment for the dry starch particles which prevents acid degradation when the starch is dry mixed and stored with an acid.

The effective amount of buffer to stabilize the starch when mixed with acid and under storage conditions can vary depending upon the buffer and the type and amount of acid the dry starch is to be mixed with, but generally the amount of buffer is within the range of 0.05 to 10% by weight of the dry starch and preferably 0.25 to 3% by weight of the dry starch. The buffer in the amounts used generally does not appreciably effect the final pH of the finished product when the dry mix is hydrated. However, the buffer effectively controls the immediate environment of the dry starch particles, protecting the dry starch from degradation by the acid in a dry mix under storage conditions.

The slurry of starch and buffer is then dried by conventional processes, such as drum drying, flash drying, or spray drying. Conveniently, the drying procedure can also be employed to gelatinize the starch thus producing a pregelatinized, stabilized starch.

Typically, the starch stabilized by this process is found to be stable and maintain its desirable viscosity and textural characteristics after one year under dry acidic storage conditions, as compared to a starch without the buffer system in which unacceptable acid degradation is evident by five months under dry acidic storage conditions. Generally, accelerated storage conditions involve 9 to 12 weeks at 110°F.

Example: Trisodium citrate was added at a level of about 1% by weight of the starch solids to an aqueous, tapioca starch slurry which had been hydroxypropylated and crosslinked with propylene oxide and phosphorous oxychloride. The pH of slurry was about 6. Following drum drying this slurry, the codried starch composition at a level of about 15% by weight was blended into an instant yogurt pudding mix base containing about 3% by weight of malic acid as well as sugar, yogurt powder, gelatin, color, oil, emulsifiers, flavor and nonfat milk solids. A control sample of drum dried hydroxypropylated and crosslinked tapioca starch without added buffer was also bended at the same level into an identical instant yogurt pudding mix base containing malic acid.

Both samples were packaged in a foil pouch and placed in an accelerated storage of 9 and 12 weeks at 110°F and evaluated after the storage by beating about 125 grams of the instant pudding mix into 2 cups of cold milk, followed by allowing the pudding to set. The finished pudding had a pH of about 4.2. At 9 weeks the control sample of pudding mix upon hydration showed a substantial loss of viscosity of from 44,000 cp prior to storage to 13,000 cp after storage, as well as exhibiting a softer, looser and gummier texture after storage. At 9 weeks the pudding mix containing the codried buffered-starch composition upon hydration exhibited a viscosity of 44,000 cp prior to storage and substantially maintained this viscosity at a level of 34,000 cp after storage, as well as maintaining its desirable texture after storage. Even at 12 weeks the pudding mix containing the codried buffered-starch composition upon hydration maintained its viscosity at 35,000 cp, as well as maintaining its desirable texture after storage.

Texture Stabilizer in Whipped Products

M. Trop and A. Livne; U.S. Patent 4,338,347; July 6, 1982; assigned to Ben-Gurion University of the Negev Research and Development Authority, Israel provides a powdered composition suitable for mixing with a liquid to obtain a mousse product. This product comprises, by weight proportions, about 30 to 50 parts of a vegetable lipid whipping agent, about 5 to 15 parts milk powder, about 30 to 60 parts sugar, about 3 to 10 parts of a texture stabilizer and about 0.05 to 10 parts flavor additives. The lipid whipping agent comprises about 15 to 35% of sugar or corn syrup solids, about 6 to 11% sodium caseinate and about 50 to 70% of a lipid component, which lipid component comprises about 75 to 85% partially hydrogenated vegetable oil, about 10 to 12% lactylated fatty acid esters of glycerol and propylene glycol and about 8 to 10% of fatty acid mono- or diglycerides.

The vegetable lipid whipping agent which forms one ingredient of the composition has been developed to replace cream for making instant aerated products without the storing and handling disadvantages of fresh cream as a manufacturing ingredient. The whipping agent has a neutral taste and whippable properties but cannot be overwhipped.

The texture stabilizers of this composition can be any known products and preferred ingredients are precooked starch, instant gelatin, calcium lactate and tetrasodium pyrophosphate. Especially preferred is a combination of texture stabilizers in which precooked starch is the main stabilizer and minor amounts of calcium lactate and tetrasodium pyrophosphate are added thereto to give further firmness to the mousse product.

The mousse product itself is simply prepared by whipping together about 200 ml of cold liquid, i.e., milk, water or juice and the powdered compositions. For preparing the cream filling for napoleon cake, the higher range of powdered components to liquid is preferred.

To prepare the vegetable lipid whipping agents, an aqueous solution containing about 50 to 70% sugar or corn syrup solids and 18 to about 22% sodium caseinate is first heated to about 48° to 50°C. There is separately prepared a lipid composition composed of about 75 to 85% partially hydrogenated vegetable oil, e.g., palm kernel, coconut, cottonseed oil, etc., having a wetting point of about 30° to 50°C, about 10 to 12% lactylated fatty acid esters of glycerol and propylene glycol and about 8 to 10% of fatty acids mono- and/or diglycerides. The aqueous and lipid solutions are then combined in a 1:1 to 3:7 ratio to form an emulsion suspension the temperature of which is raised to 75°C for 30 minutes after which the dispersion is homogenized and spray dried to form the dry lipid whipping agent used in the example.

Example 1: A firm chocolate mousse is prepared from 40 grams lipid whipping agent, 40 grams sugar, 10 grams skim milk powder, 10 grams cocoa, 7.5 grams precooked starch, 0.17 gram calcium lactate, 0.17 gram tetrasodium pyrophosphate, and 0.3 gram chocolate flavor.

The above dry powder composition was whipped together with 200 ml cold water and a volume increase of about 2½ times was noted. The resulting chocolate mousse had excellent taste and firm texture which remained under refrigeration conditions for several days.

Example 2: A cream filling for a napoleon cake was prepared from 50 grams lipid whipping agent, 55 grams sugar, 15 grams whole milk powder, 9 grams precooked starch, 0.22 gram calcium lactate, 0.22 gram tetrasodium pyrophosphate, 0.2 gram vanilla flavor, and 0.02 gram yellow color.

The above dry mixture was whipped with 200 ml orange juice and the resulting mousse cream filling was spread between layers of flaky puff pastry to produce a tangy and delicious napoleon cake.

CORNSTARCH ADDITIVES

Starch Hydrolyzate in Egg White Foodstuffs

O. Iimura; U.S. Patent 4,138,507; February 6, 1979; assigned to Kewpie KK, Japan has prepared coagulated egg white foodstuffs comprising coagulated egg white, a water-combinable material, a viscosity increasing agent, and a starch hydrolyzate. These foodstuffs exhibit little or no water separation with no significant increase in hardness when they are thawed after storage in a frozen state, and at the same time, the thawed foodstuffs have excellent palatability.

An egg white liquid usable for the production of the coagulated egg white is one which is ordinarily obtained by breaking a shelled raw egg and separating the white from the yolk. The egg white liquid may be an egg white liquid obtained by thawing frozen egg white or a solution of a dried egg white in water.

Examples of the water-combinable materials usable for this process are carbohydrates such as cereal powders, starch and other edible powder compositions which are capable of absorbing and combining with water due to their polymer structure. Individual examples of the carbohydrate are cornstarch, rice starch, potato starch, and wheat flour. Denatured starches may also be used.

In order to avoid a significant hardness of the thawed product, the water-combinable material should be preferably added in a quantity of not greater than 10% by weight.

Ordinarily, for the guar gum, xanthan gum and tragacanth gum, the addition quantity is in the range of about 0.1 to 1.0% by weight, preferably 0.2 to 0.7% by weight, with respect to the total starting materials.

As the starch hydrolyzate, use may be made of a partially hydrolyzed product of dextrin. Derivatives of dextrin such as dextrin alcohol may also be used. Particularly, a dextrin alcohol may be advantageously used. In the case where the starch hydrolyzate is a dextrin, it is desirable that the hydrolyzate have a DE value of about 5 to 25, particularly about 7 to 22.

The quantity of the starch hydrolyzate added is in the range of 2 to 15% by weight, particularly 2.5 to 10% by weight, with respect to the total starting materials. When the quantity of the starch hydrolyzate is less than 2% by weight, the desired result is difficult to attain. On the other hand, in the case where the quantity of the starch hydrolyzate exceeds 10% by weight, the palatability inherent in the egg white is liable to deteriorate.

The coagulated egg white foodstuffs may be prepared, for example, by adding the water combinable material, the viscosity increasing agent and, if desired, the starch hydrolyzate to the egg white liquid, mixing these materials together, introducing the resulting mixture into a container of any suitable shape and heating the mixture to coagulate it. Alternatively, egg white mixture may be placed in a layer around the outer periphery of a coagulated yolk in the form of a rod to coagulate the resulting composite by heating thereby to produce a hard-boiled egg in the form of a rod.

In the production of the coagulated egg white foodstuffs, heating is carried out under a condition such that the coagulation of the egg white will be attained. Ordinarily, heating at a temperature of 80° to 100°C for a period of about 15 to 25 minutes is satisfactory. Furthermore, any additives other than the water combinable material, viscosity increasing agent and starch hydrolyzate, for example, table salt and other condiments may be added, if necessary. In this case, the other additives should be desirably added in a quantity such that they cause no hardening of the egg white. In order that the resulting product retain the characteristics as an egg white food, it is desirable that the egg white liquid constitute at least about 70% by weight of the total raw materials.

The coagulated egg white foodstuffs obtained according to this process when thawed after frozen storage exhibit far less water separation than conventional coagulated egg white foods. Ordinarily, little water separation is observable in the thawed coagulated egg white foods of this process. Further, if the coagulated egg white foods contain particularly the starch hydrolyzate, they exhibit little or no water separation with no attendance of a substantial increase in the hardness of the coagulated egg white, and at the same time develop an appearance of fine structure and possess a smooth palatability.

Example: Raw chicken hen eggs were broken and separated into the yolk and the egg white liquid. The yolk was heated and coagulated into a rod having a diameter of 20 mm. The resulting rod was introduced as a core into a bag made of polyethylene which had a diameter of 35 mm. Then, the egg white liquid with wheat flour, tragacanth gum, and a dextrin alcohol (having an average degree of polymerization of 12) added thereto was introduced around the outer surface of the core. The quantities of the wheat flour, tragacanth gum, and dextrin alcohol were 6%, 0.3%, and 7%, all by weight, respectively, based on the total starting raw materials except for the yolk. Then, the bag was tied at its two ends and heated at a temperature of 95°C for 20 minutes to coagulate the egg white liquid. A hard-boiled egg in the form of a rod was obtained.

High Amylose Starch in Potato Product

Partially or wholly cooked potato products, in various shapes, are available commercially. A common type are frozen fried potato products, such as French fries and puffs, which are meant to be cooked to completion or reheated for consumption either by frying in oil or by being baked in an oven.

Such frozen products, however, are not entirely satisfactory in that they tend to exude oil when cooked to completion by baking. In addition, those products made by reconstituting heat-treated potatoes, such as potato flakes and granules, lack the desired potato flavor and texture.

A toaster French fried potato product has been provided by *G. Finkel; U.S. Patent 4,135,004; January 16, 1979.* This product can be reheated in a toaster which overcomes the problems of oil exudation upon toasting and which has excellent texture and flavor.

Briefly stated, this product comprises a dough adapted for forming into fried potato products comprising a potato base, water, oil and a high amylose product, the oil and amylose product being present in an amount sufficient to give controlled oil absorption; to give shape retention during frying, and to prevent any significant exudation of oil from the product when toasted.

It is essential that the product contain a potato base; namely, comminuted raw potatoes, dehydrated potato granules, or potato flakes, or mixtures therof with the preferred material from a taste viewpoint being comminuted raw potatoes. If comminuted raw potato solids are used, they should not be more than about 55% by weight of the potato solids in the formulation. The remainder of the necessary potato solids can be supplied by the dehydrated potato flakes or granules. This limit on the amount of raw potato solids is due to their high moisture content which would raise the mash moisture above that necessary to have a suitable product.

A second essential component is a high amylose product. Examples of such are the amylose products resulting from the fractionation of whole starch into its respective amylose and amylopectin components, or to whole starch which is composed of at least 55%, by weight, of amylose. The amylose may be further treated as with heat and/or acids or with oxidizing agent to form so-called thin boiling products. Or, the amylose may be chemically derivatized, as by means of an esterification reaction which would thus yield amylose esters such, for example, as the acetate, propionate, and butyrate; or, by means of an etherification reaction which would thus yield amylose ethers such, for example, as the hydroxyethyl, hydroxypropyl, carboxymethyl or benzyl.

Of the high amylose starches, such as those obtained from potato, corn, tapioca, rice and the like, it is preferred to use the high amylose starch from corn and most preferred is amylose acetate prepared by reacting acetic anhydride with high amylose cornstarch having an amylose content of 55% by weight.

The final frozen product should contain no more than about 15% by weight of oil, and preferably about 8 to 14% by weight.

The remaining components are water and can include the usual seasoning ingredients such as salt, pepper, and the like, as well as antioxidants and chelating agents which are added to give the final products their effect.

Example: A potato dough was prepared containing 28.46% by weight dehydrated potato flakes, 4.74% by weight high amylose product (Amylomaize), 3.42% by weight vegetable oil, 61.67% by weight water, and 1.71% by weight salt.

The Amylomaise and water were formed into a slurry and, separately, the potato flakes, salt, and oil were thoroughly admixed. The oil mixture and slurry were then thoroughly blended with agitation to form a dough in which the oil and Amylomaize were substantially uniformly dispersed.

The dough temperature was then elevated to 120°F and the dough molded into a number of shaped strips comprising several interconnected rods shaped to resemble French fried potatoes. The size and shape of the strips was such that they fit into conventional toasters. The mold was also maintained at a temperature of 120°F during molding.

The shaped strips were then fried in a conventional fryer containing peanut oil until cooked at which point the shaped strips were self-supporting. The fried products had a lower water content, 45 to 50%, and a higher oil content, 8 to 15%.

The fried products were then frozen and kept frozen until desired to be eaten. At that time they were prepared for eating simply by being placed in a toaster and toasted. There was no dripping of oil during toasting and when toasting was completed, the toasted product had the taste and the texture of conventional fried potatoes.

Extenders for Cocoa Powders

J.R. Kimberly, Sr.; U.S. Patent 4,235,939; November 25, 1980; assigned to A.E. Staley Manufacturing Company has provided a synthetic base mix which possesses the functional attributes of cocoa powders in food products. This synthetic base mix comprises:

(A) from about 25 to about 300 parts by weight granular starch characterized as having (at 95°C) a swelling power of less than 22 and critical concentration value of at least 5.0; and

(B) 100 parts by weight of a defatted vegetable seed material with the material containing at least 35% by weight vegetable seed protein, and having an NSI (nitrogen solubility index) ranging from about 20 to 80.

The proportions of starch component (A) and seed material (B) are sufficient to provide a base mix which absorbs at least its total granular starch and seed material dry weight in water at 93°C. The aforementioned starch (A) and defatted vegetable seed material (B) components (hereinafter referred to as the "base mix ingredients") may be formulated with other additives such as triglycerides, thickening agents, coloring and flavoring additives, etc.

The unmodified high-amylose cornstarches containing at least 50% by weight amylose such as Mira-Quik "C" (A.E. Staley Manufacturing Co.) and "Amylon" 55 and 75 (National Starch and Chemical Corp.) are particularly effective as the granular starches.

Defatted soy flour, grits and concentrates containing from about 40 to 75% protein (preferably about 45 to 55%), less than 0.5% fat (ether extraction), about 5 to 35% carbohydrate (most typically from about 25 to 35%) and about 1 to 20% crude fiber (preferably about 2 to 5%) are the preferred defatted vegetable seed materials.

The relative proportion of granular starch and vegetable seed material in the base mix will depend upon the particular properties of the starch and seed material used to formulate the mix. For most bakery applications, the base mix will most typically contain about 75 to 150 parts for each 100 parts by weight defatted vegetable seed material.

Natural and Dutch cocoa powders may be simulated with the base mixes of this process. Natural cocoa powders are more viscous than Dutch cocoa powders. The viscosity characteristics of natural cocoa powders may be simulated by increasing the vegetable seed material content employing a higher NSI seed material or by adding gums to the base mix formula.

Conventional gums may be used to impart the desired viscosity effect to the base mix. Such gums include edible polymeric materials which thicken or gel when dissolved or dispersed in water. Illustrative gums include natural gums (e.g., gums found in nature), modified gums or semi-synthetic gums (e.g., chemical derivatives of natural materials and gums obtained by microbial fermentation of natural materials), synthetic gums (e.g., synthetically made from chemicals), mixtures thereof and the like.

For most applications, the viscosity of the base mix formulation will range from about 5 to 20 cp for the Dutch cocoa applications and from about 10 to 30 cp for those mixes simulating natural cocoa powders (at 10% total base mix dry solids in 23°C water including viscosity additives). The gum concentration in the base mix will depend upon its efficacy as a thickener. Such gums are generally present in the base mix at a level of less than 25 (e.g., 0 to 25) and preferably less than 20 parts by weight for each 100 parts by weight defatted seed material.

In addition to starch, defatted seed material and thickeners, the base mix may also be formulated with solid edible triglyceride solids (i.e., normally solid at 20°C). The triglyceride imparts a cocoa appearance (e.g., cocoa-like sheen) and improves upon its recipe functionality.

Example: A base mix simulating Dutch cocoa powder was prepared from 100 pbw crosslinked dent cornstarch, 100 pbw defatted soy flour, 22.5 pbw partially saturated vegetable oils, 15.9 pbw cocoa flavoring, and 6.1 pbw cocoa coloring.

The base mix was prepared by initially dry-blending at 43° to 46°C the inhibited, crosslinked cornstarch and the soy flour in a dry blender equipped with a water-jacket. The partially saturated vegetable oil was melted (60°C), atomized and dry-blended (43° to 46°C) into the heated base mix ingredients. The fat-containing base mix was then ambiently cooled to 23°C to provide a dry-blend of starch granules and soy flour particles uniformly coated with solidified fat.

The natural and artificial cocoa flavor and artificial cocoa coloring were homogeneously dry-blended into the fat containing base mix at 23°C. The flavored and colored base mix was screened through an 18 mesh screen (U.S. Sieve Series) so as to provide a base mix having an average particle size comparable to Dutch cocoa powder.

A conventional devil's food control cake containing 100% Dutch cocoa powder and one in which 50% of the Dutch cocoa powder was replaced with the base mix were prepared. The cake volume, symmetry, texture, moistness, mouthfeel, flavor, color and overall eating quality characteristics of the base mix containing recipe were comparable to the control cake recipe. Due to the sensitivity of the devil's food cake recipe system, replacement of 50% by weight of the Dutch cocoa powder with the synthetic cocoa extender herein would normally be expected to severely and adversely affect the baked cake quality.

Low Calorie Esterified Starches

R. Carrington and G. Halek; U.S. Patent 4,247,568; January 27, 1981; assigned to Pfizer Inc. have found that products which are substantially nondigestible can be produced by heating starch, or partially hydrolyzed starch, in the presence of edible di- or tricarboxylic acids.

By starch hydrolysate is meant a product of partial acidic or enzymatic hydrolysis of starch, which includes products variously known as thin boiling starches, corn syrup solids, white dextrins and amylase dextrins.

Edible di- and tribasic carboxylic acids which may be used include maleic, fumaric, succinic, adipic, malic, tartaric, citric and isocitric acids. The preferred acid is citric acid. Anhydrides which can be used include maleic, succinic and citric anhydrides. The edible acid may form from 1 to 25% by weight of the mixture of starch or starch hydrolysate and edible acid before heating.

The amount of edible acid used has an important influence on the physical properties of the end product, in particular on the proportion of water-soluble material in the product and on the ability of the insoluble material to take up water. The use of smaller amounts of edible acid increases the proportion of soluble material and the uptake of water by the insoluble material, while the use of larger amounts of edible acid reduces the proportion of soluble material to very low levels, and also reduces the water uptake of the insoluble material. Preferably the amount of edible acid used is in the range from 5 to 15% by weight of the mixture before heating.

It is important that the moisture content of the mixture should be low not only prior to but also during heating. The mixture must contain less than 5% water at all times, and preferably less than 2% by weight of the mixture. The mixture of edible acid and starch or starch hydrolysate may therefore have to be dried, e.g., by heating at a temperature below 120°C, preferably in the range from 60° to 120°C to reduce its water content to below 5% by weight, before heating in the range 140° to 220°C is begun. Water is formed by reaction between the carboxyl groups of the edible acid and the hydroxyl groups of the starch or starch hydrolysate, and also by condensation reactions between glucose moieties, and this must be removed continuously from the mixture during heating, by carrying out the heating step under reduced pressure, e.g., at below 100 mm of mercury, preferably below 50 mm of mercury.

The temperature of heating in the process is in the range from 140° to 220°C. The temperature actually used will depend on the physical nature of the mixture in this temperature range. When the mixture is in the liquid state, e.g., when mixtures containing starch hydrolysates are used, temperatures in the lower part of the range, e.g., 140° to 180°C may be used. When the mixture is in the solid state, higher temperatures, e.g., 165° to 220°C may be used. A temperature of about 180°C is generally preferred however.

Samples of the product are taken at regular intervals to determine the effect thereon of amylolytic enzymes, and when the product is resistant to the action of such enzymes then a nondigestible product has been formed.

A product is taken to be resistant to the action of amylolytic enzymes when not more than 15%, and preferably not more than 5%, of the product is hydrolyzed by the enzyme under standard conditions, as measured by its dextrose equivalent relative to that of starch.

Example 1: Raw maize starch powder having an average particle diameter of about 25 microns, containing about 10% moisture, was dried in an oven at 60°C to reduce its water content to 1% by weight. The dried powder (170 grams) was then blended with anhydrous citric acid powder of average particle diameter about 150 microns (30 grams) in conventional blending apparatus to give a homogeneous mixture. The mixture was then heated in a glass flask at 180°C on an oil bath at a pressure of 50 mm mercury (maintained by a vacuum pump) for 3 hours. Samples taken at intervals during the heating period showed that proportion of the product hydrolyzable by α-amylase was progressively reduced until at the end of the period it was only 4%. The proportion of water-soluble material was also progressively reduced to 10%.

After cooling to room temperature, the product (180 grams) was suspended twice in water (820 ml) and filtered, resuspended in water (820 ml), bleached by adding 72 ml of a 10% aqueous solution of sodium chlorite, neutralized to pH 6 by addition of 10% aqueous sodium carbonate, filtered, washed with water and dried to give 160 grams of a pale cream-colored product, only 1% of which was hydrolyzable by α-amylase and which was insoluble in water.

Example 2: Raw maize starch (1,020 grams) was added to a solution of adipic acid (180 grams) in water (13.2 ℓ) and the mixture was stirred and heated to 100°C. The mixture was held at this temperature for 1 hour before being cooled and spray-dried. The resulting solid (1,040 grams) which has a water content less than 5% was heated under vacuum at 180°C as in Example 1 for 3 hours. After cooling the crude product was bleached, neutralized and dried as described in Example 1 to give 856 grams of a product only 3.4% of which was hydrolyzable by α-amylase and which was 99% insoluble in water.

The nondigestible food additives produced by this process are useful for replacing wholly or in part the carbohydrates and/or fat content of foods, thereby providing dietetic foods of low calorie content. In particular, those which are substantially insoluble or have a low water uptake may be used as substitutes for flour and other starch-containing natural products, in cakes, biscuits, cookies, pastries, and other baked products, as well as unleavened products including pastas, e.g., spaghetti. Those which have a higher water uptake may be used as substitutes for potato starch in "instantized" and "snack" products including "instant" mashed potato and crisps.

At high starch replacement levels, particularly when more than 50% of the starch is replaced by a nondigestible food additive produced by this process, it has been found advantageous in order to produce a more acceptable food product, to add a small proportion of an emulsifying agent and this helps to retain a desirable crumbly texture in the finished food product. For example, it has been found that lecithin may be added at a level of from 0.5 to 3%, and preferably at a level of 2% based on the dry weight of the composition, for this purpose.

Cold Water Hydrating Starch

A process has been used by *W.G. Hunt, L.P. Kovats and E.M. Bovier; U.S. Patent*

4,281,111; July 28, 1981; assigned to Anheuser-Busch, Incorporated for preparing cold water hydrating starch which, when hydrated in water, produces a short creamy textured paste, especially useful in instant pudding formulations. These properties are attained without the use of the classical types of polyfunctional crosslinking reagents such as epichlorohydrin, phosphorus oxychloride, acrolein, etc. This process comprises treating starch with from 0.1 to 1% chlorine (preferably 0.4 to 0.6%) as sodium hypochlorite to inhibit the starch, but not to degrade or depolymerize it, and then reacting the inhibited starch with an alkylene oxide to produce a hydroxypropyl starch having a degree of substitution of about 15.5 to 18.5%.

Starches suitable for preparing the cold water hydrating starch product may be derived from tapioca, corn, high amylose, sweet potato, potato, waxy maize, canna, arrowroot, sorghum, waxy sorghum, waxy rice, sago, rice, etc. A preferred source of the starch is dent cornstarch.

The degree of starch granule inhibition with chlorine treatment must be carefully controlled; if the granule is not sufficiently inhibited it will fragment on hydration; the sol is of less viscosity and loses its short creamy texture. If the degree of inhibition is too great, the granule will not hydrate to produce the desired viscosity.

The preferred concentration of chlorine as sodium hypochlorite is 0.4 to 0.6% as shown by the Brabender viscosity (BU) data obtained in tests. The pH is readjusted to about 3.0 after the addition of sodium hypochlorite. It is preferred to conduct the addition of sodium hypochlorite at a temperature of about 95°F to about 105°F at a pH of about 3.0. At this temperature the reaction will require about 3 to 3½ hours.

After completion of the reaction, the starch slurry is treated with sodium bisulfite to inactivate the unreacted sodium hypochlorite in the slurry. After the unreacted sodium hypochlorite has been inactivated, the slurry is neutralized with a 2% aqueous sodium hydroxide solution or equivalent. The slurry is then diluted with two volumes of water and filtered. The filter cake is resuspended in water and filtered again. The thus washed starch derivative is dried in an air oven.

The chlorine treated starch is inhibited rather than being depolymerized or degraded as are the starches of the prior art when treated with chlorine or sodium hypochlorite. This process results in the solubility of the starch being decreased by the hypochlorite or chlorine oxidation step rather than being increased, as would normally be expected from a relatively heavy chlorine treatment. The oxidation step does not bring about degradation of the starch molecule, but does, however, result in mild inhibiting of the starch.

In a preferred hydropropylation procedure, isopropyl alcohol, sodium hydroxide and water are placed in a flask equipped with a stirrer and pressure gauge. A suitable substitute for isopropyl alcohol is any water miscible organic solvent. To the foregoing mixture are added the sodium hypochlorite inhibited starch and propylene oxide.

The reaction flask is sealed and the reaction is started by increasing the temperature in the water bath. The temperature is not critical. However, it must be at a high enough level to initiate the reaction. At temperatures below about 100°F, the reaction will proceed too slowly to be economically feasible. At temperatures in excess of 200°F, product recovery will be extremely difficult. Therefore, the reaction should be conducted at a temperature between about 100° and 200°F. After the pressure in the reaction flask has decreased, additional propylene oxide is then added, and the reaction is continued. At least 25% propylene oxide on a dry solids basis is added in two portions to the alkaline slurry.

After the reaction is complete, the reaction mixture in the flask is neutralized with acetic acid or equivalent. The resulting product is recovered by means of centrifugation.

The centrifuge cake is reslurried in aqueous isopropyl alcohol, and the washed product is recovered by means of centrifugation. The washed product is dried in a fluid bed dryer at about 215°F for 1 to 2 hours. The dried material can then be pulverized to a granular form, preferably to a size capable of passing an 80 mesh screen. The hydroxypropyl content of the product is about 15 to 20%, based on the weight of the starch.

In another embodiment of the process, the starch may be modified by reaction with chlorine gas, rather than sodium hypochlorite.

Example 1: A starch slurry was prepared by mixing 1 part dent cornstarch with 1.4 parts water. The pH of the starch slurry was adjusted to about 3.0 with dilute hydrochloric acid.

Chlorine, as sodium hypochlorite solution, containing 8% active chlorine, was added to the starch slurry in amounts ranging from 0.5 to 0.7% chlorine, based on the weight of starch. The reactions were conducted for 3 hours at a pH of 3.0. The reactions were conducted at 95°F in all cases, except one. This reaction, in which the concentration of sodium hypochlorite was 0.5%, was conducted at 105°F. The pH was readjusted to 3.0 after the addition of sodium hypochlorite.

Three hours after the completion of the reaction, the unreacted sodium hypochlorite in the starch slurry was inactivated with sodium bisulfite. The starch slurry was then neutralized with 2% aqueous sodium hydroxide solution.

The slurry was then washed with two volumes of water and filtered. The filtered cake was resuspended in water and filtered again. The washed starch derivative was then dried in an air oven to 12% moisture. Maximum Brabender viscosities obtained ranged from 310 to 380 BU.

Example 2: Three pounds of 100% isopropyl alcohol, 0.18 pound of 50% sodium hydroxide, and 0.3 pound of water were placed in a 5 liter three-necked flask equipped with a mechanical stirrer and pressure gauge. Three pounds of sodium hypochlorite inhibited starch and 0.35 pound of propylene oxide were added to the foregoing mixture. The flask was sealed, and the reaction was started by increasing the temperature to 130°F. After 24 hours of reaction, the pressure in the flask decreased. Another 0.35 pound of propylene oxide was added. The reaction was continued for another 24 hours at this (130°F) temperature. After 48 hours, the reaction mixture in the flask was neutralized with 0.15 pound of acetic acid. The product was recovered by centrifugation.

The starch product was reslurried in 3 pounds of 86% aqueous isopropyl alcohol, and the washed product was recovered by centrifugation. The product was dried in a fluid bed dryer for two hours at 215°F. The dried product was then milled to a size so that 90% was capable of passing a 200 mesh screen (U.S.S.). The hydroxypropyl content of the product was 17.4%.

TAPIOCA STARCHES

Crosslinked Tapioca Starch in Foodstuffs

A cold-water dispersible crosslinked tapioca starch has been prepared by *C.-W. Chiu and M.W. Rutenberg; U.S. Patent 4,229,489; October 21, 1980; assigned to National Starch and Chemical Corporation.*

In the preparation of this modified starch, native tapioca starch in its intact granular form is reacted with any crosslinking agent capable of forming linkages between the starch molecules. Typical crosslinking agents suitable are those approved for use in foods such as epichlorohydrin, linear dicarboxylic acid anhydrides, acrolein, phosphorus oxychloride, and soluble metaphosphates; however, other known crosslinking agents such as formaldehyde, cyanuric chloride, diisocyanates, divinyl sulfone, and the like may also be used if the product is not to be used in foods. Preferred crosslinking agents are phosphorus oxychloride, epichlorohydrin, sodium trimetaphosphate (STMP), and adipic-acetic anhydride (1:4), and most preferably phosphorus oxychloride.

The crosslinking reaction itself is carried out according to standard procedures described in the literature for preparing crosslinked, granular starches.

The reaction between starch and crosslinking agent may be carried out in aqueous medium, which is preferred, in which case the starch is slurried in water and adjusted to the proper pH, and the crosslinking agent added thereto.

The crosslinking reaction may be carried out at a temperature of 5° to 60°C, and preferably 20° to 40°C. It will be recognized that use of temperatures above about 60°C will be undesirable for this purpose, since granule swelling and filtration difficulties or gelatinization of the starch may result therefrom; and the starch must retain its granular form until it is drum-dried. Reaction time will vary depending mainly on the crosslinking agent and temperature used, but is typically about 0.2 to 24 hours.

After the crosslinking reaction is complete, the pH of the reaction mixture is generally adjusted to 5.5 to 6.5, using a common acid. The granular reaction product may be recovered by filtration and washed with water and dried prior to conversion. However, such a washing step is not necessary for purposes herein, and the crosslinked product may be converted directly without isolation thereof.

As an approximate guideline, the amount of phosphorus oxychloride used for reaction generally will vary from about 0.005 to 0.05% by weight on starch, depending on the desired degree of conversion of the starch and the type of drum drier; other crosslinking agents may be employed in different amounts.

Brabender viscosities of the crosslinked starch before conversion are best measured by the peak viscosity attained by the starch when it is heated in a pH 3 buffer solution to a maximum temperature of 95°C in a viscometer. The peak viscosity of the crosslinked starches which are applicable herein may range from about 250 to 850 BU. The amount of crosslinking is not only determined by peak viscosity; a more important parameter in defining the crosslinked starch intermediates is the time required for a slurry of the starch to reach peak viscosity, starting at 50°C. Thus, from the time the starch slurry is at 50°C the starch should reach peak viscosity in about 22 to 65 minutes.

After crosslinking, but before the drum-drying step, the starch is converted to its fluidity or thin-boiling form using a suitable method of degradation which results in the modified starch defined herein, such as mild acid hydrolysis with an acid (e.g., sulfuric or hydrochloric acid), conversion with hydrogen peroxide or enzyme conversion, etc. In a preferred embodiment, the starch is converted via acid hydrolysis.

In order that the gelling instant starch of this process will be obtained, not only must the BVD of the crosslinked and converted starch be within a narrowly defined range, but also the Brabender viscosity of this starch as measured at 80°C must have a certain minimum value, i.e., it must be at least about 100 BU when measured at 7% solids using a 350 cm-g cartridge. It will be recognized that this specified minimum value is an absolute minimum, and that it may need to be higher than 100 BU, depending on the level of crosslinking in the starch.

The crosslinked and converted starch obtained by the steps outlined above must be pregelatinized to become dispersible in cold water. The pregelatinization is accomplished by using a suitable drum drier, having a single drum or double drums, to dry the starch to a moisture level of about 12% or less. The starch slurry is typically fed onto the drum or drums through a perforated pipe or oscillating arm from a tank or vat provided with an agitator and a rotor.

The crosslinking levels, BVD and minimum viscosity ranges specified above are interdependent, but they also vary to some degree with the drum drier employed. It has been found that drum driers which produce higher shear than a laboratory single-drum drier (such as a commercial single-drum drier) require that the starch have a higher level of crosslinking to obtain this modified starch with its gelling properties.

Without limitation to any one theory, it is postulated that the unique gelling properties of the products herein are related to the release of amylose during drum drying. The combined treatment of crosslinking and conversion appears to control the amount and rate of amylose release on the drum drier, with the conversion step also possibly altering the size of the amylose. Drum driers with higher shear possibly tend to disrupt the granules to a greater extent, liberating more amylose at a faster rate, which amylose then retrogrades on the drum. If, however, the starch is more highly crosslinked, it will resist this disruption and can be successfully drum-dried using higher-shear apparatus without an adverse effect on its gelling properties.

After drying, the starch product is removed from the drum drier in sheet form and then pulverized to a powder. Alternatively, the product may be reduced to flake form, depending on the particular end-use, although the powdered form is preferred. Any conventional equipment such as a Fitz mill or hammer mill may be used to effect suitable flaking or pulverizing.

The final product obtained from the drum-drying operation is a cold-water dispersible starch which forms a gel when dispersed in water. The determination of gel formation and the measurement of gel strength are accomplished by subjective evaluation and by Bloom Gelometer readings. These two methods of measurement are not always consistent (due in part to the cohesiveness of some of the products), but for purposes herein, this modified starch must form a gel having a Bloom strength (as defined herein) of at least 50 grams, and preferably at least 90 grams.

Pregelatinized Starch for Instant Puddings

J.D. O'Rourke; U.S. Patent 4,215,152; July 29, 1980; assigned to General Foods Corporation has provided a pregelatinized starch suitable for use in instant puddings by drum drying a slurry containing an ungelatinized starch in water with a protein and an emulsifier.

The resultant modified starch when ground to a fine particle size can be used in making an instant pudding which closely matches those made using chemically modified pregelatinized starches. The emulsifier acts to impart a creaminess, smoothness and gloss to the resultant puddings, while the protein adds viscosity and increases the texture and mouthfeel. The combination of the two ingredients, emulsifier and protein, provides advantages over plain starch that neither ingredient can fully produce independent of the other.

Various starches may be modified according to the process including corn, potato, rice and amioca. However, tapioca starch is preferred as it has long been considered the prime starch source for instant puddings because of its clean taste and minimal flavor-making effects.

Various emulsifiers can be employed in the drum drying process to contribute creaminess, smoothness and gloss to the resultant pregelatinized starch when employed in instant puddings. These emulsifiers include lecithin, polyglycerol monostearate, mono- and diglycerides, glycerol lactyl palmitate, glycerol lactyl oleate, succinylated monoglycerides, sorbitan monopalmitate, polyglycerol monophosphate, and phosphated monoglycerides. However, the most preferred emulsifier is polysorbate 60 [polyoxyethylene (20) sorbitan monostearate] as it acts to impart characteristics to the resultant starch which most closely mimic a chemically modified starch, such as starches chemically modified with propylene oxide and either epichlorohydrin or phosphorus oxychloride.

The protein employed in the slurry can be gelatin, egg albumin or soy protein isolate, but the most preferable protein is sodium caseinate. Sodium caseinate when combined with an emulsifier, especially polysorbate 60, imparts optimum characteristics to the resultant pregelatinized starch which most closely mimics a chemically modified starch, such as crosslinked starch modified with propylene oxide.

Preferably when a starch, such as tapioca starch, is modified with polysorbate 60 and sodium caseinate, the sodium caseinate is employed at levels of about 0.05 to 2% by weight of the starch, and polysorbate 60 is employed at levels of from about 0.2 to 1% by weight of the starch.

When a slurry is formed preparatory to drum drying, the level of starch in the slurry must be such as can be effectively dried in the particular drum drier employed, as is common in the art. Generally, the level of starch in the slurry is within about 15 to 50%, preferably 20 to 40% by weight of the slurry. After the addition of an effective amount of a protein and an emulsifier, the slurry is then drum dried under conditions tailored to optimize the textural, sheeting and drying characteristics of the particular starch being processed. Either a single drum drier or a double drum drier can be employed. A double drum drier is preferably employed under the following conditions: water 50 to 85% by weight of the slurry; starch 15 to 50% by weight of the slurry; slurry temperature 5° to 45°C; slurry pH of 4 to 8; holding time of 0 to 24 hours; drier gap of 0.001 to 0.005 cm; rpm of 0.5 to 7; and steam pressure of 0.7 to 7 kg/cm^2. Optimum conditions for a double drum drier would generally be that the starch, at a level of about 20% by weight of the slurry, is slurried in the water with a protein and an emulsifier at a temperature of about 20°C, the slurry being adjusted to a pH of 6 and the drum drier being adjusted so that the holding time is one hour, the drier gap is about 0.003 cm, the rpm is 3 and the steam pressure is 4.2 kg/cm^2.

After drum drying, the resultant pregelatinized starch is then ground and screened to a size suitable for use in a product, such as an instant pudding. Preferably, the dried pregelatinized starch is ground to a size wherein the particles pass through a 200 mesh U.S. Standard screen, optimally to a size wherein at least 95% by weight of the particles pass through a 230 mesh U.S. Standard screen.

Example 1: A slurry was formed containing tapioca starch in water, the starch being present at a level of about 40% by weight of the slurry. Polysorbate 60 at a level of 0.4% by weight of the tapioca starch and sodium caseinate at a level of 0.2% by weight of the tapioca starch were added and mixed into the slurry. The slurry at room temperature was adjusted to a pH of 6 with phosphoric acid and then dried on a single drum drier under the following conditions: steam pressure of 2.8 kg/cm^2; a holding time of 1 hour; and with the applicator roll and rpm adjusted to attain a minimum sheet thickness and a moisture content of 4 to 5%. After drying, the starch was ground and screened through a 200 mesh U.S. Standard screen.

When evaluated in an instant pudding mix, the resultant prepared pudding was found to be very near parity to puddings prepared utilizing a chemically modified starch, particularly a tapioca starch modified with propylene oxide and epichlorohydrin. The resultant prepared pudding possessed a smooth, creamy mouthfeel and a glossy surface, but without a mucid mouthfeel.

Example 2: Polysorbate 60 at a level of 0.4% by weight of the starch and sodium caseinate at a level of 0.2% by weight of the starch were added and mixed into an aqueous slurry containing 40% of amioca starch by weight of the slurry. The slurry, at room temperature, was adjusted to a pH of 6 with phosphoric acid and then dried on a double drum dryer under the following conditions: holding time of 1 hour; dryer gap of 0.003 cm; rpm of 3 and steam pressure of 5.6 kg/cm^2. After drying, the starch was ground and screened through a 230 mesh U.S. Standard screen.

When the modified amioca starch was evaluated in an instant pudding mix, the resultant pudding had a smooth and creamy mouthfeel, a firm texture and a glossy surface, but without a mucid mouthfeel. The resultant pudding was judged to be near parity to puddings prepared utilizing a chemically modified amioca starch, i.e., amioca starch modified with propylene oxide and epichlorohydrin.

In Dispersible Chocolate Liquor

Chocolate liquor is a low melting solid containing approximately 53% by weight of fat. Because of its high fat content, chocolate liquor is not water dispersible and cannot be used directly in dry mixes which must be reconstituted in water. As a consequence, no dry food formulations containing chocolate liquor have been on the market. Chocolate liquor is introduced into commercially available dry mixes in the form of cocoa, which is a powdered chocolate liquor having most of the fat removed. However, cocoa is not a good substitute for chocolate liquor because much of the true flavor of the chocolate is lost in the defatting process by which cocoa is produced.

B.H. Nappen and N.G. Marotta; U.S. Patent 4,191,786; March 4, 1980; assigned to National Starch and Chemical Corporation have provided an improved process for preparing dispersible chocolate liquor for use in dry food mix applications. This comprises the steps of:

(a)   mixing melted chocolate liquor and starch in a ratio of melted chocolate liquor to starch of from 10:2.5 to 10:15, with 5 to 25%, by total weight of the mixture, of water;

(b)   passing the mixture through a heated extruder at an elevated pressure and temperature within the range of 110° to 135°C for a period of time sufficient to partially hydrate the starch; and

(c)   extruding the mixture through an orifice.

Preferably, the ratio of melted chocolate liquor to starch is from 10:3 to 10:10, and most preferably 10:3 to 10:5.

In another preferred embodiment, the amount of water in the mixture is 5 to 10% by total weight of the mixture, whereby the product is extruded directly in powder form.

The extrusion process for preparing dispersible chocolate liquor offers distinct advantages over the analogous drum-drying process of the prior art. The extrusion process has lower energy requirements because the extrusion step is conducted at lower temperatures and no further drying of the product is necessary.

The particular starch used is not an essential feature of this process. Satisfactory results are achieved using various starches suitable for use in food products such as those derived from corn, potato, rice, sago, tapioca, waxy maize, wheat, etc., with tapioca starch being preferred due to its better taste and low viscosity characteristics. The starch may be employed in its granular or pregelatinized form. Furthermore, modified starches such as dextrins prepared by the hydrolytic action of acids and/or heat, oxidized starches prepared by treatment with oxidants, e.g., sodium hypochlorite, and fluidity, or thin-boiling starches prepared by enzyme conversion or by mild acid hydrolysis may be employed. In addition, the starch may be chemically derivatized as by means of an esterification reaction to give esters, for example, the acetates, propionates, and butyrates; or by etherification to yield, for example, hydroxyethyl, hydroxypropyl or carboxymethyl ethers.

The starch is mixed with the melted chocolate liquor and water in the desired proportions using any suitable blending or mixing equipment.

The amount of added water required to complete the extrusion mixture must be sufficient to prevent the components from scorching under the particular extrusion conditions employed. This specific amount will depend on the starch, the form of end-product desired, e.g., powder or chips, and the temperature and pressure employed in the extrusion operation. In general, concentrations of water ranging from about 5 to 25%, preferably 5 to 10%, based on the total weight of the mixture, are employed. It should be noted that the moisture which may be inherently present in the various components of the composition is not included in determining the amount of water which is to be added to the mixture. When lower levels of water, i.e., 5 to 8%, are employed, the extruded product is obtained directly in powder form. Higher amounts of water, e.g., 15 to 25%, result in a strand of product which can be cut into pieces and shaped as desired. If the mixture contains amounts of water over about 25%, however, a satisfactory product will not be obtained.

Before the extrusion operation, artificial flavorings, colorings, etc., may be added to the mixture, although this is not deemed necessary since the natural chocolate taste and aroma are retained to such a large extent in the product of this process.

When the mixture is thoroughly blended, it is then passed through a heated extruder, by means of a pump, ram, double motion ribbon blender, or any other suitable apparatus.

The temperature required within the extruder depends upon the amount and type of starch present in the mixture as well as on the moisture content thereof, but generally must be maintained within the range of about 230° to 275°F (110° to 135°C) to prevent scorching of the product in the extruder.

The temperature within the forming section of the extruder and the temperature of the extruder die itself will in most instances be kept within the range of about 75° to 200°F (24° to 93°C). The precise temperature employed within the above cited range is also directly related to the composition of the extruded mixture.

Example: A chocolate liquor product designated as Sample A was prepared as follows: A total of 22.50 pbw hot water was added to 54.25 pbw commercially obtained chocolate liquor until the liquor was thoroughly melted. Tapioca starch was then added in an amount of 23.25 pbw. The resulting mixture was stirred and then fed into a Wenger X-5 extruder at a temperature of 121°C at 500 rpm and at a feed rate so as to have a residence time in the extruder of about 30 seconds. The chocolate product was removed before one revolution was complete and was obtained as an elongated rope, which was sliced into small pieces. The product was brown in color, homogeneous, and completely dispersible in water.

To test the effect of extrusion temperature on the product, the mixture of Sample A was extruded in an identical manner as described above except that the temperature of the extruder was raised to 149°C, the rpm to 750, and the feed rate was increased slightly. The product obtained thereby was scorched due to the high temperature of the extruder and blocked the extruder barrel so that it was difficult to remove. It can be seen that extruder temperatures of over about 135°C should not be employed.

In Frozen Whipped Toppings

A process is provided by *W.J. Dell, W.E. Flango, Jr., W.H. Povall, L.H. Freed and S.D. Fencl; U.S. Patent 4,251,560; February 17, 1981; assigned to General Foods Corporation* for preparing a frozen whipped topping composition containing milk fat, which upon thawing, has an extended refrigerator shelf life.

This process makes it possible to prepare a frozen whipped topping composition containing real cream which may be distributed and solid in a frozen state and which upon thawing retains its excellent volume, smooth, continuous, light and fluffy texture, and eating properties for an extended period of time. The thawed composition may be stored at refrigeration temperatures for a period of about 5 to 7 days or longer without an apparent loss in volume, texture and eating properties. Thus, the composition may be described as having exceptional freeze-thaw stability as well as extended stability upon thawing at refrigerated temperatures.

Critical to stabilizing milk fat in this frozen whipped topping is employing a modified starch to stabilize the emulsion and provide the freeze-thaw capabilities while withstanding homogenization and high temperatures during processing of the topping. The specified modified starch is critical in that other starches or gums have not been found to impart the requisite freeze-thaw and refrigerator shelf life stability or impart the desired volume, texture and eating qualities to the frozen whipped topping composition. The starch is modified by crosslinking and hydroxypropylation with a crosslinked and hydroxypropylated tapioca starch providing the optimum stability and characteristics upon thawing the frozen whipped topping although other crosslinked and hydroxypropylated starches, such as waxy maize starch, are also suitable.

The starch is hydroxypropylated, for example, with propylene oxide, to preferably a level of at least about 0.2% with a preferred upper limit of 6%, by weight of hydroxypropyl groups by weight of the starch.

As measured on a Brabender Viscoamylograph this crosslinked and hydroxypropylated starch should have a viscosity of about 80 to 500 units at 95°C and after 10 min hold at 95°C the increase in viscosity should be within the range of about 0 to 100 units. Preferably, the crosslinked and hydroxypropylated starch has a viscosity of about 150 to 350 Brabender Units at 95°C.

The crosslinked and hydroxypropylated starch may be prepared by, for example, suspending 2 kg of tapioca starch in 3 liters of water containing 750 grams of sodium sulfate adjusted to a pH of 11. Propylene oxide (hydroxypropylation agent) at a level of 400 ml was added and the suspension stirred at room temperature for 20 hours. Phosphorus oxychloride (crosslinking agent) at a level of 0.024 ml was then added and the suspension stirred for an additional 2 hours. The modified starch was then filtered, washed thoroughly with water and air dried to 10% moisture.

The resultant modified starch had a hydroxypropyl content of 0.4% and as measured on a Brabender Viscoamylograph had a viscosity of 180 units at 95°C and a 60 unit increase during the 10 min hold at 95°C. The modified tapioca starch is preferably employed in the frozen whipped topping at a level of 0.05 to 1.0% by weight of the composition, and optimally at levels within the range of about 0.25 to 0.5% by weight of the composition.

Other ingredients which may be included in the frozen whipped topping compositions prepared by the process are emulsifiers, stabilizers, carbohydrates, flavoring agents, colorants or dyes, vitamins, minerals, and the like.

Example: The frozen topping composition was prepared containing the following in percent by weight: 62.5% heavy cream (40% fat), 20.5% sugar, 14.4% water,

1.2% dextrose, 0.5% crosslinked and hydroxypropylated tapioca starch, 0.3% polysorbate 60, 0.2% sorbitan monostearate, 0.2% flavor, 0.1% xanthan gum, and 0.1% guar gum.

The tapioca starch had a hydroxypropyl content of 0.4% and at 95°C a viscosity of 180 units and a 60 unit increase during the 10 min hold at 95°C, as measured on a Brabender Viscoamylograph having a 700 CM GMS cartridge, operated at 75 rpm with 30 grams of dry starch in a total charge weight of 500 grams. The ingredients were mixed together and then pasteurized at 160°F (70°C) for 30 min. The pasteurized mix is then homogenized in two stages to form the emulsion. The first stage homogenization employing pressures of about 7,200 psi (500 kg/cm^2) and the second stage employing pressures of 800 psi (55 kg/cm^2). The homogenized mix is then cooled for 20 min at a temperature of 32° to 36°F (0° to 2°C) to allow fat crystallization. The cooled mix is then whipped and aerated in a Votator C.R. Mixer to above 200% overrun. The whipped mix is then packaged and frozen.

The composition so prepared is characterized by its excellent freeze-thaw stability even after several freeze-thaw cycles. After thawing and storage at refrigerator temperatures about 40°F (5°C) for 5 days and longer the texture remained light, fluffy, continuous and smooth and did not become loose, soupy (no resilience), open textured (grainy, webby), or exude free liquid. Over the 5 days refrigeration storage the thawed topping composition maintained a mouthfeel, texture, volume, appearance and eating quality characteristic of freshly prepared whipped cream.

OTHER STARCHES

Cow Cockle Starch as Clouding Agent

Cow cockle starch is obtained from the seed of the *Saponaria vaccaria* plant, commonly referred to as cow cockle, cow soapwort or cow fat. The starch contains the normal amount of amylose (iodine affinity value of 4.3), as well as the normal amounts of ash, fat and protein. It gelatinizes in the range of 60° to 65°C. Cow cockle starch is a natural product and, as with other food starches, it is stable, does not affect the flavor of a product, and does not change in flavor during storage.

O.B. Wurzburg and J.M. Lenchin; U.S. Patent 4,279,940; July 21, 1981; assigned to National Starch and Chemical Corporation have found that cow cockle starch may be used to cloud fluids. It may be added directly to the fluid or, in the case of food and beverage products, to dry powder mixes, flavor concentrates, and/or syrups. When the product contains a nonessential oil flavorant, the starch provides the entire cloud. When the product contains an essential oil flavorant, the starch enhances the cloud.

If used in a liquid system, it is recommended that cow cockle starch be used in combination with a preservative in order to prevent bacterial growth. Any preservative approved for use in foods or beverages is suitable for use with the starch in the cloud blends. The practitioner will recognize that the amount of preservative needed will depend upon the pH of the final product. Typical preservatives include among others sodium benzoate and methyl- or propyl-p-hydroxybenzoate. If a food or beverage product is not involved, it may be possible to use other preservatives.

Cow cockle starch may be added directly to the finished fluid which is to be clouded. Alternatively the starch may be added in the form of a dry or liquid cloud blend. Liquid cloud blends should contain a preservative unless they will be used immediately. The fluids to be clouded include water, alcohols and/or juices. The fluids may be flavored, sweetened, carbonated and/or colored, depending upon the type of product desired. The fluids used in the preparation of beverages generally contain additional preservatives and, if a soft drink is being prepared, sufficient edible acid to provide the desired pH in the finished drink.

Example 1: This example demonstrates the use of cow cockle starch to enhance the cloud of a beverage prepared from flavor oil emulsion. An orange flavor emulsion concentrate was prepared by dissolving 10.5 pbw Purity Gum BE (a modified food starch, National Starch and Chemical Corp.) in 78.56 pbw water containing 0.1 pbw sodium benzoate, 0.3 pbw citric acid and 0.01 pbw FD&C Yellow No. 6. A total of 10.5 pbw of single fold orange oil blend was added under moderate agitation. The mixture was then passed through a two-stage (2,500 and 500 Pa) Gaulin homogenizer.

An unflavored syrup was prepared by mixing together 79.39 pbw Nulomoline 11 (an inverted sugar product, Sucreft Corp.), 16.98 pbw water, 0.48 pbw sodium benzoate, 2.42 pbw citric acid, 0.03 pbw FD&C Yellow No. 6, and 0.7 pbw cow cockle starch.

A flavored syrup was prepared by adding 1.66 pbw flavor concentrate to 98.34 pbw of the syrup and mixing thoroughly. The final orange-flavored beverage was prepared by thoroughly mixing 21.4 pbw of the flavored syrup with 295 pbw of carbonated water. The finished beverage contained about 0.047% cow cockle starch. The beverage's transmittance was 1.5% (540 nm) after 1 hr compared with 38.0% for a control beverage prepared without the cow cockle starch. The results show that the starch greatly enhanced the cloud.

Example 2: This example demonstrates the use of cow cockle starch to provide the cloud in a beverage prepared from a flavor extract. A flavored syrup was prepared by dissolving 2 pbw of an orange flavor extract in 98 pbw of a syrup prepared by mixing together 79.39 pbw Nulomoline 11, 15.68 pbw water, 0.48 pbw sodium benzoate, 2.42 pbw citric acid, 0.03 pbw FD&C Yellow No. 6, and 1.5 pbw of cow cockle starch.

The final orange-flavored beverage was prepared by thoroughly mixing 21.4 pbw of the flavored syrup with 295 pbw of carbonated water. The finished beverage contained about 0.1% cow cockle starch.

The beverage's transmittance was 1.0% (540 nm) after 1 hour compared with 74.5% for a control beverage prepared without cow cockle starch. The results show that the starch provided a good cloud.

Example 3: This example demonstrates the use of cow cockle starch in a gelatin product. A total of 0.2 g starch was dry blended with 13.3 g orange gelatin and 100 ml water were added. The product's transmittance was 3.5% (540 nm) after 1 hour.

Example 4: This example demonstrates the use of cow cockle starch in a frozen juice product. A total of 0.1 g starch was suspended in 100 ml of orange juice and the mixture was frozen. The juice's transmittance was 2.5% (540 nm) after 1 hour.

Separation of Flour from Oats

A process for the separation of a flour fraction, a bran fraction and oil from comminuted oats in which gum does not cause significant process problems and in which the flour is essentially free of gum has been found by *R.W. Oughton; U.S. Patent 4,211,801; July 8, 1980; assigned to Du Pont of Canada Limited, Canada.* Accordingly, this process provides for the separation of a substantially gum-free flour from oats by:

(a) admixing comminuted oats with an organic solvent, the solvent being capable of extracting oat oil from the oats, and

(b) separating substantially gum-free flour from the admixture of comminuted oats and solvent, the amount of flour separated from the admixture being at least 20% by weight of the comminuted oats.

The comminuted oats used in the process are preferably dehulled oats. The dehulled oats, herein frequently referred to as groat, are comminuted in order to facilitate extraction of oil and to facilitate separation of the comminuted groat so obtained into a flour fraction and a bran fraction. Conventional comminuting techniques, for example, pinmilling, hammer milling, corrugated rollers and other shearing techniques, would appear to produce an acceptable comminuted groat.

The comminuted groat is added to a solvent for the oil in the oats. The solvents used should be acceptable for use with food, for example, be nontoxic at the levels remaining in the products produced, not cause the formation of toxic materials in the product and not have a significant deleterious effect on the nutritional value of the product, and must be capable of causing separation of the flour and bran fractions.

Examples of solvents are pentane, hexane, heptane, cyclohexane and alcohols of 1 to 4 carbon atoms, and mixtures thereof; as used herein the solvents hexane and heptane include those solvents referred to in the food industry as hexane and heptane. The preferred solvent is hexane.

According to this process there are a number of techniques for separating the flour fraction from the bran fraction. In a so-called batch process the comminuted groat and hexane are thoroughly mixed for a period of time so as to extract oil from the comminuted groat. The mixing may then be adjusted to effect separation of the mixture of comminuted groat and hexane into a flour fraction and a bran fraction, or the mixing may be discontinued. If mixing is discontinued, the bran fraction tends to settle relatively rapidly, thereby allowing the flour fraction to be separated from the bran fraction of the admixture. The flour fraction should be separated as soon as practical after cessation of mixing as the flour in the flour fraction also tends to settle thereby making separation from the bran fraction less efficient and/or more difficult.

In the process involving the cessation of mixing it is preferable to repeat the above sequence of steps one or more times, for example, by adding hexane each time and remixing, in order to effect a high degree of separation of the comminuted groat into flour and bran. Alternatively the admixture of comminuted groat and hexane may be separated into fractions by sieving the admixture. The mesh size of the sieve selected will depend primarily on the degree of separation desired. Preferably a sieve having a fine mesh, e.g., 300 or finer, is used, the use of a 325 mesh Tyler sieve being exemplified hereinafter. The bran fraction is retained on the sieve and may be used as such or subjected to further comminution. Preferably the mesh size of the sieve is such that the solid component of the flour fraction, which passes through the sieve, is white and essentially free of bran.

In the so-called continuous process, separation of the admixture of comminuted groat and hexane may be effected by careful control of the mixing of the comminuted groat and hexane, especially immediately prior to and during separation of the flour and bran fractions. Such control of the mixing is essential to cause a nonuniform distribution of comminuted groat in the hexane and separation of the comminuted groat into fractions. Separation may be effected by removing a portion of the mixture of comminuted groat and hexane, separating, e.g., by using a sieve, the flour fraction and recycling the bran fraction.

Example 1: Dehulled Hinoat oats (Agriculture Canada) were ground (comminuted) in a Casella grain mill having a 2.5 mm diameter circular hole sieve. 250 g of the resultant ground groat were placed in a vertical cylinder having a diameter of 6.3 cm and a height of 40.6 cm. 600 ml of hexane were added to the cylinder and the resultant mixture was maintained as a slurry, using an agitator, for 20 min at ambient temperature. Agitation was then stopped and the upper bran-free layer was siphoned off. The procedure was repeated four times with 500 ml of hexane being added each time. The bran layer was then centrifuged to separate hexane and the bran fraction was dried under vacuum at room temperature and weighed.

The hexane fractions were passed through a 325 Tyler mesh sieve and oversize particles were added to the bran fraction. Flour was then centrifuged from the hexane solution. Oil was obtained from the resultant hexane solution by evaporating the hexane.

The centrifuged flour was admixed, as a slurry, with hexane to remove any adsorbed oil and recentrifuged. The flour was then dried under vacuum. The white flour obtained was very bland to taste. This first separation produced 89.6 g flour, 12.3 g of oil and 149.5 g of bran.

The bran obtained above was reground in the Casella grain mill using a 0.5 mm diameter circular hole sieve. 112 g of the reground bran were treated three times with hexane using the procedure described above for ground groat. The bran and flour fractions so obtained were dried. This second separation produced 28.7 g flour and 82.9 g bran.

A 55 g sample of this bran fraction was wet ground in hexane in a puck mill and retreated with hexane. The resultant bran and flour fractions were dried. This third separation produced 5.1 g flour and 49.6 g bran.

Example 2: A sample of dehulled Hinoat oat was pinmilled using an Alpine Contraplex 250 CW pinmill. It is believed, based on data from the pinmilling of wheat, that 90% of the resultant comminuted groat would pass through a 325 mesh Tyler sieve. 200 g of the comminuted groat were placed in a column that was 38.1 cm high, 6.3 cm in diameter and adapted so that hexane could be fed to the bottom of the column and removed, by means of an overflow, near the top of the column. The column was equipped with a stirrer, the blades of which were near the bottom of the column. Over a period of about 2 hours 2,000 ml of hexane was passed through the column. During this period the stirrer was adjusted so that a separation of the comminuted groat/hexane mixture into a flour fraction and a bran fraction occurred approximately 7.7 cm below the overflow.

The bran was separated from the hexane in the column and dried under vacuum. The flour, which passed with the hexane through the overflow, was centrifuged from the hexane and dried. Oil was separated from the hexane. This process produced 128.0 g flour, 62.7 g bran and 12.8 g oil.

R.W. Oughton; U.S. Patent 4,211,695; July 8, 1980; assigned to Du Pont of Canada, Limited, Canada has also provided a modification of the process of obtaining flour from oats. This process comprises:

 (a)   admixing comminuted oats with an organic solvent, the solvent being capable of extracting oat oil from the oats;

 (b)   forming a slurry of the admixture of comminuted oats and solvent; and

 (c)   subjecting the slurry to the influence of centrifugal force and thereby separating the comminuted oats in the slurry into at least two fractions, the fractions differing in composition.

To effect separation of the comminuted groats into fractions in this modification, the slurry is subjected to centrifugal force. The means used to subject the slurry to centrifugal force is a centrifugal separator, preferably a centrifugal separator capable of being operated on a continuous or semi-continuous basis. Examples of centrifugal separators are continuous centrifuges including semi-continuous centrifuges, and, in particular, hydrocyclones.

In order to effect separation of the comminuted groats into fractions in a hydrocyclone, the slurry of comminuted groats and hexane is fed to the hydrocyclone whereupon the slurry is subjected to centrifugal force. Under such conditions, fractionation of the comminuted groats in the slurry tends to occur. The operation of the hydrocyclone so as to obtain a desired fractionation of the comminuted groats in the slurry will depend on a number of process variables. Examples of such variables are the degree of comminution of the comminuted groats, the amount of solid material in the slurry, the pressure drop across the hydrocyclone, the ratio of the flows through the so-called "underflow" and "overflow" outlets, the difference in density between the solvent and the particles of the comminuted groats, the viscosity of the solvent and the like.

The operation of the hydrocyclone is adjusted so that a desired fractionation of the comminuted groats in the slurry fed to the hydrocyclone is obtained. In particular, the hydrocyclone is operated so that a bran fraction, in hexane, flows out the underflow outlet and a flour fraction, in hexane, flows out the overflow outlet. Preferably one fraction contains at least 20% and in particular at least 40% of the comminuted oats.

Example 1: Dehulled Hinoat oats (Agriculture Canada) were comminuted on an Alpine Contraplex pinmill operating at approximately 19,000 rpm. 1,000 g of comminuted groats (24.1% protein) were admixed with 6.06 ℓ of hexane and maintained in the form of a slurry for 15 minutes. The slurry was then passed through a hydrocyclone (Dorr-Oliver Doxie Type A Impurity Eliminator) using an inlet pressure of 1.4 kg/cm^2, the slurry being separated into an overflow or so-called flour fraction and an underflow or so-called bran fraction. An additional liter of hexane was then passed through the hydrocyclone so as to remove any residual amounts of the slurry. The overflow and underflow were each centrifuged so as to separate the solids. The solids from the overflow were readmixed with 0.5 liter of hexane, to remove any residual oil, and then recentrifuged. The resultant overflow solids and the underflow solids were dried in a vacuum oven and analyzed for protein content. The 92 g of white overflow solids contained 65.5% protein and the 849 g of buff colored bran contained 20.5% protein.

Example 2: 2,500 g of comminuted groats (17.5% protein) obtained by hammer milling and then pinmilling dehulled Hinoat oats were admixed with 15 liters of hexane and maintained in the form of a slurry for 15 minutes. The slurry was then passed through the hydrocyclone of Example 1 using an inlet pressure of 1.4 kg/cm². The overflow from the hydrocyclone was centrifuged to separate the solids. The solids were readmixed with hexane, recentrifuged and the solids thus obtained (protein concentrate) were dried in a vacuum oven. The underflow from the hydrocyclone was passed through a 200 Tyler mesh and then a 325 Tyler mesh sieve. The material (bran) retained on the two sieves was combined and dried in a vacuum oven. The underflow, after the sieving, was centrifuged and the solids obtained were readmixed with hexane and recentrifuged. The solids thus obtained (flour) were dried in a vacuum oven. All the hexane solutions were combined and the oil was recovered therefrom by removal of hexane in a single stage evaporator operated at 100°C and at atmospheric pressure.

This process resulted in 103 g of protein concentrates (61.3% protein), 750 g bran (21.0% protein), 1,265 g of flour (12.1% protein) and 169 g oil.

Gelatinized Wheat Flour in Food Pastes

Food pastes having characteristic pizza-like qualities provided in the past have generally not contained both meat in fairly large discrete form and cheese in the same hermetically sealed aseptic or sterile package. Thus, in many instances, such pastes providing characteristic pizza-like qualities containing meat, tomato paste, etc., have been provided as a sauce in a hermetically sealed aseptic or sterile package, and the cheese has been provided in a separate package to be sprinkled on after the sauce-like preparation paste has been applied to the bread, etc. Such two component package for providing the food paste and the eventual use of the product is considered to be undesirable.

D.H. Horner; U.S. Patent 4,206,239; June 3, 1980; assigned to The House of Paris Pate Inc./La Maison Paris Pate, Inc., Canada has disclosed a process for providing a cheese-containing, meat-containing, and tomato-containing paste having the characteristic pizza-like qualities in a unitary hermetically sealed aseptic or sterilized package or can. This process comprises the steps of:

(1) mixing about 2 to 5% by weight of the final product of wheat flour, about 0.5 to 2% by weight of the final product of tomato powder, and flavoring ingredients, with a sufficient quantity of water and for a sufficient period of time at a sufficient temperature to gelatinize the wheat flour;

(2) adding further water along with about 2 to 8% by weight of the final product of vegetable oil, vinegar and about 10 to 35% by weight of the final product of cheese and heating the mixture to a temperature of about 140° to 160°F (about 60° to 71°C);

(3) homogenizing the mixture to a fine, oil-in-water emulsion at a temperature of about 140° to 160°F (about 60° to 71°C);

(4) adding about 1 to 35% by wt of the final product of discrete sliced cured meat product having a thickness to provide sufficient pliability to withstand the rigors of pumping and filling;

(5) subjecting the food paste to heating, filling by pumping the homogenized mixture and the discrete sliced meat product, hermetic sealing and sterilizing procedures, thereby to provide a commercially sterile food paste comprising a homogenized mixture of about 30 to 50% by weight of the final product of water, about 2 to 5% by weight of the final product of gelatinized wheat flour, about 0.5 to 2% by weight of the final product of tomato paste, about 2 to 8% by weight of the final product of vegetable oil, about 10 to 35% by weight of the final product of cheese, and vinegar, and, dispersed therein, about 1 to 35% by weight of the final product of sliced cured meat product in discrete sliced form.

The preferred cured meat product is pepperoni but other similar products, e.g., smoked sausage, ham or bacon may be substituted. The preferred cheese is mozzarella, but other cheeses, e.g., parmesan or cheddar, may be used. Other ingredients are optional, but enhance the flavor of the food spread. The added spices generally consist of paprika, anise, sage, celery seed, black pepper, garlic powder, onion powder, oregano, sweet basil, thyme and marjoram. Sodium citrate is included for the purpose of firming the cheese as well as providing an appropriate gloss to the cheese. Dehydrated green peppers which become discrete pieces upon absorption of water, dehydrated mushrooms which become discrete pieces upon absorption of water and onion powder which becomes onion paste upon absorption of water can also be included to provide a paste which, when applied to the surface of a baked and/or toasted product and then heated in a grill, provides a pizza-like product. Other ingredients traditionally used on pizzas, e.g., green olives, bacon, anchovies, or onions may also be provided.

Example: Hot water, at just below the boiling point, is introduced simultaneously with wheat flour, tomato powder, spices, salt, sodium citrate, into a ribbon mixer. Mixing is carried on for 4 to 5 min to permit gelatinization of the flour. After this initial step in the process, a final portion of hot water along with ground mozzarella cheese, vegetable oil and vinegar, are added to the mixer which is operating during the entire operation.

The mixture is then passed through a homogenizer in order to form a fine emulsion. The emulsion is then pumped to a second mixer where discrete particulate sliced cured meat, dehydrated mushrooms and dehydrated green peppers are added. From this second mixer, the mix is pumped to a heating vessel and heated to a temperature of 125° to 135°F. The controlled-temperature mixture is then pumped to a filling machine where the containers are filled and hermetically sealed for subsequent sterilization. The product is sterilized for a period of 75 min at 240°F. (The time will vary with the size of container.) After the sterilization, the product is cooled and packed in shipping containers.

The food paste of aspects of this process may be spread on toasted bread, rye bread or English muffins or prepared pizza pastry and then heated under a grill to provide a pizza-like product.

Potato Pulp in Sauces

Potato pulp generally constitutes a by-product from the extraction of starch from potatoes. During this extraction, the cells of the potato are ruptured, e.g., in a grating machine which comprises a cylinder bearing teeth which revolves within a casing, or by means of a machine similar to a hammer mill, after which the largest part of the starch is separated from the resultant suspension, by example with the aid of a centrifugal sieve which retains the pulp and lets through the starch.

The pulp, which is then dried on a pneumatic drier comprises, as well as the unextracted starch, the internal and external cellular walls of the potato; these walls are of polysaccharides such as cellulose, hemicelluloses, pectin. The weight ratio between these polysaccharides, frequently called structural polysaccharides, and the starch essentially depends on the efficiency of the starch extraction. Modern starch factories generally enable extraction of at least 95% of the potato starch, such that a typical potato pulp constituting the residue from such an extraction contains, based on the dry matter, 20 to 45% starch, 45 to 65% polysaccharides and up to 10% of mineral, proteinic and fatty substances.

M. Huchette and G. Bussiere; U.S. Patent 4,160,849; July 10, 1979; assigned to Roquette Freres, France have disclosed a method for preparation of potato pulp characterized by the fact that the drying stage is carried out under conditions such that the starch fraction is gelatinized. Preferably the drying stage is carried out in an installation of the drum-dryer type appropriate for ensuring gelatinization of the starch fraction.

This dried potato pulp is in powder form and comprises less than 70 weight percent starch, 5 to 25 weight percent humidity, 1 to 7 weight percent proteins (N x 6.25), 0.5 to 5 weight percent minerals, 0.1 to 1.5 weight percent oily materials, 5 to 25 weight percent cellulosic material and 10 to 55 weight percent, by difference, of other structural polysaccharides.

This potato pulp may be used as one component for foods of the group constituted by fruit compotes, sauces and preparations based on tomatoes, fruit-juices and drinks based on fruits, confectionery and pastries comprising fruits, as well as their equivalents.

Example 1: Seasoned tomato sauces which are ready for use usually contain a viscosity agent of the starch type. Such a control sample contains 180 g of tomato concentrate with 28% dry matter, 120 g malto-dextrin (DE 40), 100 g 6° vinegar, 250 g water, 30 g salt, and 20 g Col Flo 67.

A sauce is prepared according to this process in which the viscosity agent Col Flo 67 is replaced with an equal amount of a mixture of potato pulp and Col Flo 67. In the two instances, all the ingredients are mixed and then cooked on a waterbath for 10 min at 90°C. In the control example, the viscosity is about 10,200 cp; in the case of the sample according to this process, the viscosity is about 9,800 cp. The two sauces are thus very similar in viscosity level; whereas the control has a smooth texture, this product has an agreeable pulpy texture.

Example 2: Frequently compotes, marmalades and products of the same general class occurring commercially have an insufficiently pulpy texture. The causes arise from numerous sources and are found, for example, in the severe industrial manufacturing techniques which do not sufficiently respect the fragile nature of fruits, or in the employment of fruits which have lost a part of their characteristic features following storage over a long time, for example in the presence of sulfurous anhydride, or again in the employment of fruits such as pears which are naturally unsuited for giving a pulpy texture.

To 100 g of tinned apple compote bought commercially, 12 g of potato pulp hydrated to 15% of dry matter is added, being 1.8% of dry pulp; this hydrated pulp is previously cooked on a water bath of 5 minutes at 95°C. The viscosity of the compote changes from 12,000 cp before incorporation of the pulp to 17,000 cp, and above all the texture, which for the commercial product is smooth, becomes a pulpy one after addition of the potato pulp.

With the same amount of pulp obtained on a drum, hydrated by addition of the same quantity of boiling water, the viscosity is about 18,000 cp and the texture is similar.

The same operations as with the apple compote have been effected on a pear compote; the viscosity changes from 10,000 to 15,000 cp with the usual pulp and 15,500 cp with the pulp obtained on the drum; the textures are analogous in the two instances. It is emphasized that in this case the improvement is marked, the commercial "pear compote" employed having practically no pulpy texture itself.

Cold-Water Dispersible Modified Potato Starch

C.W. Chiu and M.W. Rutenberg; U.S. Patent 4,228,199; October 14, 1980; assigned to National Starch and Chemical Corporation have prepared a cold-water dispersible, modified potato starch which forms a gel when dispersed in cold water without the necessity for conversion of the starch.

This cold-water dispersible, modified potato starch with gelling properties is prepared by drum-drying a potato starch which has been reacted with a crosslinking agent such that the crosslinked starch has a Brabender Viscosity Differential, measured between 80° and 95°C, of from about –35 to +180%, measured at 5% solids using a 700 cm-g cartridge, and has a Brabender viscosity at 80°C of up to about 3,100 BU, measured at 5% solids, or of from about 800 to 1,400 BU, measured at 7% solids, using a 700 cm-g cartridge. This crosslinked starch after drum-drying is capable of forming a gel having a Bloom strength of at least 60 g.

The step of converting the starch to a certain water fluidity before or after crosslinking is not necessary in this process, however, the potato starch may be converted to a water fluidity of up to about 60 prior to the crosslinking step, or converted after the crosslinking step, if desired.

This modified starch is useful in any food formulation where a starch which will gel without further cooking is desired, and is particularly suited for use in pie and cream fillings, puddings, spreads, jellies, and instant mixes of the type which are reconstituted with water or milk and allowed to set at room temperature or lower. A food system containing such a starch will have properties, e.g., texture, appearance, gel structure, and flavor, which closely resemble those of a food formulation which is cooked.

The starch base employed in this process is potato starch, which, as used herein, refers to potato starch in its intact granular form which is either raw or has been converted to a water fluidity of up to about 60.

In a preferred embodiment, the starch is acid-converted prior to crosslinking to a water fluidity of 20 to 50. In the preparation of this modified starch, the potato starch is reacted with any crosslinking agent capable of forming linkages between the starch molecules. Preferred crosslinking agents are phosphorus oxychloride, epichlorohydrin, sodium trimetaphosphate (STMP), and adipic-acetic anhydride (1:4), and most preferably phosphorus oxychloride.

The crosslinking reaction itself is carried out according to standard procedures described in the literature for preparing crosslinked, granular starches. After the crosslinking reaction is complete, the pH of the reaction mixture is generally adjusted to 5.5 to 6.5, using a common acid. The granular reaction product may be recovered by filtration and washed with water and dried prior to drum drying.

The crosslinked starch, whether it has been converted or not, must be pregelatinized to become cold-water dispersible. The pregelatinization is accomplished herein by using a suitable drum drier, having a single drum or double drums, to dry the starch to a moisture level of about 12% or less. The starch slurry is typically fed onto the drum or drums through a perforated pipe or oscillating arm from a tank or vat provided with an agitator and a rotor.

The water fluidity and crosslinking levels specified above are interdependent, but they also vary to some degree with the drum drier employed. It has been found that drum driers which produce higher shear than a laboratory single-drum drier (such as a commercial single-drum drier) require that the starch have a higher level of crosslinking to obtain this modified starch with its gelling properties.

After drying, the starch product is removed from the drum drier in sheet form and then pulverized to a powder. Alternatively, the product may be reduced to flake form, depending on the particular end-use, although the powdered form is preferred. Any conventional equipment such as a Fitz mill or hammer mill may be used to effect suitable flaking or pulverizing. The final product obtained from the drum-drying operation is a cold-water dispersible starch which forms a gel when dispersed in water.

Example: The test starches were prepared as follows: A total of 200 g of a raw potato starch having a peak Brabender viscosity of 1,550 was slurried in 250 ml water containing 1.0 g sodium chloride and 1.2 g sodium hydroxide. Thereafter, with good agitation, reagent-grade phosphorus oxychloride ($POCl_3$, BP 105° to 108°C, d = 1.675) was added; and the mixture was allowed to react at a temperature of about 22° to 27°C for 2 hours.

After reaction was complete, the mixture was neutralized with dilute hydrochloric acid (1 part 36.5 to 38% HCl to 3 parts water) to pH 5.5 to 6.5, filtered, washed and dried.

Each starch sample was evaluated for Brabender viscosity using a 700 cm-g cartridge at 5% anhydrous solids. Then, each sample was drum-dried by slurrying 200 g starch in 300 ml water and drying the slurry on a steam-heated steel drum, with steam pressure of 105 to 110 psi (7.38 to 7.73 kg/cm^2).

The pregelatinized starch sheets thus obtained were then pulverized using a laboratory pulverizing mill (No. 008 screen, Weber Brother Metal Works).

Starch samples treated with 0.015 to 0.030% of $POCl_3$ (based on starch) produced strong gels with Bloom strength of 185 to 219 g.

CELLULOSE-TYPE ADDITIVES

Starch and Cellulose Ethers

Baked foodstuffs which come into contact with moist toppings or fillings often absorb quite considerable amounts of liquid from the latter. These baked foodstuffs include, in particular, flan cases and pastry cases for fruit tarts, and also ice cream wafers or other baked foodstuffs which enclose ice cream or are covered with ice cream and which become moistened through very quickly when the ice cream melts.

Accordingly, it is the object of the disclosure by *G.-W. von Rymon Lipinski; U.S. Patent 4,172,154; October 23, 1979; assigned to Hoechst AG, Germany* to provide an improved additive for a baked foodstuff, which additive prevents the softening of the foodstuff caused by absorbed moisture.

In this process, the foodstuff contains a carbohydrate derivative, which has been found modified by means of heat energy, radiation or an additional chemical compound, and which is water-insoluble to the extent of at least about 25% by wt and is capable of swelling. In preferred embodiments, the modification is cross-linking and the modified carbohydrate derivative is a starch ether or cellulose ether.

The modified carbohydrate derivatives used in the foodstuff according to the process are compatible with the customary base materials and auxiliaries used in the bakery trade; for example, this applies to compatibility with fats, common salt, sugar, sweeteners, flavoring substances, preservatives and antioxidants.

The procedure for manufacturing the foodstuff can be as follows. The modified carbohydrate derivatives are worked into the mixture during the required preparation of the mixture for manufacturing these foodstuffs. This procedure succeeds not only when the modified carbohydrate derivatives are employed in a preswollen state induced by water but, surprisingly, they also can be homogeneously worked into the mixture in the dry form.

In one of the preferred embodiments for the manufacture of the foodstuff, the modified carbohydrate derivatives are added to the total quantity of the mixture, uniformly distributed therein and the whole is then baked through. Appropriately, the quantity of additive is then in the range from about 0.05 to 15% by weight, preferably from about 0.2 to 10% by weight, relative to the total weight of the mixture. In these ranges, the addition of modified carbohydrate derivative does not impair the appearance, smell and taste of the foodstuff after it has been baked through. In the case of values in the medium and/or upper part of the range, however, it is possible that, depending upon the type of mixture, a somewhat more solid textural structure is obtained.

Example 1: For comparison, a mixture is prepared from 1.35 kg wheat flour, 1.10 kg sugar, 5 g sodium bicarbonate, 2.5 g common salt, 10 eggs and 1 ℓ milk. A part of the mixture is baked through directly to give thin hard biscuits; baking time about 20 to 25 minutes, baking temperature about 180° to 200°C.

Example 2: Thin biscuits are likewise prepared under the conditions of Example 1 from a part of the mixture prepared according to Example 1, with the addition of about 2%, relative to the total amount of mixture, of a thermally crosslinked carboxymethyl starch which was uniformly worked into the mixture. With reference to the appearance, smell and taste, there is no difference compared with the biscuits prepared according to Example 1. When both types of biscuits are stored at room temperature in moist ambient air, the biscuits prepared without the additive already exhibit a noticeable softening after about half a week, while the biscuits prepared with the additive still have very good strength after two weeks.

Example 3: The procedure followed is as in Example 2 but using hydroxyethylcellulose crosslinked by epichlorohydrin. The results correspond to the biscuits prepared according to Example 2.

Example 4: The procedure followed is as in Example 2 but using carboxymethylcellulose modified by N-methylolacrylamide. The results correspond to the biscuits prepared according to Example 2.

Example 5: The procedure followed is as in Example 2 but using methylcellulose crosslinked by dichloroacetic acid. The results correspond to the biscuits prepared according to Example 2.

Pea Hull Fiber in White Bread

One of the important ingredients missing from plain white bread, which is present in whole wheat bread, is natural fibers. These natural fibers are biologically active and are highly desirable in foods, serving an important function in human digestion. For instance, they are an important aid to regularity and may be helpful in preventing functional problems associated with the gall bladder, e.g., assist in bringing down the bile acids.

Since white bread can be "enriched" by the addition of sources of missing protein, vitamins and minerals, it would seem obvious that it should also be possible to enrich white bread by the addition of natural fibers. However, the problem is that natural fibers tend to be dark in color so that these fibers are clearly visible in white bread, giving the appearance of being impurities rather than an integral part of the bread formulation.

M. Satin; U.S. Patent 4,237,170; December 2, 1980; assigned to Multimarques Inc., Canada has prepared a high fiber content white bread by employing as a portion of the conventional dough ingredients, a fiber component obtained from field pea hulls. The result is a unique white bread having an improved fiber content.

According to the process, a composition for use in the making of high fiber content white bread contains the usual white flour and about 5 to 20 parts by weight based on flour, of pea fibers having particle sizes in the range which pass a 20 mesh screen but do not pass an 80 mesh screen. The pea fibers are obtained from the hulls of yellow or green field peas.

The bread to which the process relates may be any conventional bread based on wheat flour, and may be made in any conventional way such as straight dough, sponge and dough, continuous mix and variations thereof. The wheat flour used in the formulation is conventional wheat flour for bread making, and can include blended flour of wheat and other materials. While the dough is referred to herein as bread dough, it will be appparent that the dough is also useful for making buns, rolls and the like.

In preparing a dry mix for use in making a high fiber content white bread, the constituents usually present will normally include 100 parts by weight of flour, about 1 to 10 parts of sugar, about 5 to 20 parts by weight of the pea hull fibers and leavening present in an amount sufficient to provide dough expansion. A great many minor ingredients can be employed if desired, to provide optimum performance and to impart special characteristics, although they are not necessary. These may include flavors, egg yolk for tenderizing the dough, emulsifiers which produce tenderness, gluten for the purpose of strengthening the dough and making it more resilient, yeast food, and antimycotic agent such as sodium diacetate and color among others.

A dry mix of the above type may also be prepared in which only part of the final total amount of flour is present, e.g., a mix containing about 5 to 20 parts of the pea hull fibers and 5 to 20 parts of wheat flour, with the balance of the flour being added later.

In one example of a typical formulation including 100 parts commercial baker's flour, the minor ingredients can consist of 3 parts yeast, 2 parts salt, 6 parts sugar, 3 parts shortening and 7 parts pea fiber.

There is, of course, no actual lower limit to the amount of pea fiber that can be employed since any amount of added fiber has some small effect on increasing the fiber content of the bread. However, based upon dietary requirements as well as nutritional requirements that have been carefully gathered from nutritional experts consulted concerning the formulation, about 5 to 20 pbw of the fibers is preferred.

The particle size of the fibers used is, on the other hand, very important to the process. Thus, in order to produce a bread of acceptable commercial quality and which provides an improved fecal output, the particle sizes should be such as to pass a 20 mesh screen but not pass an 80 mesh screen (U.S. Standard Sieve), with –20+60 mesh particles being particularly preferred.

Large quantities of field peas are used for split peas, the preparation of which consists of cleaning and grading, kiln-drying, splitting, and screening out the hulls and chips from the full half peas. It is the hulls obtained from such procedure which are particularly useful in this process.

The hulls obtained from field peas of either the yellow or green variety may be used in the process. Actually, the so-called yellow peas tend to be a light buff color and the hulls have this light buff color. The green field peas, on the other hand, have a kernel which is green in color even after kiln-drying, but the hulls obtained from these dried green peas tend to also be generally light buff in color and are very similar in appearance to the hulls of yellow field peas. Moreover, the characteristics of the finished breads baked with fibers from field peas of the yellow and green varieties are very similar in both chemical and physical properties, so that the fibers from the different varieties may be used either mixed together or separately.

Gum-Coated Fibrous Cellulose

In the past, efforts have been made to use conventional fibrous cellulose as a bulking agent in low calorie food compositions and in pharmaceuticals. Fibrous cellulose has the advantage, in addition to providing desirable dietary fiber, of providing desired bulk without calories. However, a principal defect of this material has been its objectionable texture.

The coated or encapsulated fibrous cellulose base product described by *B.R. Hutchison and A.M. Swanson; U.S. Patent 4,143,163; March 6, 1979; assigned to Maxfibe, Inc.* consists of discreet particulate fibrous cellulose which is coated with a gum solution to provide a surface texture which is smooth and pleasant to the taste. This material may be incorporated in natural and simulated food and pharmaceutical compositions without adversely affecting the palatability of the compositions.

The gum solution used in this material consists of one or more gums or hydrocolloids in aqueous solution. The gum solution provides a coating around the fibrous cellulose particles which results in coated particles having the bulk and dietary fiber content of the fibrous cellulose, but the taste and texture of the gum solution. Accordingly, this gum coated fibrous cellulose may be incorporated into natural and simulated food compositions as a noncaloric constituent to add bulk and dietary fiber to the composition. In the composition, the gum coating imparts to the fibrous cellulose particles a smooth texture which is pleasant to the mouth and easy to chew and swallow. The gum coating substantially eliminates the dry and gritty surface texture and taste of the refined fibrous cellulose which may otherwise be apparent and unpleasant when consumed in its unmodified condition.

It was found that the palatability of the coated fibrous cellulose product is enhanced further if the gum solution also includes a quantity of a polyhydric alcohol (sometimes referred to as polyol). The polyol enhances the mouthfeel or palatability of the gum coating and has a desirable effect on the tongue and other sensory receptors of the mouth which imparts a feeling of smoothness and moistness to the mouth, and which contributes to the palatability and pleasant texture of the coated fibrous cellulose within the mouth. The polyols also have a plasticizing effect on the gum coating which improves the mouthfeel of the product during chewing. This characteristic imparts to the coated fiber a mouthfeel which is substantially the opposite of the harsh, gritty texture of the untreated fibrous cellulose and other dietary fibers.

One commercially available source of particulate fibrous cellulose satisfactory for use in this encapsulated food product is known as Solk-A-Floc BW-300 (Brown Co.). This particulate fibrous cellulose, also known as powdered cellulose, is a mechanically disintegrated and purified cellulose generally obtained from primarily alpha-cellulose derived from wood pulp. 99.5% of this material will pass through a 33-micron screen and 99.0% will pass through a 23-micron screen. The average fiber length is 21 microns and the average fiber width is 17 microns. Such relatively fine powdered cellulose as Solka-Floc BW-300, or an equivalent finely powdered cellulose, will provide cellulose fiber which may be treated according to this method to produce a noncaloric, palatable bulking agent.

The gums which have been found to be particularly advantageous in this product are high viscosity, sodium carboxymethylcellulose (CMC), guar gum, locust bean gum, xanthan gum and alginates. These gums when mixed with water in low concentrations form thick, heavy pastes.

The polyhydric alcohol, or polyol, which is preferably incorporated in this gum solution to further enhance the palability of this coated fibrous cellulose product, includes the following polyols that are approved for use in food, i.e., glycerol, sorbitol, propylene glycol and mannitol. Of these polyols, glycerol, commonly called glycerin, was found to be particularly advantageous and preferred.

Example: A coated fibrous cellulose product was prepared from 565.4 g water, 7.1 g sodium carboxymethylcellulose (7HCF Hercules), 9.4 g guar gum, 64.8 g glycerin, and 353.3 g Solka Floc BW 300.

All of the water was heated to 170° to 180°F. The gums, consisting of sodium carboxymethylcellulose (CMC) and guar gum, were then blended together. Approximately 475 g of the heated water were placed in a Waring blender, followed by the glycerin and the blended gums, and the materials were blended to form a thick, heavy gel-like paste coating solution as the mixture solubilized. The Solka Floc BW 300 fibrous cellulose was then placed in a Hobart mixing bowl equipped with a flat beater. The thick, heavy coating solution was poured into the same mixing bowl and the beater turned on at low speed to mix and knead the mass.

The remaining hot water was then poured into the Waring blender bowl to rinse out the remaining gum solution and this rinse water was poured into the Hobart mixing bowl. The speed of the Hobart beater was increased to medium agitation to avoid build-up on the beater blades and aid in mixing and kneading the mass.

When the mass was intimately and thoroughly mixed, it was spread on a drying pan and dried with occasional agitation at a temperature of approximately 220°F until it was a dry, free-flowing granulated product. The granulated product mass was then comminuted by grinding to produce a desired finely divided, free-flowing product.

A microscopic examination of the product revealed that the fibrous cellulose particles were coated or encapsulated by the dried gum solution. When placed in the mouth, the product effected a smooth and palatable mouthfeel, with desirable moistness. When chewed and swallowed the product retained its desirable mouthfeel, did not accumulate or leave a residue in the mouth, and left no undesirable aftertaste. The taste and mouthfeel of the product appeared to be that imparted by the gum solution coating of the particles, with the fibrous cellulose particles themselves contributing their bulk only to the mouth sensation.

Chitosan as Lipid Binder

Chitosan constitutes a known material and is partially or fully deacetylated chitin. Chitin is a natural cellulose-like polymer which is present in fungal cell walls and exoskeletons of arthropods such as insects, crabs, shrimps or lobsters. The polymer structure of chitin consists of N-acetyl-D-glucosamine units linked by β-(1-4) glycosidic bonds which impart to the material characteristics similar to that of cellulose. Chitosan is conventionally prepared by the alkaline deacetylation of chitin with concentrated sodium hydroxide at elevated temperatures. Depending upon the conditions of the deacetylation, chitosan with various degrees of acetylation is obtained. In the most common products, the degree of deacetylation is between 70 and 90 percent. Although chitosan is usually obtained by chemical deacetylation of chitin, it may also be obtained by the fermentation of certain foods.

Chitosan is formulated and utilized in the process developed by *I. Furda; U.S. Patent 4,223,023; September 16, 1980* as a food additive to reduce both absorption of lipids and caloric intake.

It was found that the chitosan is capable of binding various fatty acids to form the corresponding complex salts. It is believed that the binding is induced by the number of free amino groups in the chitosan which forms an ionic bond causing a binding effect considerably stronger than that obtained in conventional absorption or adsorption. The chitosan-fatty acid complexes can be prepared by neutralization of chitosan with various amounts of fatty acids, preferably edible fatty acids, such as oleic, linoleic, palmitic, stearic or linolenic acid.

These complexes can be prepared by neutralization of the chitosan with various amounts of these fatty acids, including stoichiometric amounts, less than stoichiometric amounts, or greater than stoichiometric amounts than those required for the neutralization of the chitosan. The chitosan-fatty acid complex, after ingestion by a mammal, will bind additional lipids, most probably due to its strong hydrophobic characteristic. Such lipids include natural triglycerides, fatty and bile acids, and cholesterol and other sterols, and a great portion of these bound lipids will be excreted rather than absorbed and utilized by the mammal.

In accordance with one embodiment of the process the chitosan, as such or in the form of its fatty acid complex, is admixed with a food substance to form a composition consisting of the food substance containing a minor quantity, as, e.g., between 1 and 10% by wt, of the added chitosan, as such or as a fatty acid complex. This food composition, containing the minor quantity of chitosan, has a lower effective caloric value than the corresponding amount of food without the chitosan, as the chitosan in the composition has the ability of binding and preventing the absorption of a large amount of lipids. The food composition, when ingested, will thus reduce the fat intake of the mammal, promoting the binding and excretion of fatty materials such as cholesterols, sterols, triglycerides, etc. The chitosan, as such, is not capable of absorbing neutral oil, but once complexed with a fatty acid, the complex is capable of absorbing it. Therefore, it is often desirable to use the chitosan at least partially in the form of its fatty acid complex.

The chitosan itself, i.e., not in the form of fatty acid complex, may, for example, be utilized as a low calorie filler in connection with various food, such as breads, cakes, cookies, and the like.

In the form of the fatty acid complex, as, for example, with linoleic acid, palmitic acid, stearic acid, or linoleic acid, the same may be used as a fat extender added, for example, to margarine, spreads and dips, salad dressing, or the like.

The characteristics of the chitosan, with respect to its nondigestibility, its biodegradability, and its bland taste make the same excellent as a food additive and, due to it ability to bind the lipids, may be considered as having a negative caloric vlaue, i.e., lowering calorie effect of the foods to which it is added.

Example: *Preparation of chitosan-fatty acid complex* — Emulsion consisting of linoleic acid (or any fatty acid) and water is prepared by homogenization of by vigorous stirring (Waring blender) of 600 ml water containing 0.1 to 1.0 g benzoic acid (preservative), 1 to 7 g polysorbate 80 (emulsifier) and 300 ml linoleic acid containing 0.05 g BHA (antioxidant). The good quality emulsion is usually obtained in less than one minute of stirring.

70 g of chitosan powder (80–200 mesh) is added to the emulsion and stirring (homogenization) is continued for another one minute. The complex of chitosan-linoleic acid is then isolated by filtration.

The ratio of fatty acid to chitosan for the preparation of the complex may vary considerably, depending on specific requirements. The preferential ratio is between 0.1:1.0 and 5:1.

Other examples in the specification show the use of the chitosan-fatty acid complex in the preparation of rye bread, pound cake, salad dressing, and chicken noodle soup mix.

Microfibrillated Cellulose in Puddings

In a previous work there has been disclosed microfibrillated cellulose, distinguished from prior celluloses by a vastly increased surface area, greater liquid absorption characteristics and greater reactivity. The microfibrillated cellulose is prepared by repeatedly passing a liquid suspension of fibrous cellulose through a high pressure homogenizer until the cellulose suspension becomes substantially stable. The process converts the cellulose into microfibrillated cellulose without substantial chemical change.

A.F. Turbak, F.W. Snyder and K.R. Sandberg; U.S. Patent 4,341,807; July 27, 1982; assigned to International Telephone and Telegraph Corporation have discovered that it is possible to prepare a wide variety of food products containing microfibrillated cellulose in a single stage operation in which the microfibrillated cellulose is prepared in situ during preparation of the food product. The ingredients of the food product are added to the original slurry containing fibrous cellulose prior to fibrillation of the cellulose. The fibrillation process then converts the cellulose to microfibrillated cellulose and produces a food product in the form of a homogeneous, stable suspension containing microfibrillated cellulose. This process is useful for producing fillings, crushes, soups, gravies, puddings, dips, toppings and other food products.

Specifically, this process comprises mixing together an edible liquid which swells cellulose, a food additive and fibrous cellulose to form a liquid suspension and repeatedly passing the liquid suspension through a small diameter orifice in which the mixture is subjected to a pressure drop of at least 3,000 pounds per square inch gauge and a high velocity shearing action followed by a high velocity decelerating impact. The process converts the cellulose into microfibrillated cellulose and forms a stable homogeneous suspension of the microfibrillated cellulose, liquid and food additive.

Examples 1 and 2: A series of toppings, puddings, fillings and dips were prepared by one or both of two methods. The first method consists of premixing all of the ingredients, including cellulosic pulp, water and food components, followed by passing through an homogenizer and in situ production of microfibrillated cellulose in accordance with this process. The second method consists of the prior preparation of microfibrillated cellulose followed by admixture of the microfibrillated cellulose slurry and the food additives. Both methods used the same grade and fiber size of cellulosic pulp in a 2% slurry in water as the starting material and both methods entailed passing the respective slurries through an homogenizer having an 8,000 psig pressure drop for ten passes, at which time a stable gel point was reached. The starting temperature of the process was 25°C in each instance. In these examples, a low calorie topping (unflavored) was made by the two methods from a composition containing 8% soybean oil, 6% sugar, 2.2% cellulose and 83.8% water. By the first method, the ingredients were all admixed with cellulosic pulp and the food topping and microfibrillated cellulose prepared in one step by passing the mixture through the homogenizer. By the second method, the pulp was first microfibrillated and then the oil was added and emulsified. The sugar was added last. It was found that the first method gave a superior product; the topping made by the one step procedure was smoother and more consistent in texture.

Examples 3 through 5: In this series of examples, the same two methods were used for the preparation of a series of puddings and dips. Each of the products were prepared from a neutral pudding base composition containing 7% nonfat dry milk, 0.13% carboxymethylcellulose, 1.86% soybean oil, 2.32% cellulose and 88.69% water. In the first method, they were all passed through the homogenizer together; in the second method, the cellulose was first fibrillated and then mixed with the nonfat dry milk and microfibrillated cellulose.

Example 3 consisted of the foregoing pudding base prepared by using both the first and second methods. Example 4 was a lemon pudding prepared by both methods from the foregoing pudding base to which was added lemon juice and sugar. Example 5 was a dip prepared by both methods from the same pudding base to which was added onion soup and a party dip mix. In all examples, the results were essentially the same, regardless of which method was used. The product was firm and smooth and there was little or no perception of fibrous material.

Microcrystalline Cellulose in Custards

The freeze-thaw drawbacks and other disadvantages of conventional custard formulations are overcome by the method used by *B.A. Croyle; U.S. Patent 4,341,808; July 27, 1982; assigned to SCM Corporation.*

This freezable raw custard preparation contains a substantial proportion of whole egg or egg white based whole egg replacer. The composition also has an effective amount of food particulates and a milk derived portion. The egg and milk derived portions are present in a weight ratio of about 0.45:1 to about 0.75:1, the milk derived portion comprising fluid milk, a water absorbing quantity of a protein concentrate, and a thixotropic gum. The total protein content (by Kjeldahl analysis) from the fluid milk and protein concentrate is at least about 4% by weight of the milk derived portion. The gum content is that effective to provide a viscosity of the raw custard sufficient for homogeneous suspension of the food particulates prior to freezing.

Preferably, the fluid milk is formulated from water and nonfat dry milk solids to provide a solids content comparable with fluid milk.

Preferably, the milk derived portion consists essentially of about 82 to 84% water, about 8 to 10% nonfat dry milk solids, about 4.5 to 8.0% water absorbing bland protein concentrate, about 0.5 to 1.75% thixotropic gum, and effective amounts of flavors and seasoning components.

Preferred embodiments are those wherein the water absorbing protein is a whey protein concentrate and the gum is a thixotropic cellulosic derivative. A water dispersible microcrystalline cellulose is a preferred gum.

Other cellulose derivatives which may be employed and are thixotropic are methylcellulose, carboxymethylcellulose, hydroxypropylmethylcellulose, and sodium carboxymethylcellulose.

Microcrystalline cellulose is a well-known industrial product, and is frequently used in low-calorie ice creams at relatively high levels. To make it readily dispersible, it is combined with sodium carboxymethylcellulose and is known as Avicel RC-591 and Avicel CL 611 (FMC Corp.). These compositions are stated to be colloidal forms of microcrystalline cellulose which have been blended with sodium carboxymethylcellulose and dried. In Avicel RC-591, the amount of sodium carboxymethylcellulose is about $11 \pm 1\%$ by weight of the microcrystalline cellulose.

Other gums which may be used alone or in combination with the cellulosic derivatives, to provide thixotropy and meet the other requirements listed above, are plant extracts such as acacia gum and karaya gum; marine plant extracts such as algin and carrageenan; seed extracts such as guar gum and fruit extracts such as pectin.

Following freezing and then cooking, for instance by baking, the product exhibited the texture, flavor and appearance of unfrozen baked egg custard. After at least four freeze-thaw cycles, the product showed no discernable significant weeping.

Example 1: A quiche Lorraine filling was prepared from: (A) 43% milk portion which comprised 83 parts water, 9.5 parts nonfat dry milk, 5.5 parts of a water absorbing protein concentrate (HiSorb, Stauffer Chemical Co.), 1.0 part of Avicel RC-591 with the seasonings comprising the remainder; (B) 23% liquid whole eggs; and (C) 34% food particulate which comprised 23.5 parts cooked bacon and 76.5 parts chopped cheese.

Example 2: This example is substantially the same as Example 1, except that the milk portion contains 74% water and 9% white wine to prepare a quiche Francais filling. Also, the amount of milk portion is reduced to about 34%, while the egg portion becomes 19% and the food particulate 47%. The food particulate is characterized by roughly 44% cheese, 26% cut asparagus and 30% crab meat.

The above formulations were prepared by hydrating the gum stabilizer in the water fraction using a high-shear mixer. The remaining solids in the milk portion were then added and mixed thoroughly. In this example, the wine was added at this point. Next, the food particulate portion was dispersed in the mix. The eggs were added and mixed as a final step. Fillings were held under refrigeration until use. Measured amounts of the finished, homogeneous fillings were mechanically deposited into previously prepared pie crusts in appropriate containers. Quiches were frozen in a blast freezer to an internal temperature of $-10°F$, and packaged.

The above quiche was tested for performance (custard strength, stability, and particulate distribution) and organoleptic characteristics (flavor, mouthfeel, and appearance) against a control product (a nonfrozen cookbook quiche formulation subjected to conventional preparation and baking procedures). In organoleptic attributes, the example quiche compared favorably with the standard. In performance characteristics, the example products were superior to the control in all respects. Freeze-thaw testing showed the products described above to be stable (exhibit no appreciable syneresis) after four freeze-thaw cycles.

PROTEIN ADDITIVES

PREPARATION OF PROTEIN ADDITIVES

Encapsulated Chlorella Protein

It has been determined that the so-called Chlorella is rich in protein with high quality and contains the so-called Chlorella Growth Factor (CGF), i.e., its contents are about 55% protein, about 15% fat, about 20% carbohydrate, about 6% minerals including unknown materials and about 4% moisture. There have been various proposals to use Chlorella, not only for the purpose of preparing a nutritious diet product but also for the purpose of preparing a food additive. However, no proposal has been forthcoming since Chlorella, either in cell powder, or in water-extract, is difficult to deal with and to preserve.

Previously, there has been no attempt to combine garlic, ginseng, American aloe and Chlorella to obtain a single product.

In a process developed by *Y. Tanaka; U.S. Patent 4,143,162; March 6, 1979* an improved extremely nutritious foodstuff is provided which does not impart a disagreeable odor or flavor. The foodstuff provided is composed of: (a) Chlorella, (b) garlic, ginseng and/or American aloe, and (c) one or more vegetable oils.

The foodstuff is encapsulated and it can not only be conveniently used as a seasoning but also can be accepted per se as a nutritious foodstuff without imparting the feeling of any undesirable strong nasty smell and disagreeable taste caused by its contents.

It has been found that when a powdered water-extract of Chlorella algae is mixed with extracts of the so-called Chinese foodstuffs, such as garlic, ginseng and/or American aloe, which have been believed to be nourishing, its preserving effect is further improved and together with this, the disagreeable odors and flavors of the Chinese foodstuffs are extremely decreased.

In further studies, it was found that when the above mixture is admixed with one or more vegetable oils, the components of the above mixture may be well preserved because an oil layer is formed on the surface of the above mixture which prevents the components from being contacted with air and the odors caused by Chinese foodstuffs are further decreased.

In order to enable the mixtures to be easily handled and preserved, the mixture is encapsulated in a water-soluble and digestible capsule which is very useful not only as a food additive to season foods, but also as a daily nourishing diet supplement.

The encapsulated products are, when used as a seasoning, most effective when they are used in combination with other foods while being cooked, because the capsules are easily cleaved by the action of water and heat. Thus, the encapsulated materials are released during cooking and will spread into the foods which are therefore eventually seasoned.

The powdered water-extract of Chlorella algae is prepared by suspending about 1 kg of dried cells of Chlorella algae in about 10 kg of water, boiling it at about 100°C until the total volume has been reduced by about one-third, filtering the product under pressure conditions and freeze-drying the filtrate to obtain a powder, wherein about 20 g of the powdered product will be obtained per about 1 kg of the filtrate.

The garlic, ginseng and/or American aloe are preferred to be in powdered form and can be prepared as follows. Powdered extracts of garlic are prepared by pressing the garlic to obtain a liquid, i.e., garlic oil, which is then freeze-dried, thereby obtaining powdered garlic. The amount of water used is about ten times that of ginseng or American aloe. The oils are preferably vegetable oils, such as peanut oil, coconut oil, soybean lecithin and the like. However, peanut oil is the most preferable.

The mixing ratios of powdered water-extract of Chlorella algae, powdered extract of Chinese foodstuff and vegetable oil are, in general, respectively about 5 to 10% by weight, about 3 to 6% by weight, and about 60 to 70% by weight. The remainder may be other appropriate foodstuffs such as chemical seasonings, honey wax, vitamins or the like, in accordance with necessity and use.

Example: 35 mg of powdered water-extract of Chlorella algae was mixed with 6 mg of powdered water-extract of ginseng and 4 mg of garlic oil. To the mixture 225 mg of peanut oil and 10 mg of soybean lecithin were added and then 200 mg of purified honey wax was added.

The product thus obtained was encapsulated in a capsule with a 300 mg volumetric capacity made from starch. This encapsulated product was useful for seasoning a Chinese dish when it was applied thereto during its cooking. This encapsulated product was willingly taken as a nourishing diet product without giving any unnatural feeling.

Fractionation of Soy Protein

R.M. Davidson, R.E. Sand and R.E. Johnson; U.S. Patent 4,172,828; October 30, 1979; assigned to Anderson, Clayton & Company provide an improved method for processing soy protein from defatted soybean flakes. This process is illustrated by the following flow chart, Figure 6.1.

Figure 6.1: Schematic Diagram of Fractionation of Soy Protein

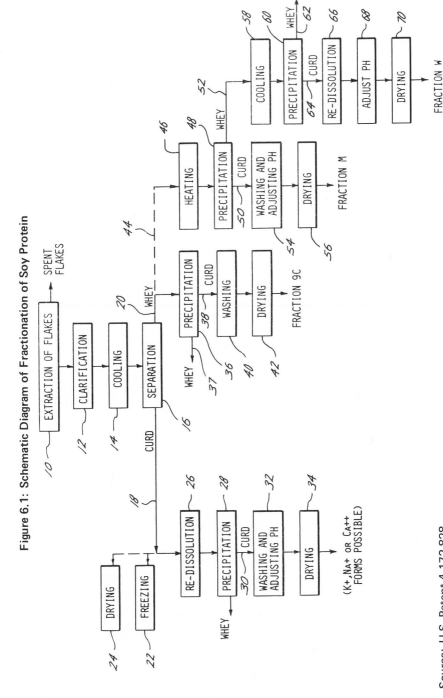

Source: U.S. Patent 4,172,828

Each fraction is a mixture of different specific soy protein molecules but, unlike conventional soy protein isolate, each fraction contains different proportions of the constituent proteins. As a result, each fraction has different properties from others, and each has different potential uses as an ingredient for foods such as imitation cheese and the like.

According to this process, defatted soybeans preferably in flake form are mixed with water and maintained at 55° to 70°C for a time sufficient to solubilize a substantial portion of the protein present in the flakes.

Insolubles are then removed from the aqueous mixture such as by clarification followed by cooling of the clarified liquor over a time period of at least one to three hours and to a temperature of from about 5° to 10°C. This results in the formation of a curd which is then separated from the whey. The resulting curd can be preserved by drying without washing or any further treatment which is quite surprising by comparison with conventional soy isolates. This curd comprises a protein composition referred to as "Fraction C." Fraction C protein material in its most natural state is a potassium salt form of protein.

A concentrated solution of Fraction C material somewhat resembles sodium caseinate solution but bears no resemblance to soy isolate solution. Surprisingly, the Fraction C material is thermoplastic in that it is softened by heat but regains its original properties upon cooling, is light in color and very bland in taste, and thus is useful in the preparation of, among other things, imitation cheese. Fraction C is thermoplastic at temperatures up to about 65°C but, if overheated much above this, e.g., above 75°C, it may convert to a thermosetting condition.

The whey remaining after separation of the curd therefrom may be processed to recover additional valuable protein fractions. Depending on the fractionation techniques employed as shown in Figure 6.1, any of three distinct protein compositions may be obtained. For the sake of clarity, these protein fractions will be referred to hereafter as "Fraction 9-C," "Fraction M" and "Fraction W."

Fraction 9-C somewhat resembles conventional soy isolate but is a stronger gelling agent than soy isolate in the sense that it is more viscous in solution and will stand by itself. Fraction 9-C material is also light in color and bland in taste, and hence has extremely useful properties for food products. Fraction M also resembles conventional soy isolate but is a stronger gelling agent.

Fraction W, depending on concentration and pH before drying, is (1) viscous, smooth and rubbery which, in simplistic terms, is somewhat analogous to bubble gum, or (2) a custard-like gel. When dried and placed in powder form, high concentrations of Fraction W material can be dissolved in water and, quite unexpectedly, the resulting solution of about 40% solids is thermosetting in that it solidifies or sets on heating and cannot be remelted to its prior state. Consequently, Fraction W protein material can be heat set (that is, it cannot be resolubilized without destroying the protein) at approximately 77°C and thus is highly useful as, for example, an egg white (albumen) substitute. However, by varying concentration and pH, Fraction W material can be placed in a thermoplastic condition with varying gel strengths.

Yeast Codried with Whey

E.J. Saunders and T.A. Lappin; U.S. Patent 4,182,777; January 8, 1980; assigned to Standard Oil Company (Indiana) have prepared protein containing food products in which yeast is codried with whey. The codried yeast-whey food is suitable for human foods. Preparations containing yeast-whey product do not have a yeast character or flavor in some food applications.

An aqueous suspension of yeast and whey can be produced by combining a yeast cream, generally having from about 10 to 20% by weight solids, and whey, generally having from about 3 to 70% by weight solids. It is preferred that the yeast-whey suspension have from about 30 to 40% by weight of the total solids as whey solids. The suspension preferably is pasteurized prior to codrying. Alternatively the yeast cream and the whey suspension may be individually pasteurized. The pasteurization step can be performed at a temperature from about 160° to 195°F. Preferably the suspension is spray dried at an outlet temperature of from 180° to 210°F to produce a product of low moisture.

Products which contain the codried yeast-whey product include cakes, puddings, space extenders, flavor ingredients, margarines, muffins, meat containing luncheon loaves, chicken coating batter, chiffon pie filling, and mushroom soup bases. Specifically the process resides in a process for making an improved food product comprising mixing a *Candida utilis* yeast and a dairy whey suspension, pasteurizing the yeast and the whey suspension, and spray drying the yeast-whey suspension.

A synergistic effect has been discovered in codrying yeast and whey. The codrying process produces a product which when used in various foods lacks objectionable yeast or whey flavors. This reduced flavor influence enhances the utility of codried yeast-whey as a food ingredient. The codrying process has been shown to impart better functionality and characteristics to food. Meat products which have codried yeast-whey ingredients shrink less, lose less cooking fluid and have better taste and mouthfeel. The whey is dried in conjunction with the yeast easily and in one step to a typically nonhygroscopic powder.

Example: 1,600 gallons of *Candida utilis* yeast cream, in aqueous suspension of about 18% solids by weight, and 551 gallons of condensed No. 2 cheddar whey of about 39% solids by weight in an aqueous suspension were blended in a 4,000 gallon vessel. The mixture was pasteurized at 174° to 179°F for a residence time of about 3 minutes. The suspension was then fed to a spray drier operating at 483° to 497°F inlet, and 201° to 203°F outlet temperatures. About 4,500 pounds of product were collected from the drying operation. The moisture content of the codried yeast-whey was about 4.0% by weight. The whey content of the codried yeast-whey product was 40% by weight. The material had a pH of 6.0, a bulk density of 25.0 lb/ft^3, and contained 44.8% water-solubles.

The codried proteins were useful in replacing 50 to 75% of whole eggs in chocolate cake, in replacing 25% of milk solids in puddings, and as a particle substitute for spices.

Wheat-Based Lipoprotein Complex

Gluten is a concentrated natural protein generally taking the form of a light tan powder having a relatively bland taste and aroma. It usually contains about 75 to 80% protein, 6 to 8% native lipids, fiber, residual starch, a small amount of mineral matter and between 4 and 12% residual moisture. Gluten, per se, is generally considered to consist, in approximately equal amounts, of the proteins gliadin and glutenin. Commercially, vital wheat gluten is manufactured by one of several washing processes in which wheat flour is kneaded with water to remove the starch and water-soluble materials from the gluten, the latter usually being obtained as a tough, rubbery, elastic mass containing a high proportion of water (about 67% by weight).

Many attempts have been made to use gluten in such nonbaking applications. However, this has not been successful primarily because the resulting gluten films are found to be lacking in the strength and maleability required, especially as regards their ability to be self-supporting and, consequently, supplementation with other film-forming materials such as collagen has been necessary and/or processing conditions must be very severe.

B.D. Ladbrooke, G.R. Quick and N.S. Singer; U.S. Patent 4,200,569; April 29, 1980; assigned to John Labatt Limited, Canada have provided a lipoprotein complex comprising gliadin reacted with a selected polar anionic lipid.

The selected "(polar) anionic lipids" to be used fall within the generally accepted definition of "anionic lipid" which refers to a class of compounds, and derivatives thereof, which is characterized by a hydrophobic hydrocarbon chain or "tail" covalently bound to a hydrophilic polar group or "head" carrying a negative charge when allowed to ionize.

Specific lipids which meet the above criteria and constitute preferred classes of lipids are: (a) alkali metal alkyl aryl sulfonates; (b) diacetyl tartaric acid esters of mono- and diglycerides; and (c) palmitoyl-1-aspartic acid.

This process also provides a protein composition comprising a lipoprotein complex of gliadin and a selected polar anionic lipid formed in situ in wheat flour by treatment of wet vital wheat gluten with the lipid.

It is more convenient, from a practical viewpoint, to work in terms of the amount of lipid per unit of gluten. When formulated in this manner, it has been found that the amount of lipid need not exceed 10% of the weight of the gluten. Products having as low as 1% by weight of lipid exhibit very useful properties and about 7% by weight based on the weight of gluten has been found to be optimal. These amounts provide the essential greater than 50% complexing of available gliadin in the gluten.

In addition, it is believed that the products formed directly from wheat flour are more effective or efficient due to the possible utilization of gliadin which would, if the wheat flour were subjected to the normal washing operation, be irreversibly bound in some manner and thus, become unavailable for complexing with the added lipid.

Turning to the reaction conditions in more detail, the dilute aqueous gluten medium contains at least about 50% moisture and generally more than 60%. For practical reasons, if wet gluten is to be treated, the direct product of the known starch washing operations is preferably used, this containing from 60 to 70% and usually about 66% moisture.

The reaction may be effected at a temperature ranging from below ambient (when cooling may be required) to a maximum less than that at which the protein loses its vitality under the conditions involved (i.e., usually about 70°C). From a practical viewpoint, temperatures of from about 15° to 35°C are satisfactory and preferred since heating the mass of gluten is thus unnecessary, especially since the wet gluten emanating from the washing processes generally has a temperature of from 20° to 25°C, usually about 22°C.

In the preferred embodiment, the lipid is converted into the corresponding hydrate prior to complexing. The lipid in hydrate form has been found to promote the desired complexing most efficiently and, moreover, in that form, the lipid is not subject to the temperature restriction referred to above when melted lipid is used. In other words, the lipid hydrates may be used at temperatures below the lipid melting point and thus, it is unnecessary to heat the protein substrate prior to complexing with the lipid and this provides obvious significant advantages.

The following example shows the in-situ preparation of the gliadin-lipid complex using Panodan AM, diacetyl tartaric acid esters of fatty acids, obtained from Grinstead Products, Inc.

Example: *Preparation in situ of the gliadin-lipid complex* — 3,000 g of freshly thawed wet vital gluten (approximately 66% moisture) at 10° to 15°C, was placed in a Prodex mixer modified by the addition of cutter blades (bowl temperature 25° to 30°C). A Panodan AM hydrate preparation (pH 3.6 to 3.9) was prepared by dispersing 110 g of Panodan AM flakes in 440 ml of H_2O, containing 4.5 g of NaOH, at 60° to 65°C. The sample was mixed vigorously using medium speed on a Waring blender for about two minutes until uniform (pH 3.6 to 3.7). The Panodan hydrate preparation was added to the wet gluten in the Prodex, vacuum was drawn and sample mixed at 2,600 rpm for 20 seconds, when mixing was stopped and the bowl is scraped down. Vacuum was drawn again and sample mixed for an additional 20 seconds, when the product obtained had a uniform soft oatmeal-like consistency. (The temperature of the product was 18° to 24°C.)

From Glanded Cottonseed

An improved process has been developed by *R.S. Kadan, D.W. Freeman, J.J. Spadaro and G.M. Ziegler, Jr.; U.S. Patent 4,201,709; May 6, 1980; assigned to the U.S. Secretary of Agriculture* for producing from glanded cottonseed a high-protein flour essentially free of free-gossypol so that the product is suitable for human consumption.

This is accomplished by an improvement in the general process comprised of the following steps: (a) flaking dehulled glanded cottonseeds having moisture contents of from 5 to 12%; (b) solvent extracting the cottonseed flakes to reduce fat content; (c) desolventizing the solvent extracted flakes; (d) milling the desolventized flakes into a flour; and (e) air classifying the flour to produce a coarse fraction and a first fines fraction.

The improvement comprises reducing the moisture content of the cottonseed flakes to 1 to 4% by weight prior to solvent extraction to produce a cottonseed flour product suitable for human consumption having a free gossypol content of 0.045% or less.

The product of the above improved process can be further processed to a cottonseed flour having even lower free gossypol content.

In this process, dehulled cottonseed kernels essentially devoid of hulls are flaked at a moisture content of about 5 to 12% in equipment preferably adjusted to yield flakes of about 0.01 to 0.03 of an inch in thickness. The drying is done immediately after flaking preferably by applying an air stream of about from 70° to 105°C temperature, in a high velocity "through-flow" type dryer to obtain a flake material with a moisture content of about 1 to 4% by weight, preferably 2% or less.

Lipids are extracted from the flakes by adding a quantity of nonpolar solvent at temperatures of about from 20° to 40°C in the extractor to obtain flakes with a lipid content of about from 1 to 3%. Flakes containing more than 3% lipids would result in air-classification yields below what is thought to be practical and would also result in product lipid contents above an undesirable level of 2.0%. Below 1% lipids the pigment glands tend to rupture more easily during milling. Desolventizing is carried out immediately and continuously in equipment designed to recover solvents by the use of heat.

The desolventized flakes are then milled in an impact-type mill, such as a pin or hammer mill, designed to produce about 60% of the material with particle size of 25 microns or less. Significant pigment gland rupture does not occur. This fact was unexpected because it is generally believed that impact milling would do excessive pigment gland damage. However, mills such as oilseed crushing rolls and particle against particle disintegrators, were found to result in excessive pigment gland damage thus causing free gossypol contents above 0.060% while the use of impact mills in this method results in products containing 0.045% or less free gossypol.

Separation of pigment glands from the milled flour preferably is achieved by a centrifugal air-classifier employing centrifugal and gaseous classification simultaneously.

Edible, that is, less than 0.045% free gossypol, protein flour products which are obtained are about from 35 to 42% by weight of the starting defatted material. Milled flours are air classified in such a manner so as to produce a fines fraction that represents about 50 to 60% of the weight of the starting defatted, unclassified flour.

Example: Fifteen pounds of Mississippi cottonseed kernels (8% moisture) were flaked to a thickness of 0.015 of an inch and divided into two lots; one lot was dried to a moisture content of 2%, and the other to a moisture content of 5%. Drying of both lots was carried out at 82°C in a "through-flow" type dryer. The dried flakes were then extracted to remove lipids. To do this, flakes were extracted with 5 passes of fresh hexane in a basket extractor using a total solvent-to-meal ratio of 2:1, employing hexane at 22°C. Extraction time was approximately 20 minutes per pass.

The extracted flakes were desolventized at 82°C for a period of 2 hours, under 15 inches of vacuum. Desolventized flakes were milled in a fixed hammer disintegrator equipped with a 0.016 inch diameter opening sizing screen. The milled flour was air classified three times in a centrifugal air classifier (Donaldson Acucut A-12), always reclassifying the fines fraction with an air flow of 70 scfm. The first fines fraction was obtained using a rotor speed of 750 rpm, the second fines fraction was obtained using a rotor speed of 850 rpm, and the third fines fraction was the final product and was obtained using a rotor speed of 950 rpm.

The fraction dried to 2% moisture produced a flour having 0.087% total gossypol and 0.040% free gossypol. The fraction dried to 5% moisture had 0.138% total and 0.082% free gossypol.

These data show it is advantageous to dry flakes to 2% moisture prior to solvent extraction in order to obtain an edible air-classified cottonseed product, containing less than 0.045% free gossypol.

Instantized Blend of Caseinate and Soy Protein

A. Schapiro; U.S. Patent 4,209,545; June 24, 1980 has prepared an instantized blend of proteins which can be readily mixed with water or milk. The term "instantizing" is intended to mean the process of converting a protein (for example soybean protein) which is water-insoluble and which is difficult to disperse in water, into a form which is readily dispersible in water or other aqueous medium such as milk.

In an earlier disclosure, U.S. Patent 3,988,511 an instantized soy protein was prepared. This earlier process starts with a small amount (i.e., small compared to the weight of final product) of a substrate which may be a carbohydrate such as a sugar (e.g., sucrose) but which preferably is a small amount (similarly defined) of the protein which is to be instantized. To this substrate is added a surfactant which is acceptable as a food ingredient, such as lecithin and the two ingredients (substrate and surfactant) are blended together. To this resulting premix or core of substrate and surfactant is added a small quantity of the protein to be instantized and the resulting mixture again is blended.

This process is repeated until the final product results. That is, the process is one of the incremental addition of a small quantity of the protein to be instantized to the blend or mixture resulting from the immediately preceding step, followed by thorough blending, and repetition until all of the protein that it is desired to instantize has been added.

This procedure is applicable with considerable advantage to the instantizing of mixtures of soy protein and one or more caseinates. (By "caseinate" is meant a sodium caseinate or a calcium caseinate or a mixture of the two in any proportions.)

By this means, a very desirable food product is produced which can be mixed with water or milk and stirred manually with no great amount of effort and at room temperatures, to produce an instantized product which is stable in the sense that phase separation does not occur to any substantial degree.

Example 1: To make a 1,000 lb batch, 400 lb soy protein were blended with lecithin in accordance with the procedure of U.S. Patent 3,988,511. The soy protein known as Promine D (Central Soya Co.) contained about 90% protein as determined by nitrogen analysis. 25 lb of this soy protein were added to 8 lb lecithin known as 3 Fub (unbleached, 3 times filtered, Central Soya Co.). This lecithin contained about 65% phospholipids, the balance being naturally occurring vegetable oil, which is a natural and ordinary concomitant of commercial lecithin derived from soybeans. These two ingredients (soy protein and lecithin) were mixed in a ribbon blender to form a uniform mix or preblend. To this preblend were added 50 lb of the soy protein and the mixture was blended. (This and subsequent blending steps were carried out in a larger ribbon blender.) The balance of the soy protein (except the last 25 lb increment) was added in increments of 50 lb each, with blending at each stage.

Example 2: To the instantized soy protein of Example 1 were added 400 lb of a mixture of equal parts by weight of sodium and calcium caseinates. This mixture was a commercial product known as No. 60 (Western Dairy Products). This mixture, as is typical of sodium and calcium caseinates and mixtures of the two, is very difficult to instantize. This caseinate was added to 50 lb increments, with thorough blending after each increment, to the blend of soy protein and lecithin. The blender was operated constantly during and after each addition of an increment of caseinate.

By this means a thoroughly and easily instantized protein was produced which can be used as is. However, it was mixed with 150 lb of whey powder (Foremost Dairy No. 35) which was also added in 50 lb increments with blending of each increment before adding the next increment.

This product can be mixed with milk, water, fruit juice and other aqueous media to produce a nutritious and pleasantly flavored beverage.

Bland Neutralized Casein

Neutralized caseins are produced by precipitating the protein fraction of milk or milk products with acids to produce acid casein and then neutralizing the acid casein with an alkali. The neutralized casein is referred to as the salt, or caseinate, of the corresponding alkali.

Whether the casein is in the form of a caseinate or acid casein, the casein contains over 90% protein and can provide an exceptionally important source of protein for foodstuffs. However, the acid caseins have a relatively unpleasant "acid" taste and the caseinates have a relatively unpleasant consistency in the mouth (referred to as mouthfeel), i.e., a somewhat soapy mouthfeel. Therefore, the amount of either acid casein or conventional caseinate which may be added to foodstuffs is considerably limited. Generally speaking, foodstuffs will not contain more than about 5% of the acid casein or caseinate or otherwise the unpleasant flavors and consistencies adversely affect consumer acceptance of the foodstuff.

The method developed by *S.B. Zavagli and R.L. Kasik; U.S. Patent 4,209,544; June 24, 1980; assigned to Beatrice Foods Company* provides a bland neutralized casein which has essentially no taste whatsoever.

The process involves forming a water slurry of casein at a temperature of between 100° to 210°F. A nontoxic water-soluble magnesium base is added to the slurry in an amount sufficient to produce the corresponding magnesium neutralized casein. The neutralized casein is then separated from the slurry to produce white, bland neutralized casein. It is preferred that the casein be produced by first mixing casein granules with a neutral salt solution, e.g., magnesium salt solution, heated to at least 150°F, e.g., at least 180°F. This causes the formation of a coagulum of the casein. In the preferred process, the casein granules are acid casein granules.

It appears that the magnesium salt step opens the casein granules so that the subsequent magnesium base step is far more effective. The magnesium salt solution seems to have the unexpected property of causing almost explosive hydration of the casein granules which dislodges or otherwise removes undesired components and places the granules in a condition for an active effect by the magnesium base.

Example 1: Into a steam heated, jacketed mixing tank were placed 865 lb water and 10 lb magnesium chloride. With heating and stirring the resulting solution reached a temperature of about 200°F. Thereafter, with vigorous stirring, were added 125 lb of commercial acid casein on a solids basis and the mixture was stirred for 3 minutes while maintaining the temperature near 200°F. The stirring was then stopped.

A coagulum formed and rose to the top of the liquid, forming a curd of bread-dough-like consistency. The pH of the curd was approximately 4.5. The curd was off-white in color, and it was apparent from the appearance, consistency and taste of the curd that significant amounts of undesired components and flavors had been removed. The jacket of the tank was heated to a temperature of approximately 200°F and the bottom draw of the tank was opened to drain the mother liquid.

300 lb of the curd (containing 120 lb of the casein solids) were resuspended in 300 lb of water heated to 140°F; with stirring the curd disintegrated into a coarse slurry.

Magnesium oxide was added to this slurry and the pH was monitored. After approximately ½ hour the pH of the liquid and curd was about 7.0. About 1.6 lb of magnesium oxide per 100 lb of curd had been added to provide this pH.

The slurry passed through a Martin-Gaulin homogenizer operated at 500 psig and spray dried in a conventional box spray drier with a gas (air) inlet temperature of about 300°F and an outlet temperature of about 185°F. The nozzle pressure of the spray dryer was about 1,500 psig.

The resulting spray dried neutralized casein powder was very white in color and had essentially no taste. An analysis of the dried neutralized casein, on a dry basis, showed 94.0% protein, 1.25% fat, 4.5% ash, less than 0.5% carbohydrates and 1.25% magnesium.

10 ml of a 5% solution of the product were dialyzed against 1 ℓ of doubly de-ionized water for 72 hours, using a cellulose dialyzer tube (Fisher Scientific Co.). The membrane retains materials with molecular weights of 12,000 and higher, and has an average pore diameter of 4.8 mμ. The dialyzed sample was dried under vacuum at room temperature and the quantity of magnesium was determined by atomic absorption using the Perkin-Elmer's method and showed that after dialysis the magnesium content in the sample was only about $5 \times 10^{-3}\%$. The loss of magnesium on dialysis shows that the magnesium is not tied to the larger protein molecule, but is free to be removed by passing through the membrane.

Example 2: The product of Example 1 is added to nonfat dry milk solids in a ratio of 1 part by weight of the neutralized casein to 3 parts of dry milk solids. Chocolate flavor and sugar are added. The ingredients are mixed with water in a volume ratio of 1 part of the mixture to 3 parts of water and with mixing a thickened chocolate milk-shake-like drink is produced.

Example 3: A conventional muffin mix is prepared from flour, corn meal, salt, baking powder, shortening, eggs, milk and sugar. One part of the product of Example 1 is mixed with 4 parts of muffin mix and 1 part water. Muffins are baked from the mixture to provide a protein rich bread.

From Sunflower Seeds

Proteins have been isolated from sunflower-seed meal by the direct extraction of proteins in the presence of the polyphenolic compounds (especially chlorogenic and caffeic acids). These compounds are one of the principal hindrances against the direct use of such meals in human feeding.

It is well known that, at the alkaline pH at which proteins are extracted, polyphenols are oxidized to quinones and the latter, by reacting with the proteins, impart to the proteins a dark green hue while concurrently depreciating their nutritional value. For this reason, the extraction of polyphenols is carried out beforehand, both in the case in which it is desired to prepare a protein enriched edible meal (a proteinic concentrate) and in order to be enable to extract the proteins subsequently in an alkaline environment in order that a proteinic isolate may be obtained.

C. Nuzzolo, R. Vignola and A. Groggia; U.S. Patent 4,212,799; July 15, 1980; assigned to Snamprogetti, SpA, Italy have used an improved method for extracting the polyphenolic compounds from the proteins. This method comprises the step of using an aluminum salt (a sulfate or a chloride) in the stage of aqueous extraction of the proteins with NaOH.

The aluminate ion $(AlO_2)^-$ or $Al(OH)_4^-$, or $Al(OH)_4(H_2O)_2^-$ which is present at the pH 10.5 of the extraction has the task of complexing the orthobiphenols (i.e., the chlorogenic and the caffeic acids) and of preventing their oxidation to quinones. From the extract, the proteins are then separated with the usual method of precipitation to the isoelectric point (pH 5) by using in the latter stage a compound which is capable of maintaining the aluminum in solution. This is because it is known that at a pH of from 3.5 to 9.5 the insoluble hydroxide, $Al(OH)_3$ is stable, which would otherwise precipitate with the proteins if the complexing agent should not be introduced.

However, in order to have a low Al content (about 200 ppm) in the end proteins (i.e., the isolate) it is required that a double precipitation be carried out, that is, that the proteins should be redissolved with NaOH and precipitated once again in the presence of the complexing agent.

The most common among the substances which complex aluminum in solution are citric, tartaric, malic and oxalic acid, the pyrophosphate, polyphosphate and EDTA (sodium salt of the ethylenediaminotetraacetic acid) and acetylacetone. The most efficient complexing agents, that is, those for which a high stability constant for the complex with Al is experienced, are EDTA, acetylacetone, citric acid and polyphosphates.

It has proven advantageous to use an acidic complexing agent, inasmuch as it unfolds a twofold action when it is added to the alkaline proteinic extract, viz: to make acidic the medium down to pH 5 while concurrently forming the soluble Al complex. Citric acid is the most suitable substance, also by virtue of the fact that there are no limitations to its use in foodstuffs both from the point of view of toxicity and of legal prohibitions.

The proteinic isolate as obtained with this method is acceptable as regards the color: It is cream-white at a pH of 5, and is slightly yellow-green at alkaline pH values on account of a small amount of residual polyphenols (0.4 to 0.5%) in the freeze-dried isolate which are strongly bound to the proteins and thus difficult to remove under the nondenaturating conditions employed herein.

Example: The meal had the following composition in percent by weight: 10% moisture, 50% proteins, 1.5% lipids, 4.2% chlorogenic plus caffeic acids and 4% crude fiber. 20 grams of meal were added to 200 ml of deionized water containing 1 gram of $Al_2(SO_4)_3 \cdot 18H_2O$.

The slurry, stirred at room temperature, was adjusted to a pH of 10.5 with normal NaOH. The solubilization of the proteins was performed for a time of from 30 to 40 minutes at this pH, which was kept constant by adding normal NaOH in increments. The slurry was then subjected either to filtration or centrifugation.

On the filtrate or the supernatant, precipitation of the proteins was effected by adding 5% aqueous citric acid (0.22 M) to a pH of 5.0. The precipitated proteins were separated from the liquid by centrifugation at a low speed. Reprecipitation of the proteins was then carried out so that the sediment was supplemented by 200 ml of water and the slurry was adjusted to pH 10.5 with normal NaOH until complete solubilization was attained (about 20 minutes). The proteins were then precipitated at pH 5 with citric acid as above. After centrifugation, the sediment was slurried in about 100 ml water, neutralized with soda and freeze dried. The freeze-dried product, 6 grams, was the end proteinic isolate having a proteinic concentration of about 95% and a content of polyphenols (chlorogenic plus caffeic acids) of about 0.5%.

Low Cholesterol Casein

Casein is a mixture of phosphoproteins naturally occurring in milk, cheese, beans and nuts. It is usually isolated from skim milk by isoelectric precipitation or by enzymatic coagulation. The isolated and recovered protein contains all of the common amino acids and is particularly rich in essential amino acids. Due to its desirable nutritional balance, casein is very useful as an ingredient in many food products such as simulated meats, cheeses and other dairy products. Unfortunately, the casein which is commercially available has the relatively high cholesterol content of between 15 and 30 mg/100 g. This high cholesterol level limits the use of casein in certain foodstuffs where a lower cholesterol level is desirable.

A process for preparing casein with lowered cholesterol has been developed by *A.J. Hefley and H.P. Furgal; U.S. Patent 4,255,455; March 10, 1981; assigned to Miles Laboratories, Inc.* The process comprises the steps of dispersing a casein-cholesterol bond-reducing emulsifying agent within an aqueous medium containing casein and cholesterol and thereafter isolating the casein from the aqueous medium.

Preferably, the emulsifier used will have a hydrophilic-lipophilic balance (HLB) of about 7 to 16 and will be soluble in aqueous media. Examples of such casein-cholesterol bond-reducing emulsifiers include the polyglycerol fatty acid esters and the sorbitol fatty acid esters. The polyglycerol fatty acid esters having an HLB between about 7.0 and 16 are preferred. The most preferred of these are octaglycerol monooleate, octaglycerol monostearate and triglycerol monostearate all of which are commercially available.

Suitable aqueous media containing casein and cholesterol which can be treated by this process are commercially available or can be easily prepared from commercially available materials. For example, commercial skim milk is an aqueous mixture containing casein and about 0.5 to 1.75 mg cholesterol per 100 ml. The aqueous medium can also be prepared from nonfat dry milk solids which contain about 15 to 30 mg cholesterol per 100 g or from commercial casein which contains about 15 to 30 mg cholesterol per 100 g. When so prepared, the aqueous medium will contain between 10 and 20 parts by weight milk solids or casein to 80 to 90 parts by weight water.

The effective amount of the bond-reducing emulsifier, such as octaglycerol monooleate or octaglycerol monostearate, which is dispersed within the aqueous medium will depend upon the amount of cholesterol in the sample to be treated and the amount desired in the isolated casein, but it can be determined empirically without undue experimentation. Preferably, the emulsifier concentration will range from 0.2 to 0.5% on a weight/volume basis based upon the total volume of the aqueous medium.

Various means may be used to disperse the emulsifier within the aqueous medium. For example, the emulsifier may be added to the aqueous medium with agitation using any high speed mixing or shearing device. The amount of agitation is not critical as long as it is sufficient to disperse the emulsifier in the aqueous medium.

The temperature at which the aqueous medium is maintained can be widely varied from about 4° to 45°C. Preferably, the temperature is maintained at about 26° to 43°C.

The time during which the emulsifier is in contact with the aqueous medium before the casein is isolated can range from about 1 to 90 minutes. Very good cholesterol reductions in the isolated casein have been effected by contact of the emulsifier with the aqueous medium for about 15 minutes.

After treating the aqueous medium with the emulsifier the casein having lowered cholesterol content can be isolated by any well-known means, such as isoelectric precipitation, enzymatic coagulation and the like. Isoelectric precipitation is the preferred means. If isoelectric precipitation is selected, the aqueous medium is heated to about 45° to 60°C and the pH of the aqueous medium is adjusted to about pH 4.4 to 4.6 to precipitate the casein. The pH of the mixture before adjustment usually ranges between 6.0 and 7.0. The pH can be adjusted by the addition of any suitable acidic means, such as acetic acid, hydrochloric acid, lactic acid, phosphoric acid, sulfuric acid, acidic buffer salts and the like.

Example: Two separate 1 liter portions of skim milk were heated in a water bath to about 45°C. To one sample portion, representing a control sample, there was then added sufficient acetic acid (20 to 35% v/v) to adjust the pH of the mixture to about 4.6 and to precipitate the casein. The precipitated casein was separated from the mixture by decantation, washed six times with about 500 ml of warm water to remove residual soluble whey components and simply pressed to remove residual water. This control sample was assayed for moisture and for cholesterol content determined on a dry weight basis using the abovedescribed assay.

Into the remaining sample portion, representing a test sample, there was dispersed by mixing agitation about 3 grams of octaglycerol monooleate (Santone 8-1-0, Durkee Industrial Foods Group, SCM Corp.; this amount corresponds to 0.3% on a weight/volume basis based upon the total volume of the sample) and the thus treated sample was maintained at about 45°C for 30 minutes. The sample was then treated with sufficient acetic acid (20 to 35% v/v) to adjust the pH to about 4.6 to precipitate the casein. The precipitated casein was separated from the mixture by decantation, washed six times with about 500 ml of warm water to remove residual soluble whey components and simply pressed to remove residual water.

The control sample contained 27.7 mg of cholesterol per 100 g of isolated casein, while the sample prepared by this process contained 2.8 mg cholesterol per 100 g casein.

Protein Fractions Obtained by Ion Exchange

D.J. Phillips, D.T. Jones and D.E. Palmer; U.S. Patent 4,218,490; August 19, 1980; assigned to Viscose Group Limited, England have found that proteins extracted from protein sources by ion-exchange generally possess good functional properties, depending on the intrinsic functional capability of the protein concerned.

This process is primarily concerned with the use of proteins which have an intrinsic functional capability in terms of one or more of the following surfactant properties: foaming, foam stabilization, water-binding, fat-binding and gelling. The isolation of protein from protein sources by a process of ion-exchange extraction will typically result in a protein isolate which, as compared with protein obtained from the same source by other methods (especially ultrafiltration, diafiltration and precipitation techniques) possesses superior and/or more widely utilizable functional properties especially as regards properties dependent on surface activity. Especially good results are obtained, in terms of foaming and gelation properties, from protein isolates produced by ion-exchange extraction from milk whey.

The protein source may be of animal, fish or vegetable origin, for example, egg white or milk; or a liquid extract or waste effluent such as soy whey or extracts from rape seed, groundnuts, palm nuts, sunflower seeds, or olives; or blood (e.g., abattoir effluent) but is preferably milk whey.

Advantageously, a cellulosic ion-exchange material is used in the protein isolation process, and the ion-exchange material is preferably a regenerated cellulose substituted with ion-exchange groups. An especially preferred ion-exchange material is one made by reaction of cellulose or a cellulose derivative with an activating substance capable of conferring ion-exchange properties thereon, followed by regeneration of the substituted cellulose reaction product into the desired physical form.

The isolation process for use in preparing functional protein may comprise, for example, the following steps:

(a) A liquid protein source (for example, milk whey) is treated with an ion-exchange material (preferably a regenerated cellulose ion-exchange material such as a regenerated carboxymethyl or diethylamino ethyl cellulose, made by effecting ion-exchange activation before regeneration) to extract protein therefrom.

(b) The protein is recovered from the ion-exchange material by a desorption process, typically yielding a dilute solution containing about 1 to 5% protein.

(c)   The solution is concentrated to a protein content of about 10 to 30%, for example, by ultra-filtration.

(d)   The concentrated solution is dried, for example, by spray-drying or freeze-drying, to give a dry protein product which will typically contain less than 1% fat (about 0.2 to 0.3%) and about 3% ash in addition to the protein material. Higher ash contents should be avoided as far as is practicable, as they may adversely affect certain functional properties.

If desired, step (a) can be preceded by an initial concentration step effected, for instance, by ultra-filtration or by a controlled evaporation technique.

Example: Protein was extracted from a milk whey by a process involving ion-exchange extraction using a cellulosic ion-exchange material, followed by recovery, concentration and drying of the protein isolate. The dry isolate was incorporated with water and the mixture beaten to form a foam. Foam volume and drained liquid volume were measured after 30 minutes. A high foam volume of 140 at 30 minutes and a low drained liquid volume of 28 after 30 minutes are both indications of good foaming properties of the agent used.

For comparison, the tests were repeated using a protein concentrate obtained from the same milk whey by ultra-filtration, and with a mixture comprising 90% of that material and 10% of the protein isolate obtained by ion-exchange extraction.

This product had zero foam volume (i.e., it had collapsed) at 30 minutes and a drained liquid volume of 50.

It will be seen not only that the protein isolate acccording to this process had superior foaming properties but also that a mixed foaming agent containing as little as 10% of the isolate exhibited properties almost as good.

The mixture containing 10% of the isolate produced a foam volume of 130 at 30 minutes and a drain volume of 30 at 30 minutes.

Products from Fava Beans

Skim milk powder has been relatively scarce on the world market, its cost has increased sharply in recent years and economic projections suggest that this situation will continue or worsen in the foreseeable future. There have thus been strong incentives to find alternatives to skim milk powder.

These problems are overcome by the process described by *I.F. Duthie; U.S. Patent 4,259,358; March 31, 1981; assigned to Agricultural Production and Vegetable Products, Ltd., England.* This process provides for preparing a food product including treating a slurry of ground or flaked legume material with aqueous liquid, the slurry having a pH of 8 or less, with amylolytic enzyme at a temperature and for a period sufficient to liquefy and hydrolyze at least part of the starch contained in the legume material to dextrins. The food products obtained are useful for young mammals including humans and ruminants.

The beans of *Vicia faba* L. of the family Leguminosae, comprising the two sub-species *Vicia faba* L. var. minor, which is being grown on an increasing scale both in Europe and in other parts of the world, and *Vicia faba* L. var. major, provide the most preferred legume material for this process.

The ability of young mammals, particularly ruminants, to digest starch is limited, and it is an important feature of this process to effect reduction of starch present in legume material into forms which can be digested by young mammals. This is accomplished by treating a slurry of legume material in aqueous liquid with one or more amylolytic enzymes capable of hydrolyzing starch to forms of carbohydrate readily digested by young mammals, in the case of young ruminants such forms being mainly dextrose and small amounts of di- and tri-saccharides composed of glucose units. This treatment prefaced by a protein-solubilization step, if so desired, is accomplished as follows.

A slurry of ground or flaked legume material, which may, if considered desirable have been heat-treated as described above, is made with aqueous liquid, preferably in a weight ratio of 1:1 to 1:15 on a dry matter basis. The pH of such a slurry may then be adjusted with a suitable alkali to a value of from the natural value to 10.0 and stirred for from 5 to 60 minutes at a temperature preferably of from 10° to 65°C to permit solubilization of the protein component.

To effect starch hydrolysis, the slurry is adjusted, if necessary, to a pH in the acid or near acid range, preferably from 5.0 to 8.0, and most preferably about 6.0 to 7.0, with, for example, sodium hydroxide, and a preparation of α-amylase in a suitably purified form is added with stirring as necessary at a rate preferably of 0.25 to 5.0 grams or more per kilogram of dry legume material of a preparation containing 1,000 to 1,500 SKB units of α-amylase activity per gram. This reaction is further illustrated by Example 1 below.

As an alternative to prolonged α-amylase treatment it is preferred to add another amylolytic enzyme, amyloglucosidase, to promote saccharification as shown in Example 2 below.

Example 1: Clean whole mature beans of *Vicia faba* L. var. minor were hulled by a suitable dry method. The resulting cotyledon fraction was ground to a fine powder passing a 1 mm mesh. Powdered cotyledon material recovered from the hulling waste may be combined with the ground cotyledons at this stage if desired.

The powdered legume material was slurried with water, employing continuous agitation at this and subsequent stages, to provide a slurry containing 35% legume material on a dry matter basis, and having an initial temperature of 30°C. Food grade sodium hydroxide was incorporated as necessary to adjust the pH of the slurry to 7.0.

0.75 gram of a suitable α-amylase preparation containing 1,000 to 1,500 SKB units of activity per gram was incorporated in the slurry per kg of legume material, and the slurry was passed through a high pressure homogenizer operating at a pressure of 1,500 psi or more. The slurry temperature was raised to 70°C at a rate of 1°C/min and then to 85°C at a rate of 0.5°C/min. The temperature of the slurry was maintained at 85°C for 45 min and then, or after a further period at a lower temperature depending on the degree of starch dextrinization required, the pH was adjusted to 7.0, the temperature being 85°C or lower. The slurry may then be passed again through a homogenizer if desired at 1,500 psi, and it was then spray dried in a spray drier equipped with a centrifugal atomizer, the temperature conditions for spray drying being inlet 180°C and outlet 80°C. The flow agent, aluminum silicate, was added during spray drying at the 1% level of dried product. The moisture content of the resultant spray-dried product was from 3 to 5%.

Example 2: The procedure of Example 1 was followed, but following the dextrinization treatment of up to 45 min at 85°C, the temperature of the slurry was lowered to 50°C, the pH of the slurry was adjusted by addition of food grade hydrochloric acid, to 4.5, and 1 g of a suitable amyloglucosidase preparation containing from 100 to 175 AG units per g was incorporated in the slurry per kg legume material. The selected pH and temperature conditions were maintained for 6 hours or longer to allow saccharification to proceed to the desired extent. Following this saccharification step the pH of the slurry was raised by incorporation of food grade sodium hydroxide to 7.5 and the temperature was adjusted to 85°C.

The slurry was then spray dried as in Example 1 and had the following composition: 33.6% protein, 46.5% dextrose and 1.90% fat on a dry matter basis. According to the procedure used the spray dried slurry may have the following range of compositions: 25 to 35% protein, 40 to 55% dextrose and 1.5 to 2.0% fat.

Trub Flour from Brewery Waste

Conventional processes for the brewing of beer produce large quantities of waste materials, some of which are rich in protein. One such material is trub, which is obtained from whirlpool extraction of the residue remaining in the wort kettle used in the brewing process. The trub obtained from a wort kettle is a highly proteinaceous matter.

The development by *P.M. Townsley and R.L. Weaver; U.S. Patent 4,315,038; February 9, 1982; assigned to The Molson Companies Limited, Canada* provides a process for preparing a flour from the trub which can be used in the preparation of human food.

A considerable quantity of wort can be recovered from hot break trub following removal of the trub from the whirlpool separator. Much of the wort can be recovered by simple pressing. For example, with a sample of 6,000 lb dry weight of trub, at least 1,064 gallons and possibly as much as 2,000 gallons of wort can be recovered.

Water is then removed from the trub by isopropanolic extraction using an azeotropic isopropanol. The solvent wet trub is passed through a roller drum drier to yield dry trub in flake form. The additional solvent recovered from the drier is vaporized and recovered in the same condenser used to condense the distilled solvent. The solvent, after separation of the trub, contains hop resins. The crude resins are subjected to evaporation to yield solvent and a crude resin extract, which is subjected to isopropanolic extraction to yield a fraction insoluble in alcohol and a trub resin extract. The trub resin extract, after separation from the solvent by evaporation, can be used in ale fermentation.

Waste brewer's grain can be used in a similar manner to produce flour for use in human food. The waste brewer's grain is fractionated by simple physical separation techniques into a protein flour suitable for human food and a husk containing material suitable for animal feed.

Trub protein contains all of the amino acids normally found in cereal proteins. However, the trub protein is unique in that the cysteine (cystine) level is high (122% of that found in egg white protein.) The sulfur amino acid level in trub is higher than that found in cereals such as wheat or barley or in seeds such as soybean or rapeseed. The nutritional value of trub protein as indicated by the essential amino acid index is 0.55 which is a value slightly lower than wheat or barley which have an index of 0.63 and 0.69 respectively. The first amino acid to be limiting in a diet utilizing trub protein as the sole source of essential amino acids would be tryptophan, followed by lysine and methionine. The limiting concentrations of methionine in trub protein would be overcome by the encouragingly high concentrations of cysteine. Threonine and arginine occur at comparatively high levels in trub protein.

The solubility of trub protein in water is approximately 2 to 2½% over the pH range of 2 to 11. When the pH of the trub in water suspension is adjusted to 12, the solubility of the protein is only 17%. The addition of NaCl to the water does not increase the solubility of the protein. Since many food applications require that the protein be soluble in order to take advantage of the functional properties of the proteins, trub protein in its present form appears to have limited application. On the other hand, because of its nutritional value, trub can be used with advantage as an additive in products such as cookies, buns and pasta.

The color of the trub protein flour presents a problem. The dry flour is a pleasant light tan color, but in the presence of water becomes much darker. Food products requiring a light color for consumer acceptance could not contain trub flour as an ingredient because of its darkening effect.

Trub flour can be used to increase the nutritional value, especially in terms of sulfur amino acid, of dark colored bakery and cereal products. Products such as bran muffins and cookies containing trub are good. The nutritional value and possibly the flavor of pasta products can be improved using trub flour. However, the products are much darker (tan colored) than the traditional food. The number of acceptable food products that can be made from trub protein is restricted for one or more of insolubility, color, neutral or slightly undesirable flavor and poor foam capability (ale). Another problem encountered with trub flour is in baking. During baking, the volatile aromatics of the flour entering the baking area are not as pleasant as those of wheat flours. The negative attribute is not noted in the finished bakery products.

USE OF PROTEIN ADDITIVES

Acetylated Proteins in Beverages

Protein fortification of acidic foods having a pH in the range of 2 to 6 is complicated by the unavailability of protein exhibiting satisfactory nutritional and taste characteristics. Oleaginous proteins, especially soy protein, when altered to be acid soluble, are characterized as having undesirable taste characteristics variously described as beany and bitter. The objectionable flavor is especially noticeable in acidic beverages that contain an insufficient number of other ingredients to mask the strong flavor at the protein levels of 2 to 8% or higher.

The disclosure by *C.V. Fulger and J.E. Dewey; U.S. Patent 4,138,500; February 6, 1979; assigned to Kellogg Company* provides improvements in taste characteristics of oleaginous hydrolyzates obtained by enzyme hydrolysis of the protein. The hydrolyzate is chemically modified by reaction with an acylating agent, preferably a dicarboxylic acid anhydride such as succinic anhydride. By this procedure, the free amino groups of the polypeptides are acylated, resulting in substantial taste improvement, particularly in acidic solution.

The seeds, such as soybean, are first treated to isolate the protein. For example, soybean meal may be extracted with a mild alkali having a pH of about 7.5 to about 9.5. The aqueous extract is separated from the solid fraction and the pH is then adjusted to about 4.5 to precipitate the proteins. The precipitated proteins may be washed and are then ready for enzyme hydrolysis.

An aqueous slurry of the soy protein may be heated to a temperature between about 200° to 265°F for a period of time without deleteriously affecting the flavor optimization of the final product. A heating period from 1 to about 25 seconds is generally sufficient, depending on the temperature.

The partially denatured protein is then treated with a proteolytic enzyme, preferably under slightly acid or basic conditions. Suitable enzymes include endopeptidases, such as neutral and alkaline plant or animal proteinase enzymes, pepsin, ficin, papain, rennin, trypsin and chymotrypsin, as well as microbial exopeptidases, such as aminopeptidase and carboxypeptidase A and B.

The enzyme hydrolysis is continued until a major portion of the protein has been hydrolyzed, preferably between 60 and 85%, and is then terminated. The resulting polypeptide mixture will be soluble in a solution having a wide range of pH values, but for this process, the mixture will be soluble in a solution having a pH between 2 and 6. After removal of any insoluble fraction, the solution may be dried.

The protein isolate is then placed in solution and acylated by conventional methods. The acylating agent is a compound capable of acylating the functional groups of the protein hydrolyzate, such as oxygen, sulfur or nitrogen, which have replaceable hydrogen atoms to provide sites for reaction with this agent. Suitable agents include carboxylic acid anhydrides, either internal or external, such as monocarboxylic acid anhydrides, including acetic anhydride and propionic anhydride, as well as dicarboxylic acid anhydrides, such as succinic anhydride, glutamic anhydride and maleic anhydride, as well as mixtures of the foregoing. Succinic anhydride is preferred and is added as a solid to the hydrolyzate solution maintained at a neutral pH of about 7.0 to 7.2. The solution is preferably cooled to a temperature of from about 1° to about 15°C during the reaction.

Inorganic salts produced by the reaction may be removed by dialysis, electrophoresis, reverse osmosis or other suitable desalting techniques. The solution may then be dried to obtain the carboxyacylated hydrolyzate in solid form.

The product thus obtained has a very mild or bland odor and flavor and does not develop objectional flavors when dissolved in acid solutions having a pH of from about 2.0 to 6.0. The product is thus particularly suitable for incorporation into acidic beverages including natural or artificial acidic fruit drinks. Suitable beverages would include, for example, citrus and tomato beverages. The product is also suitable for incorporation into carbonated beverages.

Example 1: In the preparation of bland acid soluble protein, 134 g of enzyme hydrolyzed soy protein (Type 1205, Gunther, Division of Staley) was dissolved in 1 liter of distilled water in an ice bath. The pH of the solution was 4.46, and the temperature was 37°F. To the solution was added 12.5 ml NaOH (5 N) in order to bring the pH to 7.05. Succinic anhydride (11.7 g) was sprinkled slowly into the reaction mixture while maintaining the pH between 7.0 and 7.2 by continuous titration with NaOH (5 N). About 48 ml of the sodium hydroxide was used in this titration.

When the pH remained constant, indicating completion of the reaction, the reaction mixture was dialyzed versus 5 one-gallon volumes of distilled water using a hollow fiber dialyzer and changing the water every 20 minutes. The reaction mixture was then freeze dried.

Example 2: Three different types of enzyme hydrolyzed soy protein, Gunther Type 1205, Gunther Type 1043 and Central Soya Type FS704 were subjected to succinylation in accordance with Example 1. An orange drink was prepared and divided into three portions. The same amount of succinylated protein was added to equal volumes of orange drink and subjected to taste tests. Identical samples containing the same amount of protein without succinylating were prepared for the sake of comparison. In all cases, a taste improvement was observed with the succinylated protein, including a reduction in soy flavor and bitterness.

Soy-Treated Gluten for Baked Goods

Wheat glutens, and particularly the undenatured or so-called "vital" wheat glutens employed in the production of baked goods, exhibit a marked tendency to agglomerate when introduced into an aqueous medium at or near neutral pH, and dispersions of wheat gluten thus are unduly difficult to prepare and not adequately uniform.

Prior methods for inhibiting agglomeration of wheat gluten by coating the gluten particles have achieved considerable success in the production of bread, rolls and other yeast-leavened baked goods, but the necessity for coating the gluten has increased the effective cost of the gluten to the baker. There has accordingly been a continuing need for an effective but less expensive solution to the problem of inhibiting agglomeration.

An improved and more economical method for producing stable aqueous dispersions of wheat gluten has been provided by *A.W. Kleinschmidt; U.S. Patent 4,179,525; December 18, 1979; assigned to J.R. Short Milling Company.*

The process is applicable to any finely particulate gluten which tends to agglomerate excessively when dispersed in aqueous media, and is particularly useful in connection with vital wheat gluten. The process stems from the initial discovery that vital wheat gluten, exhibiting by itself an excessive tendency to agglomerate when agitated in an aqueous liquid, is readily dispersible in water when premixed with a soy flour, the protein content of which has not been denatured, while premixing with a soy flour treated to denature the soy protein failed to reduce agglomeration. The same results are achieved with peanut flour, navy bean flour, lentil flour, cottonseed flour and rapeseed flour, so long as the protein thereof is undenatured, as well as with the proteins isolated or concentrated in undenatured form from such materials. When the aqueous medium in which the gluten is to be dispersed does not contain a substantial amount of salt, the same results are also obtainable with dairy whey solids. Though higher proportions of the additive material can be employed without serious disadvantage other than lack of economy, best results are achieved when the soy flour or other agglomeration-inhibiting additive is included in an amount equal to 3 to 15% by weight, based on the total weight of the gluten and the agglomeration-inhibiting additive.

It is of particular importance, especially when the protein material to be rendered nonagglomerating is vital wheat gluten, to employ that material in a particle size such that the average maximum particle dimension does not exceed 350 microns. When the particle size of the gluten is significantly larger, the effect of the soy flour or like proteinaceous additive in preventing agglomeration of the gluten is markedly reduced. It is advantageous to employ the gluten to be dispersed in a particle size such that most of the gluten has a maximum particle dimension smaller than 150 microns and no substantial proportion thereof has a maximum particle dimension larger than 350 microns. The average maximum particle dimension of the soy flour or other agglomeration-inhibiting additive should not exceed 350 microns in all events and, advantageously, is made to approximate that of the gluten.

Example 1: Using a Waring blender, 200 ml of water was introduced into the bowl of the blender, 25 g of New Era vital wheat gluten was added, and the wheat gluten was dispersed by operating the blender for one minute. The gluten agglomerated after the dispersion had stood at room temperature for only 30 minutes. The New Era vital wheat gluten employed was finely particulate, having 52% smaller than 105 microns.

Example 2: 25 g of full fat enzymatically active soy flour and 225 g of the same vital wheat gluten employed in Example 1 were placed in a glass jar and tumble-blended by hand for 15 minutes. Using a Waring blender, 25 g of the resulting mixture and 200 ml water were combined in the bowl and mixed for one minute. The resulting dispersion was allowed to stand at room temperature for observation. After 7 hours, no significant agglomeration or settling could be observed.

Example 3: Using a Waring blender, 200 ml water was placed in the bowl, 25 g of full fat enzymatically active soy flour was added, and the blender then operated for one minute to disperse the soy flour. 225 g of the vital wheat gluten of Example 1 was then added and the blender again operated for one minute to disperse the gluten. The resulting dispersion was set aside at room temperature for observation. No agglomeration or settling had occurred at the end of 7 hours. The procedure was repeated, except that the vital wheat gluten was dispersed in the water before the soy flour. Again, no agglomeration or settling was observed at the end of 7 hours.

Chelated Gluten for Baked Goods

A dispersion of powdered gluten once obtained is unstable because of the tendency of the individual gluten particles to coalesce. Simple mixing of vital wheat gluten with water produces a lumpy product which on vigorous agitation is transformed into a highly intractable rubbery elastic mass.

Various attempts have been made to provide an activated form of gluten which would hydrate sufficiently, easily and relatively rapidly, i.e., in the time period of the modern baking processes, and thereby become available for incorporation into the protein network.

The process used by *H.P. Johannson; U.S. Patent 4,150,016; April 17, 1979; assigned to Industrial Grain Products, Limited, Canada* provides a dry modified gluten which when rehydrated is equal to, or even superior in, functionality to rehydrated dry vital gluten or fresh wet gluten.

It was found, that washing of a dough with a medium containing a chelating agent resulted in a product which met the above objects. The gluten-chelating agent complex of this process is usually a pale yellow powder having a moisture content generally less than 8%.

The chelating agent content of the complex may vary but usually is between 0.5 and 3%, or 1 and 3%, by weight based on the dry gluten content, the most preferred concentration being about 1.5% by weight, based on the dry gluten content.

The following are chelating agents which may be used. They are placed in two groups purely for convenience in further describing this process. Group 1 consists of: (1) ethylenediaminetetraacetic acid, preferably in the form of the disodium salt; (2) phytates, such as sodium and calcium phytate; (3a) simple phosphates such as ammonium phosphate, alkali metal, e.g., sodium and potassium phosphates and alkaline earth calcium phosphates; (3b) complex phosphates or molecularly dehydrated phosphates such as salts of pyrophosphoric acid, e.g., disodium pyrophosphate, polyphosphoric acid, e.g., sodium tripolyphosphate, and metaphosphoric acid, e.g., sodium hexametaphosphate.

Group 2 consists of: citrates, such as alkali metal citrates, for example, sodium citrate and potassium citrate and ammonium citrate.

This process for the production of a powdered modified gluten comprises mixing vital gluten and a chelating agent in the presence of water until a substantially homogeneous mixture is obtained, drying the mixture and reducing the dried mixture to a powder form.

The pH during the mixing step is extremely important. For the desired product to be obtained the mixing of the gluten and the chelating must be at the effective pH of the specific chelation agent being used.

Many chelating agents which may be used are effective at a pH of about 8 to 10. Examples of such agents are those of Group 1. However, the pH of gluten is about 6 and when using such agents, it is necessary to adjust the pH of the mixture of gluten and chelating agent prior to their being reacted.

However, certain chelating agents such as those of Group 2 such as sodium and potassium citrate, have an effective pH of about 6, i.e., substantially the same as gluten itself. It has been found that no adjustment of the pH of the mixture of gluten and such chelating agent is therefore necessary in order to obtain or induce reaction between the two components. For this reason, gluten-citrate complexes constitute a preferred form of this process.

Example 1: This example shows the preparation of gluten-sodium citrate (1.5%) complex from fresh wet vital gluten by a laboratory or bath process. A solution comprising 2 g sodium citrate in a minimum of water was introduced into a standard laboratory farinograph bowl which contained 400 g of freshly prepared wet vital gluten and mixing was effected at 200 rpm for 5 to 7 minutes, i.e., until the curve had leveled off. The product was thinly spread onto a suitable surface and freeze-dried overnight. The dry product having a moisture content of from 2 to 4% by weight, was reduced in a mill so that the desired gluten-sodium citrate (1.5%) complex, i.e., containing 1.5% by weight based on dry gluten, of sodium citrate, was obtained as a powder which passed a 60 mesh screen.

Example 2: A continuous commercial process may be used by a simple modification of the well-known processes used to separate starch and gluten from wheat flour. Wheat flour dough is introduced to a starch/gluten separator and the wet gluten so obtained is passed through a gear pump to a ring drier.

To produce the modified gluten it is merely necessary to introduce the chelating agent, i.e., sodium citrate, in solution into the gear pump from a tank via a metering device. The mixing obtained by the action of the gear pump is sufficient to thoroughly mix the wet gluten and the chelating agent and the mixture is then passed to the drier and finally to a hammer mill, where it is reduced to a powder which passes an 80 mesh screen, and subsequently bagged for shipment.

Obviously, the flow rates of wet gluten and chelating agent solution are controlled to provide the desired concentration of chelating agent in the product. For example, about 250 lb/hr of dry powdered gluten-sodium citrate (1.5%) complex according to this process is produced by passing 750 lb/hr fresh wet vital gluten (equivalent to 250 lb/hr dry gluten) directly from the separator through the gear pump mixer while introducing into the gear pump 13.0 lb/hr of a sodium citrate solution having a concentration of 3.5 lb/Imp gal of sodium citrate. The so-obtained product had a moisture content of about 5% by weight.

Protein Micelles in Baked Goods

Known procedures for preparing plant protein isolates involve an isoelectric precipitation step of solubilized proteins. In most cases, the proteins are solubilized by alkaline extraction with the treatment in some cases being enhanced by increased temperature, enzyme activity and/or salt addition. However, regardless of the solubilization scheme, acid is always used for the isoelectric precipitation of the desired product. Furthermore, it is to be noted that in order to achieve a reasonable level of solubilized protein (and hence an efficient process) an alkaline pH step is normally demanded.

These known processes result in the formation of an amino acid derivative lysinoalanine (LAL) which is absorbed poorly by the gut of growing animals.

The isolation process developed by *E.D. Murray, C.D. Myers and L.D. Barker; U.S. Patent 4,169,090; September 25, 1979; assigned to General Foods, Limited, Canada* eliminates the use of acids and alkalis.

Proteins from a wide variety of sources, typically, starchly legumes, starchy cereals, and oilseeds, are extracted and recovered with gentle aqueous conditions which do not employ extremes of alkali, acid, or temperature. The ionic environment of these proteins is manipulated to firstly produce high solubility (salting-in) followed secondly by ionic strength reduction to cause the proteins to precipitate by a hydrophobic-out mechanism. This process results in the formation of proteinaceous products, referred to as protein micelles.

The first step in this process is to solubilize maximal amounts of the desired globular proteins. To do this the starting material must be broken physically and reduced or ground to a very small particle size so that a high surface area will be exposed to solubilizing solution.

The protein fraction (usually a dry flour or concentrate) is then mixed into a solution containing only water and an appropriate food grade salt (sodium chloride, potassium chloride, calcium chloride, etc.) for a time sufficient to salt-in the desired proteins. After agitation of the protein/salt/water system for a suitable time (usually 10 to 60 minutes) at a moderate temperature (usually 15° to 35°C), the insoluble particulate matter (usually cellular debris and perhaps starch granules) is removed from the solubilized proteins by settling, filtering, screening, decanting or centrifuging. In practice, the latter step is preferred. Although the salt concentration is in the range of 0.2 to 0.8 ionic strength, the actual level of salt used is selected experimentally by determining the minimal salt concentration required to yield maximal levels of solubilized (salted-in) protein.

Ideally, this extract should have a protein concentration of at least 15 to 20 mg/ml (1.5 to 2.0% w/v) and it may approach 75 to 100 mg/ml (7.5 to 10% w/v). The extract is at a preferred pH of about 6.00±0.50 which is often the natural pH of the protein/salt/water system.

The second step in the preparation of the protein isolate is to simply reduce the ionic strength of this medium exposed to the solubilized proteins. This can be done by various methods involving membrane separation techniques (e.g., dialysis) or merely by dilution of the high salt protein extract in water. In practice, the latter step is generally used.

This causes a rapid decrease in molecular weight of the very loose protein aggregates formed during salting-in and the generation of a preponderance of a comparatively low molecular weight species. This accumulation of amphiphilic globular proteins can be likened to the formation of a CMC (critical micelle concentration) encountered in detergent systems and when this CPC (critical protein concentration) is achieved, the proteins seek a thermodynamically stable arrangement whereby polar moieties on the protein surfaces are exposed to water and hydrophobic moieties cluster together in an attempt to avoid water. This stable arrangement manifests itself as small microscopic spheres containing many associated globular protein molecules. These spheres vary in size but they can be seen readily with an ordinary light microscope and have been called "protein micelles."

Example: Fava beans (*Vicia faba* L. var. minora) were pin milled to a fine particle size and then air classified to produce a concentrate of 53% protein (N x 5.85). The proteins from this starchy legume were then extracted with an aqueous sodium chloride solution at 37°C. The dry concentrate was mixed with a 0.3 M sodium chloride solution (ionic strength 0.3) at a 10% w/v level, i.e., part concentrate to 10 parts salt solution. The mixture was stirred for 30 minutes with no pH adjustments being necessary to hold the extract to pH 5.90±0.20. The system was then processed to remove cellular debris and starch granules by centrifugation using a continuous, desludging unit. The resulting high-salt protein extract (i.e., the supernatant) contained greater than 80% of the total seed protein originally in the air classified concentrate and had a protein concentration of about 45 mg/ml. This extract, which was still at 37°C, was next diluted into cold tap water in a ratio 1:3 (1 part supernatant and 3 parts water). Immediately upon dilution, a white cloud formed in the dilution system. Due to the rapidly reduced ionic strength, dissociation of the high molecular weight aggregates (formed by salting-in) is followed by reassociation into protein micelles as the CPC of the micelle forming unit is achieved. A microscopic check of this cloud showed the presence of many small spheres which bound a protein specific stain (Ponceau 2R). The dilution system was allowed to stand unagitated for about 30 minutes while the protein micelles precipitated therefrom. The supernatant was then decanted and a viscous gelatinous precipitate was found in the bottom of the vessel.

This material which was then spray dried at 100°C outlet temperature possessed protein content of 95.57%; it is apparent that the protein-protein interactions during micelle formation produced a rich protein isolate (PMM) with little contaminating material. This isolate contained 35 mols of lysine per 10^5 g of protein compared to 37 mols in the starting concentrate.

E.D. Murray, T.J. Maurice, L.D. Barker and C.D. Myers; U.S. Patent 4,208,323; June 17, 1980; assigned to General Foods, Limited, Canada have also disclosed a similar process for the preparation of protein micelles. The steps of this process comprise:

(a) extracting the protein source material with an aqueous food grade salt solution having an ionic strength of at least about 0.2 and at a pH of about 5 to 6.8, at a temperature of about 15° to 35°C to cause solubilization of protein material in the protein source material and form a protein solution;

(b) increasing the protein concentration of the protein solution while maintaining the ionic strength substantially constant;

(c) diluting the concentrated protein solution to an ionic strength below about 0.2 to cause the formation of protein micelles in the aqueous phase; and

(d) settling the protein micelles as an amorphous sticky gelatinous gluten-like protein micellar mass. The protein isolate so formed, after separation from residual aqueous phase, may be dried to a powder form.

The concentration step of this process may be effected by any convenient selective membrane technique, such as, ultrafiltration of diafiltration. The concentration step has the beneficial effect of increasing the yield of isolate which may be obtained from the process, and thereby increasing the overall efficiency of the protein isolation process.

Example: A protein concentrate (about 50 wt % protein) of field peas was mixed with a 0.4 molar sodium chloride solution at a 10% w/v level at a temperature of about 25°C. The mixture was stirred for about 25 minutes at a pH of about 6.0. The aqueous protein extract was separated from residual solid matter and had a protein concentration of about 40 mg/ml.

The extract then was concentrated on an ultrafiltration unit using two Romicon type XM50 cartridges (Rohm & Haas Co., "50" = MW cut-off of 50,000) over a processing period of about 40 minutes at a temperature of about 45°C. Concentrates at various volume reduction factors (i.e., the ratio of initial volume to that of concentrated solution) in the range of 3.0 to 5.0 were prepared and these concentrates had a pH of about 6.0 to 6.3.

The concentrates were each diluted into cold water having a temperature of about 8°C at a volume ratio of 1 to 5 (i.e., 1 part concentrate to 5 parts water) i.e., to an ionic strength of about 0.07. Immediately upon dilution, a white cloud of protein isolate observed to be in the form of protein micelles formed in the dilution system. The protein micelles were allowed to settle as a highly viscous amorphous gelatinous precipitate in the bottom of the vessel.

The wet PMM recovered from the supernatant liquid was spray dried to provide a dry powder product which was analyzed for salt, moisture and protein. Overall protein yields of about 35 to 40% were obtained, based on the initial protein over the tested volume reduction factor range, as compared with a yield of about 20% for the equivalent process but omitting the concentration step.

Protein and salt analysis results for the dry PMM demonstrate that the purity of the PMM (in terms of protein content) increases and the sodium chloride content decreases when volume reduction factors of about 3.5 to 5 are employed.

E.D. Murray, C.D. Myers and L.D. Barker; U.S. Patent 4,285,862; August 25, 1981; assigned to General Foods, Limited, Canada have used the protein isolate produced by the processes of U.S. Patents 4,169,090 and 4,208,323 in various food products.

This isolate is substantially undenatured, has substantially no lysinoalanine content and substantially the same lysine content as the protein source material from which it is derived. The isolate has substantially no lipid content and takes the form of an amorphous, viscous, sticky, gluten-like protein mass, or a dried form of the mass.

Example 1: White bread was prepared containing ½ cup milk, 3 tablespoons sugar, 2 teaspoons salt, 3 tablespoons margarine, 1½ cups warm water, 1 package active dry yeast and 6¼ cups flour.

Two batches were run, one using sifted all-purpose wheat flour and the other using pea PMM flour consisting of 15 wt % dry pea PMM and 85 wt % cornstarch. Bread was baked from each recipe and the properties compared. The PMM product was similar in properties to the wheat-based product.

Bread was baked from a further batch in which soy PMM flour at the same ratio was substituted for the pea PMM flour. The bread has very similar properties to the wheat-based product and the characteristic soy taste was absent.

PMM products, derived from various plant proteins using the procedure of Example 1 of U.S. Patent 4,169,090 were also used as a partial or complete substitute for wheat flour in a variety of other normally wheat-based products and products similar in character, properties and taste to the conventional wheat flour product were obtained. Products included bread rolls, oatmeal cookies, noodles and chocolate cake.

Example 2: Dry PMM products derived from various plant proteins using the procedure of Example 1 of U.S. Patent 4,169,090 were used as a partial or complete replacement for egg white in a variety of food compositions in which egg white is conventionally used as a binder. In each case, the cooked product was substantially the same in taste and texture to the same product formed using egg white. Products tested were cake mixes, muffin mixes, pancake mixes, a meat loaf analog and a bacon analog.

Metal Gluconates for Producing Bland Wheys for Ice Creams and Gels

R.K. Remer; U.S. Patents 4,235,937; November 25, 1980; 4,325,977; April 20, 1982; and 4,325,978; April 20, 1982; all assigned to Hull-Smith Chemicals, Inc. has processed whey to provide a product with a bland odor and taste.

During blending of whey in the presence of heat there is added a metal gluconate solution, to give the whey a bland odor and taste, as well as a colloid enhancer component to impart a colloidal type condition to the whey. Thereafter, a floc initiator may be added to separate the whey into a solids fraction floc and a liquid fraction. These products may be used in the production of ice creams and gelled products.

Whey is power blended in the presence of a blandness imparting agent and a colloid enhancer to form a colloidal type of system. Such a system is then readily separable into a solids fraction or floc and a liquids fraction by the addition of a floc initiator. Other ingredients, for example, a pH adjuster, an oxidizing agent, an enzyme, such as a lactose, a colorant, a flavoring, a natural moss, or other type of ingredient may be added in order to impart certain properties to the formulations.

An important feature of this process is the use of whey that is fresh as possible and that has not been permitted to cool down to a significant extent from the temperature at which the whey is generated in a cheesemaking process. Less than completely fresh whey can be adequately processed provided its temperature is maintained as high as possible and preferably such that it does not drop below 90°F, more preferably 100°F. Usually, the more efficient the temperature maintenance and the shorter the time lapse between generation of the raw whey and its treatment in accordance with this process, the more valuable and more advantageous is the final product of this process.

Blandness imparting agents are especially useful for masking or blanding the unpleasant odor and taste properties of natural protein sources, especially whey. They are also especially valuable within systems that exhibit colloidal tendencies. Metal gluconates are particularly useful in this regard, especially when provided in solution and in combination with an aminocarboxy acid, preferably aminoacetic acid or glycine. In an especially preferred embodiment, the blandness imparting agent is an aqueous solution containing a 1:1 molar combination of metal gluconate and glycine. Acceptable metal gluconates include iron or ferrous gluconate and copper gluconate. In the preferred embodiment, the 1:1 molar combined solution of metal gluconate and glycine is added in a total quantity of between about 5 to 60 ml for each gallon (3.785 liters) of raw whey.

Colloidal enhancers belong to the general class of cellulose derivatives, particularly those of the cellulose glycolate type, most preferably a carboxyalkyl cellulose material, including alkali metal carboxymethylcellulose compounds. The preferred colloidal enhancer is sodium carboxymethylcellulose, also known as CMC. Useful in this regard are other synthetic or natural gums which have somewhat high viscosities and exhibit general protective colloid properties such as locust bean gum, guar gum, karaya gum, gum tragacanth or their mixtures. Quantities of the colloid enhancer may generally range between about 2 and 40 g for each gallon (3.785 liters) of liquid. When the colloid enhancer is subjected to power blending conditions in the presence of raw whey, especially raw whey already having the blandness imparting agent and a metal hydroxide blended therewith, it imparts to the whey a colloidal-type physical condition which renders the whey particularly susceptible to the operation of the floc initiator of this process when it is included in order to form a solids fraction and a liquids fraction.

When floc initiator is to be used, it is preferably added after the blandness imparting agent, a metal hydroxide if used, and the colloid enhancer are all added, whereupon the whey product will, without any further treatment whatsoever, separate into the solids fraction and the liquids fraction.

The floc initiator generally belongs to the class of alkali metal silicates, most advantageously provided in the form of highly alkaline sodium or potassium solutions, especially sodium silicate solutions which typically contain approximately 40% $Na_2Si_3O_7$ and which can be represented by the formulation (SiO_2/Na_2O) = 3.22. Floc initiator solution may be added in quantities between about 5 and 60 ml for each gallon (3.785 liters) of raw whey, preferably between about 10 and 40 ml.

An optional ingredient that can be added is a source of divalent metal ions, preferably calcium ions in the form of calcium hydroxide so as to serve to raise the pH of the raw whey while adding calcium ions thereto. From about 1 to 15 g of calcium hydroxide powder may be added to each gallon (3.785 liters) of raw whey, depending generally upon the pH of the raw whey as modified by the other materials added to the system. Preferably, the pH of the system after addition of the floc initiator will be at or near the isoelectric point of the whey proteins, which is typically at a slightly acidic pH.

In some instances, it may be desirable to add an oxygen source with a view toward further enhancing the bacteriological preservation properties of the system above those provided by the blandness imparting agent. Ingredients such as hydrogen peroxide or sources of ozone bubbled into the system will generally improve the bacteriological stability thereof. Very small concentrations of these oxygen sources are adequate, for example, between about 2 and 15 ml of hydrogen peroxide can be added.

Example 1: One gallon of cottage cheese or acid whey having a pH of 4.6 and a temperature of 115°F was placed into a stainless steel beaker having a capacity of approximately two gallons, the beaker being fitted with a controllable heat supply in the form of an electric hotplate, together with a power blender having high rpm capacity and having a blending blade especially designed for imparting high shear properties to the materials being power blended. Blending and heating were commenced, and 20 ml of a 1:1 molar blend of 472 mg of ferrous gluconate and 75 mg of glycine dissolved within 4,720 cc of water were added, as were 20 ml of a 1:1 molar mixture of 472 mg of copper gluconate and 75 mg of glycine dissolved within 4,720 cc of water.

Added next while heating and blending continued, were 40 ml of sodium silicate solution [identifiable as (SiO_2/Na_2O) = 3.22, 41.0°Bé], 5 ml of hydrogen peroxide, and 10 g of high viscosity sodium carboxymethylcellulose, after which were added 3 g of calcium hydroxide dissolved within 40 ml of decanted whey water. Heating and blending were stopped when the temperature reached 130°F. The thus treated whey was then removed from the power blender and was allowed to sit under ambient temperature and pressure within a transparent bottle having a spigot. A whey floc formed, and the liquids portion of the separated whey was decanted from the solids portion by passing it through the spigot in the bottom of the transparent bottle.

Both the liquids portion and the floc portion were analyzed on an Emzymax flowing-stream lactose/glucose analyzer (Leeds & Northrup Co.). The liquids fraction at 68°F and a pH of 6.3, was found to have a lactose content of 4.4 wt % and a glucose content of 0.9 wt %, while the floc at 68°F and a pH of 6.4 was analyzed to have a lactose content of 4.2 wt % and a glucose content of 0.7 wt %. Untreated raw cottage cheese whey analyzes as having a glucose content of less than 0.1 wt %.

Example 2: A gelatin substitute was prepared from a soybean based material and a whey based material in combination with a natural gum. A commercial isolated protein (400 g of Promine D) was dispersed to a heavy paste within 2,500 ml of distilled water, after which it was washed under heavy shear action with hydrochloric acid to a pH of about 4.7 and then filtered. Into this paste was dispersed 20 ml of a 1:1 molar copper gluconate-glycine solution as well as 20 ml of 1:1 molar solution of iron gluconate and glycine. Also blended in were 3.5 g of calcium hydroxide dissolved in 100 ml of distilled water and 2 g of the enzyme papain dissolved within 100 ml of distilled water, followed by 3 ml of an emulsion of 1.5 ml of carbon disulfide with 1.5 ml of a food-grade emulsifier, after which the solution cleared into a bland solution when blending was complete.

A buffer of 10 g sodium citrate dissolved in 100 ml of water was then added, followed by the addition of 2,000 ml of the liquids fraction prepared generally in accordance with Example 1, this blending being made within the shear imparting apparatus of Example 1. While the shear blending was proceeding, about 50 g of a blend of 80% Irish moss and 20% guar gum were dispersed therein. Upon standing at room temperature, a clear, gelled product having good mouthfeel and thermomeltability was developed.

Whey Protein in Macaroni

An enriched wheat macaroni product conforming to FDA standard of identity regulations has been prepared by *D.S. Cox; U.S. Patent 4,158,069; June 12, 1979.* This product comprising a major portion of a milled wheat ingredient, minor portions of nonwheat protein ingredients including at least one milk fraction protein source, and a predetermined amount of L-lysine.

At least one milk fraction protein element is included in this macaroni formulation, and preferably two different milk fraction protein ingredients are used. Denatured dried whey protein having a 35 to 50% whey protein with the balance being essentially lactose has been found to be an excellent protein source and also has a protein efficiency role (PER) of about 3.2 to 3.4. Problems encountered with the drying of macaroni made with undenatured whey are overcome by the use of denatured whey, which does not hold onto the water molecule and does not crack and fall apart subsequent to drying. Denatured dried whey protein containing either 35 or 50% whey protein is commercially available from several manufacturers.

Another preferred source of protein is calcium caseinate which has 95% protein and a good PER of about 2.5 to 2.8 and, in addition, the calcium content of the macaroni product is supplemented and calcium caseinate adds toughness and lightens the color of the product. Therefore, a combination of denatured dried whey solids and calcium caseinate is preferred in the macaroni formulation, but soy protein isolate or soy concentrate may be substituted for one of these milk fraction proteins.

Wheat gluten is also a useful ingredient in the preferred macaroni formulation. Wheat or gum gluten while having a low PER of 0.6 to 0.8 (the same as wheat) contains about 75% protein and adds toughness and good color to the macaroni product, but soy protein isolate or soy concentrate may be substituted as indicated in the example. It should be understood that the protein enrichment ingredients may vary in amount by plus or minus 15% with the other ingredients and the semolina also varied in amount to compensate and still maintain a finished (cooked) food protein quantity of 20 to 25% and a protein quality of at least PER 2.5. The preferred formulation also includes a small amount of about 0.10% of calcium hydroxide, thereby supplementing the calcium content.

Example: A preferred formulation of enriched wheat macaroni is prepared from 88.35% semolina, 5.60% dried whey (50% protein), 4.0% calcium caseinate, 1.2% wheat gluten, 0.75% L-lysine and 0.10% calcium hydroxide.

In addition, a conventional enrichment or "nutritional package" of thiamin, niacin, riboflavin and iron is added, and permissible additional ingredients may be added. The formulation is first dry-mixed and then wet-mixed with water in a conventional manner to prepare a dough, which is then extruded by conventional techniques to produce a macaroni product. This product permits the use of commercial drying times and temperatures in conventional equipment, i.e., about 6 hours at 125° to 140°F, and produces a product that is light yellow or golden and is firm with a fine brightness and luster. When cooked, the macaroni turns white, and retains its firmness or cohesiveness and has a good texture and tastes like macaroni should taste.

The finished food product has a protein quantity of approximately 22.6% and a protein quality exceeding PER 2.50, and a total L-lysine amount below 6.4% of total protien, thereby complying fully with FDA and USDA regulations for "enriched wheat macaroni, with fortified protein" and "meat alternate" status.

Skim Milk Protein for Low Calorie Margarine

K. Wallgren and T. Nilsson; U.S. Patent 4,259,356; March 31, 1981; assigned to Mjölkcentralen Arla Ekonomisk Förening, Sweden has disclosed a protein concentrate obtained from skimmed milk useful in the manufacture of low calorie margarine of the water-in-oil type. The process comprises acidifying skimmed milk, skim milk powder or a mixture thereof to a pH of 4 to 5 to precipitate protein, heating the acidified raw material in a first heating stage to a temperature of about 35° to 65°C and maintaining the temperature for a period of at least 15 minutes, then quickly raising the temperature of the acidified milk in a second heating stage to a temperature of about 60° to 95°C and immediately after reaching the intended temperature, concentrating the precipitated protein and cooling the resulting concentrate. The process will be described in more detail in eight steps.

(1) If desired, a protein-like whey protein or reconstituted milk powder may be added to the skim milk raw material. The raw material may also be given an addition of salts like citrate or phosphate in order to stabilize the albumin and/or to reduce the influence of season variations on the milk.

(2) The skimmed milk is subjected to a pasteurizing heat treatment, which may range from a temperature of 72°C for 15 seconds to an ultra-high temperature and a corresponding temperature/time relationship depending on the season and the yield desired.

(3) The skimmed milk is preferably given a slight addition of cheese rennet to stabilize the protein particles. Preferably cheese rennet is added in an amount of 5 to 40 ml per 1,000 liters of skimmed milk.

(4) In this step the skimmed milk is made acid at a suitable temperature by means of nongas-forming acid which consequently does not result in the formation of carbon dioxide or any other as during the acidification. The milk is acidified to a pH value of between 4 and 5.

(5) The acidified milk is heated in a first heating step to a temperature of preferably 52° to 55°C, and the milk is kept at this temperature for at least 15 minutes and up to 60 minutes or more depending upon the particular temperature employed.

(6) The temperature used in step 5 is too low to enable a separation so as to get a protein having a sufficiently high dry substance content and a sufficiently reduced bacteria and enzyme content. Therefore, the skimmed milk is subjected to a momentary temperature increase up to preferably 65° to 70°C. This high temperature should be maintained for as short a period as possible, and the temperature rise (generally of at least 10°C) may be established by steam injection directly into the skimmed milk or onto the tube in which the skimmed milk is transported.

(7) In direct connection to the momentary temperature rise according to step 6 the protein is concentrated as quickly as possible, preferably by being separated. This gives a protein concentrate of 20 to 24% which mainly contains casein but which also contains some amount of whey proteins.

(8) The protein concentrate from skimmed milk is very sensitive to high temperatures and if the high temperature from step 7 is maintained for too long a period the protein quickly grows grainy, its water-keeping property is reduced and whey falls out. It is therefore important that the protein concentrate is cooled, and if the protein concentrate is not used immediately for the manufacture of low calorie margarine, it is important to cool the protein concentrate immediately after the separation to a temperature of less than 8°C. The protein concentrate ought to be used as soon as possible for the manufacture of low calorie margarine, but at a temperature of less than 8°C the protein may without disadvantage be stored for two or three days.

Example: 4,500 liters of skimmed milk was heated momentarily to 87°C and then cooled to 20°C. Thereafter 0.7% cottage cheese acid and 90 ml rennet was added. The coagulate was broken at pH 4.6 and it was thereafter heated to 55°C and was kept at that temperature for 30 minutes. Thereafter the milk was transferred to a quark separator by means of a positive controllable pump. Just before the separation the temperature of the milk was raised to 68°C by direct injection of steam. In this case the milk could not stand a higher temperature without the risk of grain formation and loss of water-keeping properties. The separation of the milk was carried out to obtain a protein concentrate of 20 to 24% which contained both casein and some portion of whey protein. The protein concentrate was then stored at a temperature of 4° to 8°C. The protein concentrate thus obtained had an even and smooth consistency without any hard lumps and it proved to have good water-keeping property and was well suited for manufacture of low calorie margarine of water-in-oil type having a high protein content and a fat content of only about 40%.

Bread Crumb Coating Composition

A protein containing adhesive is used by *J.M. Rispoli and J.R. Shaw; U.S. Patent 4,260,637; April 7, 1981; assigned to General Foods Corporation* to prepare an improved one step bread crumb coating composition.

The bread crumb composition comprises bread crumbs and an adhesive, the adhesive having been applied to and adhering to the surface of the crumbs. The bread crumbs have a particle size wherein at least a majority of the crumbs are retained on a 20 mesh U.S. Standard Screen after passing through a 5 mesh U.S. Standard Screen and wherein not more than 10% of the crumbs by weight are retained on a 5 mesh U.S. Standard Screen. The adhesive contains a protein at a level of at least about 1% by wt of the crumbs, and can also contain a starch and/or a gum. Preferably, the adhesive is applied to the surface of the crumbs with an edible oil.

The adhesive is generally present within the range of about 1 to 35% by wt of the bread crumbs. The adhesive critically comprises a protein at a level of at least about 1% by wt of the crumbs, and preferably, can additionally contain a gum and/or a starch. The protein may be whey protein, milk protein, soy isolate, gelatin, egg albumin, wheat gluten, etc., and mixtures thereof, and is generally present at a level of from about 1 to 20% by wt of the crumbs and preferably present at a level of from about 5 to 15% by wt of the crumbs. In addition to the protein, the adhesive may also contain a starch, generally at a level up to about 10% by wt of the crumbs. The starch may be a raw, modified (chemical or physical) or pregelatinized starch, for example, starches such as cornstarch, waxy maize starch and tapioca starch are all effective in enhancing the cohesive properties of the adhesive.

The adhesive may also contain a gum, generally at a level up to about 5% by wt of the crumbs. The gum may be a natural, modified or synthetic gum, for example, gum arabic, gum tragacanth, locust bean gum, or cellulose derivatives such as methylcellulose, carboxymethylcellulose or microbial gums such as xanthan gum are all effective in enhancing the cohesive properties of the adhesive. Other edible materials which enhance the cohesive properties of the adhesive may also be employed in combination with the protein in the adhesive.

Critical to this bread crumb composition is that the adhesive is applied to and adheres to the surface of the bread crumbs. Without the adhesive adhering to the surface of the bread crumbs, the bread crumbs, due to their relatively large particle size would not uniformly coat and adhere to the surface of the comestible when the moistened comestible is coated with the bread crumb composition and cooked.

Example: The protein blend used in this example comprises 50% by wt sodium caseinate and 50% by wt egg white solids. The seasoning blend contained salt, paprika, monosodium glutamate and white pepper.

The bread crumb composition was prepared from 148 g of bread crumbs, 20 g of protein blend, 22 g of seasoning blend, and 10 g of vegetable oil.

The bread crumbs consisted essentially of wheat flour, yeast and salt, had an elongated, porous and striated shape and structure and had been toasted to uniformly brown the crumbs. The bread crumbs had a particle size wherein 35% by wt of the crumbs were retained on an 8 mesh U.S. Standard Screen after passing through a 5 mesh U.S. Standard Screen, 37% by wt of the crumbs were retained on a 14 mesh U.S. Standard Screen after passing through an 8 mesh U.S. Standard Screen, 20% by wt of the crumbs were retained on a 20 mesh U.S. Standard Screen after passing through a 14 mesh U.S. Standard Screen and 8% by wt of the crumbs passed through a 20 mesh U.S. Standard Screen. The protein blend and seasoning blend were mixed together and then added to the bread crumbs in a rotating coating kettle. The protein and seasoning blends were applied to and adhered to the surface of the bread crumbs by spraying vegetable oil onto the mixture of crumbs, protein and seasoning in the rotating coating kettle.

A 2½ pound chicken was cut up into pieces and the pieces were then dipped into water. Then each moistened chicken piece was completely covered with the bread crumb composition (about 120 g of composition for the chicken pieces) and the bread crumb composition was firmly pressed onto each piece.

The coated chicken pieces were then placed skin side down on a baking pan which had ¼ cup of oil covering the pan surface. The coated chicken pieces were then baked for 25 minutes at 400°F (204°C), turned, and baked for an additional 25 minutes.

The resultant baked, coated chicken pieces were found to have a continuous, uniform, adherent and crisp coating. The baked, coated chicken was found to have the taste, texture and appearance of fried chicken.

Soy Proteins in Whipped and Frozen Juices

I.H. Rubenstein; U.S. Patent 4,293,580; October 6, 1981; assigned to Maryland Cup Corporation has prepared fruit and vegetable juice concentrates by simultaneously whipping and freezing a concentrate in the presence of whipping agents and, optionally, stabilizers. Suitable whipping agents to be employed include soy proteins and egg whites. The resulting concentrated juice is discharged from, e.g., an ice cream freezer at a temperature of 18° to 30°F. The frozen fruit concentrate has a soft whipped texture, which makes it divisible into aliquot portions for use by a consumer.

Thus, this process makes it possible to eliminate the need for defrosting an entire container of juice in order to be able to serve only a single portion.

In the procedure of this process, the juice is concentrated in the conventional manner with or without essence recovery. During this operation, a measured amount of fresh juice is prepared to which are added a quantity of a whipping protein (for example, derived from a soy protein or from egg white) and a small amount of stabilizer. While soy proteins and egg whites are preferred protein whips to be employed in this process, a whipping system based on a variety of other techniques can be employed, including the use of various emulsifiers such as glycerol monostearate and stearoyl-2-lactylates.

A soy based whipping protein, with or without the addition of phosphate salts, is readily soluble in water, and for this purpose can be used in a concentration of 0.05 to 2% by wt, depending on the source of the protein and the final product desired. Liquid egg white can be used in a concentration of 0.5 to 3% and it too is water-soluble.

The stabilizer to be employed in accordance with this process can be an all-natural gum such as locust bean gum, guar gum, carrageenans, agar, or any other material which will add viscosity to an aqueous system. A synthetic gum such as carboxymethylcellulose can be used. The choice of gums depends primarily on two factors, i.e., the need for having an all natural product and the texture desired in the finished drink.

In many cases, it is desirable to use a cold swelling gum such as guar gum or carboxymethylcellulose. However, there is no objection to dissolving a gum in, for example, a hot orange juice concentrate, prior to the addition of the single strength juice thereto.

When the whipping protein and the stabilizers have been added to either the juice concentrate or the single strength juice, the single strength juice portion (which may constitute about 0 to 90% by wt of the total) and the concentrate portion are blended, and while still hot, pass through a whipping/freezing system in a similar manner as in the making of ice cream. In this system, the finished concentrated juice is whipped, frozen and discharged at a temperature in the range of from 18° to 30°F and then packaged for consumer use.

Example 1: A typical orange juice concentrate may be manufactured in the following manner. The ingredients given in percent by weight are as follows: 1,000 parts orange juice concentrate (67 Brix) (approximately 5+1 concentration), 400 parts single strength orange juice, 10 parts whipping proteins (Gunther 157A), and 5 parts guar gum.

The guar gum and whipping protein are dissolved in the cold orange juice, added to the hot concentrate and then passed through an ice cream freezer incorporating air and increasing the volume from 10 to 100% with the final product discharging from the ice cream freezer at a temperature of 19° to 30°F, preferably 22°F. Assuming an increase of 50% in volume, it would then be desirable to reconstitute the juice by adding two parts of water to one part of the orange juice whip. The water and the concentrate are blended and stirred until the drink is ready.

Example 2: A juice drink can be made by taking the whipped frozen juice of Example 1 and diluting it with three to six times its volume of water, adding a quantity of sugar and whipping the entire mixture in a blender with the following formulation: 1,000 parts frozen whipped concentrate, 5,000 parts water and 500 parts sugar.

Cleaning Material for Foodstuffs

In many instances it is desirable to wash foodstuffs of both animal and vegetable origin to rid the same of dirt, blood, insects and insect detritus, microorganisms, such as mold, mildew and pathogenic bacteria, etc. Water alone is not always efficient and ordinary additives to water, such as surface-active compounds, may give rise to a change in the taste appeal of the washed foodstuff.

An aqueous concentrate for the cleansing of such foodstuffs without affecting the taste and aroma of the washed foodstuff has been described by *H.-J. Lehmann, R. Bietz and J. Wegner; U.S. Patent 4,177,294; December 4, 1979; assigned to Henkel KGaA, Germany.* This aqueous concentrate for the cleansing of foodstuffs consists of:

    (a)   from 0.1 to 10% by wt of water-soluble to water-dispersible proteins,

    (b)   from 0.01 to 3% by wt of water-soluble polymers having a MW of at least 10,000,

    (c)   from 1 to 15% by wt of a water-soluble sequestering agent,

    (d)   from 0.01 to 1% by wt of water-soluble food preservatives,

    (e)   from 0 to 0.5% by wt of food colors and food odorants, and

    (f)   the remainder to 100%, water.

Water-soluble to water-dispersible proteins that are suitable are obtained from animal or vegetable products, such as albumin from cattle plasma, egg albumin, sodium caseinate, gelatin, extracts from protein-containing seeds, etc. Because of the better clear solubility in water in the presence of salts, native proteins are particularly suitable, which requires, however, special precautions in their production and processing.

Particularly suitable among the proteins and protein hydrolysates obtained from vegetable products are primarily vegetable globulins, e.g., legumin and vivilin from peas, glycinin from soybeans, or phaseolin from beans.

The sequestrants or water-soluble sequestering agents employed are, for example, the hydroxyl-carboxylates normally used in washing agents, such as the alkali metal citrates, lactates, tartrates, the amino-carboxylates, such as the alkali metal salts of ethylenediaminetetraacetic acid, nitrilotriacetic acid, as well as inorganic sequestering salts, such as the alkali metal phosphates and polyphosphates, preferably as sodium salts.

Furthermore, the cleaning solutions contain water-soluble polymers, particularly those having a molecular weight of at least 10,000. These are primarily polyvinyl alcohols with molecular weights of 10,000 to 500,000, preferably 50,000 to 100,000, with a degree of hydrolysis of 50 to 100%, as well as polyvinyl pyrrolidones with molecular weights of 10,000 to 1,000,000, preferably 500,000 to 1,000,000 or mixtures of these polymers.

Other suitable water-soluble polymers are polymer-saccharide compounds, such as cellulose carboxylates or vegetable gums, also cellulose ethers, such as methylcellulose, methylhydroxypropylcellulose, hydroxyethylcellulose, methylhydroxyethylcellulose, etc.

Finally the cleaning mixtures contain preservatives, such as alkali metal and alkaline earth metal salts of sorbic acid, benzoic acid, formic acid, boric acid, or other preservatives which are particularly suitable for foods.

The cleaning liquors are normally adjusted to a pH value of 6 to 8, but they can be used in a wide pH range of 5 to 10.

Example 1: To prepare a vegetable protein, 150 kg of yellow, dried peas were crushed, suspended for 30 minutes in 300 kg of a 10% common salt solution or sodium sulfate solution. The liquid was next separated in a sieve centrifuge, heated for 30 minutes to 70°C with stirring, and the albumins were precipitated. About 10% filter aids (diatomite) were added and the product was filtered under pressure. Yield: 150 kg pea extract, about 5% protein (legumin, vicilin), in the form of a clear solution.

Example 2: To prepare a cleaning formulation, the following ingredients were diluted with water to a final weight of 100 kg: 15.0 kg pea extract of Example 1, 1.5 kg methylhydroxypropylcellulose, viscosity of 1% solution = 20,000 cp, 7.5 kg sodium citrate, 0.1 kg of a preservative (Kathon 886). After mixing this concentrate with water in a ratio of 1:100, the cleaning preparation obtained is used for cleaning fruits and vegetables.

The preservative Kathon 886 is a mixture of: (1) 5-chloro-2-methyl-4-isothiazolin-3-one calcium(II) chloride; and (2) 2-methyl-4-isothiazolin-3-one calcium(II) chloride.

S. Herbst and R. Bietz; U.S. Patent 4,244,975; January 13, 1981; assigned to Henkel KGaA, Germany have described another cleaning agent which does not contain the water-soluble polymers. This cleaning agent consists of:

(a)   from 0.1 to 20% by wt of water-soluble to water-dispersible proteins;

(b)   from 0 to 15% by wt of a water-soluble sequestering agent;

(c)   an effective amount of a preservative selected from the group consisting of (1) from 0.01 to 1% by wt of water-soluble food preservatives and (2) from 5 to 30% of ethanol;

(d)   from 0 to 0.5% by wt of food colors and food odorants; and

(e)   the remainder to 100%, water.

It has been found that not only combinations of protein compounds and water-soluble synthetic organic polymers can be used for the cleaning of food and feed, as described in U.S. Patent 4,177,294, but that a good cleaning effect is obtained with the use of single protein compounds as well as of combinations of different protein compounds as effective agents.

The advantage of these cleaning agents is the fact that they consist practically completely of substances of natural origin, which has special significance with respect to their area of application. Beyond this, aqueous, preferably aqueous-alcoholic concentrates with a much higher content of active substance, protein compounds in this instance, can be prepared, which leads to products that are characterized by a reversible gel formation, even after storing at 4°C for days and rewarming to room temperature. The products consequently can be prepared in an advantageously concentrated form and diluted to the ratio 1:500, preferably 1:100 to 1:250, for application.

Sequestering agents are not absolutely necessary to achieve the cleaning effect. However, they can have a favorable effect on the storage life of the protein solutions and are, therefore, included as a rule. Also suitable can be the addition of preservatives, coloring agents and fragrances.

These preparations are characterized by the complete absence of surfactants or surface-active compounds and are physiologically harmless. They are low sudsing and do not cause skin irritations when used by hand.

Example 1: 3.0 kg of gelatin hydrolyzate (MW about 10,000); 7.0 kg of sodium citrate; and 0.1 kg of a preservative Kathon 886 (Rohm & Haas Co.) were dissolved in 90 kg of water. Kathon 886 is a mixture of 5-chloro-2-methyl-4-isothiazolin-3-one magnesium chloride and 2-methyl-4-isothiazolin-3-one magnesium chloride.

This clear, low-viscosity concentrate was used as cleaning agent, after diluting with water at a ratio of 1:100.

Example 2: 2.0 kg of gelatin; 1.0 kg of sodium caseinate; 2.5 kg of sodium citrate; and 0.1 kg of Kathon 886 were made up to 100 kg with water.

The slightly turbid, low-viscosity concentrate was used as cleaning agent, after diluting with water at a ratio of 1:100.

Example 3: 1.5 kg of sodium caseinate; 1.5 kg of soybean protein; and 5.0 kg of sodium citrate were made up to 100 kg with water.

The slightly turbid concentrate was used as cleaning agent, after diluting with water at a ratio of 1:100.

PROTEIN SUBSTITUTES

FOR MEAT AND MEATLIKE PRODUCTS

Egg White in Gellied Fish Paste

When salt is added to raw fish meat undergoing grinding, myosins, which are soluble, in salt solution are dissolved out from the meat to form a "sol," which is very adhesive. When this adhesive raw ground meat is heated, the sol converts to gel, which forms a network construction, and this gel imparts elasticity to the fish meat paste. Such elastic fish meat is called "neriseihin" in Japan and fish meat which forms a strong gel is referred to as fish meat having a good gel-forming capability.

However, fish infested with certain sporozoa harmless to humans quickly form a product called jellied meat or milky meat, which is spotted or soft. The fish is edible and not putrefied.

Attempts have been made to utilize the fish meat of fish containing jellied meat as a raw material for neriseihin. However, all of these attempts have as yet failed to produce satisfactory results. In general, in order to make a high-quality "kamaboko" which has high elasticity, chopped fish meat must be soaked in cold water for several hours to remove fat, blood, odorous substances, and water-soluble proteins. The cold water is changed two or three times during the soaking. In general, the less fresh the fish meat is, the longer is the soaking time required.

In the case of meat obtained from fish containing jellied meat, even when the frequency of the process of soaking in water is increased, or the soaking and dehydration procedures are carefully conducted, or the addition ratio of sugar or condensed polyphosphates is increased, the dehydrated fish meat has little gel-forming capability and is unsuitable for use as a raw material for fish-based neriseihin, such as kamaboko and chikuwa, which require a high gel-forming capability.

H. Haga, R. Shigeoka and T. Yamauchi; U.S. Patent 4,207,354; June 10, 1980; assigned to Nippon Suisan KK, Japan provide a method for processing fish meat of fish containing jellied meat characterized by the addition of egg white in a quantity of the order of from 4 to 40 parts, calculated on the basis of fresh white, to 100 parts of the fish meat.

The egg white is added in a quantity preferably on the order of 10 to 20 parts of fresh white to 100 parts of raw material fish meat. Where condensed egg white or dried egg white is used, it may be added in a quantity corresponding to fresh egg white (moisture content: about 89%) in consideration of their respective moisture content. For example, in the case of dried egg white (moisture content: about 9.5%), the addition in a quantity of about 0.5 to 5 parts corresponds to the above stated addition ratio. In the case where the addition ratio of the egg white is lower than the above range, the desired result cannot be achieved and it is impossible to obtain a raw material suitable for producing a neriseihin having a great gel-forming capability.

On the other hand, when the white is added in a ratio greater than the given range, the resulting neriseihin possesses the property of heat-coagulated white, and the feel of resilience and resistance to chewing kamaboko which is characteristic of neriseihin, is reduced. As a result, the neriseihin becomes brittle and gives off a strong smell of the egg white, which reduces the organoleptic quality of the neriseihin.

The egg white may be added to fish meat collected by a meat separator at any time during the subsequent processing steps. Moreover, the addition of the white results in advantageous effect irrespective of whether or not the fish meat is soaked in water. For example, when fish meat is processed, the white may be added before or immediately after soaking in water or at the time of grinding. In the case where soaking is excluded from the following procedure, the white is preferably added immediately after the fish meat is put through the meat separator or at the time of grinding. Moreover, when freezing and thawing process steps are adopted after meat separation or soaking, the white may be added before freezing or at the time of grinding after thawing.

Example: Frozen Chilean hake (*Merluccius guyi*) containing spotted jellied meat was thawed. Then, fish meat was collected from the thawed hake by means of a meat separator. The fish meat was soaked in water and drained to obtain 50 kg of dehydrated meat. 5 kg of fresh egg white was added to the dehydrated meat before it was ground with addition of table salt. Thereafter, 2.5 kg of sugar, 1.5 kg of table salt, 4.0 kg of starch, 1.2 kg of "Mirin" and 0.5 kg of sodium glutamate were added to the ground meat, and the mixture was ground and formed into "itatsuki-kamaboko" according to a conventional method.

Mirin is a kind of wine used as seasoning in Japan. This seasoning, made from waxy rice by fermentation, is rich in glucose and various amino acids. Sodium glutamate is a most popular seasoning in Japan, being a kind of amino acid and used widely in cooking and various food processing.

When shaping kamaboko, the seasoned ground meat is made to adhere by its own adhesiveness to a thin wooden board, usually in the shape of a quonset hut, for which reason it is called itatsuki-kamaboko or kamaboko on wooden board.

A control itatsuki-kamaboko was prepared according to the procedure described above, except that egg white was not added.

Tests show that by the addition of about 10% of the white to the hake containing spotted jellied meat, it is possible to produce neriseihin having a strong gel-forming capability. The product prepared in this example was equivalent to a conventional kamaboko on the point of flavor.

The control product without the egg white had a much lower gel point, and poor biting qualities and folding properties compared to the product of this example prepared with egg white.

Soy Isolate in Curing Hams

The pumping of hams with brine, sometimes referred to as "pickle," probably antedates recorded history. The most common salt employed is sodium chloride which provides curing (color), preservative (shelf life) and organoleptic (taste) functions.

Starting in the mid 1960s, soy protein isolate was viewed as an especially attractive supplement to the brine to permit the introduction of more fluid while maintaining the nutrition level, particularly relative to protein.

A typical soy isolate-augmented brine employed over the years in Europe included 4% isolate, 10% salts including the chloride and nitrite, 3% phosphate and 3% sugar including monosodium glutamate with the remainder water.

It was felt desirable to be able to increase the concentration of isolate in the brine, increasing the effective weight of hams with less costly ingredients. However, to be acceptable, the water-isolate relationship should be such that after cooking, the isolate was present in the remaining water at a level comparable to the percentage of protein actually present in the ham, viz, 17-20%. The isolate concentration had to be stepped up as more water was employed.

The deterrent to this increase has been the viscosity of the isolate solutions at higher concentrations. Even with stitch pumping, the brine could not be used advantageously—plugging the needles and failing to diffuse completely through the muscle structure.

The process used by *V.V. Kadane, E.W. Meyer and R.W. Whitney; U.S. Patent 4,164,589; August 14, 1979; assigned to Central Soya Company, Inc.* overcomes this difficulty by using a unique isolate that provides a relatively low viscosity brine even with isolate concentration above 12%.

In general, the steps of processing soybeans so as to obtain edible isolated soy protein include screening, cracking, dehulling until full fat flakes are developed. These are solvent extracted to provide white flakes which are then further extracted and spray dried as mentioned previously so as to provide edible isolated soy protein.

To produce a soy protein isolate which is especially advantageous in this process, the isolated soy protein produced as above is combined with a minor amount of a lipid material. This is achieved advantageously by placing the isolate in a blender and then spraying onto the isolate from about 0.2% to about 2.0% of melted lipid material. In one preferred form, advantageous results were obtained using 0.35% commercial lecithin heated to about 135°F on the spray-dried particles of nongelable soy protein isolate.

Contrary to the teaching of the prior art, the soy protein isolate especially useful in the process is not characterized by a gel formation upon heating.

A conventional gelling soy protein isolate had a viscosity of about 5,000 poises. The lipid-augmented nongelling protein of this process had less than 50 poises.

Example: The procedure for brine preparation was as follows:

> (1) To produce 500 lb brine 4.875 lb of phosphate was dissolved in 38.11 lb of hot tap water.
>
> (2) The remainder of the water (350 lb) was added to the brine tank and mixing started (two moderate agitation mixers, i.e., portable, propeller type, were used to provide a rolling action).
>
> (3) The 61.11 lb of Lot #9950 of pumping isolate was added to the water and allowed to wet and disperse for 30 minutes. Manual mixing was employed to initially hydrate all of the material.
>
> (4) The polyphosphate suspension was added to the protein dispersion.
>
> (5) The 39.0 lb of salt (NaCl) was added dry and mixed in.
>
> (6) The 6.375 lb of (sucrose) sugar was added dry and mixed in.
>
> (7) The 0.13 lb of sodium nitrite was added and dispersed.
>
> (8) The 0.4 lb of sodium erythorbate was added and dispersed.

The viscosity of the brines was measured using a Brookfield viscometer, Model RVT. After step #3, the viscosity for Lot #9950 was 10.2 poises and the viscosity for the final brine was 1.3 poises.

The 500 lb batch of brine was pumped to the stitch injector which was operated at a pressure of 65 psi through 0.032-inch-opening needles. The hams (25°F) which possessed a green weight of 78.875 lb were injected 3 times per side in an attempt to attain a proposed weight of 118.32 lb (150% pump). The actual pumped weight was 106 lb, or 134.4% increase over green weight.

The pumped hams were weighed and placed in the pilot massager. Since 150% pump was not achieved, 13 lb of brine was placed in a Lynggaard massager Model 900E with the hams. Four ounces of clove oil (0.05%) was added to the massager. The hams were massaged at 3 rpm for 30¾ hours with a 20 min massage, 10 min rest, 20 min reversed massage and 10 min rest cycle. The ham weight after massaging was 116 lb representing a total pump of 147.2%.

The hams were removed from the massager, weighed and trimmed. The trimmings amounted to 23.5 lb resulting in hams weighing 92.8 lb, so the trimming loss was 20%. The hams were manually stuffed into casings, air pockets removed and placed in individual spring-loaded presses. The hams were cooked in the smokehouse for 4 hours to an internal temperature of 156°F. The dry bulb temperature was 180°F and the wet bulb temperature was 160°F. The ham weight after cooking was 86.6 lb. The hams exhibited good sliceability besides having the substantially greater weight due to isolate pumping.

The time required for dispersion, although 30 minutes, was substantially less and with the expenditure of considerably less energy and using much simpler equipment, than that required for other commercially available isolates—and even where these were at lower concentrations.

Heat-Coagulable Viscous Protein

Protein isolates prepared by conventional processes are useful as emulsifiers, binders and as the chief materials of meat analogs prepared by fiber spinning and heat gelation techniques. Although widely used for these purposes, the rigorous alkaline treatment required to solubilize the proteins in the seed material or subsequently required to resolubilize and gel the proteins for fiber spinning or extrusion results in protein degradation and loss of physioelastic properties. The proteins which are recovered by precipitation in acid salt solutions are invariably contaminated by large amounts of absorbed salts which are only partially removed by extensive washing.

M. Shemer; U.S. Patent 4,188,399; February 12, 1980; assigned to Miles Laboratories, Inc. has prepared a protein product from oleaginous seed materials which has excellent functional and heat-coagulating properties and without the disadvantage of the absorbed salts.

This process comprises forming an aqueous slurry of an oleaginous seed material having a pH of about 5.1 to about 5.9, separating the liquid from the solids portion of the slurry, and recovering the protein fraction from the separated liquid.

The process is best accomplished by forming an aqueous slurry of the full fat, low fat or defatted oleaginous seed meals, flakes, or flours, the slurry having a pH in the range of about 5.1 to about 5.9. Preferably, the pH is adjusted to about pH 5.5.

The aqueous slurry is then agitated for a time sufficient to extract the desired acid-soluble proteins from the oleaginous seed material. The particular time will depend upon the temperature and concentration of the slurry. Optimum extraction is achieved at about 40°C.

The concentration of oleaginous seed material in the slurry is usually less than about 15% (weight/volume basis) based upon the total volume. Preferably, the concentration ranges between about 8 to about 12% (w/v basis).

To protect the oleaginous seed material from undue oxidation, it is desirable to add an antioxidant to the aqueous extraction media prior to suspending the oleaginous material. Preferably, an antioxidant such as sodium sulfite is added at a concentration of about 0.1 to 1.0 weight percent based upon the weight of oleaginous material.

Following the extraction of the desired acid-soluble proteins, the liquid portion of the slurry is separated from the insoluble portion by any convenient means such as filtration or centrifugation. The insoluble portion, which consists of proteins, carbohydrates and cellular matter, insoluble in the acid pH range, is set aside for concentrate and isolate processing.

The desired protein product is recovered from this separated liquid by such convenient means as isoelectric precipitation, ultrafiltration, reverse osmosis and the like.

Example 1: Fifty pounds of defatted soy flour and 0.5 pound of sodium sulfite were suspended in 500 pounds of 40°C tap water adjusted to pH 5.5 by the addition of phosphoric acid. This was a solids concentration of 10% soy flour on w/v basis. The aqueous suspension was maintained at pH 5.5 and agitated for 30 minutes to extract soluble proteins and carbohydrates. The liquid portion of the suspension was then separated from the insoluble solid material by centrifugation.

The separated supernatant, containing extracted proteins and carbohydrates, was adjusted to pH 4.5 by the addition of phosphoric acid to precipitate the protein. The precipitated protein was then separated from the supernatant liquid by centrifugation. The precipitated protein product obtained in a 10.9% yield on solids basis was a viscous liquid. The product contained 50% moisture, 50% solids, 4.18% ash, 14.84% carbohydrates and had a Kjeldahl nitrogen of 80.6%. The heat coagulation temperature was 90°C.

Viscosities were measured at pH 4.39 at 22°C, 37°C and 53°C and were found to be 200,000 cp, 41,600 cp and 17,600 cp respectively at these temperatures. The viscosity characteristic is clearly different from the alkaline, heat or salt produced properties of prior art protein products.

Example 2: This example illustrates the preparation of spun protein fibers using the protein product of this process. A portion of the viscous liquid protein product produced by the procedure of Example 1, having about 50% protein solids at a pH of 4.5, was forced through a 0.5 inch diameter spinneret and into water heated to above 90°C to produce fibers. The spinneret contained about 60 holes having a uniform orifice diameter of 0.008 inch. The resulting fibers were extremely resilient and elastic. When placed in the mouth and masticated, the fibers exhibited chewy characteristics similar to real meat.

Example 3: This example illustrates the use of the protein product as a binder in the preparation of a meat analog. A portion of the viscous liquid protein product produced by the procedure of Example 1, adjusted to about pH 5.5, was incorporated into a meat analog formulation containing 42 g texturized vegetable protein, 99 g water, 22 g fat and emulsifiers, 10 g starch, and 22 g of the protein fraction of Example 1.

The mixture was stuffed into a mold and retorted for 30 minutes at 104°C. The material was then uniformly sliced and fried.

Two control mixtures were similarly prepared. The first mixture was prepared as a negative control containing no binder. The second mixture was prepared as a positive control containing the conventional binder, egg albumen, instead of the desired protein product. Each mixture was then molded, heated, sliced and fried in the same manner described above.

The fried products were compared on the basis of mouthfeel and texture. The fried analog prepared with the protein product of this process and the positive control analog containing egg albumen were similar in mouthfeel and had excellent texture. The negative control analog containing no binder had poor mouthfeel and a mushy texture. The ease and low cost of obtaining the protein product of this process as compared to the prior art binders represents a considerable advantage in preparing meat analogs.

Whey-Caseinate Meat Binder

R.M. Lauck and N. Melachouris; U.S. Patent 4,259,363; March 31, 1981; assigned to Stauffer Chemical Company have found that the extension of comminuted meats can be improved by using as binder a blend of deproteinized or delactosed whey by-products and casein or its sodium or potassium salts. The flavor of the comminuted meats is also enhanced by the use of these blends.

Further, there is provided an improved process for drying delactosed or deproteinized whey by-product solutions which comprises admixing with the solution from about 5% to about 50% by weight of casein or its salts such as the sodium and potassium salts thereof prior to drying. The solutions are particularly adapted for spray drying. Reduced hygroscopicity has also been found.

The dried products of the process are blends of delactosed or deproteinized whey solids with from about 5% to about 50% by weight of casein or its salts. These blends can be prepared by dry blending the ingredients. Preferably, the blends are prepared by codrying a delactosed and/or deproteinized whey by-product solution with the casin or its salts. The codried blends are more easily handled and show a reduced hygroscopicity over the dried delactosed or deproteinized whey itself.

These by-product solutions can be derived from either acid or sweet cheese whey which has been processed to remove all or a part of the lactose or protein content. The term "whey by-products" is used here to encompass the second fraction obtained from the molecular sieve separation of cheese whey, the permeate obtained from the ultrafiltration concentration of protein from whey, and delactosed permeate.

Also effective in the process is the permeate obtained from the ultrafiltration of cheese whey. Ultrafiltration membranes are utilized to separate the high molecular weight fraction of the whey (the protein) from the liquid and low molecular materials, i.e., the lactose and ash in the whey solution. The protein-enriched solution is retained on the membrane and it is called the retentate. The water and low molecular weight fraction passes through the membrane and is called the permeate. The liquid permeate is then used in the process.

In accordance with this process, casein or its sodium or potassium salts are added to the liquid whey solution or the dry whey solids in an amount of preferably from about 5% to about 40% by weight based on the total weight of the dry solids in the whey solution or the weight of the dry solids. The casein salts can be added as the preformed sodium or potassium salt or by dissolving casein by pH adjustment of the water suspension of the whey protein by-product solution. Casein itself can be used though this is less preferred. If a product for a low sodium diet is required, potassium caseinate can be used.

The blend of the caseinate and the whey protein solution can be dried by any known means. Preferably, an atomizing-type dryer is utilized.

If desired, one can also include a small proportion of a drying agent or a flow control agent selected from the group consisting of tricalcium phosphate, dicalcium phosphate, kaolin, diatomaceous earth, silica gel, calcium silicate hydrate and mixtures thereof.

These blends can be used in food products as a flavor enhancing agent, flavor agent or a binding agent. More specifically, it has been found that the blend derived from the process can be used in meat products, for example, soups, stews, gravies, breadings, batters, beef patties and imitation sausages. Also, the product can be used in chip dips, cheese spreads, process cheese foods, spray-dried cheeses and the like.

Example: Various amounts of the second fraction obtained from the gel filtration concentration of whey protein (ENR-EX, Stauffer Chemical Co.) and sodium caseinate were mixed and dissolved in 300 ml of water. The solutions so obtained were freeze dried. The liquefying of dry ENR-EX was a laboratory approximation of the liquid second product from the gel filtration of whey since the dry ENR-EX was the second product in dry form. All amounts are on a dry solids basis.

Combinations were prepared which contained 0%, 5%, 10%, 15% and 20% (by weight) of sodium caseinate. The remainder was the ENR-EX mixture.

After freeze drying, the ENR-EX sample alone could not be powdered and could not be handled. It was hygroscopic and caked. Overnight, it collapsed and turned to paste. At 5% caseinate, the dried product could be powdered in a mortar and pestle and was handleable. The dried samples containing 10%, 15% and 20% sodium caseinate could be powdered in an Osterizer blender. These samples flowed easily, were essentially nonhygroscopic and were more easily handled.

Sodium Caseinate in Comminuted Meats

It is known that water-soluble proteins play an important part in the proper preparation of good sausages. Meats contain myosin which is soluble in brine. When meats are comminuted in the presence of water and salt, the myosin dissolves in the brine. The resulting protein solution coats the meat fibers and the comminuted fats to help form a dispersion and a stable emulsion or batter. When this batter is formed and heat processed, the myosin along with other heat-sensitive meat proteins, sets to a solid, gelled or semisolid condition.

In recent years, high speed comminuting machines have come into use to make sausage batters. These processors reduce roughly ground meats and the other sausage ingredients to a batter in a fraction of a second. The conventional cutter-chopper that has been and still is used for reducing ground meats to a batter requires as much as 15 minutes to accomplish the same meat and fat particle size reductions. It is believed, that better sausages are made by the slower cutter-chopper system than when the high speed comminuting machines are used. It is reasoned that the high speed machines do not allow sufficient time for maximum myosin extraction to occur and the subsequent even coating of the fat and meat fiber particles.

A.E. Poarch; U.S. Patent 4,202,907; May 13, 1980 has used nonreversible gels formed during cooking of enzyme-activated sodium caseinate and calcium ion.

The cation exchange treatment of milk converts the calcium phosphocaseinate complex of milk to sodium caseinate. Calcium phosphocaseinate in milk is not soluble in water and therefore milk is an opaque fluid. Sodium caseinate, on the other hand, is soluble in water. A sodium-for-calcium exchanged skim milk is hazy and semitransparent to light passage. When dried it is commercially marketed as calcium-reduced dried skim milk. This product is used by the sausage industry at a multimillion-pound annual volume.

When rennet enzyme is caused to act on sodium-for-calcium exchanged milk under certain circumstances, the low calcium skim milk is altered. The alteration is not apparent to the eye since no clotting occurs. However, the enzyme-treated ion exchanged milk is chemically different and is very sensitive to calcium ion addition plus the application of heat to form a gel structure. When calcium is added to a solution of calcium-reduced dried skim milk and the mixture is heated, a white opaque milk-like fluid results but no gel formation occurs. Calcium ions may be added to the solution of enzyme-treated low calcium milk at low temperatures without the formation of curds or a gel.

However, when the mixture is heated to temperatures used to cook sausages, a firm rigid nonreversible gel is formed. When the gel structure is cooled, even to refrigerator temperatures, the gel remains firm and intact.

The characteristics of the special gel formed by the reaction of enzyme-treated low calcium milk and calcium are quite different than those of the regular gels formed by concentrated solutions of gums, vegetable proteins, starches, water-soluble milk proteins or animal gelatins as they each may be used as sausage ingredients. The special gels are "brittle" in nature as opposed to the elastic or rubbery and plastic texture of the regular gels. This is of particular significance in the manufacture of firm but still succulent meat products such as frankfurters and meatballs and when the described product is a component of canned meat spreads.

Example: Fresh fluid skim milk is the desirable starting material. It may be raw or pasteurized. The latter is preferred since it introduces fewer living bacteria into the subsequent operations. Reconstituted skim milk made from nonfat dry milk and water can be used.

The ion exchange resin material is a high capacity cation type nuclear sulfonic polystyrene resin (Duolite C-20, Diamond Shamrock Corp.) which has been preconditioned with a sodium chloride solution. Similar exchange resins are available from different sources. From experience it has been learned that this particular resin is stable, has long life, and is predictable in performance.

The degree of preconditioning with sodium chloride solution and the rate at which milk passes through the resin bed determines the extent of the sodium for calcium exchange that is accomplished. From the practical viewpoint it is difficult, expensive and unnecessary to remove all of the calcium from the skim milk. However, the quality of the gel resulting from the reaction of calcium ions with the final rennet treated milk product depends to a considerable extent on the calcium content of the exchanged milk.

In this example the calcium ion content of the skim milk which was passed through the ion exchange resin was held to not more than 8 mg of calcium per 100 ml of skim milk. A calcium content of 8 mg or less is preferred but not necessary.

Various samples were run with calcium contents of the samples ranging from 8.0-92.1 mg of Ca per 100 ml of milk.

It was found that gels are formed when the calcium content of the exchanged skim is between 8 and about 70 mg per 100 ml. In view of the factors affecting gel formation, it is to be expected that when calcium is present at less than 8 mg per 100 ml, gels will be formed.

When a sodium for calcium exchange of this magnitude occurs in skim milk, the milk shifts in pH value from about 6.8 to a pH of 7.4-7.6. The preferred pH of the low calcium skim milk for rennet treatment is about 6.5. It becomes necessary to adjust the treated skim milk substrate to attain this pH 6.5 range.

The preferred method is by the direct addition of acid to the exchanged milk to shift the pH to the desired range.

Either hydrochloric or lactic acid are edible acids and are satisfactory choices for pH adjustment of the exchanged substrate. Of these two, hydrochloric acid is preferred because of its relatively low cost and ready availability.

The preferred procedure is to rapidly agitate the exchanged skim milk and slowly add 1.0 N edible hydrochloric acid to the exchanged milk to a pH 6.5.

The preferred enzyme for use in this process is Standardized Cheese Rennet Extract (Chr. Hansen's Laboratory, Inc.) with a substrate pH of 6.5.

The temperature range of rennet milk clotting activity is 5° to 55°C. The maximum activity of rennet occurs at about 40°C and that is the preferred temperature for preparation of the milk product. The preferred pH range is from about 6.0 to about 7.0.

No clots form when rennet is added to the low calcium milk. An arbitrary choice of 30 minutes treatment was employed as a matter of convenience. The time of rennet treatment should be sufficient to alter the protein of the low calcium skim milk to effect the maximum potential gel formation in the final product.

It is important, at the end of the rennet treatment step, to inactivate the enzyme and stop its action on the milk protein. The temperature-time relationship used in the preparation of the milk product was 60°C for 30 minutes. Pepsin is inactivated at temperatures above 40°C. The above microbial enzyme requires temperatures above 80°C for inactivation. It is common practice to inactivate rennet enzymes at 75°C for 18-20 seconds by passing rennet-containing products through high-temperature/short-time processors.

Calcium lactate is the preferred source of calcium. The ratio of 125 mg calcium per equivalent of 100 ml of treated milk was chosen as being commercially practical for the manufacture of the sausage and meatball products. However, from the tests it is evident that the range of calcium may vary from 100 to 150 mg per equivalent of 100 ml of treated milk.

The milk product can be used in varying amounts in sausage products. In connection with specific sausage products it can be used in amounts up to 3.5% of the final sausage weight. In connection with nonspecific sausage products it can be added up to about 20% of the final sausage weight.

Fibrous Milk Protein Product

S. Ohyabu, S. Kawai, H. Akasu, T. Akiya, K. Matsumura, N. Yagi, K.Y. Kim and T. Nakaji; U.S. Patent 4,251,567; February 17, 1981; assigned to Kuraray Co., Ltd., and Minaminihon Rakuno Kyodo KK, Japan have provided an improved fibrous milk protein food having a high tensile elongation and an excellent stability to hot water cooking at a high temperature.

This fibrous milk protein food comprises bundles of fibers comprising mainly a milk protein, each fiber having a diameter of less than 10 μ, the bundles of fibers having a tensile elongation of 115 to 380%, preferably 160 to 380%, and without fusing of microfibril by cooking in hot water of 135°C for 4 minutes.

These workers have previously disclosed various processes for preparing milk fibers for use in food products. This process is an improvement of prior processes.

The process for producing the fibrous milk protein food of the process comprises forming the starting milk protein into fibers, and treating the resultant fibrous milk protein composition in an aqueous solution containing at least one salt selected from the potassium salt, sodium salt and calcium salt in an amount of 1 geq/ℓ or more (as the total concentration of cations) and at least one compound containing aldehyde group or aldehyde type reducing group and having a ratio of molecular weight (M) to number of aldehyde groups in one molecule (n) of 120 to 360 (M/n = 120-360), the aldehyde compound being present in an amount of 5 to 200 g/ℓ at a solution pH value of 2.5 to 6.5 and at a temperature of 100° to 140°C for 20 minutes to 3 hours.

The compound containing aldehyde group or aldehyde type reducing group is incorporated into the aqueous solution as a crosslinking agent. The compound includes reducing monosaccharides or disaccharides (e.g. glycerose, erythrose, xylose, arabinose, ribose, glucose, fructose, galactose, mannose, rhamnose, fucose, maltose, lactose), and derivatives thereof, such as uronic acid, phosphate, sulfate, fatty acid esters or condensates of these saccharides, (e.g. glucuronic acid, mannuronic acid, galacturonic acid, glyceraldehyde-3-phosphoric acid, an oxidate of glycerine monofatty acid ester e.g., glyceraldehyde monopalmitate, glyceraldehyde monostearate, condensate of glycerose, condensate of erythrose), which have the ratio (M/n) of 120 to 360.

Besides, it has experimentally been found that when the ratio of molecular weight to number of aldehyde groups (M/n) of the reducing saccharides or derivatives thereof is increased, the elasticity and tensile elongation of the final fibrous milk protein product are also increased.

The ratio (M/n) of the representative reducing saccharides or derivatives thereof are as follows: xylose = 150, glyceraldehyde-3-phosphoric acid = 170, glucose = 180, fructose = 180, glyceraldehyde monopalmitate = 329, maltose = 342, lactose = 342, glyceraldehyde monostearate = 357. The crosslinking agent has the ratio (M/n) of 120 to 360, preferably 180 to 360. When the ratio (M/n) is lower than

120 as in the case of formaldehyde, glycolaldehyde, glyceraldehyde or glutaraldehyde, there cannot be produced the desired fibrous milk protein product having a tensile elongation of 115% or more. For instance, when a dialdehyde type oxidized starch having the ratio (M/n) of 80 is used as the crosslinking agent, the fibrous milk protein product tolerates the treatment in hot water of 135°C for 4 minutes, that is, it has a sufficient stability to hot water cooking, but it does not satisfy the condition of tensile elongation of 115 to 380%.

Example: To a suspension of acid casein (25 g) in warm water (100 ml) at 50°C is added a 28% aqueous ammonia (1.3 ml) to give a solution. To the solution is added a 25% aqueous calcium chloride solution (10 ml) to form a micellar structural composition. The micelle is treated with rennet-substitute protease (20 mg) to form a gel composition. The gel composition is orientated and fibrillated by drawing to give a fibrous composition. The fibrous composition thus obtained is pretreated by dipping it into an aqueous solution (1 ℓ) containing 10% by weight of sulfuric acid at room temperature for 1 minute and then actually stabilized by dipping into a saline bath of pH 5.0-5.1 containing 5% (60 g/ℓ) by weight of lactose and 28% (5.8 geq/ℓ) by weight of sodium chloride at 108°C for 2 hr. After washing with water and draining off the water, there is obtained a fibrous protein product having a water content of about 70% by weight (80 g).

The fibrous product maintains about 90% of the microfibrillar structure. When the product is treated in a hot water of 135°C for 4 minutes in an autoclave, the microfibrillar structure is almost maintained and the product has an excellent stability to hot water cooking. Moreover, the product has a toughness (chewiness) of 3 g/mm^2 (before actual stabilization) and 53 g/mm^2 (after actual stabilization) and has a tensile elongation of 170%.

In the same manner as described above except that a dialdehyde type oxidized starch having the same equivalent aldehyde group (1.2% by weight = 14 g/ℓ) is used instead of the lactose, a fibrous protein product is prepared. While fibrous protein product has the same stability to hot water cooking by the treatment in a hot water of 135°C for 4 minutes as that of the above product, the product shows such a small tensile elongation as 22% and a toughness of 7 g/mm^2 (after actual stabilization).

Gluten as Binder for Meatlike Products

A physical form of gluten which has self-binding properties has been developed by *M. Shemer; U.S. Patent 4,238,515; December 9, 1980; assigned to Pedco Proteins and Enzymes Development Co. Ltd., Israel.*

This gluten may be utilized as a self-binder for meat-like products excluding the need of egg albumen. The physical form contains an inert material bound within its matrix and is characterized by its net-like fibrous structure, which fibers have a diameter smaller than 2 mm diameter and a viscosity of at least 50,000 cp.

This gluten is obtained by the steps of: (a) agitating vital wheat gluten with a reducing agent at a temperature below 70°C, and (b) incorporating during the agitation the solid inert material having particle sizes below 5 cm in diameter.

During the first step of agitation in the presence of a reducing agent, the gluten particles are softened moderately and a net-like fibrous structure is gradually formed. This fibrous structure is subsequently stabilized by the incorporation of the solid inert material. The mechanism which will explain how the stabilization of the net-like fibrous structure occurs is not yet fully elucidated. As known the intramolecular bonds in gluten are complex and probably many forces are in action besides hydrogen bonding and the disulfide groups. It may be assumed that after a prior reduction of the gluten, which has been found to be absolutely required in order to obtain the new gluten form, the incorporation of the inert material (i.e. textured vegetable protein) to the incipient fibrous form of gluten, contributes to the interaction which occurs between the gluten and gliadin as a result of the difference in the electric charge, thus causing the stabilization of the net-like fibrous structure. According to a preferred embodiment, the solid inert material added during agitation in the second step is textured vegetable protein. The physical form of gluten thus formed has been found to possess outstanding binding properties for the ingredients in meat-like products, being thus capable to replace the expensive egg albumen. Furthermore, the presence of gluten improves the "meaty" structure of the final product.

Although the physical form of gluten can replace even completely the relatively expensive egg albumen, there are cases when egg albumen may also be added but at lower levels for persisting the meaty bound texture of the meat-like product, especially after being heated.

The entire process for the manufacture of the physical form of gluten is very simple. Fresh gluten or rehydrated is thoroughly mixed with a solid or an aqueous solution of a reducing reagent. Examples of such reducing agents are: tocopherol, ascorbic acid, butylated hydroxyanisole, butylated hydroxytoluene, sodium sulfite, sodium bisulfite etc. An incipient fibrous structure can be observed but this completely collapses when the mixer is stopped. While continuing the vigorous agitation, a textured vegetable protein as inert material is added and the physical form of gluten is achieved. A typical example of textured vegetable is hydrated extruded soy flour.

Example 1: An amount of 600 g rehydrated wheat gluten was mixed with 0.3 g of sodium sulfite. After 10 minutes of vigorous agitation an amount of 100 g of rehydrated textured soy (particle size between 1 to 2 cm) was added while the agitation was continued. Additional ingredients added were: 10 g of vegetable oil, spices and flavor. The final mix was prepared as meat-like hamburger in which all ingredients were bound in a homogeneous mass, although no egg albumen was present.

Example 2: An amount of 100 g of rehydrated wheat gluten was mixed with 0.01 g of ascorbic acid. After about 10 minutes of continuous mixing an amount of 600 g of rehydrated textured soy flour was added while the mixing continued. The following ingredients were further incorporated: 10 g of vegetable oil, 35 g of egg albumen, spices, flavor and caramel. The final mix was prepared as meat-like steak in which all ingredients were bound in a homogeneous mass. It was of an excellent quality as tested by a panel group.

Example 3: The same procedure described in Example 2 was repeated with the exception that 200 g of rehydrated wheat gluten were mixed with 0.1 g of sodium sulfite and 500 g of rehydrated textured vegetable protein. The same amounts of flavoring agents, spices and fat were added, the difference being the requirement of only 25 g of egg albumen in order to obtain the same bound homogeneous mass as in Example 2.

Example 4: This example illustrates the utilization of the physical form of gluten as meat extender. An amount of 300 g of rehydrated gluten was mixed with 0.2 g of sodium sulfite and after about 10 minutes of continuous agitation, 400 g of hydrated textured vegetable protein were added. It was found that up to 3 parts of the gluten form to 1 part meat can be successfully utilized (as meat extender), without impairing the binding or texture properties of the mass obtained.

Gluten-Seroprotein-Albumin Mixture as Binder

A. Fabre; U.S. Patent 4,265,917; May 5, 1981; assigned to Rhone-Poulenc Industries, France has prepared binder compositions for texturized proteins, characterized in that the binders are comprised of at least gluten, dairy seroproteins and albumin, optionally together with a consumable liquid, i.e., an edible (potable) liquid diluent.

One of the advantages of the use of the binder composition for binding spun proteins to one another resides in the pleasant texture obtained. The consistency and cohesion of the products thus obtained are remarkably similar to those natural products of animal origin.

The gluten used in the binder is preferably wheat gluten because this is the commercially available product and it is low in cost. It is obviously possible to use rye gluten, or gluten from any other cereal, provided it has the same amino acid constitution as wheat gluten. Regardless of the form of gluten employed, it is particularly valuable to use "vitalized" gluten, which has neither been modified nor denatured by prolonged heating or by an alkaline treatment. It is desirable to use freshly prepared gluten, but it is possible to employ the gluten in its dehydrated form, which is preferably reconstituted by adding an amount of water of from 50 to 75% by weight.

The dairy seroproteins employed can be obtained in known manner by ultra-filtration of milk, or by thermal coagulation.

The albumin preferably originates from fresh eggs or from dried egg white, but it is possible to envisage other sources, such as milk, fish, elastin and keratin.

By a "consumable liquid" there essentially is intended milk or water, and preferably the latter. If one or more constituents of the binder composition is or are in the liquid state, the amount of consumable liquid to be added can be reduced, or the liquid can even be omitted.

The percentage of gluten, calculated as dry material, relative to the total weight of the binder composition can, for example, vary from 10 to 20% by weight, and is preferably between 13 and 17%. The dairy seroproteins can represent, as dry material, from 10 to 20% by weight of the total binder composition, and preferably from 13 to 17%. The albumin, again calculated as dry material, can be present to the extent of 1 to 5% by weight in the binder composition, and preferably 1.5 to 3%. The consumable liquid is used in sufficient amount to make up the binder composition to 100% by weight.

The meat substitutes using this binder are characterized in that they consist of spun proteins of vegetable and/or animal origin, and of the binder composition. Preferably, spun proteins consisting of filaments having a diameter of 10 to

300 microns and possessing, in their cross section, particles of fatty matter having a diameter of 1 to 60 microns, and a degree of saponification of less than 5%, are used; this means that there are not more than 5% of ester groups of the fatty matter employed which can be saponified at the time of contact with the alkaline protein gel.

The binder is used in such amount that it yields a product having fibers bonded to one another in a structured and stable manner, in order that the product shall acquire a cohesion which withstands subsequent processing treatments, such as the various methods of cooking or of preparation. The weight ratio of the wet spun proteins to the composition can vary from 4/1 to 1/2 and is preferably about 1/1. This ratio corresponds to protein fibers containing from 60 to 70% by weight of water.

A mixture of the spun proteins, which may or may not contain fatty matter, and of the binder composition can be prepared in a malaxating device, such as, for example, planetary malaxators, sigma malaxators, ribbon mixers, double-paddle malaxators and Hobart mixers. The binder composition has the consistency of a fluid paste. In order to homogenize it well, it is preferable to introduce it first into the malaxator. The mixture of protein fibers and binder composition is malaxated for a period which can vary from 15 minutes to 1 hour but preferably from 30 to 40 minutes. In general, this impregnation is carried out at ambient temperature, without that being essential.

It is possible to add flavorings, colorants and spices and, optionally, an emulsifier or fatty matter, in cases where these are not yet present in the proteins, either by introducing the additives into the binder composition or by impregnating the spun proteins separately with such ingredients; preferably, they are introduced directly into the binder composition.

The impregnated fibers are then subjected to a heat treatment by heating them to a temperature at which the binder composition coagulates.

Example: 100 g of the following binder composition:

>Drinking water: 66.6%
>Vitalized gluten (wheat): 15.0%
>Dairy seroproteins (obtained by ultrafiltration of milk): 15.0%
>Dried egg white (ovalbumin): 2.2%
>Natural chicken flavoring: 1.0%
>Spices: 0.2%

and 100 g of soya fibers obtained by spinning and having the following characteristics:

>Diameter of the fibers: from 115 to 130 microns
>Water content: 70%
>Proteins isolated from soya: 25.5%
>Beef suet incorporated into the fibers: 4.5%
>pH: 5.8 to 6.0

were successively introduced into a mixer. The mixture was homogenized for 30 minutes at ambient temperature. The mixture was then spread in the form of a layer about 20 mm thick and was cooked at a material temperature of 95°C for 30 minutes.

After cutting, pieces having the following composition were obtained:

> Water content: 68.2%
> Protein isolated from soya: 12.8%
> Beef suet: 2.3%
> Vitalized gluten (wheat): 7.5%
> Dairy seroproteins: 7.5%
> Ovalbumin (expressed as solids): 1.1%
> Chicken flavoring: 0.5%:
> Spices: 0.1%

FOR EGG WHITE AND/OR EGG YOLK

Blood Albumin as Whole Egg Substitute

J. Heine, R. Trama and E. Penedo; U.S. Patent 4,167,586; September 11, 1979 have disclosed a process for preparing a high protein material simulating whole egg material.

This product can be obtained by means of an adequate processing of the following elements: animal albumin obtained from phosphated or citrated blood, fatty materials of animal or vegetable origin; surfactants used as emulsifiers, or wetting agents, and potable water, and as optional additives, carbohydrates, antioxidants, and/or stabilizers.

Fatty materials of animal and/or vegetable origin, partially or totally hydrogenated, obtainable from peanut, cotton, corn, coconut or grape oil and/or cow, pig, sheep and fish and similar materials may be used in the range of 4.0 to 17 weight percent of the composition.

Generally, the following are preferred as surfactants: mono- and diglycerides of edible fats and oils, acetylated and/or lactylated mono- and diglycerides, propylene glycol monoesters, and sorbitol esterified fatty acids, all of them alone or combined totally or partially with each other.

Carbohydrates if added may include sugars obtained from sugar cane, sugar beet, corn and others, alone or combined with each other.

If antioxidants are added gallic acids, propyl gallate, guaiacum gum, norhydroguaiaretic acid, tocopherols, phospholipids, butylated hydroxyanisole, hydroxybutyltoluene, ascorbic acid, citric acid, thiopropionic acid, ascorbic palmitate and dithiolauryl propionate etc. may be used.

The stabilizing agents if used may be various phosphates, cellulose esters, carrageenates, guar gum, xanthan gum, arabic gum, tragacanth gum, simple starches, hydrolyzed or modified starches, alginates and others.

In the case that natural egg or similar color is desired, the natural dyes contained in the raw materials added may be used, or synthetic-natural food dyes as beta-carotene, cryptoxanthin, beta-apo-8'-carotenal and carotenoids etc. may be added.

Example: A protein product of this process is prepared from 11.0% hydrogenated peanut oil, 3.0% soy lecithin, 34.4% potable water, 1.0% glyceryl monostearate, 0.1% cane sugar, 0.5% sodium benzoate and 50.0% liquid phosphated animal albumin.

The manufacturing of the product comprises the following steps, which are carried out in one reactor that is, for example, of the Pfaudler type with an "homomizer" stirrer, with high stirring capacity.

The glyceryl monostearate is melted and is added to all of the hydrogenated oil, previously melted in the stirrer, and the antioxidant is added while trying to maintain the mixture at a melting temperature at least 70°C with continuous strong stirring.

Maintaining the mixture under intense stirring, the temperature is lowered to 40°C, with the purpose of preserving the basic properties of the soy lecithin, which is added once this temperature has been reached, so that the phospholipids represented in this case by this lecithin may react as such in the mixture.

Immediately after, water in which is dissolved the cane sugar and sodium benzoate, the stabilizer is added to the reaction mixture. The dye may be added in this phase or in the fatty phase, as preferred.

Previously the water has been boiled, to become sterilized, and then cooled to a temperature of 60° to 70°C.

With the addition of this aqueous phase (60°-70°C) which is done while maintaining the intensive stirring of the mixture previously made, a low but progressive increase in its temperature is achieved, which does not entail any risk for the properties of the phospholipids. A certain crystallization of the lipids of the fatty acids in the mixture takes place and, as the intense stirring continues, the crystals remain dispersed in the aqueous phase due to the emulsifiers which have been added.

In this way, as the water is added, an emulsion of the water in fat type is formed, until equilibrium is attained. At that time, a certain coagulation takes place and the mixture thickens acquiring the physical characteristics of a pasty gel.

As water continues to be added, an inversion in the emulsion takes place and it becomes of the fat-in-water type.

Once the temperature of 40°C is obtained, the liquid animal albumin is added; at first a small part of the total volume is added slowly, e.g. 20%, at the same temperature of the emulsion; then the rest of it is added rapidly. In this way it becomes part of the aqueous phase and, due to the intense stirring, a perfect emulsion is obtained.

Even though due to the intense stirring there is a true homogenization of the emulsion, the particles suspended in the fatty phase do not attain the size of two or three microns. Therefore the mixture at a temperature of approximately 40°-50°C is subjected to the action of an homogenizer, which can be a one or two stage type or a colloidal mill.

The dough is then subject to a regeneration process, passing it through an "APV" or similar regenerator, where the temperature drops to 25°C and then must be dropped as quickly as possible to a temperature not over 10°C and held at that level.

The emulsion obtained with this procedure has been cooled slowly, so that germs of fatty crystals are formed. The growth of these crystals is regulated with the temperature, causing them to develop to a limit; their development finishes when the temperature drops to 10°C at the end of the process, even though the growth stability is only achieved when the finished product is stored at a low temperature in a refrigerator, to be preserved until its use.

The emulsion must be kept in a refrigerator at a low temperature preferably at 5°C or less.

If a powdered product is desired, the emulsion at 50°C, may be fed to a conventional spray-dry or fluid bed drying system.

Modified Soy Protein in Mayonnaise

A process for preparing mayonnaise-like foods by using a specifically refined soybean protein instead of eggs as an emulsifier has been used by Y. Mikami, H. Kanda and A. Uno; U.S. Patent 4,293,574; October 6, 1981; assigned to The Nisshin Oil Mills, Ltd., Japan.

According to this process, with alcohol denaturation, soybean protein is refined to one free of an odor and color tone peculiar to soybean, and mayonnaise-like foods prepared therefrom are by no means inferior in flavor and organoleptic properties to the conventional mayonnaise starting from egg. Further, the synergistic effect of the alcohol denaturation and the solubilization treatment by enzymolysis with protease results in a remarkable increase in the emulsifying ability of protein, particularly at an acid pH. Details of the process are shown in the following example.

Example: 20 kg of a slightly denatured, defatted soybean were washed with 200 kg of an aqueous 60 w/w % solution of ethanol in a sealed tank at 50°C for 30 minutes. After filtration 32 kg of cake were obtained and transferred into a pressure-reduced drier. After drying at 70°C, 13.5 kg of a powdered, concentrated soybean protein were obtained which contained 6.5% water, 67% protein, and had an NSI of 11% on a dry basis.

Next, the whole amount of the above product was added to 150 ℓ of warm water (60°C) in a tank provided with jacket and dispersed while stirring. Ammonia water was added to adjust a pH to 9.0 and temperature of the solution was elevated to 55°C and maintained at this temperature.

When 50 g of Bioprase (20,000 U) were added and reacted at 55°C for two hours, a pH was lowering slowly up to 7.5 and a 10% TCA solubilization rate was 20%. This solution was heated by means of a plate-type heat exchanger and cooled. The highest temperature of the solution which was reached was 120°C. After centrifuging with a decanter, 120 ℓ of a solution of 7.3% in solid content were obtained, of which 5.0 ℓ were sampled (Sample A) and the remainder were concentrated and spray-dried to obtain 7.5 kg of refined soybean protein (Sample B). The sample B in form of powders was free of a beany flavor, having a bright white color and a high water solubility and contained 7.0% moisture, 83.5% protein (on a dry basis), had 95% NSI and a 10% trichloroacetic acid solubilization rate of 26%.

Each of 4.0 ℓ of a protein solution having a 5% concentration was prepared from Samples A and B. 100 g of sugar, 200 g of salt and 50 g of seasonings mixture were added and dissolved. Next, 5 kg of soybean salad oil were added slowly while maintaining the temperature at 30°C and emulsified. Then, 1.0 kg of vinegar was added slowly and emulsified. Finally, the finishing was conducted using a colloid mill to obtain mayonnaise-like foods A and B. As for these two mayonnaise-like foods, the emulsification state and flavor were both good and both had viscosities of 45,000 cp as recorded on a Brookfield Viscometer.

In comparison tests mayonnaise-like foods prepared from the conventional separated soybean protein powders and the further enzymolyzed, isolated soybean protein powders are inferior in the flavor and emulsification stability to the food product obtained by enzymolyzing the alcohol-denatured soybean protein as in this example.

Whey Protein-CMC for Replacement of Albumin

A whey protein concentrate composition has been developed by *P.K. Chang; U.S. Patent 4,214,009; July 22, 1980; assigned to Stauffer Chemical Company* as a substitute for the egg albumen requirement in a food product.

It has been found that the difficulties of replacing egg albumen with a whey protein concentrate can be overcome by replacing a portion of the albumen (dry solids basis) depending on the application up to 100% with a quantity of a substantially non-heat-denatured whey protein concentrate containing at least 29% and preferably at least 35% protein prepared substantially from and preferably from at least 80% acid, e.g. cottage cheese whey in such an amount that the total protein content provided by the whey protein concentrate (WPC) and albumen is no less than about 50% and preferably no less than about 38% below the protein content of the egg albumen (on a dry solids basis) originally present in the recipe and from about 0.5% to about 15% by weight carboxymethylcellulose based on the weight of the whey protein concentrate. If desired, the composition can include from about 25% to about 0% of another protein-containing whey-based product.

The use of the carboxymethylcellulose in combination with the whey protein concentrate in baked goods such as cakes overcomes the deficiencies in cake texture and volume normally encountered in replacing egg albumen with whey protein concentrate. The composition of the process also effectively replaces egg albumen in other food areas such as puddings, sauces, soups, batters and the like.

The acid cheese whey used is derived from the acid coagulation of milk protein by the use of lactic-acid-producing bacteria or by the addition of food-grade acids such as lactic or hydrochloric acid. In either case, acidification is allowed to proceed until a pH of approximately 4.6 is reached. At this pH, casein becomes insolubilized and coagulates as cheese curd. The whey obtained in this manner is commonly called cottage cheese whey.

The whey protein (WPC) concentrate as used in this process is preferably derived from 100% acid cheese whey though minor amounts of other cheese wheys of up to 20% can be utilized. Such other cheese wheys include but are not limited to cheddar cheese whey which is produced by the rennet coagulation of protein and is commonly called sweet whey. It is preferred that the whey source be at least 90% cottage cheese whey.

The whey protein concentrate must be substantially non-heat-denatured, i.e., at least 40% of the protein as determined by solubility at pH 4.6 has not been denatured by heating such as in pasteurization and drying. Thus, freeze-drying would denature less protein than spray drying. The use of a food-grade sulfite such as sodium sulfite, sodium bisulfite, cysteine, cystine and the like in from about 0.1% to about 0.5% by weight based on the whey protein concentrate is particularly advantageous in reducing the coagulation temperature of the whey protein from about 80°C to about 70°C. This more nearly approximates the coagulation temperature of egg.

Carboxymethylcellulose or CMC is a water-soluble cellulose ether generally available as the sodium salt. CMC is known to have molecular weights which range from about 21,000 to 500,000. CMC is commercially available in viscosities ranging from above 10,000 cp in 1% solution to 25 cp in 2% solution and even lower viscosities. It is preferred that the CMC used provides a viscosity within the range of from about 1,000 cp to about 10,000 cp in a 1% solution at 25°C.

The carboxymethylcellulose is used in an amount ranging from about 0.5 to about 15% and for use in baked goods preferably from about 0.5 to 5% and more preferably from about 1-3% based on the weight of the whey protein concentrate. In other areas higher levels as high as 15% based on the whey protein concentrate may be used.

The albumen replacer can be used to replace up to 75% by weight of the albumen requirement in baking applications such as white, yellow, sponge and devil's food cakes, sweet doughs, biscuits, pancakes, doughnuts, muffins and the like. The replacement should be on a basis sufficient to provide at least 50% and preferably at least about 62% of the protein replaced (the combined protein content of the whey protein concentrate and albumen is no less than 38% below the protein content originally added to the recipe by the egg albumen).

Example: A control sponge cake was prepared from 300 g cake flour, 360 g sugar, 22.5 g nonfat dry milk, 9.4 g salt, 8.5 g baking powder, 15.0 g Atmos G-2462 emulsifier, 37 g dried egg yolk, 16 g egg albumen, 2 cc vanilla and a total of 360 cc water.

Test cakes were prepared using whey protein concentrates of this process as substitutes for the nonfat dried milk and/or the egg albumen. Test cakes were judged for flavor, strength, structure, color and overall appearance against the control cake. When the egg albumen was substituted with 50% WPC containing 5% CMC the cake was similar to control cake. When 50% of the albumen and 73% of the NFDM were replaced with WPC containing 1.63% CMC, the cake was also similar to the control. However, when the above replacements were run with WPC containing no CMC, the cake batters had high specific gravities and the cakes had lower volumes.

Whey-CMC-Shortening for Whole Egg Replacement

It is known to provide an egg albumen whole egg partial replacer for bakery goods which comprises a whey protein concentrate and from about 0.5% to about 5% carboxymethylcellulose (CMC). However, this product is not effective in partially replacing (up to 50%) the whole egg solids in baked custard. While the texture of the custard is not affected, the custard does not brown properly.

In attempting to overcome the problem various fats and oils were added to the blend. Butter while providing some browning is disadvantageous as it leaves the surface of the baked custard oily.

These problems can be overcome by the process developed by *C.R. Corbett; U.S. Patent 4,214,010; July 22, 1980; assigned to Stauffer Chemical Company.*

It has been found that up to about 75% of the whole egg requirement of whole egg-containing baked custard can be replaced with a composition comprising:

    (a) from about 40% to about 60% of a blend of a whey solids product and from about 0.5% to about 5.0% based on the weight of the blend of carboxymethylcellulose, the whey solids product comprising:

        (1) from about 75 to 100% of a substantially non-heat-denatured whey protein concentrate having at least 35% protein wherein at least 50% of the protein is prepared from acid whey, and

        (2) from about 25 to 0% of another protein-containing whey-based product; and

    (b) from about 60% to about 40% of the composition of a lactylated shortening.

The composition replaces at least 90% of the whole egg solids replaced on a w/w basis, all percentages herein being on a dry weight basis.

The use of the above composition effectively replaces whole egg in baked custard providing good browning and taste characteristics while at a definite cost advantage based on market prices.

The whey protein concentrate (WPC) and carboxymethylcellulose (CMC) used in this process are described in the previous process of U.S. Patent 4,214,009.

The whole egg replacer composition also includes from about 40% to about 60% by weight of a lactylated shortening. Shortening is intended to include those materials which are naturally solid at room temperature as well as hydrogenated and plastic shortenings. The shortening contains from about 5% to about 20% glyceryl lacto esters (lactylated mono- and diglycerides). Preferably, the shortening contains from about 5% to about 20% of glycerol lacto esters of fatty acids. These are well-known compositions formed by esterifying glycerol with lactic acid and fatty acids.

The shortening can contain small quantities (from about 5% to about 10%) of other emulsifiers. It is preferred that the shortening contain from about 5 to about 10% mono- and diglycerides.

The shortening preferably contains a small quantity of lecithin in an amount up to 5%.

The preferred shortening is a partially hydrogenated vegetable oil with mono/di-glycerides, glyceryl lacto esters of fatty acids and lecithin added. A product of this type is distributed by Durkee Food Service Group of SCM Corp. (D-21).

Example: A baked custard pie control was prepared by mixing in the usual manner 4.9% whole milk solids, 3.8% whole fresh egg solids, 4.9% NFDM, 7.7% sugar, 0.2% salt, 0.9% instant starch, 0.8% vanilla flavoring and 76.8% water.

A second custard was prepared where 1.9% of a whey protein was substituted for half of the whole fresh egg solids. A third custard was prepared with 1.9% of whey protein concentrate/emulsifier shortening, and CMC was substituted for half of the whole egg solids.

The custards were placed in ordinary 9" pie shells and baked in a 218°C oven for 30-35 minutes. The pies prepared from the control were brown and had the acceptable appearance of a custard pie. The pie made from the second custard was white and had an unacceptable appearance. The pie prepared from the third custard in accordance with the process had a brown and an acceptable appearance closely approximating that of the control.

Defatted Soy Flour in Egg Yolk Replacer

W.B. Chess; U.S. Patent 4,182,779; January 8, 1980; assigned to Stauffer Chemical Company has found that an egg yolk replacer can be prepared from defatted soy flour which provides the same beneficial qualities as a similar composition prepared using full fat soy flour.

In accordance with this process, an improved egg yolk extender can be provided by combining defatted soy flour having a protein dispersibility index of less than about 60, an edible food grade oil in a ratio to defatted soy flour of from about 18:100 to about 40:100, grain flour in a ratio of grain flour to the combined weight of defatted soy flour and oil within the range of from about 1:10 to about 1:1, and lecithin in a ratio of lecithin to the combined weight of the defatted soy flour and the oil within the range of from about 1:100 to about 20:100. This composition can effectively extend egg yolks when used in an amount of up to about 75% and preferably up to about 50% egg yolk replacement depending on the area of use of the yolk.

These compositions can be used as such in the dry state as an egg yolk extender for dried egg yolks. Also, these compositions can be mixed with equal amounts by weight of water to form a liquid egg yolk extender.

These compositions more nearly approximate the functional characteristics of egg yolk replaced when used on a per weight replacement basis.

The defatted soy flour is blended with a grain flour in a ratio of grain flour to the combined weight of defatted soy flour and oil within the range of from about 1:10 to about 1:1. The flour can be from any commercial source as long as the flour is of fine particle size (less than about 150 mesh). Preferably, the grain flour is derived from wheat. Short patent wheat flour is preferred.

If it is desired to cut the amount of flour used, a filler can be used for that purpose. Though the use of a filler is not preferred, one can use a material such as corn syrup solids as a filler. Up to 50% of the flour may be replaced with the filler without seriously harming the results achieved.

The oil used in the process can be any edible food grade oil such as corn oil, safflower oil, sesame oil, soy oil and salad oil. It is preferred that the oil be soy oil though any food grade oil can be used as dictated by availability and price.

While the blend of defatted soy flour, an oil, grain flour and lecithin can be used as such as an effective egg yolk extender, it has been found desirable to include additional ingredients to expand the range of usefulness and provide a product more fully adapted to extend egg yolk on a per weight basis of the yolk replaced.

These additional additives include humectants, gums, additional emulsifiers, other than lecithin and spice mixes.

The egg yolk extender is generally prepared by mixing the defatted soy flour with the wheat flour followed by blending therewith a dispersion of the lecithin in the oil. As an alternative, the lecithin can be made fluid by heating. The lecithin/oil dispersion is added incrementally so that it can be fully absorbed into the flour as it is mixed. Mixing can be accomplished at ambient temperature. If desired, all of the additional emulsifiers can be added with the lecithin. If the emulsifier is a solid other than a finely divided material defined as less than about 60 mesh, or a viscous fluid, it can be liquefied in the heated lecithin and added to the blend at this point with the lecithin and oil.

Example 1: *Preparation of egg yolk extender* — Into a ribbon blender was placed 119 kg (53%) defatted soy flour and 48.27 kg (21.5%) short patent wheat flour. 29.2 kg (13%) soy oil in which 13.47 kg (6%) lecithin had been dissolved was incrementally added while mixing. After the oil and lecithin mixture were thoroughly blended with the flour, 6.8 kg (3%) glycerine (USP) were slowly added with mixing. To this blend was added with mixing 4.5 kg (2%) polysorbate 60 which is a polyoxyethylene sorbitan monostearate (Tween 60). After the emulsifier was thoroughly blended, there was added with mixing 3.37 kg (1.5%) carrageenan (Carastay 26, Stauffer Chemical Co.). The product was then milled in a Fitzmill hammer mill and packaged in bags or drums. The total yield of product was 224.61 kg of dry product.

Example 2: Using the egg yolk extender prepared by the method of Example 1, a salad dressing is prepared containing 80.55% soybean oil, 6.66% water, 6.66% cider vinegar, 1.74% egg yolk solids, 1.74% egg yolk extender of Example 1, 1.40% salt, 1.10% sugar and 1.15% dry mustard powder.

The water is placed in a mixer. The dry ingredients are added and mixed until just blended. A small portion of oil is added very slowly and then the vinegar is added with mixing. The remaining oil is then added very slowly with mixing.

A salad dressing characterized by acceptable body, flavor, and appearance is provided.

Example 3: A Hollandaise sauce was prepared by mixing 7.14 parts of the dried egg yolk replacer of Example 1 with 14.5 parts of boiling water in a double boiler under heat. The contents of the boiler were whipped with a wire whisk until thickened. Then 12.66 parts of boiling water and 4.41 parts vinegar or lemon juice were whipped in.

After removing from the heat, the sauce was whipped well. While under heat, the 56.15 parts of melted butter was added and the sauce was heated until thick. A Hollandaise sauce equivalent to an all egg yolk control was obtained using either vinegar or lemon juice.

Protein Modified with Cysteine-Enriched Plastein

The so-called "plastein" is a protein-like high molecular substance which is produced by hydrolyzing proteins with an enyzme and then reconstituting the resultant. The plastein is considered to be a complex mixture of high molecular polypeptides having a molecular weight of about 4,000 to about 20,000 and protein-like low molecular substances, but the arrangement of the amino acids and the distribution of the molecular weight are not yet sufficiently made clear.

In the fields of nutrients and medicines, there has been an increasing interest in the plastein, because protein-like substance having an interesting nutritious or immunological effect may be produced by incorporating various amino acids into the plastein when it is reconstituted, or by eliminating a certain amino acid from the plastein.

M. Fujimaki, S. Arai, M. Watanabe, Y. Hashimoto and A. Kurooka; U.S. Patent 4,145,455; March 20, 1979; assigned to Fuji Oil Company, Limited, Japan have provided a protein composition which is modified with a cysteine-enriched plastein (CySH-P).

In earlier work, the workers have synthesized various CySH-Ps by reacting an enzymatically hydrolyzed product of various proteins with a cysteine ester (e.g., ethyl cysteinate) and have mixed the CySH-P thus obtained with a water-soluble protein, such as albumin and globulin. The products show very interesting properties. For instance, when a small amount of the CySH-P produced from soybean protein is admixed with soybean protein or egg-white, the mixture shows an increased foamability with the lapse of time while the CySH-P per se has no such foamability. Besides, when the CySH-P produced from soybean protein is heated together with soybean protein in an aqueous medium, the mixture shows an extremely increased viscosity while the CySH-P per se has no coagulation property by heating.

Taking into consideration the fact that the CySH-P is an entirely harmless protein-like substance, in addition to these newly found properties thereof, it will be useful in the field of foodstuffs.

Example 1: *Preparation of CySH-P from soybean protein* — A 2% aqueous solution of soybean protein (Fujipro-R, Fuji Oil Co., Ltd.) is adjusted to pH 10 with NaOH and is allowed to stand overnight and thereby is hydrated and modified. To the resultant is added Bioprase (Nagase Sangyo KK) in the ratio of 1% on the basis of soybean protein and the mixture is hydrolyzed by heating at 55°C for 5 hours. The hydrolyzate is adjusted to pH 4.5 and the resulting precipitates are filtered off, and the supernatant liquid is concentrated. To the resultant is added ethyl cysteinate in the ratio of 10% by weight on the basis of the concentrated product (solid component), and the mixture is controlled to the concentration of the substrate: 50% (w/v) and the pH value: 5.5. To the mixture is added papain (Difco Co.) in the ratio of 1% by weight, and the mixture is incubated at 37°C for 48 hours. After the incubation, the mixture is adjusted to pH 10, and the unreacted ethyl cysteinate is hydrolyzed and then the mixture is again adjusted to pH 7.0. To the reaction mixture is added ethanol in a concentration of 90% (w/v), and the resulting precipitates are collected to give the desired CySH-P.

Example 2: *Preparation of a modified soybean protein* — To a 5% aqueous solution of the soybean protein as used in Example 1 is added the CySH-P produced from soybean protein (half cysteine content: 5% by weight or more). The mixture is adjusted to pH 7 and heated at 80°C for 30 minutes and then lyophilized. The product thus obtained has excellent properties as a foaming agent.

Example 3: *Preparation of a modified egg white* — To egg white is added 5% by weight of the CySH-P produced from soybean protein. The mixture is mixed well by agitating and then lyophilized. The lyophilized egg white thus obtained has superior foamability in comparison with the conventional dried egg white.

Example 4: A mixture of milk casein (160 g), isolated soybean protein (40 g) and cheese whey powder (50 g) is added to a food cutter (Hanaki Seisakusho KK) and thereto are added a molten palm oil (MP: 36°C, 200 g) and hot water (40°C, 600 g). After the mixture is completely mixed and dissolved, CySH-P (10 g) of Cheddar cheese flavor (0.1 g) are added thereto, and the mixture is mixed well with a food cutter. The mixture is filled into a synthetic resin-made casing and sterilized by immersing it in hot water of 90°C for 30 minutes to give a cheese-like foodstuff.

Milk Serum Proteins as Whipping Agents

An egg white substitute is prepared by *D. Pâquet, K.S. Thou and C. Alais; U.S. Patent 4,226,893; October 7, 1980; assigned to Agence Nationale de Valorisation de la Recherche (ANVAR), France* from milk serum. This process comprises:

> (a) bringing the pH of a milk serum based material having a protein content of the order of 2 to 15%, to a value comprised between about 2 and 6;
>
> (b) bringing for about one to ten minutes the temperature of this solution to a temperature comprised between about 45° and 80°C and of maintaining the temperature for about 5 to 20 minutes;
>
> (c) rapidly cooling the solution to a temperature comprised between about 15° and 25°C;
>
> (d) alkalinizing the solution to a pH comprised between about 7 and 9, which leads to a liquid product directly suitable for foaming.

The milk serum based raw material may be of various origin; such as for example, milk serums or mixtures of milk serums from cheese making or from casein manufacture directly obtained in the manufacture of cheese or of casein. The process is also applicable to milk serum mother liquors.

It is also possible to use a milk serum concentrate or in powder form which is diluted to obtain the protein concentration desired comprised between 2 and 15% weight by volume.

From the organoleptic point of view, the best results are obtained by using ultra-filtration and dialysis retentates of the milk serum.

The duration of the heat treatment (b) must be relatively short whatever the temperature comprised between 45° and 80°C and preferably between 50° and 80°C. For a preheating period comprised between one and ten minutes, it must not exceed 20 minutes. Beyond this time the stability of the foam obtained becomes less than 50% of maximum stability.

Example 1:   Into a tank of 500 ℓ equipped with a double wall, filled with the ultrafiltration retentate of milk serum, at concentration of 5% of protein, is added concentrated hydrochloric acid to bring the pH to 2.

The temperature is brought to 55°C by the injection of steam into the double wall, in 10 minutes with average stirring, and this temperature is held for 5 minutes. It is rapidly cooled to 20°C and then concentrated soda added to a pH of 7.

The product obtained was converted into foamed form by whipping in a manner similar to the treatment of white of egg. The foam obtained had good stability and was used after the addition of a suitable amount of sugar for the manufacture of meringues which were baked in a low-temperature oven. The meringues obtained had the appearance and taste of meringues produced from egg white.

Example 2:   In a continuous process, raw material (cheese milk serum or casein manufacture milk serum or milk sera mother liquors) at a concentration of 5% protein, was introduced continuously into the apparatus. At this level, a pH meter controlled the operation of a pump which injected hydrochloric acid to bring the pH of the solution to 2. The solution then passed through the heat exchanger where its temperature was brought from 20° to 55°C in 2 minutes; it then passes for 10 minutes through a heat-insulated chamber which was regulated to a temperature of 55°C.

At the outlet from the heat-adjusting chamber, the solution passed through the heat exchanger in which its temperature passed from 55° to 20°C.

At the end of the cycle, another pH meter controlled the operations of another pump which injected concentrated soda to bring the pH of the solution to 8.

The product thus obtained, supplemented with sucrose at a concentration of 50% (w/v) was convertible, after whipping, into meringues by baking in an oven at moderate temperature.

Protein Micellar Mass as Egg White Substitute

E.D. Murray, T.J. Maurice and L.D. Barker; U.S. Patent 4,247,573; January 27, 1981; assigned to General Foods, Limited, Canada have used a protein micellar mass to replace or extend egg white normally used in various food compositions, as a binder. The protein micellar mass exhibits similar binding and heat coagulation properties to egg white.

Protein micellar mass used in these compositions is a unique protein isolate, the production of which from various protein sources is described in detail in U.S. Patent 4,169,090.

The procedure described in this patent involves a controlled two-step operation, in which, in the first step, the protein source material is treated with an aqueous food grade salt solution at a temperature of about 15° to 35°C, a salt concentration of about 0.2 to 0.8 ionic strength and a pH of about 5.5 to 6.3 to cause solubilization (or salting-in) of the protein, usually in about 10 to about 60 minutes, and, in the second step, the aqueous protein solution is diluted to decrease its ionic strength to a value less than about 0.1.

The decrease of the ionic strength of the aqueous protein solution preferably is achieved by feeding the concentrated solution into a body of cold water containing sufficient volume such that the ionic strength is decreased to a value less than about 0.1. Preferably, the body of cold water has a temperature of about 5° to about 15°C.

The dilution of the aqueous protein solution, which may have a protein concentration, for example, up to about 10% w/v, causes association of protein molecules to form discrete highly proteinaceous micelles which settle in the form of an amorphous highly viscous, sticky, gluten-like mass of protein having a moisture content of about 60 to about 75% by weight. The amorphous protein mass so formed is referred to as "protein micellar mass," and is abbreviated to PMM. The wet PMM may be dried to a powder for use in that form. Drying of the wet PMM may be achieved using any convenient drying technique, such as, spray drying, freeze drying or vacuum drum drying.

The protein materials from which the wet PMM is formed may vary widely and include plant proteins, e.g., starchy cereals and legumes, and animal and microbial proteins. Preferably, the protein source is a plant protein owing to the readily available nature of the materials.

The production of protein micellar mass as described above results in a wet product having occluded water which also contains dissolved food grade salt, mainly sodium chloride. Upon drying, the sodium chloride remains entrapped in the PMM. It has been found that, as this concentration of sodium chloride increases, the binding capacity of the PMM decreases and, overall, the binding capacity of the PMM in any given food composition decreases with the total concentration of sodium chloride present, arising from the PMM and any additional salt necessary in the composition.

In view of the desire for as high a binding capacity as possible and thereby use of the lowest concentrations of PMM to achieve the desired binding capacity, it is preferred in most cases to minimize the concentration of sodium chloride which is entrapped in the PMM.

Improvements in the basic process of PMM production have been developed to control the concentration of sodium chloride entrapped in the PMM. The initial extraction of protein is effected over a pH range of about 5 to about 6.8, the protein solution is concentrated to increase the protein concentration while maintaining the ionic strength substantially constant, and thereafter is diluted to an ionic strength below about 0.2 to cause the protein micelle formation. This procedure may be used to decrease the entrapped salt concentration in the PMM for use in this process.

Centrifuging and dialysis of the PMM precipitate may also be used to lower the salt content.

Example 1: This example illustrates the use of PMM samples in various food mixes as a substitute for egg white.

Several commercial dry food mixes conventionally used in one case for baking cakes, in a second case for baking muffins and in a third case for cooking pancakes and normally requiring the addition of water and egg white were taken. The egg white and at least part of the water were replaced by a 15% w/v aqueous solution of PMM samples from field peas and soybean on the basis of 33 g of PMM for each egg white required by the recipe.

In each case, the cooked product was substantially the same in taste and texture to the same product formed using egg white.

Example 2: This example illustrates the formation of a white layer cake using PMM in place of egg white.

A cake mix was made up from 2 cups of flour, 1⅓ cups of sugar, 1 teaspoon salt, one-half cup of vegetable shortening, 1 cup milk, 1 teaspoon vanilla, 3½ teaspoons baking powder and 132 g of PMM.

The PMM was added as a 15% aqueous solution thereof which also provided all the water requirement and mixes were made from PMM samples from field peas and soybeans. The mix, after processing to a smooth consistency, was placed in a bake tin and baked in an oven at 350°F for 30 minutes.

The resulting cake was tested and found to exhibit similar taste and texture characteristics to a cake baked from a mix using 4 egg whites in place of the PMM.

Whey Products as Albumen Extenders

P.K. Chang; U.S. Patent 4,238,519; December 9, 1980; assigned to Stauffer Chemical Company has found that effective egg albumen extenders can be prepared using derived protein-containing compositions such as the permeate and delactosed permeate resulting from the ultrafiltration of whey.

Egg albumen extenders are provided comprising at least 65% by weight on a dry solids basis of a derived protein-containing composition from plant or animal sources wherein the derived protein-containing composition has a molecular weight of less than 20,000, a total Kjeldahl nitrogen content of from about 0.45% to about 2.1% of which at least 60% of the nitrogen is nonprotein nitrogen, and form 0% to about 30% of a whipping aid, in combination with a member selected from the group consisting of gelatin, gelatin and a water-soluble polyphosphate, a gum and mixtures thereof.

The products are useful as egg albumen extenders in whipped products such as meringue, nougat candy and divinity candy, and cakes such as yellow or sponge cake.

The gelatin used can be either of the alkaline or preferably the acid prepared type. Gelatins ranging in Bloom strength from about 100 to about 300 and preferably from about 200 to about 250 Bloom can be used. The gelatin can be predissolved in water to facilitate incorporation. Preferably "cold-water dispersible" gelatin is used.

The preferred derived protein used is the permeate and preferably the delactosed permeate resulting from the ultrafiltration of acid or cottage cheese whey.

The amount of permeate or delactosed permeate used is generally dependent on the amount of additives used. Generally, the permeate and/or delactosed permeate comprise at least about 65% by weight (dry solids basis) of the blend, the remainder of the blend upon which the percentage is based being made up with gelatin, gelatin and polyphosphate and/or a gum and, optionally, a whipping aid such as the enzymatically hydrolyzed wheat protein including any starch or sugar additives to the gluten. The gelatin can be used in an amount ranging from about 1% to about 15%, preferably from about 3% to about 12% and more preferably from about 3% to about 5% by weight.

When using gelatin alone (no polyphosphate) it is more preferably used in an amount ranging from about 3% to about 12% by weight. When used in combination with the polyphosphate, results equivalent to those obtained using gelatin alone can be obtained using less gelatin. In some cases, the amount of gelatin used with the polyphosphate can be reduced by as much as 50% over the quantity of gelatin used alone while providing substantially equivalent results. When the gelatin is used with the polyphosphate, the preferred amount of gelatin is from about 3% to about 5%, the preceding broad and intermediate ranges being applicable.

The polyphosphate used is preferably sodium hexametaphosphate. Sodium hexametaphosphate has been found to be usable within the range of from about 5% to about 25% by weight, preferably from about 9% to about 20% and more preferably, from about 18% to about 20%. In general, as the amount of polyphosphate increases, the amount of gelatin decreases.

In one whippability study at a 10% level of sodium hexametaphosphate, the stability of the foam is affected by the level of gelatin. The foam is more stable at gelatin concentrations of 8% and 12% than 4%. At a 19% level of sodium hexametaphosphate, the stability of the foam is not affected by the level of gelatin though a slight increase in the specific gravity of the foam was noticed as the gelatin content increased. It is preferred that the additive total of gelatin and polyphosphate not exceed 35% and preferably not above 25%.

If desired, sodium aluminum sulfate can be added in an amount up to and including about 5%, and preferably from about 1% to about 2% by weight of the dry egg albumen extender to further improve the stability of the foam.

Also, it has been found desirable to include from 0% to about 5% and preferably from about 1% to about 2.5% by weight of the dry egg albumen extender of an acidifying agent in the form of anhydrous monocalcium phosphate. This agent contributes calcium ion to the system.

Other optional ingredients such as flavorings, i.e., sugar, salt, vanilla extract and the like, and colorings, fillers, and the like can be added if desired.

The egg albumen extenders of this process can be prepared by dry blending the ingredients in the proportions desired. Liquid formulations can also be used but these require refrigeration. The products can be preblended for shipment to user or preblended in user's plant. Blending can also be accomplished in situ in the final use product. The dissolution of gelatin in a liquid system is facilitated by heating to 40°-50°C followed by cooling. Since permeate and delactosed permeate are hygroscopic, liquid formulations are preferred. The extender is used in an amount sufficient to provide a solids content of from about 9% to about 14% and preferably from about 10% to about 12% in water. The solid content of liquid egg albumen is about 11.5%.

Seroproteins as Gel Formers

A gel having rheology characteristics comparable to those of an egg-white gel or a gelatin gel is prepared by the process developed by *C.G.G.R. LeGrand and R.A.E.C. Paul; U.S. Patent 4,251,562; February 17, 1981*. This process comprises forming a mixture of a sol of seroprotein such as whey protein, glucides such as saccharose or hydrolyzed lactose and water, and heating the mixture under pressure at a temperature and for a time sufficient to convert the mixture into a gel.

Two of the available by-products are (1) a sol of seroprotein derived from sweet lactoserum and (2) ultrafiltrates of whole or sweet skimmed milk containing lactose. The sol of seroprotein is prepared by subjecting the sweet lactoserum of pH, preferably higher than 5.8 and lower than 7.0 to ultrafiltration, reverse osmosis, dialysis, electrodialysis, etc. The sweet lactoserum (whey) from which the sol of seroprotein is derived is obtained, for example, as one of the components resulting from cheese making processes.

The lactose-inorganic salt fraction, sometimes called the "ultrafiltrate" is subsequently hydrolyzed to a degree of between about 75-95% lactose completion. Such hydrolyzed lactose may be used as the glucide, or it may be used with saccharose. Lactose when hydrolyzed converts to a glucose and galactose which have a sweet taste, as compared to the blend or nonsweet taste of the unhydrolyzed product.

The behavior of seroproteins when subjected to heat is of significance in a better understanding of this process. Casein functions as a protective colloid for seroproteins (lactalbumins and lactoglobulins) in milk thus making it possible to heat milk without forming a flocculate. The seroproteins in the absence of substantial amounts of casein, e.g. 6 to 10 parts of casein to 1 part of seroprotein, will foam and coagulate to form a flocculate when heated to more than 70°C.

It was found that, when present in sufficient concentration, a glucide such as saccharose will act as a protector of the seroproteins and that a mixture of seroprotein containing saccharose will retain its original opalescent appearance. However, upon merely heating a sol of seroprotein containing saccharose (to 100°C and even higher) without subjecting the mixture to pressure, one obtains a poorly organized coagulate having enormous syneresis.

When a sol of seroproteins, mixed with water, a glucide such as saccharose, or an unfiltrate of sweet milk is subjected in a sealed vessel, to a temperature equal to or greater than 80°C to a pressure greater than atmospheric, a pure well-defined gelled product of a rheology somewhat comparable to that of an egg-cream or a heated egg white is obtained.

Example 1: 442 parts of a sol of seroproteins having a 13% dry solids content, 59% of the solids being protein, was mixed with 320 parts of saccharose and 238 parts of water. The mixture, thus formed, was placed in hermetically sealed containers being small cans of about 250 g capacity and heated in an autoclave for about 5 minutes at about 110°C. Gelling started during this period and, upon removal from the autoclave, continued until a homogenous, well-organized gelled product was obtained. It should be noted that according to the practice of this example, small containers were employed.

This was done to insure adequate heat transfer from the exterior of the containers through the static mass contained therein, thus producing a homogenous gel formation throughout the composition.

Example 2: The procedure of Example 1 was followed except for substituting for the saccharose an equivalent amount of an ultrafiltrate of sweet milk having a dry solid content on a weight basis of 70% hydrolyzed lactose (glucose plus galactose) and 2% minerals. The same type of homogenous, well-organized gelled product was obtained.

Example 3: A mixture of the ultrafiltrate and saccharose was employed as the glucide component for this example. The seroprotein sol, glucide component and water was treated in a sterilizer according to the previously described continuous or dynamic method employing a temperature of about 140°C and a pressure of about 2 bars. The gelling commenced in the sterilizer, and upon transferring the initially gelled material to open containers, the gelling was completed at ambient temperatures. Again the product was homogenous and of a well organized gelled consistency.

Improved Drying of Albumin Extenders

A starch is added by *D.B. Hosaka; U.S. Patent 4,297,382; October 27, 1981; assigned to Stauffer Chemical Company* to protein compositions used as extenders for egg albumen.

The drying of solutions of derived protein-containing compositions and particularly mineral-containing deproteinized whey by-products and additives such as gelatin and sodium hexametaphosphate can be improved by mixing starch, preferably, a thin-boiling modified starch, with the solution prior to drying. The dried product exhibits reduced hygroscopicity as well as improved physical and chemical properties.

These dried products can be used broadly as food additive agents, i.e., flavor enhancers. The products are particularly adapted for use as egg albumen extenders in soft meringues.

The derived protein-containing composition can be prepared from legumes, oil bearing seeds, milk or milk derived products. The derived protein-containing compositions are usually by-products of a previous procedure used to extract an ingredient from the main source.

The derived protein-containing composition is preferably obtained from a dairy source, i.e., milk and milk derived products.

The molecular weight for substantially all matter in the derived protein-containing composition is less than 30,000 and preferably less than 20,000. A material which has been ultrafiltered through a membrane having a molecular cut-off of 20,000 is considered less than 20,000.

The total Kjeldahl nitrogen (TKN) content in the derived protein-containing composition preferably ranges from about 1.1% to about 2.1% providing a total Kjeldahl protein content of from about 7% to about 13%.

The dried products of this process are preferably based on certain deproteinized whey by-product solutions. As used herein, the term "whey by-products" is intended to encompass the low molecular weight second fraction obtained from the molecular sieve fractionation of whey, the permeate obtained from the ultra-filtration concentration of protein from whey, and delactosed permeate.

The gelatin used can be either of the alkaline or preferably the acid prepared type. Gelatins ranging in Bloom strength preferably from 200 to 250 Bloom can be used. Preferably, "cold-water-dispersible" gelatin is used.

The water-soluble polyphosphates are medium chain length sequestering agent polyphosphates. Representative compositions within this group are sodium or potassium tripolyphosphate, sodium or potassium tetrapolyphosphate, sodium or potassium hexametaphosphate, the more preferred being sodium hexameta-phosphate (SHMP) with an average chain length 6-18, and the most preferred 9-12.

The gums include any of the edible gums or protective colloids. The preferred gum is carrageenan which is used in amounts ranging from 0.5% up to and including about 3% (exclusive of starch).

The starch can be any starch or blends thereof, modified or unmodified which is water-soluble (swellable) under the conditions of drying and which provides a whey by-product/starch solution of a viscosity at a level sufficient to be dried in the particular drying apparatus used.

The starch can be derived from cereal grains, i.e., corn, waxy corn, wheat, sorghum, rice; tubers or roots of such plants as cassava (tapioca), potato or arrowroot and the pith from the sago palm. For the atomization drying, starches with an amylose content of below about 25% are preferred. Tapioca starch is most preferred since it has both low amylose content and the high molecular weight amylose.

In general, the starch is used in an amount ranging from about 25% to about 50%, the percentages being based on the dry weight of the protein component and optional gelatin, polyphosphate, gum and whipping aid. For instance, a film forming starch can be used effectively at 50% but not at 25% whereas a tapioca dextrin can be used effectively at 25%.

In general, the blend of whey by-product, gelatin and/or phosphate or gum can be directly blended with the starch and spray dried. The use of slightly elevated temperature assists in the solubilization of the starch, and, if used, the gelatin. The temperature used in this stage should not cause gelatinization or thickening of the starch prior to spray drying. A temperature within the range of from about 60° and 70°C is suggested for tapioca dextrins and for gelatin.

Example: 37.85 ℓ of delactosed permeate (DLP) (31-34% TS) is blended with 15 ml of catalase to destroy any peroxide preservative. 54.25 ℓ of water is then added to the DLP and the mixture is heated to 65.6°C. A dry mixture of 9.08 kg starch (tapioca dextrin, DE less than 30, K-4484, National Starch and Chemical Corp.), 3.405 kg sodium hexametaphosphate (SHMP) and 0.68 kg gelatin is prepared by dry blending. The dry blend is added to the diluted DLP to provide a solution about 25% total solids and spray dried. The product spray dries well with scant build-up on the drier walls. 27.9 kg of a dense slightly hygroscopic powder is obtained. The ratio of DLP/SHMP/Gelatin to starch is about 2.1.

FOR MILK PROTEINS

Yeast Proteins for Baked Goods

Particular attention has been directed to the use of single-cell protein materials, such as yeast, as a replacer for egg solids and nonfat dry milk (NFDM). For example, in the bakery industry, 2 to 3% nonfat dry milk is normally used as an additive to improve the physical and nutritional quality of bread. However, in view of the increasing cost and decreasing availability of milk, many bakers are looking for a substitute. Although certain products derived from soy protein have gained some acceptance, the active search by food technologists for a suitable substitute for milk in food products continues.

K.C. Chao, J.A. Ridgway, Jr., P.G. Schnell and J.H. Pearce; U.S. Patent 4,178,391; December 11, 1979; assigned to Standard Oil Company (Indiana) have used a process for treating protein materials such as single-cell protein material, plant protein material, whey solids, or mixtures thereof in a manner whereby their color, flavor, nutritional value, and functional properties are improved for food use.

Where a mixture is being used, the amount of the single-cell protein component can vary from about 1 to about 99%. Moreover, the aqueous slurry can be treated with a basic compound, preferably a calcium compound, and fortified with an amino acid such as methionine or cystine. An aqueous slurry of the protein material is prepared and heated to a temperature of from about 75° to about 100°C and the pH of the heated protein material is adjusted within the range of about 7.2 to about 7.6, by adding a pH adjusting compound. The pH adjusting compound can be selected from among the group consisting of anhydrous ammonia, ammonium hydroxide, calcium hydroxide, sodium hydroxide, sodium bicarbonate, calcium sulfate, potassium carbonate, potassium bicarbonate, calcium carbonate, sodium carbonate, potassium hydroxide, magnesium hydroxide, and mixtures thereof, especially mixtures of calcium hydroxide and calcium carbonate or calcium sulfate.

Additionally, the pH adjustment can be accompanied by the agitation and oxidation of the single-cell protein. The pH adjusted solution is maintained at temperature for a period of about 1 to about 120 minutes and then dried. Alternatively, the pH adjusted slurry may be separated into (1) a protein extract and (2) a base-treated protein material, particularly with a basic calcium compound, wherein the base-treated protein material is removed, water washed and dried with or without the addition of amino acids. The protein extract can be heated to an increased concentration and dried for use as a seasoning ingredient.

This process can be applied to microbial cell material produced by various bacteria, yeasts and fungi, but *Candida utilis, Saccharomyces cerevisiae, Saccharomyces fragilis,* or *Saccharomyces carlsbergensis* are suggested single-cell starting component materials for this process, because each is approved by the U.S. Food and Drug Administration as suitable for use in food products.

The plant protein material is advantageously selected from oil seed protein materials such as soy flour, defatted soy flour, soy flakes, soy protein isolates and concentrates, cottonseed flour, cottonseed protein isolates and concentrates, peanut flour, peanut protein isolates and concentrates, sesame seed flour, sesame seed protein isolates and concentrates, corn grits, corn protein isolates and concentrates, gluten, cereal protein isolates and concentrates, rapeseed flour and rapeseed protein isolates and concentrates.

The whey material can be whey solids in the form of an aqueous solution, condensed suspension of crystals, or a dried powder. The whey may be derived from the processing of Cheddar, brick, Edam, Parmesan, Gouda, Emmenthaler (Swiss), or other cheeses.

Example: The following three testing samples were prepared from a 10% solids yeast cell slurry:

(a) untreated spray-dried cells

(b) heated at 95°C and pH 5.9 for 30 minutes and then spray-dried

(c) heated 95°C and pH 7.5 (0.88 g NaOH/100 g dry cell) for 30 minutes and then spray dried.

The samples were submitted for bread-baking test. The best result, as it is comparable to that of NFDM, is from the sample (c) prepared by heating at pH 7.5. The most significant improvement is in its dough handling. When used as a 2% additive in dough, sample (c) and the control containing NFDM were good in mixer, rounding and molding and good in the oven. However, samples (a) and (b) were sticky and stringy in the mixer and flat in the oven.

Untreated yeast cells have a high content of thiol groups. Soluble compounds such as glutathione and cystine, as well as the thiol group in the water-soluble protein are active materials which will weaken the gluten structure by the sulf-hydryl-disulfide interchange reaction during the dough mixing and proofing. The thiol group is readily oxidized, especially under heating at increased pH with trace amounts of metal ions. Analytical determination of thiol in yeast cells treated as above showed a loss of —SH groups of 0%, 30.5% and 61.0% for samples (a), (b) and (c) respectively.

Lactalbumin-Whey-SHMP Product in Cheese

N. Melachouris, B.B. Fracaroli and C.R. Corbett; U.S. Patent 4,163,069; July 31, 1979; assigned to Stauffer Chemical Company have provided a nonfat dry milk substitute product for use in cheese products which is low in cost and does not adversely affect the properties of the cheese.

This nonfat dry milk substitute comprises lactalbumin, a cheese texture improving amount of sodium hexametaphosphate and a modified whey solids product. The whey may be derived from a process selected from the group consisting of:

> (a) adding a divalent metal ion to a raw whey feed and adjusting the pH to a value above about 6 at a temperature below 140°F thereby causing precipitation of the modified whey solids product, and

> (b) adjusting the pH of a raw whey feed containing at least 20% acid whey to a value between 6.5 and 8, thereby causing the precipitation of the modified whey solids product.

The protein content of the nonfat dry milk substitute product is from about 17% to about 30%, (by weight, dry basis).

It is preferred that the nonfat dry milk substitute product contain about 9% to about 15% by weight, dry basis, lactalbumin.

A preferred nonfat dry milk substitute product contains a weight ratio (dry basis) of modified whey solids to lactalbumin of from 2:1 to 5:1.

A cheese texture improving amount of sodium hexametaphosphate (SHMP) is incorporated in the nonfat dry milk substitute product. The weight ratio (dry basis) of sodium hexametaphosphate to lactalbumin is preferably from about 0.05:1 to about 0.20:1.

Optionally the nonfat dry milk substitute product has contained therein dried sweet whey. The dried sweet whey is preferably present in a weight ratio (dry basis) of dried sweet whey to lactalbumin of from about 2:1 to about 5:1. It is particularly preferred that the quantity of dried sweet whey in the nonfat dry milk substitute product be about the same as the quantity of modified whey solids.

Additionally the nonfat dry milk substitute product can have contained therein a dried deproteinized whey. Preferably the dried deproteinized whey is present in a weight ratio (dry basis) of dried deproteinized whey to lactalbumin of from about 2:1 to about 4:1. It is particularly preferred that the weight ratio be about 3:1.

Optionally the nonfat dry milk substitute product may have contained therein a delactosed-deproteinized whey. Preferably this whey is present in a weight ratio (dry basis) of delactosed-deproteinized whey to lactalbumin of from about 3:1 to about 4:1. More particularly it is preferred that the weight ratio be about 3.5:1.

Preferably the nonfat dry milk substitute product completely replaces or replaces a substantial portion of the expensive nonfat dry milk or conventional whey products in the cheese products. Particularly preferred cheese product compositions have been formulated wherein at least 75%, by weight (dry basis) of the nonfat dry milk in the cheese product has been replaced by the nonfat dry milk substitute product of this process.

The following nonfat milk substitutes were prepared according to this process.

Example 1: A composition containing 20.6% protein was prepared from 9.8%

lactalbumin, 44.6% modified whey solids, and 1.0% SHMP and 44.6% dried sweet whey.

Example 2: A composition containing 20.0% protein was prepared from 14.8% lactalbumin, 39.6% modified whey solids, 1.0% SHMP and 44.6% dried deproteinized whey.

Example 3: A composition containing 21.8% protein was prepared from 10.9% lactalbumin, 49.5% modified whey solids, 1.0% SHMP and 38.6% delactosed-deproteinized whey.

Example 4: Cheese products were prepared from 61.0% raw cheese (mix of 20% aged and 80% new), 25.4% water, 6.0% dried sweet whey, 0.5% salt, 3.0% disodium phosphate emulsifier, and 0.1% potassium sorbate. The control cheese also contained 4.0% nonfat dry milk, while the test products contained a mixture of 1% nonfat dry milk and 3% of a substitute as prepared in the above examples.

The cheese, emulsifiers, dried sweet whey, nonfat dry milk, salt and potassium sorbate were blended in the bowl of a Brabender Plastograph which had been preheated to 185°F (85°C). Approximately ⅓ of the cheese was added, then the dry ingredients. The remaining cheese was added and then the water. About 5% of the water was added by steam being applied throughout the entire process by a distillation flask. Cheese temperature at the end of the processing was approximately 180°F (82°C).

The mixing was at 75 envelopes per minute for about 5 minutes. The cheese was cooled and stored in the refrigerator overnight. The next day, the cheese was allowed to warm to room temperature and evaluated for melt, moisture, pH, melt character, hardness, brittleness, adhesiveness, cohesiveness and flavor.

Evaluations showed that cheese products containing the nonfat dry milk substitute products of Examples 1, 2 and 3 were equal to or similar in functionality to the control containing only the nonfat dry milk. Taste was substantially the same for each cheese product sample.

When cheeses were prepared with the additives of Examples 1 and 2 without the addition of SHMP, the cheeses were inferior in texture and body.

Water-Soluble Vegetable Protein Aggregates

A process for increasing the water solubility of a vegetable protein has been developed by *P.A. Howard, M.F. Campbell and D.T. Zollinger; U.S. Patent 4,234,620; November 18, 1980; assigned to A.E. Staley Manufacturing Company*. This process comprises the steps of:

> (A) supplying an aqueous vegetable seed feed stream to a homogenizer with the feed stream containing on a dry solids basis at least 30% by weight vegetable seed protein and a sufficient amount of base to maintain the feed stream within the homogenizer at a pH between about 6.5 to 9.0;

> (B) increasing the water solubility of the vegetable seed protein by subjecting the aqueous feed stream in the homog-

enizer to successive pressure and cavitation cycling at a temperature between about 50°C to about 150°C; and

(C) recovering the vegetable protein product having an improved water solubility.

The successive and cavitation cycling in the centrifugal homogenizer is typically achieved by a rotor (e.g., a conical impeller, rotating rings, etc.) which accelerates the aqueous stream past a stator or stationary rings. The rotor and stationary rings are typically comprised of a series of projections or depressions such as teeth, holes, pins and the like, operatively arranged at a relatively close tolerance (e.g., 0.25 to 1.3 mm). The successive pressure and cavitation cycling causes rupturing, molecular rearrangement and aggregation of the aqueous suspended dry solids constituents therein. Steam is normally injected into the feed port to heat the aqueous feed to the proper processing temperature and to assist in the cavitation cycling by steam condensation.

The low NSI concentrate used in the following example is Procon 2000, A.E. Staley Manufacturing Company, Protein Division having 6.0% moisture, 71.5% protein (moisture-free basis), 0.3% fat (ether extraction), 3.5% crude fiber, 5.3% ash, 17.7% carbohydrates (by difference), a 6.8 pH and NSI of 8, water absorption of 3.0-3.5 to 1 and oil absorption of 1 to 1.

Example: A low NSI protein concentrate was processed into a 70 NSI product. The centrifugal homogenizer employed in this example was a Supraton Model 200 Series (Supraton F.J. Zucker KG, Germany) equipped as a Model 247.05 with a fine grinding head, and inlet pipe fitted with a steam injection unit for temperature control and a discharge pipe (4 ft) having a terminal control ball valve for back-pressure regulation with internally positioned pressure and temperature gauges. The inlet pipe to the centrifugal homogenizer was connected to a mixing vessel for slurry make-up and pH adjustment. The discharge pipe was connected to a neutralizing mixing vessel for pH adjustment and then spray-dried.

In this example, an aqueous feed slurry was prepared by uniformly admixing together in the mixing vessel 1000 parts by weight Procon, 7,000 parts by weight water and 6 parts by weight sodium hydroxide (dsb). The aqueous feed slurry was pumped to the centrifugal homogenizer at a flow rate of 5 gal/min with the steam injection unit being adjusted to 20-40 psig steam pressure. The centrifugal homogenizer was operated at 6,150 rpm and 0.9 mm clearance. The back-pressure in the discharge pipe was maintained at about 30 psig and the temperature to 104°C. The discharge product (pH 7.8) was neutralized to a pH 6.4 at 71°C with 10 N HCl. The neutralized product was then conducted through a high pressure piston pump operated at 2500 psig to a concurrent-flow spray drier having a capacity (water) of approximately 1,000 lb/hr, equipped with a No. 51 nozzle and a No. 425 flat top core by Spraying Systems, Inc. In the dryer, the inlet air temperature was maintained from 210° to 225°C and the outlet temperature from 92° to 98°C.

The water solubilities of the products are characterized by their NSI and PIS. These tests are applicable to products which form solutions as well as colloidal dispersions (e.g. milk).

The resultant spray-dried soy protein concentrate product (100% particles through a 100 mesh screen) had a 70.1 NSI, a pH 6.7 upon reconstitution with water (5% solids) and contained 67.3% protein (dsb), 4% fiber, 0.372% sodium, 6.5% moisture and 6.0-6.6% ash.

The spray-dried product was employed as a substitute binder for milk protein in a variety of comminuted meat products (e.g., frankfurters, liver sausage, weiners, luncheon meats) containing from about 1 to about 15% spray-dried product. The characteristics of the resultant comminuted meat products were equivalent in quality and workability to the milk protein control formulations. Conventional layer cake dry mixes were prepared by replacing NFDM and a portion of the egg albumin in the formulation with the spray-dried product. The quality of the soy protein recipes was comparable to the control recipes.

Soy Protein Isolates for "Creme" Fillings

There are certain food product applications of protein isolates where a high degree of solubility is not desirable. An example of this is a nonaqueous, "creme" filling of the type normally found in cookies or snacks. These fillings usually comprise a mixture of a milk coprecipitate, shortening and sugar. The filling must be very spreadable, yet retain its softness for a prolonged period of time. The use of a highly soluble protein isolate or one having a relatively high NSI results in the formation of a very hard and brittle filling when used as a replacement for a portion of the milk protein.

I.C. Cho, C.W. Frederiksen and R.A. Hoer; U.S. Patent 4,278,597; July 14, 1981; assigned to Ralston Purina Company have therefore provided a low solubility protein isolate having the ability to function in a nonaqueous filling material.

The low solubility protein isolate of the process having a nitrogen solubility index (NSI) below 20 is produced by: forming an aqueous slurry of an isolated soy protein; controlling the pH of the slurry to between about 4.5 to 5.8 by the addition of a monovalent alkali reagent; heating the slurry to a temperature of between about 170°-240°F; and neutralizing the slurry to a pH of 6.8 to 7.2 in the presence of an alkaline earth cation. The temperature range for heating of the slurry, the type of alkali used to adjust the pH prior to heating, and the stage of the process at which the alkaline earth cation is present in the heated slurry are all critical parameters which collectively provide the protein isolate with the desired degree of insolubility to serve as a partial or complete replacement for milk protein in products that require good softness, smoothness, and spreadability.

Certainly, the application of more heat to the slurry can also insolubilize the protein; however, even if the nitrogen solubility index is reduced to below 20 by the use of more heat, the protein tends to form a "gritty" texture in a nonaqueous creme filling. Likewise, if the alkaline earth is added prior to the application of heat, even when the slurry is heated to within the above temperature range, the addition of alkaline earth cation at this point in the process results in a filling which is also handicapped by a gritty texture.

Example 1: An aqueous slurry of isolated soy protein was formed having a pH of 4.5 and a solids level of 17% by weight. The slurry was maintained at a temperature of 70°F and the pH of the slurry was adjusted to 5.5 by the addition of sodium hydroxide.

Steam was injected into the slurry until the slurry reached a temperature of about 200°F. The slurry was held at the noted temperature for a period of 15 minutes. The heated slurry was homogenized at 2,000 psi and adjusted to a pH of 6.8 by the addition of 1.5% by weight of the total solids of calcium hydroxide. The slurry was spray dried to a moisture level of below about 3% by weight.

Analysis of the spray dried product showed it to have 89.6% protein, 2.49% moisture and an NSI value of 9.7.

To evaluate the effectiveness of the above isolate as a partial replacement for milk coprecipitate in creme type fillings, a sample of the above product was used in preparing such a filling, followed by an examination of the textural properties of the filling material.

The formula used in preparing the nonaqueous creme filling contained (in percent by weight) 14.86% isolated soy protein, 14.86% milk coprecipitate, 34.69% shortening, and 35.59% powdered sugar.

The filling was prepared by forming a creme of the sugar and shortening by mixing for 3 minutes with a paddle type mixer. The temperature of the creme was raised to 80°F, with the isolated soy protein and milk coprecipitate being immediately added, followed by mixing for 3 minutes.

The filling was evaluated subjectively for spreadability and was determined to spread well. The filling had an overall satisfactory color and appearance with a smooth mouthfeel. The filling also had a softness which corresponds to a filling prepared with milk coprecipitate as the only proteinaceous ingredient. On a subjective evaluation scale of 1 to 3 with 1 being the very best and 3 having the poorest properties, the filling containing the isolated soy protein had a rating of 1.

Example 2: For comparison, 400 lb of isolated soy protein having a pH of about 4.5 was formed into a slurry having 15% solids. The pH of the slurry was adjusted to 6.1 by the addition of a 50% solution of sodium hydroxide.

The slurry was heated by steam injection to a temperature of 305°F, followed by neutralization of the heated slurry with calcium hydroxide to a pH of 6.8. The neutralized slurry was spray dried to a powder containing 89.4% protein, 3.69% moisture and with an NSI of 70.9.

To evaluate the effectiveness of the above product in a nonaqueous creme type filling, such a product was prepared as set forth in Example 1. The filling was rated as 3 for textural properties.

It may be seen that when the temperature range for heating of the slurry prior to neutralization is exceeded that the isolate obtained has poor textural properties when employed in a nonaqueous creme filling.

Modified Rennet Casein in Puddings and Ice Cream

Rennet casein is greatly preferred to acid-precipitated casein for many uses because of the bland flavor of rennet casein. However, rennet casein has poor water holding properties and it has not been possible to use rennet casein as a replacement for nonfat milk solids in many food products requiring water holding properties.

A method for treatment of rennet casein to provide a casein product with desirable water binding properties has been developed by *P.F. Davis; U.S. Patent 4,213,896; July 22, 1980; assigned to Kraft, Inc.* This method produces a casein product which can provide a milk replacement for use in various food products.

In general, the method includes the following steps. Rennet casein is added to water to provide a casein dispersion. An orthophosphate salt of a monovalent cation is added to the casein dispersion and the casein dispersion is agitated to solubilize casein. A magnesium salt is added to the casein solution and a rennet casein product is then recovered by any suitable method, such as spray drying.

The rennet casein is added to the water at a level sufficient to provide from about 1 to about 22% by weight of rennet casein in the solution. The orthophosphate salt is added to the rennet casein dispersion and the dispersion is agitated. This generally takes from about 5 to about 20 minutes to hydrate the rennet casein and to solubilize the rennet casein and provide a casein solution.

It is known to use complex phosphates to solubilize casein. Complex phosphates cannot be used in the method since they produce casein solutions with high viscosity at the levels of casein used. These casein solutions tend to gel at ambient temperature and require the continuous heating to maintain a fluid solution. The use of an orthophosphate salt of a monovalent cation in this process provides a rennet casein solution with relatively low viscosity which remains liquid at ambient temperature. Rennet casein which has been solubilized with complex phosphates is sensitive to magnesium ion. Insolubilization of casein solubilized with complex phosphates begins with the addition of small quantities of magnesium and proceeds rapidly as further quantities are added. The curd formed is not suitable for purposes of the process.

The preferred orthophosphate salts are monosodium phosphate, disodium phosphate, trisodium phosphate, potassium phosphate and ammonium phosphate and particularly preferred orthophosphate salts are disodium phosphate and trisodium phosphate.

After forming the casein solution by addition of an orthophosphate salt, the casein is treated by the addition of a magnesium salt. The use of a magnesium salt results in a rennet casein product with unique properties. The magnesium salt is added at a level sufficient to provide from about 0.6 to about 1.9% by weight of magnesium ion based on the weight of the casein. A preferred magnesium salt is magnesium chloride. Other suitable magnesium salts are magnesium acetate and magnesium lactate.

After addition of the magnesium ion the rennet casein product may be recovered from the solution by any suitable method or may be used in its solubilized form.

Example 1:   100 kg of rennet casein is added to 400 kg of water which is at a temperature of 74°C. 10 kg of anhydrous disodium phosphate is added to the rennet casein mixture and the mixture is agitated for 15 minutes to hydrate and solubilize the rennet casein. 8.4 kg of magnesium chloride decahydrate are added to the solubilized casein to provide a treated rennet casein solution. Thereafter, the heated rennet casein is recovered by spray drying.

A pudding mix is prepared containing (in weight percent): 18.9% sugar, 15.2%

whole eggs, 6.7% nonfat dry milk, 4.8% cornstarch, 0.25% salt, and 54.15% water.

Pudding is prepared using the above formulation containing the nonfat milk solids and a second formulation wherein 3% of the treated rennet casein described above is substituted for the nonfat milk solids. The two pudding samples are tasted by a panel of experts and no flavor variation between the two samples is detected. A third pudding formulation containing sodium caseinate from acid precipitated casein is prepared. The pudding formulation containing the treated rennet casein prepared in accordance with this example has a flavor preferred to the pudding formulation containing sodium caseinate from acid precipitated casein.

Example 2: An ice cream mixture is prepared containing (in weight percent) 25.00% cream (40% butterfat), 14.2% condensed skim milk (30% solids), 4.86% sweet whey (96% solids), 17.91% liquid sugar (67% solids), 6.25% liquid corn syrup (80% solids), 0.30% stabilizer/emulsifier, and water qs 100%.

Ice cream is prepared using the above formulation containing nonfat milk solids from condensed skim milk and sweet whey solids. Other formulations wherein 2-10% of the dried rennet casein product of Example 1 and 0-8% of a dried neutralized whey is substituted for the condensed skim milk and sweet whey prepared. Water is added to compensate if required. The ice cream samples are tasted by a panel of experts and no flavor variation between the samples can be detected.

Whey-Sodium Caseinate in Caramels

Caramels are prepared by caramelizing sugar in the presence of milk solids. Milk protein is a major contributor to the texture, body and flavor of the caramel. The browning reaction which takes place during the manufacture of the caramel arises from a reaction between the milk protein and reducing sugars during the cooking of the caramel mix. The casein in the milk protein contributes body to the caramel.

A milk replacer for caramels is provided by *D.J. Chirafisi and N. Melachouris; U.S. Patent 4,269,864; May 26, 1981; assigned to Stauffer Chemical Company*. This replacer comprises 13.75 to 17.25% whey protein from a whey protein concentrate, 35-45% dry whey solids, and 8-12% sodium caseinate. The milk replacer is an effective dry substitute for the milk solids normally used in preparing high quality caramels. Caramels of equivalent quality are provided.

In a typical caramel containing 45.95% sweetened condensed milk, 39.9% corn syrup of dextrose, 9.2% fat and 5.47% sucrose, of which 12.87% of the total recipe is the milk solids from the sweetened condensed milk, the milk solids can be replaced with the milk replacer of the process. The sucrose which is present in the sweetened condensed milk in an amount of 42% of the sweetened condensed milk or 19.3% based on the total recipe is added to the total amount of sucrose which is needed in the recipe. The 30% water of the sweetened condensed milk or 13.97% based on the total recipe is added as an ingredient in the caramel manufacture.

In this example, the final caramel recipe would comprise 12.87% of the milk replacer of the process, 39.39% dextrose, 9.2% fat, 24.77% sucrose and 13.9%

water. The milk replacer is generally utilized to replace the total milk solids of a caramel recipe on a gram for gram basis. In general, this means that the milk replacer is utilized in an amount within the range of from about 10% to about 14% by weight based on the total weight of the caramel recipe, the variation in percentage relating to the caramel recipe itself.

It has also been found desirable to include with the milk replacer formulation of the process from 1-5% by weight of an antifoaming agent in the form of a food grade emulsifier. This additive reduces the foaming which has been encountered in using the milk replacer in continuous as well as batch manufacturing techniques. Such food grade emulsifiers are typically mono- and diglycerides, propylene glycol fatty acid esters, polyglycol fatty acid esters, sorbitan monostearate, poly-oxyethylene sorbitan fatty acids such as a polyoxyethylene sorbitan mono-stearate, sodium stearoyl-2-lactylate, dioctyl sodium sulfosuccinate and the like and mixtures thereof. The preferred emulsifiers are the mono- and diglycerides of fatty acid esters.

In preparing caramels with whey protein, two different types of foaming are encountered. Foaming during the cooking stage can be overcome by adding anti-foaming agent as described. The second type occurs at the start of cooking if the kettle temperature is raised rapidly as is common in commercial manufacturing which uses high-pressure steam to heat the kettles. This foaming can be overcome by blending the milk replacer with the water and preheating the mixture to about 55° to 72°C. Remaining ingredients can then be added and the mix heated normally.

Example 1: A laboratory duplication of a commercial (control) caramel was prepared from 5.187 kg of milk solids, 0.0415 kg of dried whey, 2.043 kg of glucose, 0.478 kg of fat, 1.29 kg of sucrose, and 0.72 kg of water. The milk solids, dried whey, part of the sucrose and the water are added as part of a condensed milk comprising 28% milk solids and dried whey, 42% sucrose, and 30% water.

The ratio of the milk solids and dried whey in the sweetened condensed milk is the same as in the kettle recipe above. The caramel was prepared by mixing all ingredients and homogenizing at 154.84 kg/cm². Then the mix was cooked to 117°C in 34 min in a steam-jacketed kettle equipped with a scraper. The scraper speed was set on a medium speed sufficient to prevent buildup on the kettle wall. The cook appeared normal and provided an excellent caramel.

Example 2: Caramels were prepared by a commercial caramel manufacturer using the recipe of Example 1, replacing the 28% milk solids with an equivalent weight of the composition containing 47% whey protein concentrate (31% protein), 40% dried sweet whey, 10% sodium caseinate, and 3% of a mono- and diglycerides emulsifier, Atmul 84, Atlas Chemical Industries. The manufacturer ran a sample in preparing a caramel cook in his kettle. After blending, a steam line having 49.94 kg pressure was opened to heat the kettle. The mix foamed out.

The example was repeated. The milk replacer was blended with the water and heated to 60°C. All other ingredients were added and the steam line turned on. No initial foaming was noted. The caramel cook proceeded normally.

Vegetable Protein in Leavened Baked Goods

In spite of the extensive use of nonfat dry milk substitutes containing vegetable protein, a continuing need exists for a nonfat dry milk replacer employing materials such as a vegetable protein isolate which is widely suitable for baking applications particularly in the preparation of leavened baked products. It is also necessary that such a replacer not only have the same desirable baking characteristics as nonfat dry milk solids, but that no characteristic flavor of the soy protein is carried through to the baked goods.

A nonfat milk replacer for baking containing an isolated vegetable protein ingredient is disclosed by *I.C. Cho; U.S. Patent 4,279,939; July 21, 1981; assigned to Ralston Purina Company.*

This nonfat dry milk replacer is highly suitable in the production of leavened baked goods, especially chemically leavened baked goods wherein the resultant texture of these baked goods is considered to be as acceptable as those obtained using nonfat dry milk solids.

The milk replacer is produced by forming a slurry containing 94-48% by weight dairy whey, 6-52% isolated vegetable protein, the slurry having a pH of 5.8-7.5 and an added alkaline earth cation concentration of 0.1 to 2.0% by weight of the solids. The slurry is heated within a critically defined temperature range of 190°-230°F for a period of time sufficient to partially insolubilize the protein, followed by cooling and drying of the slurry to form a nonfat milk replacer which has comparable baking properties to nonfat dry milk.

Example: 501 lb of dried dairy whey having a solids level of 95% by weight and 137 lb of dried isolated soy protein also with a solids level of 95% by weight were added to 1,134 lb of water which had been preheated to 130°F. The slurry had a solids content of 36.0% by weight. The slurry was agitated while the pH was adjusted to 6.4 using a 20% slurry of calcium hydroxide. The added calcium content of the slurry was estimated to be about 0.35% by weight of the solids. The slurry was homogenized at 2,500 psi and divided into five separate batches. Each of the batches which were identified with the numbers 1-5, were then heated with a jet cooker to different temperatures and were held at these different temperatures for 7 seconds. The temperatures used were: Batch #1, 170°F; #2, 190°F; #3, 210°F; #4, 230°F; and #5, 270°F.

Each batch was then discharged into a vacuum chamber which was at a negative pressure equivalent to 20 inches of mercury during which the slurry was cooled to a temperature of 150°F. The pH of each batch was adjusted to 6.8 with 50% solution of sodium hydroxide. The slurry in each batch was homogenized at 2,500 psi and spray dried at an exhaust temperature of 220°F.

The spray dried skim milk replacers from each of the five batches were evaluated in the production of layer cakes in accordance with the following formulation and baking procedure. The batter formulation contained 521.1 g of cake mix, 99.0 g of liquid egg whites, 18.9 g of one of the above milk replacers and 320 g water (70°F).

All of the above ingredients including water at 70°F were placed in a 3 quart mixing bowl and mixed with a Hamilton Beach Mixer Model #C-100 for one minute at the No. 2 speed. After this the mixing was continued for an additional two minutes at the No. 7 speed. The batters formed from each of the five batches of dried milk replacers and made pursuant to the above formula were measured for any temperature increase since the batter temperature should not have appreciably increased over the temperature of the water used for the batter.

Each batch of batter was used to make 2-3 layer cakes according to the following procedure. 445 g of cake batter was placed in a 9 inch greased round cake pan and baked at 350°F for 33 minutes in an oven. The cakes were then cooled for 20 minutes.

After a period of 2 hours, each of the cooled cakes was evaluated for texture. The textural examination consisted of a visual examination of the surface of the cake for a desirable, rounded contour without visible dips or depressions on the surface of the cake. Each of the cakes was also broken in half to generally evaluate whether or not the cake had any significant horizontal cracking, since a significant amount of cracking would be undesirable.

It was found that cakes produced from milk replacers which were heated to within the critically defined temperature range had the highest volume, the lowest weight and the best visual characteristics. These cakes compared favorably to the control which was prepared with conventional nonfat dry milk. It may be seen that the temperature of heating is an important step in the production of a dried nonfat milk replacer which has comparable baking properties to nonfat dry milk.

ACIDS AND SALTS

SALT ADDITIVES FOR MEATS

Meat Tenderizing Salt Mixtures

R.C. Gooch and J.J. Guenther; U.S. Patent 4,224,349; September 23, 1980 have provided a meat tenderizing composition comprising an aqueous solution of two or more chloride salts, namely KCl, NaCl, $MgCl_2$, and $CaCl_2$ in sufficient but controlled concentrations and amounts to measurably improve the tenderness of the meat without otherwise affecting the meat in any deleterious manner.

The tendering solutions are prepared simply by dissolving the appropriate amount or weight of potassium chloride, sodium chloride, magnesium chloride or calcium chloride, in water so that the concentration of each of the compounds in the solution is of the molarity desired with respect to the total volume of the solution. Either distilled water or tap water may be used as the solvent. The tendering solutions may be prepared immediately prior to use, or prepared in advance and strored at room temeprature or in a refrigerator. Temperature control is not a critical factor and the tendering solutions require no technical monitoring by skilled personnel, as is necessary with some of the tenderizers containing proteolytic enzymes.

The tendering solutions may be introduced or injected into the meat by any suitable means, such as through multineedle injection devices equipped with multiapertured needles. Injection pressure does not appear to be a critical element, but rather the objective is to obtain a uniform distribution of the tendering mixture in the meat tissue. The tendering solutions are harmless to plant personnel, create no particular meat handling problems for the packers, retailer, or consumer, and are easily controlled as they do not necessarily need to be altered according to the breed, sex, age or grade of the animal from which the meat is obtained, as is usually the case when proteolytic enzymes are used as tenderizers. The tendering solutions may be injected into the meat in varying amounts. Injection amounts ranging from 2 to 10% of the weight of the meat to be treated have been tested. With the 2% injection distribution of the solution occasionally is a problem, although the treatment is effective. Injections over 10% are not readily accepted by current USDA regulations. About 3% of the solution based on the weight of the meat to be injected are preferred and excellent results have been obtained with this procedure.

In numerous tests, the concentration of the potassium chloride or the sodium chloride when either was used individually or in combination with other of the aforementioned compounds was varied from about 0.3 molar to about 2.0 molar, while the concentration of the magnesium chloride or the calcium chloride when used individually or in combination with other of the aforementioned compounds was varied from about 0.05 molar to about 0.6 molar. Although the exact mechanism is imperfectly understood, test results suggest that when magnesium chloride or calcium chloride is combined with potassium chloride or sodium chloride in these tendering solutions, a synergistic effect is attained which lowers the molar concentration required for each ion, particularly the divalent cation to achieve the desired tenderization.

Example 1: The right and left eye of the round muscles were removed from a freshly slaughtered "D bone" cow carcass. Each of the muscles was divided into two equal sections, dorsal and ventral, giving a total of 4 sections. Using a table of random digits, one of 4 treatments was assigned at random to each of the above muscle sections. Section No. 1, left dorsal, was injected with 3% of its original weight of an aqueous solution containing 0.6 M KCl. Section No. 2, left ventral, was injected with 3% of its original weight of an aqueous solution containing 0.6 M KCl and 0.12 M $MgCl_2$. Section No. 3, right dorsal, was untreated and served as the control. Section No. 4, right ventral, was injected with 3% of its original weight of an aqueous solution containing 0.12 M $MgCl_2$.

All 4 muscle sections were stored at approximately 38°F for 72 hours. Then, 4 one-inch thick steaks were cut from each section. The steaks were cooked in an oven at 300°F to a center temperature of 160°F. Five ¾ inch cores were removed from each steak and sheared on a Warner-Bratzler shear machine. The average shear value for the 0.6 M KCl injected steaks was 0.5% lower than that of the controls. The average shear value for the 0.12 M $MgCl_2$ injected steaks was 36.2% lower than that of the controls. Thus the decrease in the average shear value of the 0.6 M KCl injected steaks added to that of the 0.12 M $MgCl_2$ injected steaks was 36.7% lower than the average shear value of the controls.

However, the average shear value for the 0.6 M KCl and 0.12 M $MgCl_2$ injected steaks was 45.1% lower than that of the controls. Accordingly, the additive value, 36.7%, when subtracted from the decrease in average shear value with respect to the controls, of the steaks injected with 0.6 M KCl and 0.12 M $MgCl_2$, 45.1%, resulted in a tendering effect which was 8.4% greater than the additive total of the single chloride salt treatments. A demonstrably clear synergistic effect was thus obtained when these two chloride salts were combined into a single injection.

Example 2: The right and left boneless strip loins were removed from a freshly slaughtered bull. The right side strip loin was injected with 3% of its weight of an aqueous solution containing 0.5 M KCl and 0.5 M M NaCl. The left strip loin was untreated. The loins were stored for 90 hours at 36°-38°F. Then five one-inch thick steaks were cut from each loin and cooked in an oven at 300°F to a center temperature of 160°F. Five ¾ inch cores were taken from each steak and sheared on a Warner-Bratzler shear machine. The treated steaks averaged 31.5% lower in shear value than did the untreated controls.

Electrolyzed Salt Solution to Reduce Shrinkage

Several processes are known which are said to minimize or reverse carcass shrinkage. Among these are processes in which meat is sprayed or fogged with an aqueous solution, or in which meat is dipped into a water bath. In addition, it is well known to use aqueous curing compositions or pickles which assist in reducing meat shrinkage while preserving and flavoring the meat.

According to the process used by *P.W. Rose; U.S. Patent 4,276,313; June 30, 1981,* a meat food product with reduced shrinkage is prepared by treating meat with an edible chloride salt-containing solution which has been electrolyzed by passage of direct current. By reduced shrinkage, it is meant that the meat loses a smaller amount of weight during processing and/or cooking than is usual at these steps. The treatment entails the steps of providing meat and providing an aqueous solution containing an edible chloride salt present at a concentration of at least about 0.2 molal (0.2 mol per 1000 g of liquid water). A direct electric current is passed through the aqueous solution at about 2 to about 20 amperes and until at least about 25 coulombs per liter of solution have passed therethrough to form an electrolyzed solution. The meat and the electrolyzed solution are combined to contact the meat with the solution and form an admixture within about 40 minutes after passage of direct current through the electrolyzed solution has ceased. Contact between the meat supply and electrolyzed solution is maintained for a time period sufficient for the electrolyzed solution to penetrate the meat, enhancing the moisture content thereof and forming moisturized meat. The moisturized meat is then recovered.

The process has been practiced with or without inorganic phosphate enhancement for the meat. A boned meat food product of this process having reduced shrinkage on cooking and which is free from inorganic phosphate enhancement measured as phosphoric acid, gains at least about 10 weight percent of its boned weight when 20 lb of boned meat are heated to an internal temperature of about 150°F and maintained at that temperature for about one hour.

Example: *Manufacture of pork sausage links* — Pork sausages (60 weight percent lean, 40 weight percent fat) were prepared in two batches using equal amounts of total meat, water, sodium chloride, sugar and spice. For one batch of sausages, the usual process was followed, while for the other batch, the present process was used.

Thus, a solution of 25 lb of hot tap water containing sodium chloride (1 lb) and sucrose (72 g) was provided in a stainless steel pan. The pan was used as the anode and a carbon electrode was used as the cathode for passage of a direct electric current at 10 amperes and 15 volts through the solution. This current passage raised the pH value of the solution from about 6.5 to about 9 in about 5 minutes.

The solution, at a temperature of about 110°F, was then combined with 120 lb of super lean pork chunks (9% fat by weight) with agitation. Ice particles (5 lb) were added after take-up of the solution by the pork chunks and the resulting admixture was agitated further. The aqueous solution was taken-up by the meat within about 10 seconds from the time agitation began to form moisturized lean pork chunks. 880 lb of less lean pork chunks (having a higher fat content) were supplied and admixed with the moisturized lean pork chunks to form a second pork chunk admixture which was 60 weight percent lean and 40 weight percent fat.

Usual sausage spices were then added to this second pork chunk admixture to form a spiced pork admixture. Grinding of this spiced pork admixture under usual conditions formed a ground, spiced pork product weighing about 1,030 lb and produced no water run-off from the meat. The ground spiced pork product was placed in collagen sausage casings and the casings bound as usual to form sausage links, with each link weighing about one ounce.

The sausage links were cooked for 10 minutes at 340°-350°F with continual rolling of the individual sausage links. The cooked yield weighed about 68% of the uncooked sausage weight (about 700 lb after cooking). The cooked sausages were straight, showed little if any curvature and air pockets between the casing and meat were not evident.

Ground spiced pork prepared in the usual manner with the same amount of meat, total water, sodium chloride, sugar and spices yielded about 3% less product after grinding (about 1,000 lb), due primarily to water loss. Sausages prepared from this ground pork averaged a cooked yield of about 62 to 57% (about 620 to 570 lb) of the uncooked sausage weight, were bent after cooking and showed air pockets between the casing and meat.

Thus, the process produced two increases in yield by reduction of shrinkage at two stages. First, more salable product was produced per pound of starting materials by a reduction in processing shrinkage. Second, more edible product was produced after cooking by a reduction in cooking shrinkage. The difference in cooking shrinkage between the product of this process and that normally produced (about 80 to about 130 lb) represents a greater weight differential than that due to the weight of aqueous solution used during processing.

It was also found that sausages prepared from ground pork of this process were more tender than sausages made with usually prepared ground pork. This effect was noted during normal taste and chewing tests and by breaking the sausages between the fingers and observing how the sausages snap.

NONSTICK ADDITIVES

Anticaking Dough Conditioner

Fatty acid lactylates such as calcium or sodium stearoyl-2-lactylate have been employed in bread doughs for a number of years to improve the quality. The specifications for sodium stearoyl-2-lactyalte (SSL) are set forth in Food Additive Regulations 21 CFR, Section 121.1211, while similar specifications for calcium stearoyl-2-lactylate (CSL) are found at Section 121.1047 of the same Regulations. SSL is a mixture of sodium salts of stearoyl lactylic acids and minor proportions of other sodium salts of releated acids and is manufactured by the reaction of stearic acid and lactic acid followed by conversion to the sodium salts. CSL is prepared in a similar manner but comprises the calcium salts of stearoyl lactylic acids. In general, the dough conditioning properties of such fatty acid lactylates serve to enhance mixing tolerance of doughs, while increasing loaf volumes, improving the texture of the finished baked goods, and in general producing final products of improved taste and appearance characteristics.

Although the use of SSL and CSL either alone or in combination is well established, a number of problems remain. Specifically, both of these additives have a tendency to lump or agglomerate during storage and shipping thereof, either alone or in a combined additive. This stems from the inherent tendency of the materials to agglomerate, and is aggravated when the materials are subjected to relatively warm temperature and humidity conditions such as those encountered druing shipping or storage in warehouse or in a commercial bakery.

L.F. Marnett and V.D. Barry; U.S. Patent 4,164,593; August 14, 1979; assigned to C.J. Patterson Company have disclosed anticaking dough conditioning agents which combine respective quantities of (1) a sodium salt of at least one of the acyl lactylates of C_{14-22} fatty acids, and (2) a calcium salt of at least one of the acyl lactylates of C_{14-22} fatty acids. These salts are commingled in a molten condition using essentially any feasible means and thereafter cooled to provide a solid which can be ground or otherwise treated by conventional equipment to provide a powdered product which is substantially resistant to caking or agglomeration.

In more detail, the dough conditioning agents should broadly include from about 10 to 85% of the calcium salt, and correspondingly from about 15 to 90% of the sodium salt (the combined weight of sodium salt and calcium salt taken as 100%). In preferred forms the amount of calcium salt should be at least equal to the amount of sodium salt used. For example, the calcium salt may comprise from 50 to 85% by weight (most preferably 65 to 80% by weight), with the fraction of the additive made up of calcium salts and sodium salts taken as 100%. Correspondingly, the sodium salts should be present at a level of from 15 to 50% by weight (most preferably, 20 to 35% by weight) on the same basis.

Example: In order to demonstrate the anticaking properties of the combined salt compositions the following tests were undertaken.

Equal portions (approximately 1 lb each) of seven samples of powdered CSL were blended together to form a master blend of powdered CSL. A similar method was followed to prepare a master blend of powdered SSL. Test samples were compounded from the respective master blends of CSL and SSL to produce seven test blends containing 100%, 85%, 75%, 65%, 50%, 25% and 0% CSL with the remainder being SSL.

Four hundred grams of each test blend was next melted in a water bath maintained at 95°C. As each test blend was melted it was stirred with a glass rod to thoroughly mix and commingle the components without incorporating large amounts of air. After stirring each test blend was poured into a shallow aluminum foil dish to cool. At the point at which the material solidified on the surface, the foil container was placed inside a moisture resistant plastic bag and allowed to cool overnight.

After cooling overnight each test blend was ground in a hand operated food grinder and screened. Care was taken to minimize absorption of ambient humidity and heat of friction. Duplicate samples were prepared from the ground material, and the latter was passed through a 60 mesh screen and retained on a 140 mesh screen. Twenty-five grams of the screened material was then sealed in 4 ounce jars with bakelite lids. Sample jars were placed in a constant temperature cabinet maintained at 38°C.

The storage samples were examined daily for the first week and thereafter twice weekly until it was determined there were no significant changes in the material between examination periods.

A scale of 0 to 10 was used to rate the samples, with 0 corresponding to a completely free-flowing powder and 10 representing a fairly solid mass which did not crumble with moderate pressure.

Blend No. 3 which contained 75% CSL and 25% of SSL showed the best anticaking properties with a flowability rating of 3 after 31 days elapsed storage time.

Condiments Coated with Zinc Salts

The importance of zinc in animal and human nutrition is receiving increased attention. Zinc has been shown to play an important role in taste acuity, enzyme reactions, and other physiological responses such as wound healing. Many modern diets do not provide optimal amounts of zinc to mammals. Supplementation of animal and human diets with zinc is, therefore, very important.

One supplement dietary source of zinc is "sea salt," which is obtained by evaporating natural salt water. However, sea salt is very hygroscopic and inconvenient to handle. Like table salt (sodium chloride), sugar and other granulated or powdered condiments, sea salt tends to "cake" or form large clumps on storage under ambient conditions. In order to keep these condiments free flowing and easy to dispense under humid conditions, anticaking agents such as magnesium silicate or calcium silicate are used. However, such agents add little, if anything, of nutritive value to the condiments and certainly do not provide a dietary source of zinc.

W. Jakinovich, Jr.; U.S. Patent 4,220,667; September 2, 1980; assigned to The Procter & Gamble Company has provided a palatable condiment which will not cake and which also provides the daily requirement of zinc.

This encompasses noncaking, palatable and nutritional compositions, comprising a granular condiment, especially table salt, the individual particles of the condiment being substantially coated with a safe and effective amount of a saliva-insoluble, physiologically acceptable zinc salt in the form of a finely divided powder. A process for preparing the nutritional composition by mixing the finely divided zinc salt with the granular condiment particles is also disclosed.

Many salts of zinc are nutritionally available, but due to other undesirable properties, they are not useful as anticaking agents. Zinc sulfate, for example, is an emetic. Other zinc salts have an astringent effect in the mouth and therefore are unpalatable. In addition, many zinc salts such as zinc sulfate, and the like, are quite hygroscopic and are totally unsuitable for use with free flowing granular condiments such as table salt.

Zinc salts which are saliva-insoluble, nonhygroscopic and physiologically acceptable are used in these compositions. Typical zinc salts with these properties include zinc oxide, zinc carbonate and zinc phosphate. These zinc salts are non-emetic, nonhygroscopic and, because they are insoluble in saliva, are nonastringent and do not have a bitter taste. These salts can be used alone or in combination in the practice of this process. Indeed, zinc oxide reacts with carbon dioxide in the air, and therefore may contain some zinc carbonate.

Condiments which can be kept from caking, and yet act as a vehicle for delivering the nutritional amount of zinc to the animal or human, are preferably those which form a part of a daily diet. Sugar; common table salt; dietetic salts, i.e., mixtures of from about 20% to about 80% sodium chloride and from about 20% to about 80% potassium chloride; and seasoned salts, for example, garlic salt, onion salt and other flavored salts containing mixtures of herbs, spices and flavorings, are especially useful in this process. For aesthetic purposes, those condiments which are white, or which contain white particles or granules, are most preferred, since the saliva-insoluble, nonhygroscopic zinc salts are commonly white.

Example 1: A palatable zinc supplemented table salt is prepared by mixing 3.13 g of zinc oxide and 96.87 g of granular, common table salt. The zinc oxide particles are less than 177 microns in diameter, and table salt granules are greater than 250 microns in diameter. The table salt and zinc oxide are placed in a glass bottle which is stoppered and then shaken vigorously for 2 to 3 minutes to coat the individual table salt particles.

This zinc supplemented salt provides about 25 mg of zinc per gram of sodium chloride. When stored under conditions of 73% relative humidity and 90°F, the salt does not cake and remains free-flowing.

The table salt prepared in Example 1 is sprinkled on cooked eggs. No difference in taste is perceived between the eggs "salted" with the zinc supplemented salt and those flavored with ordinary table salt. No aftertaste is perceived.

When the zinc oxide of Example 1 is replaced with 3.85 g zinc carbonate, similar results are obtained.

When the zinc oxide of Example 1 is replaced with 5.95 g of zinc phosphate, similar results are obtained.

In contrast with the procedure of Example 1, 96.87 g of granular, common table salt and 3.13 g of the finely-powdered zinc oxide are admixed by simply sprinkling the zinc oxide on the sodium chloride to form a heterogeneous mixture. No coating step is used. When stored at 73°F, and 50% relative humidity for one week, the sodium chloride cakes and is no longer free flowing.

Example 2: Sucrose (100 g commercial granulated table sugar) and 6.5 g zinc phosphate are placed in a mechanical mixer and mixed until the zinc phosphate substantially coats the sugar particles. When this mixture is stored under ambient conditions for about one week, no caking is observed.

This mixture will supply about 25 mg of zinc per gram of sugar.

Silicon Dioxide Coated Citric Acids

Dry beverage mixes containing flavor, food acid, color, sucrose and minor amounts of various other additives are well-known articles of commerce. In the production of such articles, it would be desirable to replace part of the sucrose of the mixes with the less-expensive sugar, dextrose. This presents a problem since the dextrose of commerce is the monohydrate containing 8 to 9% water of crystallization. When all or part of the sucrose in beverage mixes is replaced with dextrose hydrate, the mixes tend to cake and lose their free-flowing properties. Color and flavor changes often result.

One way to overcome this problem is that the dextrose hydrate is dried to remove nearly all of the water of crystallization. Beverage mixes are then prepared in which part of the sucrose is replaced with the dried dextrose. Care must be taken to handle the dextrose under controlled relative humidity packaging and it is often necessary to have packets of a drying agent mixed with the dried dextrose to prevent its adsorption of moisture during the mixing process. This method requires the expensive step of drying of the dextrose monohydrate. Furthermore, the hygroscopic dehydrated sugar must be handled under special conditions adding to the cost.

V.S. Velasco; U.S. Patent 4,278,695; July 14, 1981; assigned to CPC International Incorporated disclose a method for preparing a dry, free-flowing beverage mix which comprises coating the particles of a food acid with a desiccating agent and then mixing the coated food acid with a saccharide material which comprises dextrose hydrate.

This process permits the use of dextrose monohydrate as a replacement for from about 5 to about 25% by weight of the saccharide component of the mix. The unexpected discovery is that if the food acid is first coated with a very finely divided desiccating agent, and this coated acid is mixed with the beverage mix containing dextrose monohydrate, severe caking of the mixture and migration of the color is avoided. The ingredients can be mixed in any order as long as the acid is first coated with the desiccating agent.

The food acids can consist of any edible solid food acid such as citric, malic, adipic, tartaric or fumaric acids. The amount of food acid used in the preparation will vary with the desired flavor of the beverage mix. It is usually within the range of 1 to 6% by weight of the total mix.

Minor amounts of desiccating agents, i.e., from about 0.3 to about 2.0% by weight of the beverage mix are employed. Suitable desiccating agents include tricalcium phosphate, magnesium carbonate, sodium silicoaluminate, calcium silicate, silicon dioxide, and a starch hydrolyzate of a dextrose equivalent (DE) less than 20. Sodium silicoaluminate and silicon diioxide with mean particle diameters of less than 100 mμ are the preferred desiccating agents with silicon dioxide being the most preferred desiccating agent.

Example 1: A mixture of 100 g of citric acid and 20 g of desiccating agent (Zeofree 80, a hydrated silicon dioxide from J.M. Huber Corp.) was tumbled manually in a 3.8 ℓ sealed jar for 5 minutes. The resulting mixture was added to 1,602 g of sucrose, 178 g dextrose monohydrate, 15 g color and 8.5 g flavor base in a Patterson-Kelley Twin Shell Dry Blender, Model LB-4226. Mixing was continued for 20 minutes before the solid was stored in glass-stoppered bottles. Mixing operations were carried out at 21°C, at 50% or less relative humidity. Caking tests were carried out on the material stored in glass-stoppered bottles at 32°C. Test results show that dextrose hydrate can be used in beverage mixes without drying if the citric acid present is first treated with a desiccating agent before it is mixed with the major ingredients of the mixture. The above mixture which contained 5% citric acid coated with 1% silicon dioxide, 80% sucrose, 9% dextrose monohydrate, 0.75% color and 4.25% flavor base remained free-flowing after 8 weeks of storage.

Example 2: For comparison, a mixture of 90 g of sucrose, 5 g of dextrose monohydrate and 5 g of citric acid was mixed thoroughly and stored in a glass-stoppered bottle for 12 weeks at 32°C. The mixture became caked in the bottle. The procedure was repeated except that 0.7 g of finely divided tricalcium phosphate was added to the mixture. Severe agglomeration of the particles was observed. This example shows that the addition of the desiccating agent, tricalcium phosphate, to a mixture of sucrose, dextrose monohydrate and citric acid, fails to prevent agglomeration of the particles of the mix.

Silica Modified Yeast Products

At least several weeks will frequently intervene between the time bulk yeast is packaged and when it is used. In common with all living organisms, the yeast cell respires during storage, one of the end products of such respiration being water. The yeast cell membrane permits free passage of water into and out of the cell and, depending upon the relative conditions of osmolarity on either side of the membrane, the extracellular water content of bulk yeast may increase sufficiently under storage conditions to adversely affect the flowability characteristics of the yeast.

One approach to the problem of imparting improved flowability to bulk yeast is to dry the yeast to a lower initial water level, e.g., 65% or below. Although this procedure is fairly effective in terms of providing a satisfactory product, it entails additional energy and labor expenditures and thereby adds to the cost of the yeast. Moreover, a decrease in the water content of the yeast obviously mandates a concomitant increase in the amount of yeast solids in the bulk product.

A free-flowing particulate yeast product is produced by the process developed by *S. Pomper and E. Akerman; U.S. Patent 4,232,045; November 4, 1980; assigned to Standard Brands Incorporated.* The product comprises baker's yeast and a minor amount of a solid, substantially nondeliquescent drying agent having a particle size of less than about 100 mμ dispersed throughout the yeast. The amount of drying agent is sufficient to bind or absorb a portion of the extracellular water without substantially deleteriously affecting the leavening activity of the yeast. The drying agent maintains the flowability of the yeast over extended periods.

A variety of materials may act as "nondeliquescent drying agents" in the composition and method of the process. Submicronized wood pulp, cellulose, fumed or precipated silica and various other silicates have been found to be effective in achieving these objectives.

While the typical particle size of the drying agents will be less than about 100 mμ it is preferred to utilize an agent having a particle size of from about 10 to about 30 mμ.

Typically, prior to contacting or treating the yeast with the drying agent, the yeast will be granulated to average particle size not greater than about 0.25 inch in diameter and preferably will be granulated to a particle size of from about 0.04 to about 0.15 inch in diameter. Granulation may be effected by any suitable means such as grinding, pulverizing, screening, etc. The above-noted particle sizes approximate those attained by passing the yeast through U.S. No. 3, 16 and 6 screens, respectively.

Example 1: This example illustrates the effects of treating bulk yeast with submicronized wood pulp (Solka-Floc 200 BWNF) and a submicronized cellulose product (Avicel pH 101).

Bulk yeast having an average particle size of about 0.05 inch in diameter and a moisture content of 69.5% was treated with the wood pulp at levels of 1, 2 and 3% and with the cellulose product at a level of 2% based on the weight of the bulk yeast. The yeast and the above materials were combined in mason jars and the jars shaken vertically thirty times to distribute the material throughout the yeast.

The treated yeast and the untreated bulk yeast serving as control were subjectively assessed for flowability and tactile impression or "feel." The yeasts treated with both Solka-Floc and Avicel were judged to have markedly improved flowability over the untreated yeast and to feel dry to the touch. The leavening activity of the treated and untreated yeast was determined by means of sweet dough punch tests utilizing freshly prepared yeast and yeast which has been stored for three days at 77°F.

In these tests the leavening activity of the treated yeasts was comparable to that of the untreated yeast.

Example 2: This example illustrates the preparation of a bulk yeast product comprising baker's yeast and a solid, substantially nondeliquescent drying agent. Baker's yeast having a moisture content of from about 68 to about 70% by weight was granulated through a paddle-type mixer to a particle size not greater than about 0.05 inch in diameter. Two percent by weight powdered hydrophilic silicon dioxide (Zeofree 80) was metered onto the granulated yeast as the yeast moved along a screw conveyor which provided a thorough mixing action so that the yeast particles were substantially coated with the silicon dioxide. The bulk yeast product which was judged to have excellent appearance and color was free-flowing and dry to the touch.

LEAVENING AGENTS

Potassium Bromate with Reduced Fire and Explosion Hazards

Although 100% potassium bromate or potassium bromate mixed with small amounts of anticaking additives, such as 5% magnesium carbonate, can be considered a stable and relatively inert product, the fire and explosion hazard increases tremendously as soon as the product is mixed with starch or flour. Fires and explosions have occurred in mills when concentrated potassium bromate was diluted for proper feeding purposes with flour or starch and a flame or spark came into contact with this mixture. In addition, potassium bromate dust is very susceptible to ignition and combustion and this susceptibility is aggrevated by the presence of flour or starch.

It is a principal object of the disclosure by *F.D. Vidal and A.B. Gerrity; U.S. Patent 4,183,972, January 15, 1980; assigned to Pennwalt Corporation* to provide a potassium bromate product for use in the milling and baking industries which has an acceptably reduced fire and explosion hazard.

Accordingly, this composition for improving flour and dough consists essentially of a powdery mixture of potassium bromate with, for each 100 parts by weight of bromate, from, preferably about 30 parts, up to about 125 parts by weight of an inert, edibly acceptable phosphate salt, and an inert edibly acceptable hydrated salt which provides no more than about 50 parts of water per 100 by weight of bromate. Examples of these phosphate salts include the following: $CaHPO_4 \cdot 2H_2O$, $CaHPO_4$, $(NH_4)_3PO_4$, $(NH_4)_2HPO_4$, Na_3PO_4, $Na_2HPO_4 \cdot 7H_2O$, $(NaPO_3)_3$, $NaH_2PO_4 \cdot H_2O$, $(NaPO_3)_6$, $Na_4P_2O_7$, KH_2PO_4, and the like. The useful phosphate salt may or may not be hydrated. The hydrated companion salt includes the edibly acceptable hydrated alkali metal and hydrated alkaline earth metal salts of organic or inorganic acids. Examples of these hydrated salts include the hydrated phosphate salts included above as well as $CaSO_4 \cdot 2H_2O$; $MgSO_4 \cdot 7H_2O$; KNa tartrate$\cdot 4H_2O$; sodium citrate$\cdot 2H_2O$; calcium citrate$\cdot 4H_2O$; $FeCl_3 \cdot 6H_2O$; $CuSO_4 \cdot 5H_2O$, $KAl(SO_4)_2 \cdot 12H_2O$ and the like.

Potassium bromate formulations tested herein were prepared by the following blending method: All ingredients are sifted through a No. 80 mesh sieve before adding to blender. The potassium bromate is mixed with the phosphate and hydrated and hydrated salt by blending in a P-K Twin Shell Blendor for 5 to 10 minutes. The balance of ingredients, if any, is added to the mixer and blended for an additional 20 minutes.

Where potassium bromate is shown herein as potassium bromate (95%), the additional 5% of the composition consists of magnesium carbonate. To ascertain the extent of the problem of the tendency of potassium bromate to ignite when blended with flour and the conditions under which this hazard develops, tests were conducted with the bromate alone and in specified mixtures. The test results showed little difference in hazard between a 95% potassium bromate (5% magnesium carbonate anticaking agent) a 90% potassium bromate and a 50% potassium bromate.

Example: One of the formulations tested comprised 52.6% of 95% potassium bromate, 26.4% of dicalcium phosphate dihydrate, 13.0% calcium sulfate dihydrate, 5.0% magnesium carbonate, 2.5% of tricalcium phosphate and 0.5% Cab-O-Sil. This formulation had negative impact senstivity at 40 inches, a Trauzl value of 10.5 ml, showed no burning or discoloration after 600 seconds on the hot plate test, and had excellent resistance to flame showing only mild decomposition with no charring.

When the above formulation was mixed with equal portions of flour (50-50 weight percent blend), the mixture also showed good reaction to the flame test with only moderate decomposition and a slight black residue.

Calcium Sulfate Modified SALP

Sodium aluminum phosphate (SALP) has several inherent deficiencies, the most serious of which is dusting and hygroscopicity. Sodium aluminum phosphate dust is very light and rapidly permeates the air in food processing plants, creating cleaning and sanitation problems and unsatisfactory working conditions for the employees. Additionally, the finely divided particles of sodium aluminum phosphate do not flow easily.

Sodium aluminum phosphate is also an inherently hygroscopic material which will absorb a large quantity of atmospheric moisture, usually about 28 to 29% by weight. Originally produced, SALP is a dry, white crystalline product. If permitted to stand exposed in a hot, humid atmosphere, it rapidly absorbs moisture, first forming water droplets or caking at the surface, then becoming what may be termed as a viscous semifluid. Commercially, this phenomenon is minimized somewhat by the use of sealed, air-tight containers.

R.E. Benjamin and T.E. Edging; U.S. Patent 4,260,591; April 7, 1981; assigned to Stauffer Chemical Company have developed a sodium aluminum phosphate having improved handling characteristics and useful as a leavening agent. The preparative method comprises contacting a slurry of a complex alkali metal aluminum phosphate with calcium sulfate prior to drying, the drying preferably being accomplished by granulating the calcium treated product while drying under such conditions that a majority of the granulated particles are less than 840 μ (through 20 mesh) and at least 90% less than 2,000 μ. The slurry prior to calcium sulfate addition contains less than 1% free acid. It is theorized that there is provided granulated complex sodium aluminum phosphate granules combined with calcium sulfate, possibly as a coating thereon. This process can also be included as part of the process for preparing potassium-modified SALP.

The process has significant advantages in providing a product which granulates easily and dries rapidly; a longer drying zone (wet zone) in the dryer/granulator is possible thus increasing production rate and less heat is required to dry the product during milling again. This results in lowering the energy consumption. The product does not blind screens during screen analysis. The process also has provided a cleaner reactor which can be a significant factor in the economics of the process.

The products of the process are less dusty and evidence a greater density than a comparable SALP prepared without calcium sulfate and untreated SALP. The products are easier to handle, and bag in smaller bags because of the increased density and decreased dusting.

The compositions of the process are useful as leavening acids in such areas as biscuit mixes, pancake mixes, waffle mixes, cake mixes, doughnut mixes, muffin mixes, self-rising flour and the like. Various other materials can also be added to these compositions to adjust rate, such as monocalcium phosphate.

Mixed Leavening Agents

Refrigerated canned doughs are generally compositions packaged in foil surfaced fiber containers having vent holes or other means to allow gas to escape therefrom. As the dough is proofed in the container, carbon dioxide is generated from the leavening system which expands the dough in the container and drives out the oxygen. The expanded dough seals the container. An internal gas pressure is required to maintain the seal and to keep the oxygen out of the container. The gas pressure must remain after refrigeration to maintain the seal. Failure to maintain the pressure will cause the dough to spoil due to bacteriological action which can spoil the biscuits and, in some instances, cause excessive gas pressure to be generated sufficient to cause the containers to rupture. The dough must also retain sufficient leavening to allow the product to rise when baked.

Sodium acid pyrophosphate (SAPP) has been found to be especially well suited to the needs of preleavened packaged doughs and is widely used for that purpose. The addition of minor amounts of K^+, Ca^{2+}, and Al^{3+} to the SAPP during manufacture permits the controlled retardation of the rate of reaction of the SAPP with the sodium bicarbonate in the baking system. A SAPP/bicarbonate leavening system fulfills the gas generation requirements for canned doughs. However, the so-called "pyro" taste of SAPP generally is considered objectionable.

Another well-known leavening agent in the baking industry is sodium aluminum phosphate (SALP). It finds use in baking powders, self-rising mixes, preleavened pancake flours and mixers, prepared biscuit mixes, and prepared cake mixes. The SALPs of commerce include 1:3:8 SALP and 3:2:8 SALP, where the 1:3:8 and 3:2:8 refer to the $Na:Al:PO_4$ mol ratios of the leavening agent.

R.M. Lauck; U.S. Patent 4,230,730; October 28, 1980; assigned to Stauffer Chemical Company provided a new leavening acid composition which can be used in leavening canned refrigerated dough. This leavening agent comprises a potassium modified 1:3:8 sodium aluminum phosphate or the calcium modified derivatives thereof or mixtures thereof in combination with an alkali metal acid pyrophosphate. Unexpectedly, this leavening acid in combination with sodium bicarbonate produced gas at a rate slower than at least the fastest ingredient and, in many cases, slower than either leavening acid ingredient separately. When using a slow sodium acid pyrophosphate, the combination meets the criteria for leavening acid for canned refrigerated doughs, namely low gas development upon mixing, ability to develop gas to seal the cans, and the ability to maintain pressure in the cans when refrigerated. This biscuit leavening acid combination is relatively insensitive to temperature and can be used at a dough temperature of up to about 26°C (80°F) thus eliminating the need for the extensive refrigeration of the mixing bowl as presently required. Since the leavening systems of this process provide a gas release rate which is slower than at least the fastest, and since 1:3:8 SALP and SAPP are available in various grades ranging from fast to slow, it is possible to provide tailor-made leavening systems to satisfy the particular needs of a baking system.

The alkali metal acid pyrophosphate used can be either sodium or potassium acid pyrophosphate and mixtures thereof. The preferred pyrophosphate is sodium acid pyrophosphate, hereinafter SAPP, prepared by the controlled thermal decomposition of monosodium phosphate. By varying the conditions of humidity and temperature as well as the amount of cations added and the particle size during processing, SAPPs for varying reactivities can be prepared. Most commercially prepared SAPPs contain minor amounts of K^+, Ca^{2+}, and Al^{3+} which permit the controlled retardation of the reaction of SAPP with the $NaHCO_3$ in the baking system.

It is preferred to utilize the potassium modified 1:3:8 SALP prepared from food grade phosphoric acid having a concentration of about 85.0 to about 88.0 weight percent H_3PO_4 which is contacted with a sufficient amount of potassium ion to provide an analysis of about 0.5 to about 1.2 weight percent K_2O in the final product, and a sufficient amount of sodium ion to provide an analysis of from about 2.4 to about 3.2 weight percent of Na_2O in the final product. This mixture is then treated with a sufficient amount of alumina to provide a concentration of from about 15 to about 17% by weight of Al_2O_3 in the final product. The slurry thereby formed is cooled to a temperature within the range of from about 60° to about 75°C. The product is dried and granulated simultaneously.

The potassium modified SALP is used in a ratio to the SAPP within the range of from about 3:1 to about 1:3. Preferably the ratio is within the range of from about 1.5:1 to about 1:1.5. Most preferably for refrigerated canned doughs, the ratio is 1:1. Other ratios may be more effective in other baking applications. Variation in the ratios are also possible depending on such factors as the reaction rate of the potassium modified SALP, and the amount of potassium present, as well as the reaction rate of the SAPP. For example, in doughnuts, a ratio of 50/50 SALP/SAPP using a SAPP with a fast reaction rate has been found to be effective. The ratios are based on the titratable neutralizing value of SAPP and of the potassium modified SALP leavening acids. The neutralizing value, sometimes called neutralizing strength, of a leavening acid represents the number of kilograms of sodium bicarbonate which will be neutralized by 100 kg of the leavening acid.

The leavening acid compositions of the process can be formed by dry blending the potassium modified SALP and the SAPP. This insures the intimate association of the materials. The compositions can also be prepared in situ in the baked goods by blending the potassium modified SALP and the SAPP with the flour present and other ingredients of the baked goods.

The leavening acid compositions can be used to leaven any baked product presently using SALP or SAPP. In particular, the leavening acid system can be used to replace, on a one for one basis, the SAPP presently used in baking applications such as biscuits, doughnuts and the like. The gas producing agent used in the formulations is generally sodium bicarbonate.

Dione-Hydrogen Peroxide Systems

The acidogen system disclosed by *F.L. Metz; U.S. Patent 4,328,115; May 4, 1982; assigned to Mallinckrodt, Incorporated* is used to prepare raised bakery products of fine crumb characteristics.

The acidogen system is a combination of at least one of a selective group of diones and hydrogen peroxide (H_2O_2). The acid necessary for the desired acidification and ensuing protein coagulation or bakery dough leavening is generated by the oxidation of the dione "in situ." Specifically, the diones used are acyclic and cyclic aliphatic compounds containing from 2 to 6 carbon atmos per molecule, and wherein the carbonyl groups are vicinal to each other.

The acidogen system is satisfactorily employed as a substitute for yeast in leavening bakery products. In this embodiment, the system is added to prepared flour mixes in combination with a CO_2-releasing compound, particularly sodium bicarbonate. The leavening ability of the acidogen system is comparable to or improved over yeast. The products prepared possess excellent texture equivalent to yeast-leavened products.

The particular acidogen system preferred is hydrogen peroxide and pyruvaldehyde in combination with sodium bicarbonate, due to the texture, taste and overall appearance of the bakery products obtained. The percentage of each ingredient of the leavening system, based on the weight of flour, may range generally from 0.1 to 3.0% of the vicinal dione component; from 0.3 to 3.0% for the hydrogen peroxide component; and from 1.0 to 3.5% for the sodium bicarbonate. Within these ranges, based on the weight of the flour, 0.5 to 1.5% of vicinal dione, 1.0 to 2.0% of hydrogen peroxide and about 2% of sodium bicarbonate are preferably employed at present to make pizza dough. For biscuit application, from about 0.2 to 1.0% dione, from 0.5 to 0.7% hydrogen peroxide and 1.5 to 2.0% sodium bicarbonate, based on the flour weight, are preferred.

This acidogen system may also be incorporated in powdered baking mixtures so as to provide leavening thereto when the powder is moistened and heated. It, likewise, is satisfactorily incorporated as the leavening agent in doughs which are to be frozen and then subsequently baked as desired. In these and other applications, the leavening system herein is more economical than the fat encapsulated glucono-delta-lactone now used.

This acidogen system is applicable for the preparation of bakery products such as bread, rolls, biscuits, pizza dough and various refrigerator doughs.

Example 1: *Preparation of Biscuit Dough* – Using a KitchenAid table-top mixer, 115 g flour and 3.5 g salt were sifted into the mixer bowl for each biscuit batch prepared. Shortening (33 g) was cut into 4 to 5 pieces and blended into the flour-salt mixture using the No. 1 mixer speed. To 82 g portions of skim milk, the leavening agent components were added. One leavening agent-milk aliquot was added to the flour-shortening mix and blended until the dough started to cling to the beater. The dough was placed onto a lightly floured board and kneaded lightly. It was then rolled to the desired thickness and cut with a 2-inch biscuit cutter. The biscuits were baked at 450°F for 12 minutes. After baking, they were evaluated for degree of leavening (height of biscuit shape measured before and after), texture and evenness of browning.

This evaluation indicated essentially equivalent results as to texture and degree of rising are obtained using the acidogen system of this process in an amount less than half that of the conventional baking powder.

Example 2: Pizza dough was prepared following a general procedure and recipe except that the glucono-delta-lactone leavening agent was substituted by 2.0 g of 30% H_2O_2 and 2.0 g of 40% pyruvaldehyde, per each 100 g of flour.

The baked product was of fine crumb structure with a light tan, smooth crust. It had a bread-like appearance and smell. As the baked thickness was approximately 3 times that of the original unbaked dough, the leavening action of the acidogen was excellent.

OTHER BREAD ADDITIVES

Encapsulated Acetic Acid in Sourdough Bread

The standard procedure for making sourdough bread involves fermentation of a wheat flour-based dough by bacterial organisms which have the ability to produce lactic and acetic acids. This fermentation process is maintained by daily additions of flour and other nutrient materials such as nonfat dry milk and occasionally potato flour. Portions of this "ferment" are then used by adding them to standard French bread or sourdough bread formulae.

The bacterial organisms commonly go through changes which frequently results in lowered production of acetic and lactic acids. The baker, therefore, is faced with considerable difficulty in maintaining a constant flavor level in sourdough breads produced in this fashion on a day-to-day basis.

The flavor of sourdough bread is based on its high acidity due to acetic and lactic acids. The total acidity of sourdough bread is roughly ten times that of conventional bread. Of this acidity, the acetic acid is the more important of the two acids in providing the characteristic sourdough flavor.

A sourdough composition is provided by *W.H. Ziemke and E.F. Glabe; U.S. Patent 4,141,998; February 27, 1979* in the form of a flowable powder for the introduction of encapsulated acetic acid into bread doughs and other bakery products. The composition is stabilized against premature loss of acetic acid by intimate dispersion with a minor quantity of a normally solid edible fatty material.

This is prepared comprising minor quantities of acetic acid, with or without fumaric acid, tartaric acid and/or citric acid, absorbed on a major quantity of a finely divided solid edible absorptive starchy polysaccharide or cellulose and is stabilized against premature loss of acetic acid or other acids, if present, by intimate dispersion of the components of the composition with a minor quantity of a normally solid edible fatty material.

The quantity of acetic acid used is preferably within the range of 2 to 15% by weight of the composition. The quantity of fumaric, tartaric and/or citric acid, if used, is preferably within the range of 1 to 8% by weight of the composition. The weight ratio of acetic acid to fumaric, tartaric and/or citric acid, if used, is preferably at least 1:1 and is preferably within the range of 1:1 to 7:1.

The stabilizing fatty material used is preferably a fatty material having a melting point of at least 90°F and the amounts employed are preferably within the range of 2.5 to 10% by weight of the composition.

The acetic acid is usually employed in the form of glacial acetic acid. The other acids, if used, are usually employed in dry form. The fumaric, tartaric and/or citric acids can be mixed with the acetic acid after the acetic acid is encapsulated.

The starchy polysaccharides can consist of extracted starches such as wheat and corn as well as tapioca starch and potato starch and the corresponding flours, such as wheat flour and corn flour as well as starches and flours derived from waxy maize, rice, rye, and other cereal starches and flours. The starchy polysaccharide can be gelatinized or ungelatinized. A preferred type of starchy polysaccharide is pregelatinized wheat starch.

If finely divided cellulose is used as the absorbent, good results are obtained by the use of an edible wood flour. One such type of flour which is available commercially is Solka-Floc SW40. Other cellulosic absorbents which can be employed are methyl cellulose and carboxy methyl cellulose. In general, the starchy polysaccharides and the cellulosic absorbents will have a particle size within the range of 20 to 350 mesh (Standard Sieve series). The Solka-Floc has a particle size of approximately 30 mesh.

Examples of fatty materials which may be used are lecithin, hydroxylated lecithin, mono- and diglycerides and highly hydrogenated vegetable oil shortenings, and similar fatty compositions. These fatty materials serve to hold the acetic acid and the other acids, if used, in the absorbent, thereby acting as stabilizing agents, and also reducing, if not completely preventing, the evaporation of these acids during the oven baking stage of the sourdough bread process. The fatty materials employed as stabilizing agents in these compositions are normally solid at ambient temperatures of 25°C and have a melting point of at least 90°F (about 32°C).

Example: A flowable sourdough composition is prepared from the following ingredients.

Ingredients	Percent by Weight
Glacial acetic acid	12.5
Mono- and diglyceride fat compound (Atmul 500)	2.5
High melting point vegetable oil fat (Kao-Rich)	5.0
Gelatinized wheat starch	80.0

The Atmul 500 and Kao-Rich are warmed to a point where they are completely liquid. The acetic acid is then added, forming a dispersion. This dispersion is then blended with the starch in a standard ribbon blender by pouring it into the gelatinized starch while the mixer is in operation, or by spraying it into the starch with the mixer in operation. The resulting mixture is then placed in polyethylene-lined bags or cartons for use in making sourdough bread.

L-Ascorbic Acid-Dibasic Acids in Bread Doughs

There have been attempts to increase the volume of a bread to improve the nature of the inner and outer phases of the bread and to simultaneously improve its taste. For such a purpose, L-ascorbic acid is added to dough. Although it increases the volume of bread to a certain extent, such an additive is not satisfactory in terms of the quality of bread to be obtained and production process.

K. Tanaka and S. Endo; U.S. Patent 4,296,133; October 20, 1981; assigned to Nisshin Flour Milling Co., Ltd., Japan have found that breads of extremely good quality can be obtained by employing in combination (a) L-ascorbic acid and (b) an additive selected from divalent carboxylic acids and their salts, cystine, methionine, alum or nicotinic acid.

The term "an L-ascorbic acid" used in this specification means to include L-ascorbic acid, dehydroascorbic acid or their salts. Such is incorporated in a dough at a proportion of 3 to 30 ppm of the weight of wheat flour, more suitably, 5 to 15 ppm. No desirable effect will be brought about with a proportion less than the above range, and no substantial increase in effect will be expected even if more L-ascorbic acid is added than the above range.

Among the dicarboxylic acids used are malic acid, α-ketoglutaric acid, tartaric acid, asparagic acid, glutamic acid, hydroxyoxalic acid, oxo-succinic acid, di-amino-succinic acid, γ-hydroxyglutamic acid and their salts such as for example sodium salts and potassium salts. Asparagic acid, glutamic acid and tartaric acid are particularly preferred. Such a dicarboxylic acid is suitably added in a proportion of 5 to 60 ppm of the weight of wheat flour, more preferably 10 to 40 ppm.

Cystine or methionine can be suitably used in a proportion of 5 to 80 ppm, and more preferably 15 to 50 ppm, to the weight of wheat flour. Suitable as an alum is potassium alum, burnt alum or ammonium alum, which can be suitably used in a proportion of 10 to 60 ppm and more preferably 20 to 40 ppm to the weight of wheat flour.

Among nicotinic acids suitable for the process are included nicotinic acid and the salts thereof and nicotinic acid amide. Such nicotinic acid in the general term is used in a proportion of 5 to 70 ppm and preferably 20 to 50 ppm.

The above additives may be mixed and kneaded sufficiently during the kneading of a dough. If a sponge dough method is employed, it is preferable to add to the sponge dough at least either one of L-ascorbic acid and the other additive, and more preferably both of the additives.

According to the method, it is possible to obtain bread of which volume is sufficiently large, and of which the inner phase (crumb grain, color of crumb) outer phase (color and nature of the crust) and texture (feeling obtained by pressing the texture with a finger) are satisfactory. In addition, the handling of the dough is easy as the dough is not excessively sticky. The effectiveness of such additives becomes more apparent where no oxidizing agent is employed in the dough.

Example 1: Materials in the sponge formula were kneaded and then 20 mg of L-ascorbic acid and 60 mg of asparagic acid were added and kneaded further. The sponge was allowed to ferment at 27°C for 4 hours. To the fermented sponge was added the mixture of the dough formula and kneading was carried out. Loaf bread was produced in accordance with the following bread-making conditions.

Sponge formula	
Wheat flour	1400 g
Bread yeast	40 g
Yeast food	2 g
Water	800 cc
Dough formula	
Wheat flour	600 g
Salt	40 g
Sugar	120 g
Margarine	40 g
Shortening	60 g
Powdered milk	40 g
Water	520 cc
Bread-making conditions	
Floor time	20 min
Bench time	20 min
Final proof	37°C, 35 min
Baking	200°C, 30 min

Example 2: To the sponge of Example 1, was added and kneaded 14 mg of dehydroascorbic acid. The sponge was allowed to ferment for 4 hours at 27°C. To

the sponge was added the mixture of the bread-making formula of Example 1 and 34 mg of sodium glutamate. After kneading the mixture and following the bread-making conditions of Example 1, loaf bread was obtained.

Stearoyl Lactylate Salts

Stearoyl lactylate salts have been sold for many years, being used especially, in the edible forms, as additives in bakery products to improve quality. Sodium and calcium stearoyl-2-lactylates are the most common and commercially used of the salts and their addition to products for human consumption is controlled by the FDA.

The commercial grade stearoyl lactylate salts encompass various lactylates including those with a wide range of lactyl groups and various fatty acids of the acyl group; e.g., stearoyl lactylate salts when used as baking additives are generally written as sodium stearoyl-m-lactylate and calcium stearoyl-n-lactylate where m and n are the average number of lactyl groups (polylactyls); each m and n is an average of a range from 0 to 11. Lactylates having 1 to 3 lactyl groups are considered most functional in baking with an average of 2 preferred. In normal nomenclature each m and n is rounded to the nearest whole number, so that 2 may equal 1.51 to 2.50. Commercial grade stearoyl lactyls, too, may contain a wide range of acyl fatty acid radicals, including those of C_{14-22} fatty acids, the most common of which are C_{18} and C_{16} fatty acids. Thus, for example, a particular stearoyl lactylate may be made from stearin fatty acid containing 50% stearic acid and 50% palmitic acid. Therefore as used herein, stearoyl lactylate salts are understood to include the wide range of lactyl groups and various fatty acid substitutions for stearic acid which may be present in commercial stearoyl lactylate salts.

The conventional stearoyl lactylate salts are characterized by their relatively low softening and melting points and are normally highly hygroscopic, particularly the sodium salt, which will be shown below. Such salts are often used, especially in the baking industry, as powders such that 100% of the powder will pass through U.S. 40 mesh screen. The low melt and high hygroscopicity characteristics of such compositions present serious storage and warehousing problems, particularly in summer months because of the tendency for the powdered materials to coalesce into lumps and hard cakes.

It has been discovered by *C.J. Forsythe; U.S. Patent 4,264,639; April 28, 1981; assigned to Top-Scor Products, Incorporated* that stearoyl lactylate salts can be produced with improved handling and storage properties as measured by hygroscopicity, resistance to humidity and increase in melting or softening point by the thermal addition of hydrogenated stearin to such salts. The most functional stearins are those having high percentages of C_{18} fatty acid, although those having as low as about 50% (by weight of fatty acid) C_{18} fatty acid are functional. The most functional commercial stearin would seem to be hydrogenated soybean oil stearin. Other stearins such as palm oil or cottonseed oil stearin are also functional. The stearin preferably should be fully hydrogenated, having an iodine number of less than 7.

As little as 2% (by weight) edible stearin is functional in the process. At about 30% (by weight) edible stearin, the cost of further stearin addition outweighs benefits derived therefrom; however, higher percentages of stearin are functional. An optimum percentage of stearin addition should be based upon a current cost versus benefit analysis for each situation.

The stearoyl lactylate salt may be that of sodium, calcium, other alkaline earth or alkali metals, or mixtures thereof. The salt must be acceptably nontoxic if it is to be used in food products.

The process is functional with stearoyl lactylate salts having a wide range of poly-lactyls, although stearoyl-2-lactylate salt is preferred, is functional in any of the commercial stearoyl lactylate salts which may comprise a variety of acyl fatty acids radicals, and is not limited to stearic fatty acid.

Example: A sample of Emplex (a commercial brand of sodium stearoyl-2-lac-tylate, C.J. Patterson Co.) was tested for degree of hygroscopicity, resistance to caking at 110°F and melting point range. A sample of sodium stearoyl-2-lac-tylate (SS2L) was prepared by conventional methods in the laboratory without any additions. Three additional samples of sodium stearoyl-2-lactylates were also prepared wherein hydrogenated soybean oil stearin, having an iodine value of less than 7 and a melting point range of 147.7° to 151.3°F was added to the reaction charge at the same time as the stearin fatty acid in amounts of 2.5, 10 and 30% (by weight) in each sample respectively. These laboratory prepara-tions were powdered for testing.

The Emplex and untreated SS2L on testing were extremely hygroscopic and had extremely poor resistance to caking at 110°F storage temperature. SS2L products treated with the hydrogenated soybean oil stearin in the above amount showed improved properties, for example SS2L with 30% stearin showed very slight hygroscopicity, had very good resistance to caking and melted at 137.5° to 148.6°F compared to 118.4° to 122.7°F for SS2L without stearin.

IN GELATINS AND PUDDINGS

Acidulants in Cooked Chocolate Puddings

Virtually all cocoa flavored and colored cooked pudding products have been formulated with natural or lightly-dutched cocoas having a pH below 7.0. Cooked puddings previously prepared containing highly-dutched cocoas (i.e., having a pH above 7.0) exhibited a loss of appearance, texture and mouthfeel in that the puddings prepared had a soft texture, a lower viscosity and were not able to main-tain a clean pie cut.

Thus, it is a main feature of the process by *J.R. Carpenter, W.L. Steensen and C.R. Wyss; U.S. Patent 4,262,031; April 14, 1981; assigned to General Foods Corporation* to prepare a cooked pudding composition containing a highly-dutched cocoa which maintains the appearance, textural and mouthfeel attri-butes associated with cooked pudding products prepared with natural or lightly-dutched cocoas.

It is also a feature of this process to prepare a cooked pudding composition which can utilize reduced levels of cocoa to obtain optimum color and flavor.

This is achieved by incorporating in the pudding mix containing a highly-dutched cocoa (i.e., having a pH above 7.0) an amount of an acidulant effective to impart a desired texture and viscosity and a pH up to about 6.7 in the resultant cooked pudding. Preferably, the pH of the resulting cooked pudding is adjusted to within the range of about 6.3 to 6.6. Upon hydrating, cooking and cooling this pudding mix, the resultant pudding possesses a firm texture and a sharp pie cut characteristic of cooked puddings prepared utilizing natural or lightly-dutched cocoa. The preferred starch in the pudding composition is corn starch, while the preferred acidulant is acid whey.

Critical to the process is the addition of an amount of an acidulant effective to impart a desired texture and viscosity and adjust the pH of the resultant cooked pudding up to about 6.7, preferably addition of an amount effective to adjust the pH of the cooked pudding to within the range of about 6.3 to 6.6, optimally 6.4 to 6.5. It is theorized that the alkalinity of the highly-dutched cocoa interferes with the gelling of the starch in a milk-containing cooked pudding system, thus resulting in a decrease of its textural, mouthfeel, viscosity and appearance characteristics.

pH of the resultant cooked pudding is adjusted by the addition of an amount of an acidulant effective to obtain a pH of 6.7 or below and to obtain a desired pudding texture and viscosity. Preferably, the amount of acidulant is effective to adjust the pH of the resultant cooked pudding to within a range of 6.3 to 6.6 to provide a pudding with a firm texture, a clean pie cut, and a desirable viscosity and mouthfeel.

The preferred acidulant is acid whey (liquid or powder); however food grade acids such as citric, fumaric, adipic, tartaric, malic, ascorbic, lactic, acetic, etc., are also suitable to neutralize the residual alkalinity of the highly-dutched cocoa in the cooked pudding composition. Acid whey is derived from milk, contains about 5 to 9% lactic acid, and about 2 to 8 g of acid whey would be needed to neutralize about 12 g of highly-dutched cocoa. When a food grade acid is employed in a dry pudding mix for preparing the cooked pudding composition, preferably the acid is encapsulated (e.g., by fat, sugar, etc.) to prevent any interaction between the starch (or other ingredients) and the acid and thus prolongs the shelf life and the storage stability of the dry pudding mix.

Generally the viscosity of a cooked pudding composition is within the range of about 60,000 to 115,000 centipoises, while the viscosity of a cooked pie filling is within the range of about 80,000 to 130,000 centipoises. Upon the addition of a highly-dutched cocoa to a cooked pudding composition or pie filling without an acidulant there is observed a reduction in viscosity ranging from 8,000 to 30,000 centipoises, as well as the texture becoming softer and soupy and the inability of the pudding to maintain a sharp pie cut. The addition of an acidulant, such as acid whey, will increase the viscosity of the resultant cooked pudding by about 8,000 to 30,000 centipoises and will impart a firm texture (smoother, firmer mouthfeel) and will lengthen the time that a pudding will retain its shape after cutting (i.e., maintains a sharp pie cut).

Example: A dry pudding mix was prepared by mixing together 60 g sugar, 17 g raw cornstarch, 12 g highly-dutched cocoa powder, 8 g modified cornstarch, 3 g acid whey powder, 1 g salt, 0.25 g vanilla flavor, 0.2 g calcium carrageenan and 0.2 polysorbate 60.

The dry pudding mix was hydrated by mixing it with 2 cups (474 ml) of cold milk. The hydrated pudding mix was then cooked over medium heat until the pudding came to a boil, followed by chilling the pudding for about 1 to 2 hours to set the pudding. The pH of the resultant cooked pudding was about 6.4.

To prepare a pie filling the dry pudding mix is hydrated with 2¼ cups (533 ml) of cold milk, followed by cooking the hydrated mix to a full bubbling boil, then chilling for 3 hours. The pH of the resultant cooked pie filling was about 6.4.

Both the cooked pudding composition and the cooked pie filling exhibited a sharp pie cut (retained shape after cutting with a spoon), a firm texture and a desirable color, flavor, mouthfeel, appearance and viscosity. These physical properties obtained were judged to be equivalent, if not slightly superior (especially in color and flavor), to those obtained with cooked pudding products or pie fillings containing natural or lightly-dutched cocoas.

Substitution of the acid whey powder in the above example with 0.2 g of encapsulated fumaric acid gave an excellent pudding mix which could be used in the preparation of a pie filling.

Carbonated Gelatin Preparations

The disclosure by *H. Ono and Yoichi Akino; U.S. Patent 4,197,325; April 8, 1980; assigned to Taiyo Fishery Company Ltd., Japan* relates to gelatin foodstuffs containing carbon dioxide. A polysaccharide such as carragheenan may replace gelatin in these products.

In the range of temperatures between which the gelatin is dissolved in hot water and coagulates upon cooling the solubility of carbon dioxide relative to water is about 0.1% by weight at 35°C and about 0.2% by weight at 10°C. Consequently, even though the gas is dissolved in the gelatin by absorption at this stage, the jelly obtained upon solidification due to cooling does not possibly reach a state exhibiting a refreshing quality due to carbon dioxide.

In fact, although it is possible to prepare jelly by pouring cold soda drink into a hot gelatin solution when the solution reaches a temperature just before it coagulates, i.e., about 10°C, then gradual cooling and agitating moderately followed by further cooling, the obtained jelly virtually does not possess a refershing quality and merely produces a so-called "vapid" lemonade jelly.

To prepare the foodstuffs according to the process, 5.3 g of edible gelatin, 44 g of sugar and 0.75 g of citric acid are charged along with flavoring and coloring if desired into a can suitable for carbonated drink. 53 ml of water is also added to the can. There ingredients are mixed together and are made to hydrate and swell sufficiently. Thereafter, the ingredients are dissolved by heating them to about 50°C by means of a hot water bath. The resulting solution is appropriately solidified by cooling it to 5° to 10°C. 156 ml of carbonated water having a gas volume (hereinafter referred simply to as Vol) of about 3.2, which is previously prepared by a carbonator, is poured into the resulting product at 7° to 10°C by means of a filler for carbonated drink. The can is seamed immediately. The can is immersed in hot water of 50° to 70°C for 10 to 20 minutes, while rotating and shaking the can if desired, to change its contents to a homogeneous solution. After the heating, the resulting solution is sufficiently maintained at 4° to 10°C by cooling it with water. The jelly thus obtained contains carbon dioxide of about 2 Vol.

The receptacle used should be of inner pressure resistance such as a can for carbonated drink or beer can or pressure bin.

Example 1: 13.5 g of a powdery mixture of 10 parts of gelatin, 87 parts of sugar, 1.5 parts of citric acid and small quantities of orange flavor and natural coloring was charged into a specific taper can of 75 ml in volume. To the mixture was added 9 ml of water. The resulting product is made to hydrate and swell under agitation, and was then heated to about 50°C to dissolve these ingredients. The resulting solution was solidified by cooling it to about 10°C in water. 48 ml of carbonated water at 8°C and 3.2 Vol was poured into the can, which was then seamed.

In a similar manner, carbonated water of 3.8 Vol and also of 4.5 Vol was filled in other cans. These cans were seamed. The cans thus seamed were heated under rotation to 70°C for 20 minutes to obtain gelation.

The gas volumes of the product thus obtained correspond to 2.0 Vol, 2.4 Vol, and 2.9 Vol, respectively. As a result of the panel tests, the products were found to have virtually the same intensity of carbon dioxide and possess a good refreshing quality.

Example 2: 2.4 kg of gelatin and 0.24 kg of carragheenan were suspended in 22.4 ℓ of water and made to hydrate and swell in water. The suspension was then heated to 60°C to dissolve the ingredients. 30 ml of the obtained solution was poured into each specific pressure-resistant taper can each having a volume of 150 ml. The cans were then seamed and cooled to obtain gelation.

Apart from this, 10 ℓ of plum liqueur of (B 32, specific gravity 1.233) and 16 kg of sugar were agitated together with 65 ℓ of water in a tank to dissolve these ingredients. With the use of a carbonator, a solution of 8° to 10°C and 2.8 Vol was prepared. 105 ml of this solution was poured in each of the cans containing the jelly.

After filling about 250 ml cans in this manner, repeating the process a solution of 3.6 Vol and also of 4.3 Vol was poured in the remaining cans. All the cans were seamed and heated under rotation to 70°C for 20 minutes, and were then cooled to obtain carbonated plum liqueur jelly.

The gas volumes of the obtained jelly correspond to 2.0 Vol, 2.5 Vol, and 3.0 Vol. All the jelly samples were found to have the aroma characteristics of plum liqueur and possess a refereshing quality as well as a good taste.

Stable Liquid Gelatin Formulation

It is well-known that the strength of the gel formed by dissolving a given amount of gelatin in water is a function of the pH of the system, with the maximum being reached at a pH of about pH 8.5 for either type A or type B gelatin. This behavior is such that a gelatin which exhibits a Bloom of 300 at a pH of 4 will exhibit a Bloom of about 312 at pH 5, 325 at pH 6, etc. This variation of Bloom with pH is more precipitous below pH 4, but the effect is totally reversible, except for any incidental hydrolysis which might occur when the gelatin is held for long periods of time at very low pH, especially as temperature is raised.

T.V. Kueper and T.H. Donnelly; U.S. Patent 4,224,353; September 23, 1980; assigned to Swift & Company have discovered that the effect can be used advantageously to produce cold-water-soluble gelatin desserts, and other gelatin-based products.

While any acid which will provide the proper pH could be used to formulate such syrups, it is advantageous to use citric, malic, ascorbic and erythorbic acids. Food grade inorganic acids, such as phosphoric, hydrochloric and sulfuric, also give the effect, but are limited in applicability by esthetics, taste, and special ionic effects. Food acids such as succinic, fumaric and glutaric are not sufficiently soluble to be attractive candidates. The more flavorful food acids such as acetic and lactic give the effect, but are limited by considerations of taste. Tartaric acid would be as desirable as citric, except that it forms relatively insoluble acid tartrates at the pH of gelatin dessert, thus causing an undesirable turbidity. Malonic and aconitic acids are generally too expensive. It is preferred to use citric and malic acids in equal amounts, although any percentage combination is suitable.

Preferably, the syrup composition will comprise about 20% gelatin, about 20% edible acid, and about 60% water by weight. The important weight relationship is that of acid to gelatin. Generally, 30 to 200% acid should be present based on the weight of gelatin, and preferably from about 75 to 150% acid based on the weight of gelatin.

A sweetener of humectant, such as glycerine, sucrose or fructose, may be added to the syrup to provide a presweetened syrup. If sugar is not added to the gelatin/acid syrup, it will be incorporated with the buffer salts. The amount of sugar will range anywhere up to 1000% based on the weight of gelatin. Also, suitable flavorings and/or colorings may be added to the syrup.

The aqueous gelatin/acid syrup will be packed in a suitable container, e.g., a plastic tube, to form one component of a multicomponent dessert package. The other required component of the package will consist of a container of buffer salt or salts. The buffer salt cannot be incorporated within the syrup, prior to use in forming a dessert, inasmuch as the buffer would adjust syrup pH so as to enable the syrup solution to gel. Suitable buffer salts include sodium citrate, diammonium phosphate and disodium phosphate, with sodium citrate being greatly preferred. Sodium sulfate may be added to speed the rate of set. Also, the separate container of buffer salt will preferably contain the sugar, although it is possible to incorporate the sugar into the syrup. Like the sugar, some coloring and/or flavoring ingredients may be combined with the buffer salt.

The buffer salt will be present in an amount to adjust final pH of the gelatin dessert product to from 3.6 to 4.6, preferably 4.0. Generally, depending on type and amount of acid used, proper final pH can be achieved by using 50 to 400% by weight buffer salt based on the weight of acid, and preferably 100 to 250%. If incorporating the buffer salt and sugar as a tube of syrup within the dessert package is desired, then from about 10 to 15% buffer salt by weight is combined with 40 to 70% sugar and 20 to 50% water. As little water as possible should be used to form a flowable syrup.

The gelatin/acid composition tends to deteriorate at room temperature over extended periods of time due to hydrolysis of the gelatin in an aqueous acid solution. Accordingly, it is preferred to store and distribute the syrup in refrigerated or frozen form. Since the process contemplates a package comprising two physically separate components, the gelatin/acid component and the buffer component, the complete gelatin dessert product package will normally be stored and distributed in a refrigerated or frozen form. Most preferably, both components will be syrups, stored and distributed under refrigeration.

Example: A gelatin/acid syrup was prepared by soaking 2 parts of gelatin in 2 parts of cold water and melting when fully swelled. In this, one part of citric acid (monohydrate) and one part of malic acid were dissolved. Suitable flavor and color were added, and the clear syrup placed in a plastic tube and sealed.

A neutralizer-sweetener powder was prepared by mixing four parts of sodium citrate dihydrate with fifteen parts of sugar. A cup of dessert was prepared by dissolving 30 g of the above syrup in a cup of water, and dissolving 95 g of the powder in this solution.

Frozen Gelatin Desserts

The disclosure by *H.H. Topalian, C.R. Wyss, R.E. Kenyon and A.F. Dec; U.S. Patent 4,297,379; October 27, 1981; assigned to General Foods Corporation* relates to a frozen aerated gelatin composition which is storage stable, smooth and creamy without gummy or icy characteristics. The resultant frozen aerated composition is able to maintain its shape on a stick while remaining soft, creamy and ready-to-eat at freezer temperatures. As well, the frozen composition holds its shape and does not weep (release water) during thawing and thus avoids dripping during consumption.

The gelatin mix used to prepare the frozen aerated gelatin composition critically contains gelatin at a level of 0.6 to less than 5%, preferably 0.8 to 2.5%, by weight of the frozen composition. This level of gelatin is critical in obtaining a stable composition with the creamy and smooth texture desired while avoiding excessive gummy, rubbery and icy characteristics which are common with different levels or other gums. The gelatin also enables the frozen aerated composition to maintain its shape on a stick when frozen and even during thawing, and minimizes or eliminates any weeping or release of water from the frozen composition during thawing.

The gelatin mix also contains an acid for process and microbiological stability and taste in an amount effective to impart to the frozen composition a pH of 2.5 to 7, preferably 3.8 to 6.5. Suitable acids include adipic acid, fumaric acid, malic acid, citric acid, tartaric acid, etc. A buffer may also be included to maintain a desired acidic pH during all phases of processing. Suitable buffers include sodium citrate, disodium phosphate, potassium phosphate, sodium tartrate, etc.

The gelatin mix further contains soluble solids at a level of 5 to 50%, preferably 15 to 30%, by weight of the frozen composition. The soluble solids are essential in rendering the frozen aerated composition ready-to-eat at freezer temperature (0° to 20°F, -20° to -5°C), while imparting a creamy and smooth texture which is continuous and uniform. The solids level when combined with the gelatin and the critical processing parameters are important in controlling ice crystal size and overrun, which results in the extended stability and desired texture of the frozen aerated composition. The soluble solids can include dextrin, hydrolyzed cereal solids and preferably a sweetener such as sugar or nonnutritive sweetener, e.g., saccharin or aspartame. Appropriate sugars include sucrose, dextrose, glucose, and fructose.

The soluble solids can further contain other freezing point depressants such as glycerol, propylene glycol, maltodextrin, salts, etc., which will contribute to smaller ice crystal size and impart to the frozen aerated composition a smoother, softer, texture at lower temperatures.

The gelatin mix preferably additionally contains a stabilizer which aids in retarding the growth of ice crystals during storage at freezer temperatures. The stabilizer is preferably employed at levels up to 0.5% by weight of the composition. Appropriate stabilizers include algin, carrageenan, xanthan, locust bean gum, low methoxyl pectin, guar, hydroxypropylmethylcellulose, carboxymethylcellulose, etc., and blends thereof.

Fat can be incorporated, preferably with an emulsifier, into the frozen aerated composition, although the fat and emulsifier can readily be excluded from the composition while still obtaining a frozen aerated composition which is smooth and creamy.

Example: A frozen aerated composition was prepared which contained (in parts by weight) 76.3 parts water, 7.7 parts high fructose corn syrup (80% solids), 13.0 parts sucrose, 1.9 parts gelatin (250 bloom), 0.4 part adipic acid, 0.1 part sodium citrate, 0.1 part fumaric acid, 0.3 part blend of locust bean, xanthan and guar gum, 0.1 part flavor, and 0.1 color.

The dry ingredients were blended, then added and mixed with water heated to 160°F (70°C). The high fructose corn syrup was then mixed in and the dissolved mix was cooled to 90°F (32°C) in a heat exchanger. The dissolved mix was then cooled to about 24°F (-4°C), while simultaneously being aerated and agitated in a swept surface heat exchanger. Air was incorporated to create an overrun of about 40% by volume. The aerated mix was substantially frozen at this point and was then extruded into slices, cut, and sticks were inserted, followed by freezing to a core temperature of about 0°F (-18°C). The resultant aerated composition had an ice crystal size of about 100 mμ.

The frozen aerated composition possessed a creamy and smooth texture, without gummy or icy characteristics upon being eaten at freezer temperatures (about 0° to 20°F, -20° to 5°C). During thawing the composition maintained its shape on the stick and did not weep or drip. After thawing a soft tasty gel was produced. This composition possessed extended storage stability (over one year at 0°F, -18°C).

FOR OTHER FOOD PRODUCTS

Buffered Lactic Acid Mixtures in Candies

In the preparation of sweets with a sour taste (for example, sour hard candy or fruit drops), caramels, jelly fruit, marshmallows, jelly beans, chocolate candy fillings and acidulated suckers so-called buffered acids are preferably used in order to obtain on the one hand an agreeable, acidulous taste and, on the other hand, to avoid disagreeable side effects of acid additive, such as inversion of the saccharose as well as hydrolytic degradation of gelatin or other polymeric thickeners. On the other hand, such solutions must be as concentrated as possible in order to exclude hydrolytic effects as far as practicable; on the other hand, it is necessary to keep them sufficiently liquid, so that they can be pumped in the course of the modern continuous manufacturing methods.

A known product often used for such purposes is the so-called buffered lactic acid, which is an aqueous mixture of lactic acid and sodium lactate having a dry substance content of about 75 to 80% and an average pH value of 2.7 to 3.2, preferably 2.9 to 3.0. As a liquid acid with a lesser tendency toward inversion than, for example, tartaric or citric acid, lactic acid comes closer to the requirements to be fulfilled, i.e., that it may be pumped easily and has a low hydrolytic effect. However, its disadvantage is its relatively mild acidulous taste, as well as its higher price.

It has already been tried to produce a cheaper buffered lactic acid with a more pronounced and more refreshing taste by adding to it more or less large quantities of solid edible organic acids (for example, tartaric or citric acid). However, when more than 5% of such acids were added, crystalline precipitations resulted which substantially interfered with the processing of the buffered lactic acid; because the concentration of the acid solution was changed in an uncontrolled manner, the pumps were plugged up, crystal agglomerations could form in the candy.

H.E. Bisle; U.S. Patent 4,292,339; September 29, 1981; assigned to Boehringer Ingelheim GmbH, Germany have discovered that by neutralizing 8 to 20%, preferably 12 up to 18%, of the acid equivalents present with preponderantly, or preferably even exclusively potassium ions, stable (i.e., protected from crystallizing out) aqueous solutions of (solid) edible acids, such as citric acid, tartaric acid, malic acid or gluconic acid, or mixtures thereof, with a dry substance contact of more than 55% by weight, preferably about 70 to 80% by weight, are obtained. The pH value of such potassium-buffered, concentrated edible acid solutions is from 2.7 to 3.3; pH values refer always to measurements in a dilution of 1:10 with water.

Here, the possibility of producing technically interesting mixtures of lactic acid with other edible acids in the indicated range of concentration is of particular value. In this manner, one succeeds in admixing considerable quantities of organic acids, especially citric acid but also tartaric acid, with concentrated lactic acid solutions. For example, it is readily possible to obtain a stable, concentrated buffered acid solution by means of potassium-buffering to pH 2.7 to 3.3 (preferably 2.9 to 3.0) of the acid present in an aqueous solution of a mixture of lactic acid and citric acid comprising up to 25% citric acid with a dry substance content of about 70 to 80% by weight.

This buffered acid has the advantage that it possesses a more refreshing acidulous taste than the conventional buffered lactic acids, that its production is cheaper and that, while the increase in the inversion rate remains minimal, other disadvantages such as crystallizing out or difficult pumping are avoided. Of course, its content of edible organic acids (for example, citric acid) may also be higher than 25%, and the dry substance content may be lower than 70% by weight; but this leads to a further increase in the inversion rate.

The buffered acids may be produced either by the addition of potassium hydroxide to a concentrated aqueous edible acid solution up to a pH of about 2.7 to 3.3, preferably 2.9 to 3.0, or by admixing the quantity of edible acids required for obtaining a dry substance content of more than 55% with the quantity of their potassium salts required for neutralizing 8 to 20% of the acid equivalents present, and filling up with water to 100%.

Example: An aqueous solution of lactic and citric acid, buffered with sodium salt, contains in 100 g of finished product 60% lactic acid (= 667 mval), 11% of citric acid (= 171 mval) and 4.6% of NaOH (= 116 mval or 13.8% of acid). The dry substance content of the total mixture was 73.5%.

Acid solutions of various concentrations were prepared from the composition mentioned above and stored at room temperature as well as in the refrigerator. It was found that even an only 55% solution did not remain sufficiently stable, but formed crystals both at room temperature and when refrigerated at 6° to 8°C. When the entire NaOH was replaced with approximately 6.0 to 8.5 g of KOH, the solutions were stable and did not crystallize. When mixtures of NaOH and KOH were used, even at a K/Na ratio of 4:1, the solutions were unstable and crystallization occurs.

Thickened Acid Compositions for Salad Dressings

The work disclosed by *K. Maerker, K. Bezner, F. Biller and H. Bohrmann; U.S. Patent 4,252,835; February 24, 1981; assigned to CPC International Incorporated* is directed to an acidic, aqueous dispersion containing an improved thickening system. The dispersion may be used either directly as or as a base for a salad dressing. It contains at least one edible acid and has a pH from about 2.5 to 6.7. At least one of the thickeners contained in the dispersion consists of amorphous and/or microcrystalline glutamic acid in a concentration of about 6 to 60% by weight, based on the total weight of the dispersion.

When selecting the content of glutamic acid it should be observed that the thickening effect of a specific amount of glutamic acid as a rule decreases as the pH value rises, and vice versa, and so with pH values ranging between 4 and 5.5 the glutamic acid content of the dispersions should preferably range between 14 and 38, and in particular between 15 and 30% by weight, based always on the total weight of the essential components, i.e., the edible acid and, optionally, salts thereof, water and glutamic acid. On the other hand, in the case of pH values from 5.5, and in particular from 5.7 upward, it is recommended, using the same basis of reference, to select glutamic acid contents of at least 28, preferably at least 32, and in particular at least 34% by weight, it being advisable not to exceed a glutamic acid content of 52, preferably 47 and in particular about 42% by weight.

In a preferred embodiment one or several optional components are incorporated in the dispersions in addition to amorphous and/or microcrystalline glutamic acid, which optional components, possibly beside other characteristics, e.g., seasoning properties desired to be featured by such products, have a bodying effect. Examples of such optional components are salt, sugars, maltooligo- and/or maltopolysaccharides, vegetable, fruit and/or nonfat milk solids, powdered spices and/or soylike proteins of animal and/or vegetable origin.

It is a particular advantage of the dispersions of this process that virtually all flavoring components derived from plants or plant parts, such as herbs, onions and garlic, as well as all kinds of vegetables, in particular peppers, tomatoes, celeriac and leek, not only in fine-grain form, but also in the form of relatively large pieces, may be incorporated without sedimenting even after long storage. A particularly important point in this context is that fresh products may be used without any risk of such products perishing rapidly.

The dispersions may be produced in different ways. One possibility is to grind glutamic acid, optionally in the presence of minor amounts of water, in a colloid mill into a microcrystalline powder, then to mix with water an edible acid and, optionally, one or several optional component(s), and homogenize.

A simpler and safer way is to produce the amorphous and/or microcrystalline glutamic acid in situ by precipitating with the edible acid(s) in the presence of water from a glutamate which is water-soluble at least to a limited extent.

Although both components of the reaction may be present in the form of solutions, it is recommended to use the glutamic acid in the form of a powder which is fine-grain as possible, and the edible acid(s) in the form of an aqueous solution. While this embodiment generally provides placing the glutamate powder in the mixer and then incrementally mixing it with the edible acid solution, it is much safer and also simpler to place the edible acid solution in the mixer first, then add the glutamate powder and disperse it in the solution as quickly and homogeneously as possible.

Example: A 15% vinegar (made from an 11.3% vinegar and an 80% vinegar essence) is placed in an Eirich mixer and homogeneously mixed with mustard and a pasty premix of garlic and common salt (50:50). Another premix of salt, castor sugar, malto-dextrin, Aromat and IG is added to the running mixer and stirred in. A premix is thus obtained to which dustlike sodium glutamate is slowly added in amount equivalent to 10% by weight, based on the total weight of the finished mixture, and is quickly dispersed in it homogeneously, whereupon amorphous and/or microcrystalline glutamic acid precipitates after a short time and the mixture thickens to a semiliquid salad dressing.

Acidulants for Yeast Cultures

Compressed baker's yeast and dry baker's yeast prepared by the usual processes generally have a good activity or leavening power in both unsugared and sugared neutral doughs but their leavening power is unsatisfactory in acid doughs in the presence of organic acids such as acetic acid, propionic acid or lactic acid. Below certain pH values, these acids considerably inhibit the fermentation produced by the usual baker's yeasts.

The process discolsed by *F. Hill; U.S. Patent 4,318,991; March 9, 1982; assigned to Henkel KGaA, Germany* comprises the preparation of compressed baker's yeast or dry baker's yeast with improved activity or leavening power in an acid medium.

The compressed yeast with improved acitivity or leavening power in an acid medium is cultivated and worked up in the usual manner but from 0.1 to 10 g of aliphatic short chain carboxylic acids are added per liter of culture both in the last stage of propagation. More particularly, the process involves the production of baker's yeast with improved activity or leavening power under acid leavening conditions of the compressed yeast or dried yeast type, comprising cultivating fresh yeast conventionally to the last propagation stage, propagating the fresh yeast in the last propagation stage in the presence of from 0.1 to 10 g/ℓ of culture broth containing a source of carbon and nitrogen, of an aliphatic carboxylic acid having from 2 to 4 carbon atoms, and recovering baker's yeast with improved activity or leavening power; as well as the yeast produced by the above method.

The fresh baker's yeast obtained by this process is found to have a high leavening activity even in doughs with a high degree of acidity. This yeast can be converted into a durable dry baker's yeast which, like fresh baker's yeast, has an improved leavening activity in highly acid doughs.

Fresh baker's yeast is prepared by multistage fermentation. A stage of pure culture is followed by several stages of propagation with or without formation of ethyl alcohol. The last stage, the so-called transportable yeast fermentation, is generally controlled to prevent the formation of ethyl alcohol from the substrate. By the addition of from 0.1 to 10 g of aliphatic short chain carboxylic acids, especially those having from 2 to 4 carbon atoms, such as alkanoic acids like acetic acid and propionic acid, hydroxyalkanoic acids like lactic acid, hydroxyalkanedioic acids like tartaric acid and tartronic acid, etc., in particular propionic acid, per liter of culture broth in the last stage of propagation of the baker's yeast, a yeast is obtained which, when subsequently used as fresh baker's yeast or dry baker's yeast in doughs with a high degree of acidity, is found to have a substantially increased leavening activity. This yeast has a solids content after pressure filtration of about 25 to 30% by weight.

The increased leavening activity is preserved even after the fresh baker's yeast has been converted into dry baker's yeast by a drying process. This drying process is particularly advantageous, and the initial activity is substantially preserved, if the fresh baker's yeast is not obtained with the addition of sodium chloride as dewatering agent or hypertonic agent but with the addition of glycerol and/or urea or other nonionic dewatering or hypotonic agents. When the fresh yeast is processed with the addition of from 0.1 to 10%, preferably from 1 to 3%, based on the dewatered yeast, or a nonionic dewatering agent and pressure filtered, a yeast is recovered with a solids content of from 30 to 37% by weight.

Example 1: 4 ℓ of tap water were introduced into a 10 ℓ fermenter and 250 g of commercial fresh baker's yeast containing 29% of dry substance were suspended therein. 15 g of potassium dihydrogen phosphate, 1 g of calcium chloride and 1 mg of calcium chloride and 1 mg of biotin were also added. The temperature was maintained at 30°C and the mixture was stirred with a stirrer at a speed of 600 rpm. The supply of air to the culture broth was adjusted to 5 ℓ of air per minute. The addition of the carbon source and the nitrogen source to the culture broth was controlled by a programming device. 1.0 kg of beet molasses, diluted with water, and 70 ml of a 25% ammonia solution were added within 12 hours. The rate of addition of nutrient was adjusted so that at no point in time during the fermentation did the culture broth contain more than 0.1% of ethanol. The pH was maintained at 5.0 by an automatic titration apparatus. The development of foam was regulated by the addition of a commercial antifoam agent.

After 12 hours, the quantity of substance in the fermenter had risen to 6.6 ℓ and the yeast biomass had increased to 1,280 g (yeast containing 27% of dry substance).

The supply of air to the culture broth was continued for 30 minutes at the elevated temperature of 35+°C and the yeast was then separated from the culture broth by centrifuging, washed twice with water and filtered from the washing liquor through a pressure filter operating at 2.5 bars. The dry substance in the fresh baker's yeast was determined and the appropriate quantity of yeast was used for preparing the doughs. The leavening power of the fresh baker's yeast obtained is substantially equal to that of ordinary fresh baker's yeast available on the market since the usual method of fermentation was also used for its preparation.

Example 2: Yeast fermentation was carried out as in Example 1. However, 6 g of acetic acid were introduced into the fermenter during fermentation (in the 9th hour after onset of the addition of nutrient). The procedure was then continued as described in Example 1. The yeast obtained shows significantly improved leavening activity in doughs containing added carboxylic acid salts.

Example 3: Fermentation of yeast was carried out as in Example 1. During the fermentation (in the 9th hour after onset of the addition of nutrient), 6 g of propionic acid were introduced into the fermenter. The procedure was then continued as described in Example 1. The yeast obtained has an even better leavening activity in doughs containing added carboxylic acid than the yeasts obtained in Examples 1 and 2.

Dione-Hydrogen Peroxide Acidogen Systems for Dairy Products

F.L. Metz; U.S. Patent 4,264,636; April 28, 1981; assigned to Diamond Shamrock Corporation has provided a chemical acidogen for use in the production of acidified dairy products, for the production of "tofu." This acidogen is also used as part of the chemical leavening system in the production of raised bakery products as disclosed in U.S. Patent 4,328,115.

The acidogen system is a combination of at least one of a selective group of diones and hydrogen peroxide (H_2O_2). The acid necessary for the desired acidification and ensuing protein coagulation or bakery dough leavening is generated by the oxidation of the dione "in situ." Specifically, the diones used are acyclic and cyclic aliphatic compounds containing from 2 to 6 carbon atoms per molecule, and wherein the carbonyl groups are vicinal to each other.

Acyclic diones which are useful include ethanedial, more commonly designated as glyoxal; pyruvaldehyde; diacetyl; and 2,3-pentanedione. A specific cyclic dione used is 1,2-cyclohexanedione. In the preparation of cheese products and soybean curd, glyoxal is presently preferred for use, while in the preparation of the raised bakery products, pyruvaldehyde is the preferred component of the leavening system.

In the preparation of cheese products, the process generally may be carried out at a temperature of 4° to 40°C, with temperatures of 25° to 40°C preferred, and temperatures of 28° to 35°C being most preferable. In practice, the acidogen components may be added to the cold milk and the resulting mixture heated to the selected reaction temperature. Alternatively, the milk may be heated to the desired temperature prior to adding the acidogen thereto. Incorporation of the acidogen is effected with gentle agitation of the milk, vigorous agitation generally being unnecessary.

Throughout the acidification process, the pH of the milk is progressively lowered at a controlled rate to the protein coagulation stage (usually at a pH below 5.0) by the acid generated from the concomitant oxidation of the dione component.

To prepare soybean curd, i.e., tofu, soybean milk is first extracted from soybeans as conventionally practiced. The acidogen system of this process is added to the extracted milk to coagulate the protein, the acidogen components in aqueous solution preferably being added separately to the milk. The curd which develops is separated from the whey and is gelled (firmed) completely prior to being cut. The tofu prepared in accordance with this method has excellent gel strength by comparison to that obtained by using conventional coagulants such as calcium chloride or calcium sulfate.

From about 0.3 to 0.5 part by weight of dione in combination with from about 0.2 to 0.3 part by weight of hydrogen peroxide per 100 parts by weight of the soy milk generally will provide the desired protein coagulation rate. After boiling the soy milk for a sufficient time period to reduce the intensity of the bean flavor and to destroy antinutritional factors, the acidogen system components are added with mild stirring to the milk while it is still hot, i.e., at a temperature of at least 85°C. As mentioned previously, it is preferred to incorporate the acidogen components in increments over a time period ranging from about 20 to 30 minutes.

After acidogen addition has been completed and the curd has formed, it is allowed to settle for at least 30 minutes prior to being separated from the whey. Finally, the separated curd is pressed to the desired moisture content which typically is 80%, or slightly greater. As shown hereinafter by specific example, tofu prepared in accordance with this process possesses much improved gel strength compared to that prepared by a conventional, normally used coagulant such as calcium sulfate.

The process system is satisfactorily employed as a substitute for yeast in leavening bakery products. In this embodiment, the system is added to prepared flour mixes in combination with a CO_2-releasing compound, particularly sodium bicarbonate. The leavening ability of the acidogen system is comparable to or improved over yeast. The products prepared possess excellent texture equivalent to yeast-leavened products.

OTHER FOOD ADDITIVES

FAT SUBSTITUTES OR REPLACERS

High Solids Fats in Spreadable Syrups

D.E. Miller and C.E. Werstak; U.S. Patent 4,226,895; October 7, 1980; assigned to SCM Corporation have developed a nonpourable, flavored, spreadable emulsion comprising 8 to 15% of a high solids fat, the balance being water and sweetening agent with lesser amounts of stabilizing agent, emulsifier, and protein. The use of a high solids fat makes the emulsion temperature stable over a wide temperature range.

The process is particularly applicable to the preparation of a spreadable, maple syrup flavored topping and will be described with reference thereto, although it will be apparent that this process has other applications, for instance in the preparation of a spreadable honey.

This emulsion contains as major ingredients 8-15% high solids fat, water and sweetening agent; with less than 10% of minor ingredients including a water-dispersible stabilizer, emulsifier, a water-dispersible protein and flavoring, the proportion of sweetening agent to water being in the range of 3:1 to 4:1. It is important that the fat be a high solids fat preferably having a low lauric acid content. Suitable such fats are mostly derived from what are termed domestic oils such as soybean oil, cottonseed oil, corn oil, ground nut oil, sunflower oil and safflower oil. Other suitable oils are olive oil and palm oil. These oils characteristically have lauric acid contents of about 0.5 or less, as contrasted with typical lauric oils having lauric acid contents of about 45% or more. Thus the fats of the process preferably have a Wiley Melting Point in the range of 90° to 115°F, as determined by AOCS method Cc 2-38; and an approximate solid-fat index, as determined by AOCS method Cd 10-57 of 65±10 at 50°F and 11 max. at 100°F.

A preferred fat useful in the composition is Kaomel (SCM Corp.) having a Wiley Melting Point of 97°-101°F, an Iodine Value of about 59, and a solid-fat index of 5 max. at 100°F.

The sweetening agent employed in the composition can be any of those conventionally used in the production of sweetened topping compositions. Preferably a substantial portion of the sweetener is corn syrup solids to add bulk to the spreadable emulsion, without excessive sweetness. One suitable such syrup is Aunt Jemima Syrup (Quaker Oats Co.) containing about 75% corn syrup, 21% sugar syrup, 2% maple sugar syrup, additional corn syrup solids, cellulose gum, artificial flavor, sodium benzoate and sorbic acid (preservatives), and caramel color.

A critical aspect is using the sweetening agent and water in the range 3:1 to 4:1. Above 4:1 sweetening agent to water, sugar crystallization is apt to occur. Below 3:1 sweetening agent to water, bacterial growth is possible.

A preferred stabilizing agent is Gelcarin HWG, a carrageenan gum (Marin Colloids, Inc.). It is readily water dispersible and provides a desirable short-grained non-stringy product.

The formulation also employs a normally (at room temperature) water-dispersible or soluble protein such as sodium caseinate, soy protein, nonfat milk solids, whey solids, fish protein, calcium caseinate and cottonseed protein.

The emulsifying agent of the process preferably is a soft partial glycerol ester, as contrasted to a hard or fluid emulsifier, having a capillary melting point (as determined by AOCS method Cs 1-25) below about 140°F, preferably in the range of about 110°-140°F. A soft emulsifier of plastic consistency gives the emulsion viscosity desired and maintains the fat in a dispersed form. One suitable such emulsifier is Dur-em 204 (SCM Corp.), a cream plastic mono- and diglyceride from hydrogenated vegetable oil having about 52% minimum alpha-monoglyceride, an IV of 65-75, and a capillary melting point of about 120°-130°F.

Example: The formulation was prepared from 10.1% Kaomel as fat (percentages are by weight), 85.6% maple flavored syrup (Aunt Jemima), 2.5% skim milk solids, 0.4% of Gelcarin as stabilizer, and 1.4% of Dur-em 204 as emulsifier.

The entire amount of the protein (skim milk solids) and stabilizing agent (Gelcarin) is added to a melted mixture of the fat and mono- and diglyceride at 130°F. Simultaneously the syrup is heated to 130°F. Mixing then is carried out by adding the oil mix and syrup mix together and heating that mix to 160°F. The mixture is then homogenized at elevated temperature and cooled to room temperature. No refrigeration is necessary.

The Aunt Jemima Syrup, with its corn, sugar, and maple sugar syrups each about 80% solids, is more than 98% sweetening agents and water, in the ratio of about 4:1. Thus the above formulation in effect contains about 69% sweetening agents and about 17% water. Roughly the same formulation could be prepared by adding the sweetening agents (e.g., a mixture of corn syrup solids and dextrose) and water separately.

The flavored spread is unlike jelly or jam in that it contains fat, so that no butter or margarine is required on toast, bread, English muffins, or the like to which the spread is applied. Also the solids are high in the syrup phase so that the oil-in-water emulsions may be used as sundae toppings with freezing of the syrup. It is understood that the actual viscosity of the emulsion of this process, within limits, can be controlled by the amount of fat used, by the melting point of the fat, by the solid-fat index of the fat, by the type of emulsifier, by the amount of emulsifier, and by the amount of stabilizer employed.

Conventional products containing sweetening agents other than syrup, such as honey, jelly, jam, and the like, can be employed. In each case, the formulation would be appropriately modified to take into account proportions of sweetening agent and water present, plus gums and emulsifiers in the product used.

Synthetic Wax Ester for Parting Oils

Parting oils or releasing agents are employed in baking to coat molds before addition of the dough. Their use aids in the separation (parting) of the baked goods from the molds. Customarily natural, hardened sperm whale oil has been employed as a component of such parting oils. The reduction of the catching quotas for whales however, has led to an increasing scarcity of sperm oil, which is, among others, an important component of parting oils for baked goods.

The disclosure by *C. Heine, U. Ploog and R. Wüst; U.S. Patent 4,315,040; Feb. 9, 1982; assigned to Henkel Kommanditgesellschaft auf Aktien (Henkel KGaA), Germany* relates to a component for parting oils or releasing agents for baked goods that is physiologically safe, meets food regulations and can be used in place of natural, hardened sperm whale oil.

This component of parting oils comprises an ester prepared by combining:

> (a) a mixture of C_{12-20}-fatty acids having an iodine number of 48-96 and a content of C_{16-18}-fatty acids of at least 90%; with
>
> (b) a mixture of C_{12-20}-fatty alcohols having an iodine number of 50-95 and a content of C_{16-18}-fatty alcohols of at least 90%.

The synthetic wax ester has an iodine number of 40-110; a saponification number of from 100 to 140; an acid number of less than 1; an hydroxyl number of 5 or less; and a solidification range of from 10° to 30°C. A parting oil for baked goods contains from 20 to 60% by weight of the above wax ester and from 80 to 40% by weight of edible liquid triglycerides of natural fatty acids with a content of polyenic fatty acids of less than 16% by weight.

Starting products for the preparation of the wax ester are, for component (a), the fatty acids component, technical grade olefin fractions such as are obtained by saponification and fractionation, if desired, from fatty acids or tallow, palm oil or hardened vegetable oils of suitable composition. The fatty acid mixtures obtained therefrom contain fatty acids with the chain length of C_{12} to C_{20} and a content of fatty acids of the chain length of C_{16} to C_{18} of at least 90% and an iodine number of 48 to 96.

Suitable as the fatty alcohol component (b) are fatty alcohol mixtures as they are obtained by reduction, particularly hydrogenation of fatty acid mixtures derived from animal fats. These fatty alcohol mixtures contain fatty alcohols with the chain length C_{12-20}, with a content of at least 90% of fatty alcohols with the chain lengths C_{16-18} and an iodine number of 50-95.

The wax ester is prepared in a well-known manner by esterification of molar amounts of the corresponding fatty acid component (a) with the corresponding fatty alcohol component (b), for example, in the presence of esterification catalysts such as isopropyl titanate. The wax ester can, subsequently to the esterification, be separated from unesterified components, particularly alcohol components, by vacuum distillation and then be freed from acid by extracting with an aqueous alkali metal hydroxide, bleached and deodorized in the usual manner, to obtain a product that is completely neutral with respect to taste. Thanks to the selection of the starting components the wax esters have the following chemical or physical properties based on the analytical data:

> acid number: <1
> saponification number: 100-140, preferably about 110
> hydroxyl number: <5
> iodine number: 40-100, preferably about 70
> solidification range: 10° to 30°C.

Such wax esters possess special advantages for the intended use as sperm oil substitution product for parting oils for baked goods. For example, they have an increased resistance to oxidation due to the adjustment to a relatively low iodine number of about 70 and the low content of fatty acids or fatty alcohols with multiple olefinic bonds. Furthermore, the solidification point is easily adjusted to the seasonal temperature.

Example 1: A 2 liter three-neck flask equipped with an agitator, thermometer and distillation head was charged with 522 g of oleyl-cetyl alcohol (OH number 215, iodine number 55) and 544 g of technical grade oleic acid (saponification number 200, iodine number 92). The mixture was dehydrated at approximately 100°C with agitation for about 30 minutes. Then, 1 g of a 28% solution of isopropyl titanate was added and the mixture was slowly heated to 140°-200°C, over a period of about 10 hours. Excess oleyl-cetyl alcohol was then distilled off at 225°C/2-3 torrs, and, after raffination with 50% sodium hydroxide solution, the product was bleached by the addition of 1% activated charcoal and 1% aluminum oxide activated with acid, and then heated to and maintained at 100°C for about 30 min under water jet vacuum. After filtration, approximately 1 kg of a liquid wax ester was obtained, which congeals partially at room temperature and has an acid number of 0.12, a saponification number of 106, and an iodine number of 76.3.

Example 2: A parting oil with excellent parting qualities was prepared from 24% by weight of wax ester according to Example 1, 24% by weight of palm oil fraction, liquid, 48% by weight of medium-chain triglyceride (C_8:C_{10} = 60:30) and 4% by weight of soy lecithin.

Fluid Shortening for Batter Coating

R. Meyer and S.H. Lee; U.S. Patent 4,330,566; May 18, 1982; assigned to Armour-Dial, Inc. has developed an edible batter coating for foods which is applied to the food prior to cooking and readily adheres to the food.

This batter coating is shelf stable without the need for preservatives and provides food with the taste and appearance of deep fat fried foods. These coatings are provided by mixing together a combination of a particular type of fluid shortening, a food grade binder or adhesive, a suitable breading material, and optional flavorings and colors.

The breading materials are substantially distributed throughout the coating composition and this is accomplished without the need to incorporate emulsifiers. In short, although the coating composition is in a pourable or semipourable state, the breading material remains evenly distributed throughout and the product does not require shaking or mixing, except in the concentrated version, by the consumer prior to use. In addition, the product is shelf stable without the use of any heat treatment in the preparation of the coating composition or through the use of preservatives.

The fluid shortening is a principal and critical constituent of this coating composition, since it performs not only its usual function of shortening or tenderizing food, but also serves to uniformly suspend the breading material, food grade adhesives and any of the optional ingredients such as flavorings throughout the coating composition. The term fluid shortening means a shortening which is viscous fluid in the temperature range of 70° to 100°F and which is normally prepared from a liquid oil such as cottonseed or soybean oil. It is important that the shortening contain from about 10 to 35% by weight of solid triglyceride, largely in the beta crystal form. The solids content is measured at 92°F by the "Dilatometric Method"—AOCS Cd 10-57. Shortenings having from 20-30% of such solid triglyceride are preferred. If the percentage of beta crystals is less than about 10%, the fluid shortening will not adequately suspend the other materials present in the coating composition over an extended period of time.

Thus the viscosity of the liquid shortening should range between 1000 and 50,000 cp or about 0.2-17 Bostwick units and such viscosity is primarily dependent on the amount of beta crystals present in the fluid shortening. The total amount of fluid shortening present in the edible coating composition ranges from about 20 to 60% by weight, preferably about 45 to 55% in the ready-to-use form and about 26 to 31% in the more concentrated composition.

As food grade binder such materials as starch, sodium caseinate or spray dried egg white may be used. Starch which has been modified by the addition of sodium hypochlorite to the aqueous starch slurry—so-called oxidized starch is preferred. The amount of adhesive present in the ready-to-use composition should range between about 7 and 21%, preferably about 11 to 17%. In the concentrated form the amount ranges from about 7 to 30%, preferably about 8 to 28%.

The breading materials also form an important aspect in that they not only provide a "crunchiness" to the food product when baked, they also seem to contribute to the stability of the edible coating composition. The term "breading materials" includes all of the customary farinaceous materials such as bread crumbs, cereals, puffed carbohydrates, flours prepared from wheat, corn or oats or mixtures of these grains. Although the particular type of breading material used is not critical, the size of it is important. The breading materials used should fall within a U.S. Standard sieve size of about 3-12, preferably about 7-10. The amount of such breading may range from about 10 to 35%.

The batter coating is easily prepared by slowly adding the adhesive to the fluid shortening using low-speed mixing. Mixing is continued until lumps are no longer visible. Thereafter any other liquid ingredient—color(s), flavorings, are thoroughly mixed in. This is followed by the addition of any remaining dry ingredients other than the breading materials. Finally the breading materials are added with mixing until no visible lumps are present. Thereafter, the product is packaged in suitable containers.

Example: A ready-to-use batter coating was prepared from the following ingredients (% is by weight): 54.581% fluid shortening (30% beta crystals), 0.046% antioxidant, 0.023% citric acid, 5.690% water, 0.800% microcrystalline cellulose, 14.790% oxidized starch, 5.870% spices, and 18.200% corn flake crumbs. The foregoing product was formulated to have a so-called Southern fried chicken flavor.

Substitutes for Confectioner's Hard Butter

Hard butters are widely used in confectionary and various other edible products because they have good "stand up" characteristics at 50°-70°F, a very rapid melting beyond that point, and substantially total disappearance of any solid fraction at about human body temperature. In actual practice hard butter usually has a melting point (usually measured as a Wiley Melting Point, abbreviated WMP) between about 90° and 105°F or even higher (the high melting ones—above 103°F—generally being regarded as quite inferior substitutes for cocoa butter). The odor is nil and taste of hard butter ordinarily is bland. Such fat should be brittle at temperature up to about 75°F, this brittleness sometimes referred to as "snap." It should stand up to such temperatures without "sweating" or "bleeding" out to the surface of droplets or any visible liquid film.

J.J. Jasko and R.J. Zielinski; U.S. Patent 4,234,618; November 18, 1980; assigned to SCM Corporation have provided a process for making and using a hard butter useful as a substitute for cocoa butter. This hard butter comprises:

> at least about 96% C_{16-18} triglycerides with about 40-50% by weight of the total combined acids being unsaturated acids in the trans configuration, a substantial proportion of the remaining acids being also unsaturated and in the cis configuration, and having
>
> an SFI profile of 78-85% at 50°F and at 70°F, 70-80% at 80°F, 31-45% at 92°F, 3% maximum at 100°F, and 0% at 110°F; and
>
> a Mettler Dropping Point (MDP) between about 36° and 39°C.

This hard butter is particularly useful in a composition comprising about 15-40 parts cocoa butter, about 0-25 parts butterfat, with the sum of the parts of cocoa butter and of butterfat being not in excess of 40, and about 60-85 parts of the hard butter of this process.

The hard butter of this process is obtained by an improvement in process described in U.S. Patent 2,972,541 for the systematic, successive fractional crystallization of fat, the fat being a mixture of saturated and unsaturated triglycerides, from a fugitive solvent for the fat, the fat containing about 35-60% unsaturated triglycerides whose fat-forming acids exhibit a trans configuration, being substantially devoid of polyunsaturated triglycerides (that is, not more than about 5% by weight), and containing about 35-45% saturated triglycerides, into at least one saturated triglyceride-rich solid fraction and a solid fraction less rich in saturated triglycerides, but rich in triglycerides whose fat-forming acids are in a trans configuration and which fraction exhibits characteristics of a hard butter.

The improvement comprises subjecting the above solid fraction less rich in saturated triglycerides to (a) dissolution in fugitive solvent, thereby making a new solution, (b) fractionally crystallizing between about 75 and about 95% of the resulting solute fat as crystal from the new solution, (c) recovering the fat crystal, and (d) stripping solvent from the recovered crystal.

A fugitive solvent for this process is one that will dissolve fat to at least about 10 grams per 100 cc at a temperature not substantially above 50°C, which solvent has less solubility for the fat and particularly for the saturated parts thereof as temperature of the solution is reduced, and can be stripped away from the fat essentially completely if not entirely at a temperature and within a time that will not cause significant fat degradation (e.g., 5 to 60 minutes at a temperature not exceeding about 500°F using very low absolute pressure, e.g., 1 mm Hg or less, if necessary). Typical solvents having this property include aprotic solvents such as the nitropropanes and particularly 2-nitropropane, ketones such as acetone and 2-butanone, lower paraffins in the liquid phase such as hexane down to propane, acetamide, carbon tetrachloride, chloroform, hexamethylphosphoramide, benzene, dimethylformamide, tetrahydrofuran, and dimethyl sulfone.

Example: The starting material was a so-called "B" fraction cropped by systematic fractional crystallization of high trans selectively hydrogenated cottonseed oil as produced by the process of U.S. Patent 2,972,541.

One part of this B fraction hard butter starting material was used. Such feed was dissolved in 10 parts of 2-nitropropane at 120°F. This solution then was cooled at an average rate of about 1°F per minute to the final slurry temperature indicated. The refrigerant used was aqueous ethylene glycol on the jacket side of a scraped-wall heat exchanger. Each slurry was filtered by vacuum filter. The resulting prime cake in each instance was repulped with 175 parts of about 24°F 2-nitropropane, then refiltered under vacuum. Each resulting washed cake was stripped of 2-nitropropane and steam deodorized at 240°C and below 1 mm Hg absolute pressure for 60 minutes to yield the improved hard butter. Such product had between 14 and 15% of its combined acids in the cis configuration versus about 17% for the starting hard butter.

This hard butter obtained had an SFI profile of 79.1% at 50°F, 79.5% at 70°F, 76.7% at 80°F, 39.3% at 92°F, 0.7% at 100°F and 0.0% at 110°F. The MDP was 37.8°C and the product contained 46.1% trans acid and had an iodine value of 55.0.

When used as a replacement for cocoa butter, the products had better hardness, better snap, better gloss characteristics, and better organoleptic properties than like compositions would have had, had all the hard butter used been the starting material of U.S. Patent 2,972,541.

Fats for Marzipan Substitute

Edible compositions suitable for use an an alternative to marzipan and in particular to such a composition in the form of a free-flowing powder which converts to a marzipan substitute by the addition of water have been developed by *R.H. Barents and J.B. Rossell; U.S. Patent 4,278,700; July 14, 1981; assigned to Lever Brothers Company.*

Marzipan is a cooked mixture of ground almonds and sugar, sometimes known as almond paste where extra sugar is added. It is very expensive owing to the high cost of almonds. Chocolate fillings containing either marzipan or cheaper substitutes based on other comminuted drupes such as peach stones or soybean meal, promote bloom in the couverture, due to the incompatibility of the couverture glycerides with those in the marzipan or substitute.

The process provides a marzipan substitute composition, the glyceride content of which comprises an edible vegetable fat having dilatations at 20° and 35°C of at least 800 and at most 500 which is compatible with couverture fats to minimize blooming.

The compositions containing sugar and comminuted drupe or meal in addition to the fat may be provided in the characteristic plastic consistency for use in fillings and general confectionery purposes, or in the form of a free-flowing powder requiring only the addition of water or other aqueous additive.

Meal is the edible part of any grain or pulse, ground to a powder and is the finer part as distinct from the bran or coarser part. Preferably soybean meal is used, particularly containing at least 60% protein and previously heated to eliminate enzyme and bacterial action.

Preferably at least 3 to 30% each of fat and meal are present, especially roughly equal amounts, the residue in the free-flowing compositions being sugar of which preferably at least 40% but not more than 90% is present. The sugar may be in powdered and/or crystallized form according to the consistency required in the final product.

The fat preferably exhibits a dilation (D) at 35°C no more than 400, especially a maximum of about 200. At 20°C its dilatation is preferably at least 1000. It should be compatible with couverture fats and for this reason it should preferably consist, like most couverture fats themselves, predominantly of C_{16-18} triglycerides. Indeed the fat is more preferably itself a couverture fat. Of course, where the marzipan substitute is intended as the filling mass for a chocolate consisting of a predominantly lauric couverture then the fat in the composition should itself consist of a lauric fat, e.g., babassu, coconut or palm kernel oil. Hard butters including hardened or fractionated vegetable oils may be used.

The compositions may be in the form of a plastic mass ready for use, or a powdered mixture to which water, milk or sugar syrup including corn and potato starch syrups, may be added, preferably from 5 to 30%, in an amount according to the consistency required in the plastic product. They may be used alone or in admixture with conventional marzipan or almond pastes. The plastic composition may be used as a coating or to ornament cakes, pastries and other confectionery.

Example: A free-flowing powder was obtained by mixing 74% crystal sugar with 13.8% fat and 12% soy meal containing 65% protein. A little sorbic acid as antioxidant and almond flavor were added. Two samples were prepared containing 13.8% couverture fat, one of D_{20} 1450 and D_{35} 400 comprising a blend of hardened vegetable oils having a melting point of 38°C and iodine value 55, and the other of D_{20} 1900 and D_{35} 125 comprising a blend of palm midfraction and stearine.

Each sample was made up to a plastic consistency with the addition of 5 ml of water and 10 g of sugar syrup per 100 g of sample and enrobed in the form of bars, with a cocoabutter milk chocolate coating containing in one case a cocoabutter extender fat. In another example the couverture was based on the same fat as the filling.

These bars showed no blooming after four months' storage during which the temperature was alternated at daily intervals from 12° to 25°C. Similar bars containing only half the quantity of fat and soy meal, supplemented with 12.4% crushed almonds, were as successful, but comparative examples omitting only the added fat showed clear blooming phenomena after only two months' storage under these conditions.

Hard Fat Replacers

It has been found by *C.J. Soeters, C.N. Paulussen, F.B. Padley and D. Tresser; U.S. Patent 4,283,436; August 11, 1981; assigned to Lever Brothers Company* that the hard fat replacer properties of palm mid-fraction can be surprisingly improved by the incorporation of SOS, 1,3-distearyl-2-oleyl glycerol, or POS, 1-palmityl-2-oleyl-3-stearyl glycerol, of at least 85% purity.

It has been found that the incorporation of at least 85% pure SOS in palm mid-fraction gives a fat that is a useful and general partial replacer for cocoabutter, particularly useful in the preparation of normal plain chocolate. The fat can be used at higher levels in plain and in milk chocolate than can palm mid-fraction. Alternatively, requirements for palm mid-fraction when using such a fat can be less strict than when using palm mid-fraction alone. The incorporation of small amounts of the at least 85% pure SOS has the further advantage that more palm mid-fraction can be incorporated in plain chocolate than when palm mid-fraction alone is used as replacer.

A further and very important effect is that the antiblooming property is even better than that of palm mid-fraction alone. The incorporation of SOS in palm mid-fraction advantageously raises the dilatations and melting point and improves the crystallization and tempering characteristics. The effects are best appreciated when the mixture of the SOS and palm mid-fraction formed by incorporation of the SOS consists of between 5 and 95% of palm mid-fraction and 95 and 5%, particularly between 95 and 20%, of the SOS. 20% is the level above which the antibloom properties are particularly significant.

Further a fat consisting of a mixture of palm mid-fraction and the SOS and containing up to 42% of palm mid-fraction can be used as a partial replacer for cocoabutter to give a hardened milk chocolate. A fat consisting of palm mid-fraction and at least 85% pure SOS and containing less than 70% of the SOS, has been found to enable more palm mid-fraction to be incorporated than when palm mid-fraction alone is used as replacer.

When the amount of the at least 85% pure SOS incorporated in the palm mid-fraction is such that the fat obtained consists of 10 to 20% of the SOS and 80 to 90% palm mid-fraction, the fat is an excellent full fat replacer; excellent plain chocolate can be prepared using this fat instead of cocoabutter. Some adjustment of tempering conditions must be considered. Also, when the fat consists of 20 to 30% of the SOS and 70 to 80% palm mid-fraction, the fat can be used as a full fat replacer to prepare a plain chocolate suitable for tropical use. Some adjustment of tempering conditions must be considered.

It has also been found that a fat consisting of a mixture of palm mid-fraction and, based on the fat, at least 40%, preferably 60% of at least 85% pure POS is a useful hard fat replacer. When it contains 40 to 60% of the POS it can if necessary be used as an adequate replacer for Coberine and can be used at high levels in chocolate. When it contains more than 60% of the POS it can be used at high levels in chocolate to give plain chocolate suitable for tropical use. These replacers are preferably used up to a total of 50%, particularly at 30% of the cocoabutter normally used.

Coberine is a partial cocoabutter replacer. Such products contain other fats as well as palm mid-fraction.

A fat formed by the incorporation of at least 85% pure POS with palm mid-fraction and that consists of palm mid-fraction and up to 80% of the at least 85% pure POS is a useful partial replacer for cocoabutter. It enables more palm mid-fraction to be incorporated in plain chocolate than when palm mid-fraction alone is used as replacer.

Also it has been discovered that such a fat consisting of at least 85% pure POS and up to 45% of palm mid-fraction can be used as a partial replacer for cocoabutter to give a plain chocolate suitable for tropical use.

Such a fat consisting of 15 to 25% palm mid-fraction and 75 to 85% of the at least 85% pure POS is an excellent full fat replacer; excellent plain chocolate can be prepared using this fat instead of cocoabutter.

Further such a fat consisting of at least 85% pure POS and up to 50% palm mid-fraction is an excellent full fat replacer for use in the preparation of milk chocolate. When the amount of palm mid-fraction is less than 35%, hardened milk chocolate can be obtained.

Particular advantageous combinations of SOS, POS and palm mid-fraction are described below. It should be noted that the SOS and the POS can be incorporated separately or together. In the first case, each must fulfill the requirement of being at least 85% pure; in the second case, the mixture must fulfill the requirement of being at least 85% pure.

Advantageous hard fat replacers are those that can replace all the added cocoabutter used in the preparation of chocolate, i.e., that can be used as the additional fat to that present in the cocoa-liquor. The amount of such added fat is usually about 30% of the total fat. It has been found that a hard fat replacer consisting of a mixture of palm mid-fraction, at least 85% pure SOS and at least 85% pure POS and containing less than 30% of the SOS and less than 80% palm mid-fraction can fulfill the advantage.

It has been discovered that a fat formed by the incorporation of either at least 85% pure SOS or at least 85% pure POS or both and consisting of up to 75% palm mid-fraction, from 0 to 60% of the at least 80% pure POS and any remainder of the at least 85% pure SOS is useful as a partial replacer for cocoabutter to give a plain chocolate suitable for tropical use. When the percentage of the SOS is greater than 30, there is additional advantage that the tempering conditions need not usually be altered.

Further such a fat consisting of 3 to 35% of the at least 85% pure SOS, 36 to 54% of the at least 85% pure POS and 15 to 35% palm mid-fraction has been found to be an excellent full fat replacer in the preparation of plain chocolate. Usually the tempering conditions need not be altered.

Also such a fat consisting of 5 to 35% of the at least 85% pure SOS, 30 to 80% of the at least 85% pure POS and 10 to 30% palm mid-fraction has been found to be an excellent full fat replacer in the preparation of a tropical plain chocolate.

Also it has been discovered that a fat consisting of a mixture of 27 to 42% of the at least 85% pure SOS, 30 to 50% of the at least 85% pure POS and 17 to 33% palm mid-fraction is a useful full fat replacer in the preparation of milk chocolate. The tempering conditions usually need not be altered.

F.B. Padley, C.N. Paulussen, C. Soeters and D. Tresser; U.S. Patent 4,276,322; June 30, 1981; assigned to Lever Brothers Company have also provided a hard fat similar to that described in U.S. Patent 4,283,436.

This composition comprises a mixture of a natural, high POP/POS/SOS-fat and a narrowly defined SOS/POS-fat. Palm mid-fraction is a preferred natural, high POP/POS/SOS-fat.

Such formulations can be defined by the following Figure 9.1 where **A** represents POS fat; **B** represents SOS fat; **C** represents palm mid-fraction; **E** represents illipe; **F** represents shea stearin; and **G** represents POP.

Figure 9.1: Hard Fat Composition Useful in Chocolate

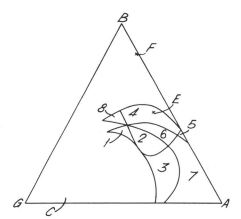

Source: U.S. Patent 4,276,322

The formulations shown in Figure 9.1 are found to be particularly useful in the preparation of milk chocolate.

The compositions defined by areas **1, 2** and **3** have been found to be useful for preparaing normal milk chocolate. Compositions defined by area **2** are preferred because usually tempering conditions need not be altered. The compositions defined by areas **4, 5, 6, 7** and **8** have been found to be useful for the preparation of hardened milk chocolate.

Compositions defined by areas **6** and **7** are preferred because hardened milk chocolate can be obtained with surprisingly sharp-melting characteristics. Compositions defined by areas **4** and **6** are preferred because usually tempering conditions need not be altered.

In describing this hard fat composition, the following symbols are used for fatty acid residues: A for Arachidonic, E for Elaidic and double-bond isomers, G for Saturated, Ln for Linoleic, M for Myristic, O for Oleic and double-bond isomers, P for Palmitic, S for Stearic, and U for Unsaturated. Triglycerides are designated as SOS, 1,3-distearyl-2-oleyl glycerol; SPO, 1-stearyl-2-palmityl-3-oleyl glycerol; and SSO, distearyl-oleyl glycerol, position of the acid residues unspecified. In detail, the hard fat of this process has a POP, POS, SOS-composition falling within an area defined in Figure 9.1 and comprises a cocoabutter-compatible hard fat with a POP, POS, SOS-composition outside the area and an SOS/POS-fat consisting of

   (A) from 80 to 98% of SOS and POS,

   (B) other O, P and S triglycerides consisting, by weight of the SOS/POS fat, of

      (a) less than 5% of GGG,
      (b) less than 5% of PPP,
      (c) less than 3% of each of other GGG,
      (d) less than 10% GGO consisting, by weight of the SOS/POS fat, of

         (1) less than 5% of each of PSO and SPO,
         (2) less than 3% PPO, and
         (3) less than 8% SSO,

      (e) less than 10% OOG,
      (f) POP and

   (C) (a) less than 10% of each SES and SLnS,
      (b) less than 5% of each of PES, PLnS, PEP and PLnP, and

   (D) partial glycerides of O, P and S, consisting, based on the SOS/POS fat, of

      (a) less than 10% of each of total monoglycerides, of total diglycerides, of diglycerides containing at least one O, and of unsaturated monoglycerides, and

      (b) less than 6% of each of saturated monoglycerides and of disaturated diglycerides, in (B), (C) and (D) at most 15%, based on the total fatty acid residues in the SOS/POS fat, of P and S being randomly replaced by M and A.

Stabilizing Fats for Peanut Butter

The disclosure by *S.A. McCoy; U.S. Patent 4,341,814; July 27, 1982; assigned to The Procter & Gamble Company* relates to peanut butter stabilizers suitable for preventing the separation of oils from peanut butter. These stabilizers improve the texture of the peanut butter and reduce the stickiness, and provide a fluid consistency over the normal range of temperatures encountered, about 50° to about 100°F.

The peanut butter stabilizer system of this process comprises:

> (a) an intermediate melting fat having an IV of 25 to 45 and a solids content of from about 80 to about 95% at temperatures of from 50°F, and a solids content of less than 50% at 115°F;
>
> (b) a hydrogenated fat or oil having an IV of not greater than 10, the fat or oil having a high proportion of fatty acids having from 20 to 24 carbon atoms.

The iodine value (IV) of an oil or fat indicates the number of grams of iodine equivalent to the halogen absorbed by a 100 g sample. In general, the lower the iodine value of a given fat or oil, the greater will be its content of solids at a given temperature. That is, as the triglyceride molecules become more saturated by the addition of hydrogen (i.e., the double bond content decreases), the consistency of the fat or oil becomes more solid. The iodine value can be readily determined by the well-known Wijs method.

The amount of stabilizer should range from about 1 to about 4%, by weight of peanut butter, and preferably from about 1.5 to about 3.5%, most preferably from 2 to 3%. If less than about 1% of the stabilizer is present the peanut butter will have an undesirable amount of oil separation due to insufficient stabilization when the product is stored at room temperature. If more than about 4% of the stabilizer is used, the product will be too firm and will have a waxy taste.

The intermediate melting fat fraction of the stabilizer is prepared by hydrogenating a triglyceride fat oil to an iodine value of from about 25 to about 45, preferably from about 30 to 40. The triglyceride oil should be hydrogenated under conditions such that the intermediate melting fat which is produced has a solid content of from about 80 to about 95% at temperatures of from 50° to 90°F. The solids content should decrease rapidly to about 35 to 50% at 115°F.

Preferred intermediate melting fractions are cottonseed oil, hydrogenated to an iodine value of about 36, and blends of soybean oil and palm oil, the ratio of soybean oil to palm oil being in the range of from about 80:20 to about 60:40. These oils are blended to an IV of about 35.

The substantially completely hydrogenated triglyceride (fat or oil hardstock) of the stabilizer should have an iodine value not exceeding about 10. The hydrogenated fat or oil is formulated to consist essentially of a hydrogenated fat and oil having a high proportion of fatty acids containing 20 to 24 carbon atoms such as hydrogenated rapeseed oil, mustard seed oil, salmon oil, herring oil, pilchard oil, menhaden oil, and sardine oil. The highly preferred hardstock component herein is hydrogenated rapeseed oil. This portion of the hardstock component having the high proportion of fatty acids having 20 to 24 carbon atoms is a non-beta-phase tending hardstock.

Optionally, a second hardstock component can be added to the stabilizer. It is derived from substantially completely hydrogenated fats and oils such as lard, sunflower seed oil, cottonseed oil, safflower seed oil, linseed oil, sesame seed oil, hazelnut oil, palm oil, soybean oil, peanut oil, olive oil, and corn oil. Highly preferred for use is hydrogenated cottonseed oil having an iodine value of about 4. This second component is added to the hardstock having a high proportion of C_{20-24} fatty acids in the ratio of 3:1 to 1:3.

Example: Peanuts were roasted, blanched and cleaned to remove all debris. The peanuts were then fed into a peanut butter mill consisting of two vertical parallel circular plates, one fixed and one rotating. The inner surfaces of the plates were equipped with burs or teeth with a small clearance between the plates. The peanuts were fed into the center of the plates and were forced out the sides in the form of fine peanut particles and oil. In separate runs various stabilizers in ground solid form, each with about 0.4% emulsifier, about 1% salt and 5.5% sugar, by weight of the final peanut butter, were mixed with the peanut paste after milling. The ground product of a temperature of about 160°F was fed into a recirculating system to thoroughly mix the ingredients.

The peanut butter with stabilizer is then fed into a Manton-Gaulin homogenizer where it is further ground to finely divide the peanuts, salt and sugar. During this grinding step the temperature rises to about 180°F. The peanut butter is then passed through a scraped wall chiller in which it is rapidly cooled down to temperatures between 90° and 100°F. Immediately following this freezing step, the product is gently agitated and placed into jars and stored at temperatures within the range of 50° to 90°F.

The following stabilizer systems were used where the amounts listed of the additives are in % by weight of the final product: (A) 2.2% cottonseed (IV 36), 0.35% cottonseed hardstock, and 0.35% rapeseed hardstock; (B) 1.0% cottonseed (IV 36), 0.55% cottonseed hardstock, and 0.55% rapeseed hardstock; (C) 2.2% 70/30 soy/palm (IV 35), and 0.7% rapeseed hardstock; (D) a combination of soybean hardstock, rapeseed hardstock, and soybean hardstock totalling 1.8%.

When 99 panelists were asked to compare products A and D, product A was preferred by 76 for being less sticky, by 79 for creaminess and by 81 for faster melting.

When product C was compared with product D by a panel of consumers, C was preferred 55 to 25% (rest no preference) for creaminess, 43 to 18% (rest no preference) for less sticky, and 45 to 15% (rest no preference) for faster mouth melt. C was also compared to a commercial creamy peanut butter. Again it was preferred for creaminess 53 to 22% (rest are no preference), for less sticky 30 to 19% (rest are no preference) and for mouth melt 30 to 19% (rest are no preference).

When product B was compared to product D, B was preferred by 60% of the panelists for texture, which is a measure of mouth melt, stickiness and creaminess.

ADDITIVES FROM CITRUS WASTES

Juice Production by Enzyme Treatment of Whole Oranges

A process for producing orange juices from whole oranges or their rinds has been developed by *T. Mouri and H. Kayama; U.S. Patent 4,299,849; Nov. 10, 1981; assigned to Toyo Seikan Kaisha Limited, Japan.* More specifically, this process for producing orange juice decomposes and softens cellulose and hemicellulose in the tissues and cell membranes of these raw materials not by any mechanical treatment but by the action of cellulase produced by specified microorganisms.

The rind of a citrus fruit such as orange is composed of flavedo layer (yellow portion) containing oil gland cells at its outermost side and an albedo layer (white portion) inwardly of it. The unpleasant smell and bitterness of juices prepared by the conventional methods are due to the disintegration (decomposition) of the flavedo layer by the action of enzyme. Investigations led to the discovery that enzyme components having the activity of disintegrating plant tissues, which will be eluted at a pH of between 8 and 12, particularly between 9 and 11, by fractionation exhibit the action of disintegrating the flavedo layer, and therefore that to achieve the object of this process, it is necessary to use cellulase which decomposes polysaccharides such as cellulose to monosaccharides and which do not have the activity of disintegrating plant tissues.

According to this process, there is provided a method for producing a juice which comprises steaming and crushing a whole citrus fruit or its rind, adding to the crushed material an enzyme produced by a microorganism of the genus *Trichoderma* or the genus *Aspergillus* and being free from those components having the activity of disintegrating plant tissues (i.e., the activity of disintegrating the flavedo layer) which will be eluted at a pH between 8 and 12 by fractionation, reacting the mixture at a temperature of 30° to 55°C, deactivating the enzyme, filtering the reaction mixture, and separating the residue.

Commercially available C-TAP (an enzyme derived from a microorganism of the genus *Trichoderma,* Amano Pharmaceutical Co., Ltd.) and C-AP (an enzyme derived from a microorganism of the genus *Aspergillus*) can be used.

In a preferred embodiment of this process, a whole citrus fruit or its rind which has been steamed and crushed is mixed with at least 0.01% by weight, preferably at least 0.05% by weight, of the specified enzyme, and subjected to enzymatic reaction with stirring at a pH of about 2.5 to 6.5 and a temperature of 30° to 55°C for 2 to 24 hours. The reaction mixture is heated at a high temperature to deactivate the enzyme. The reaction mixture is then filtered, and the residue is separated. The upper limit to the amount of the enzyme is determined depending upon its cost. Heating may be unnecessary, or even when it is necessary, the heating temperature is low. Accordingly, the loss of vitamins during the reaction is small. The juice produced by this process is free from offensive odors and bitterness, and is nutritionally excellent.

Example 1: Mandarin orange of Unshiu variety (whole fruit) having a crude fat content of 0.3% by weight was steamed at 100°C for 2 minutes, and crushed to a size of less than 5 mm. C-TAP (an enzyme produced by *Trichoderma viride*) was added to the crushed material in an amount of 0.2% based on the weight of the raw material. The mixture was reacted with stirring in an enzymatic reaction tank at a pH of 3.9 and a temperature of 45°C for 3 hours. After the reaction, the reaction mixture was heated at 100°C for 1 minute to deactivate the enzyme. The reaction mixture was filtered by a pulper of 50 mesh to produce a product having a crude fat content of 0.1 to 0.15% in a yield of 93%.

Canned juices were produced from this product, and subjected to sensual tests. The offensive smell (molasses-like smell) and bitterness of the rind were not noted at all.

The residue obtained by the filtration was found to contain a flavedo layer (containing oil gland cells).

Similar results to the above were obtained when the above procedure was repeated using C-AP enzyme derived from *Aspergillus niger* instead of C-TAP.

Example 2: The procedure of Example 1 was repeated except that doriselase (an enzyme prepared by the hyphae of mushrooms) was used instead of C-TAP. The flavedo layer containing oil gland cells was consequently decomposed, and there was hardly any filtration residue. This afforded a yield of 95%. However, the product contained 0.3% of the crude fat.

When canned juices were produced from this product and subjected to sensual tests, the offensive odor and bitterness of the rind were noted, and the juices were not fit for drinking.

Use of maserozyme, produced from a microorganism of the genus *Rhizopus*, also produced juice similar to that of Example 2.

Citrus Fiber Additive

A process has been developed by *C.C. Lynn; U.S. Patent 4,225,628; Sept. 30, 1980; assigned to Ben Hill Griffin, Inc.* for preparing an edible fiber from citrus fruit wastes. The citrus fiber is useful as an additive in various food products. Citrus fruits are normally washed, passed through an oil extraction process and then halved and reamed to extract the juice.

In the preferred embodiment of the process, fresh citrus waste, resulting from normal juice extraction, i.e. the peel, most of the pulp, membranes and seeds, is promptly subjected to oil removal, chopping into comminuted chunks, neutralized in acidity and pressed to minimize enzyme reactions and to remove about 15% of the water. The use of fresh citrus waste and the prompt processing of it is important in that it minimizes enzymatic reactions which produce undesirable flavors and/or color as is well known. The waste, constituting approximately 80% orange and 20% grapefruit, is then ground into ¼ inch uniform particle size and blended with a solvent-extracted sesame flour, containing about 50% protein, in a ratio of one part sesame flour to fifty parts citrus waste, as by weight. The mixture is then dried in a convection dryer with high velocity air flow at preferably 210°-230°F with a thickness layer of the material in the dryer being approximately 1". Less thickness decreases dryer efficiency while increased thickness tends to inhibit thorough drying of the material.

The lack of any elevated temperature previous to this drying step and the moderate temperature at which the mixture is to be dried is most important in that fast and efficient dehydration is a practical requirement, while providing for partial sterilization and enzyme deactivation to achieve stability to the final product are also important. Concurrently, the adverse effect on the valuable and heat sensitive substances is minimized, such as the denaturing of the proteins, destruction of vitamins, and carmelization of the sugars in the carbohydtate fraction of the mixture. These factors are not only important from a nutritional point of view but also are vital for obtaining a material with unique functional properties.

After drying the mixture to less than about 7% moisture, and preferably 3-5% moisture, such mixture is milled into a desired particle size having a flour consistency which may be bagged and shipped to baking plants and the like food processors for use as a fiber additive, for adding to or replacing a portion of other ingredients such as sugar, eggs, flour, extending meat products, etc. The product is intended to be an intermediate or a semifinished material which is incorporated into other common human foods to result in a nutritionally improved human food product.

While it is contemplated that the citrus waste may consist of all orange, the addition of grapefruit enhances the final product by providing increased pectin fiber, desirable color, and modifying the functional properties i.e., without grapefruit the final product would be, for example, browner in color and more readily detectable in white flour baked products, and the functional properties would be somewhat different. Different varieties of citrus fruits exhibit considerably different analytical compositions, and thus provide different functional properties to the citrus fiber additive product of this process. Therefore, it is necessary to continually monitor the product and change the proportions of orange, grapefruit, and sesame flour as needed in order to produce a product with the desired functional properties.

The citrus fiber can be used with grain flour to produce cakes and breads with no loss in volume or change in grain or color. In bread prepared from 25% whole wheat flour, 70% white flour and 5% of the additive, the resulting bread would have the volume and mouthfeel similar to white bread and resemble the texture thereof, while the total fiber content would be about the same as total whole wheat bread.

The additive can also function as an egg extender because of the high protein content of the additive. In recipes where several eggs are required the additive can replace 15-35% of the requirements of whole egg solids.

Another use for the citrus fiber additive is as a pie dough conditioner, particularly in the reconditioning of scrap pie dough resulting from trimming the dough extending beyond the edges of the pie pan.

Still another use is in flavoring preparations. Flavor concentrates are usually in the form of oils that must be transformed into some type of dilute emulsion in order to be used in food products. This frequently entails the use of a gum or an emulsifier to get the oil to be dispersed in tiny droplets in water for addition to flour for cake or bread. The citrus fiber additive serves as a carrier for the oil and eliminates the necessity of preparing an emulsion, and in addition, it provides some flavoring of its own to produce a more natural flavor when the basic flavor is to be orange, lemon, or grapefruit. As a carrier for oily flavor concentrates, the citrus fiber additive can absorb up to about 50% of its weight in oil or other liquid flavor concentrate and maintain the character of a dry flour, which is very convenient for use in cake and pastry recipes. Yet another use of the citrus fiber additive is as an additive in sausage or other ground meat. At levels of about 1% by weight of the meat, the additive serves to reduce the loss of water and fat during cooking and thus preserves the flavor and texture of the original product.

Juice Vesicle Solids in Cake Mixes

The development by *P.E. Hart and T. Nichols; U.S. Patent 4,275,088; June 23, 1981; assigned to General Mills, Inc.* relates to improved dry cake mixes which produce improved finished layer cakes. The finished layer cakes exhibit enhanced moistness initially and for several days thereafter compared to conventional finished layer cakes prepared from dry mixes containing various hydrophilic colloids. The dry mixes comprise from about 35 to 45% by weight of the dry mix of wheat flour, from about 40 to 54% by weight of sugar. The sugar to flour ratio in the dry mix ranges from about 1.20 to 1.35. The dry cake mixes additionally comprise from about 1 to 16% by weight of an emulsified shortening and from about 0.5 to 4% of a chemical leavening agent. The dry mixes also comprise from about 0.2 to 10% by weight of processed juice vesicle solids. The processed juice vesicle solids have (1) a moisture content of from about 5 to 12% by weight of the processed juice vesicle solids, and (2) particle size of from about 50 to 400 microns.

Preferably, the juice vesicle solids are prepared by drum drying raw juice vesicles. Drum drying is an established method for drying a variety of food product material and is recognized as both an economical and continuous method of dehydration. The raw citrus juice vesicle material is spread into a rotating single or double roll of a drum dryer in a thin, uniform layer.

After a partial revolution, a scraper knife removes the dried juice vesicle solid material from the drum. The rotating dryer can be either steam or gas heated to provide a roll surface temperature ranging between 300° to 350°F. Required steam pressures needed to provide such roll temperatures range between 80 to 100 psig. The roll speed of the drier depends on the temperature, the feed rate of the raw juice vesicle material and the static gap setting of the rolls. The static gap is set at approximately 0.002-0.005 inch (i.e., 1-5 mils). The feed rate is controlled by a moocher, which is a movable rake which oscillates between the rolls at the nip formed by the rolls. This insures a uniform surface application of the raw juice vesicle material on the roller surfaces. When removed from the drum drier, the dried juice vesicle solid material has a final moisture content of approximately 6-12%. The sheeted dried juice vesicle material is then conveyed to a grinder or other size reduction apparatus.

The vesicle solids useful herein are also essentially characterized by their particle size. Vesicle solids useful herein have a particle size distribution such that at least 90% by weight of the vesicle solids have a particle size of less than 400 microns (i.e., through U.S. Standard Sieve No. 40). Preferred vesicle solids have particle sizes such that at least 90% by weight of the vesicle solids have a particle size less than about 200 microns. Maintenance of vesicle solid particle size within these limits is essential for the realization of finished cakes having acceptable visual and textural attributes.

Example: A dry mix for a yellow layer cake is prepared in the following manner:

Approximately 170 g of raw juice vesicles having a moisture content of about 95% are fed to a Werner Lahara two-foot double cylinder drum drier set at a static gap of 0.5-1.3 mil. The drum drier is operated at a drum surface temperature of about 300°-350°F by 80-100 psig steam and operated at approximately 1 rpm. The raw juice vesicles are fed to the drum drier at a feed rate of 230 lb/hr. Upon removal from the drum, the juice vesicle material forms a sheet approximately 4 mil in thickness having a moisture content of about 6%.

The sheet of 6% moisture processed juice vesicle solids is then fed into a granulator or flaker at a feed rate of 25 lb/hr. Upon exiting the granulator or flaker through a screen having a size opening of a No. 10 U.S. Standard Size Wire, the juice vesicle solids are fed into an Alpine American Corp. alpine-pin Kolloplex mill of 10 hp at a rate of 6-8 lb/min. Upon exiting the alpine-pin mill, the juice vesicle solids have a particle size distribution within the prescribed limits.

The juice vesicle solids material so prepared is then mixed together with 240.3 g sucrose, 18.9 g dextrose, 205.2 g flour, 10.8 g dry milk solids, leavening and optional ingredients in a ribbon blender to form a relatively uniform mixture. Then 37.8 g of emulsified shortening is blended in the dry mix in the ribbon blender. The dry mix is then fed into a conventional finisher and worked until a sample has an oil lump size of less than 10% on a No. 10 screen, U.S. Standard size. The dry mix is then packaged in a conventional manner.

A cake batter is prepared in conventional manner by mixing the cake mix prepared as described to 3 whole eggs (150 g), ⅓ cup (75 ml) of liquid shortening, and 1 cup (240 ml) of water. The mix and liquid additives are initially mixed in a Hamilton Beach stand mixer at a low speed (approximately 200 rpm) for 30 seconds to moisten the mix. The mixer speed is then increased to a medium setting (approximately 500 rpm) and mixed for two minutes. The batter is then divided equally and poured into two greased 9" round cake pans. The batter is then baked at 350°F for 30-35 minutes.

The finished cakes so prepared are characterized by enhanced moistness and fine grain structure. When tasted, the finished cake failed to taste gummy.

Juice Vesicle Solids in Snack Foods

Intermediate moisture food products which contain major amounts of orange citrus juice vesicle solids have been described by *J.R. Blake; U.S. Patent 4,205,093; May 27, 1980; assigned to General Mills, Inc.* The food products also contain from about 12 to 40% of a carbohydrate nutritive sweetener and sufficient amounts of edible organic acid to provide a pH of about 2.5 to 4.5. The food products have a moisture content of from about 8 to 30% and a water activity ranging from about 0.30 to 0.85. The disclosed food products exhibit prolonged moisture retention and microbial stability.

Juice vesicle solids can be supplied from stabilized raw juice vesicles pasteurized by heating at 180°F or higher for 0.25 hr or longer. Typically, the solids concentration of raw, pasteurized or "stabilized" juice vesicles is from about 2 to 5%. Accordingly, when raw, stabilized juice vesicles are used to supply the juice vesicle solids, a drying step is required in the food product preparation method in order to control the food product moisture content within the given essential ranges.

In another embodiment, the juice vesicle solids are supplied from processed juice vesicles. As used herein, the term "processed juice vesicle(s)" refers to juice vesicle material whose water content has been reduced from the level of raw juice vesicles by either thermal and/or mechanical means. The term "drying" is used herein to characterize any thermal or radiation means for reducing the water content of the juice vesicle material or the present food products. Processed juice vesicles are thus characterized by the level of moisture remaining after moisture reduction. By controlling the moisture content of the processed juice vesicles, the present food products can be prepared without the necessity for a drying step.

It has been unexpectedly discovered that the food products of this process which have been prepared containing highly processed juice vesicles are particularly shelf stable against moisture loss and microbial attack. "Highly processed juice vesicles" are defined as juice vesicle material which ranges from about 3 to 12% moisture on a juice vesicles basis.

Of course, the food products can be prepared from mixtures of raw citrus juice vesicles and highly processed juice vesicles. In such case, the weight ratio of solids provided by the raw juice vesicles to the solids provided by the highly processed juice solids ranges from about 10:1 to 20:1.

Example 1: A formulation is prepared by first mixing together the heavy wet items including the 20 g corn syrup and the 1000 g raw orange vesicles in a Hobart Paddle Mixer. To this wet mixture is added a mixture of the 6.72 g glycerin and 0.185 g flavoring and color. Mild agitation is continued until the wet mixture is well mixed. A dry mixture is prepared by hand-mixing the 36.7 g sucrose, 7.2 g citric acid, 3.1 g starch, 1.56 g gelatin and 0.76 g xanthan gum. Then, the dry mixture is added to the wet mixture in the Hobart Paddle Mixer and the formulation is well mixed with mild agitation. The resulting formulation has a pudding-like consistency.

The formulation is extruded into a ½-inch diameter roll through a Selo stuffer at a pressure of 0 to 50 psig. The roll extrudate is sectioned into two pieces, each of approximately five inches. The cylindrical pieces weighing approximately 30 g are then placed into an oven and held at 140°F for 14 hr. After drying, the food products which originally totalled 1076.22 g now measure approximately five inches in length and weigh 200 g.

The food products so produced have a pH of about 4.2 and a water activity of about 0.68. The food product when consumed exhibits a tender texture similar to licorice.

The food products are then packaged in a suitable cellophane wrapper. The combination of low water activity and low pH renders the samples shelf stable for extended periods. Moreover, little moisture loss is observed.

Food products of substantially similar physical and organoleptic character are realized when in the Example 1 formulation all or part of the citric acid is replaced with an equivalent amount of ascorbic, phosphoric, tartaric, malic acids or mixtures thereof, or the sodium or potassium salts thereof.

Example 2: A food product in the form of a cylindrical food stick weighing about 100 g is prepared from 8 g corn syrup, 15 g sucrose, 7 g fructose, 64.2 g of processed orange vesicles having 26% moisture content, 3 g anhydrous citric acid, 3 g glycerin plus strawberry flavor and color.

Such a food product is prepared in a manner similar to that described in Example 1. However, approximately 235 g of raw orange vesicles (95% moisture) are first drum dried at 268°F for a residence time of 1 minute until the processed orange vesicles have a moisture content of about 25%. Then, the processed orange vesicles are mixed together with the corn syrup, glycerin, flavor and color to form a wet mix. A dry mix of the sugar, fructose and citric acid is prepared. The dry mix is thereafter blended into the wet mix to form the full formulation. The full formulation is extruded into a long, cylindrical shape and cut into sections to produce the food product.

The food product so produced has a pH of about 4.0 and a water activity of about 0.65. The food product when consumed exhibits a chewy texture similar to dried fruit.

Food products of substantially similar physical and organoleptic character are realized when in the Example 2 composition all or part of the sucrose is replaced with an equivalent amount of dextrose, glucose, maltose, maple syrup solids or apple juice powder.

Comestible Bases Containing Citrus Juice Vesicles

Citrus juice residual juice vesicles are a by-product of commercial citrus juice preparation. Juice vesicles can be obtained from the finer materials (i.e., pulp) associated with the juice which is segregated from the juice by screening. A variety of terms have been loosely used to refer to this pulp material or parts thereof in its untreated forms. It has been called at various times in its untreated state, "juice vesicles," "juice sacs," or "finisher pulp." The juice vesicles are the membranes forming the juice sacs of the citrus fruit. During juicing operations, the juice sacs rupture and release their juice. Thus, juice vesicles is used synonymously for the residual citrus juice sac materials remaining after the release of the juice from the juice sac.

A typical analysis of citrus juice vesicles useful in these food products comprises 36% pectin, 33% crude fiber, 16.5% sugar, 10% protein, 2.75% ash and 1.75% fat.

Although the moisture contents of juice vesicles are quite high, the water is tightly bound and not readily removed by mechanical means such as conventional filtration. Juice vesicles having moisture contents below about 89% obtained by partially thermally drying are not contemplated for use herein. While the precise phenonemon is not understood, it is believed that drying of the juice vesicles irreversibly alters their functional properties rendering them unsuitable for use in the comestible bases as prepared by this preparation method.

Food products useful as comestible bases and methods for their preparation have been developed by *J.R. Blake; U.S. Patent 4,232,053; November 4, 1980; assigned to General Mills, Inc.* More particularly, the process relates to comestible bases containing citrus juice vesicles and to the methods of cooking by which the bases are prepared.

The comestible bases are useful in a variety of food products. The comestible bases are prepared by first blending from about (1) 25 to 65% citrus juice vesicles having a moisture content between about 89 to 96% by weight of the vesicles; (2) from about 7 to 45% by weight of a nutritive carbohydrate sweetening agent; (3) sufficient edible nonvolatile organic acid to provide the blend with a pH of between about 2.5 to 5.5; (4) from about 1 to 5% of an ungelatinized starch; and (5) from about 8 to 60% water. The process mixes prepared by blending these essential components range in viscosity between 3,000 to 6,000 cp at 190°F. The process mix is cooked at a temperature between 180° and 280°F to form the comestible bases. The cooking step results in the comestible base having a moisture content of between 30 to 60%, a viscosity of between 7,000 and 10,000 cp at 190°F or Bostwick flow viscosity of between 7.5 and 9.5 cm at 190°F and a water activity of between 0.6 and 0.95.

In certain uses of the comestible bases, it is highly desirable to employ homogenized juice vesicles. Homogenized juice vesicles can be prepared using conventional homogenization method and apparatus. Generally, homogenizers are divided into two groups according to the kind of energy introduced to the medium homogenized: (1) rotor or rotostator systems, e.g., agitators, emulsifying pumps and colloid mills, and (2) pressure systems, e.g., wedge resonators and pressure homogenizers. The pressure homogenizer is predominantly used in food processing since it has the best homogenizing effects. Preferably, such units which are used in the preparation of the homogenized juice vesicles, usefully employed in the comestible bases, are those homogenizers which are constructed to prevent contamination. Typically, juice vesicles are easily homogenized employing low to high homogenization pressures, e.g. 1,000 to 8,000 psig.

Example 1: A comestible base is prepared in the following manner: An uncooked blend is prepared from 830 g orange juice vesicles having an average moisture content of 90%, 620 g sucrose, 41 g ungelatinized modified starch "Thin Boil 60" produced by Hubinger Co., 1 g citric acid and 450 g water.

Such a formulation is prepared by first preparing a dry mix of the sugar, starch and the acid. Then, the unhomogenized juice vesicles are mixed with water in a sauce pan with mild agitation to form a wet mix. The agitation is continued while the dry mix is slowly added to the wet mix. Agitation is continued until the dry mix ingredients are completely dissolved thereby forming the uncooked blend. The blend is then heated with mild agitation to 190°F. A 250 g sample is taken and analyzed for acidity and for both Brookfield and Bostwick viscosity, and returned to the blend. The pH of the uncooked blend is 4.0 while the Brookfield viscosity was 4,000 cp and the Bostwick viscosity was 14 cm (1.0 min flow).

The blend is then heated to about 200°F and cooked in open air to form the comestible base. Samples are periodically withdrawn, analyzed and returned to the blend until the viscosity of the comestible base is about 9,000 cp at 190°F. Viscosity measurements are taken using a Brookfield HA Viscometer with a No. 3 spindle at 10 rpm. A cooking time of about 40 minutes is required to obtain this viscosity. The comestible base is found to have a Bostwick consistometer value of 9.5 cm at 190°F (1.0 min flow time). The moisture content of the cooked base is determined to be about 50%.

About 1500 g of the comestible base is prepared by the abovedescribed procedure. The comestible base so prepared is useful in the provision of a wide variety of food products including jams, jellies and fruit toppings for baked goods or ice cream.

Comestible bases of substantially similar physical and organoleptic properties are realized when the citric acid in Example 1 is replaced with an equivalent amount of malic acid, lactic acid and succinic acid.

Example 2: An uncooked blend having the following formulation is prepared from 89.82 lb orange juice vesicles of 91% moisture content, 25.88 lb of sucrose, 52.82 lb of corn syrup of 36.7°Be, 4.22 lb of Aytex, an ungelatinized starch (General Mills Chemicals), 0.72 lb of cream of tartar, 0.1 lb of citric acid, 0.1 lb of potassium sorbate and 26.52 lb of water.

The juice vesicles are first homogenized at about 6,000 psi using a Cherry Burk homogenizer at 60°F at a rate of 12 lb/hr. The homogenized juice vesicles have a consistency similar to fruit puree.

The homogenized juice vesicles are charged into an open Groen Kettle (approximately 40 gal capacity) equipped with a swept surface-type agitator and a steam jacket. Water is then added to the kettle using moderate agitation. Thereafter, the corn syrup is added to form a wet mixture.

A dry blend comprising the starch, sucrose, cream of tartar, citric acid and potassium sorbate is added to the wet mixture with moderate agitation to form the uncooked blend.

The kettle is heated with low-pressure steam until the uncooked blend is 190°F. The viscosity of the uncooked blend is determined to be about 4,200 cp.

The blend is then heated to about 260°F and cooked with moderate agitation to form the comestible base. Samples are periodically withdrawn, analyzed and returned until the viscosity of the comestible base is about 9,500 cp measured at 190°F. A cooking time of about 2½ hours is required to reach this viscosity. The comestible base is found to have a Bostwick value of 9.5 cm (60 sec flow at 190°F). The moisture content of the base is determined to be about 48% while the pH is about 4.0. The water activity as measured by a Beckman Model SMT is found to be 0.90. About 150 lb of comestible base are produced by this procedure. Thus, about 50 lb of water are removed from the blend during cooking to realize the comestible base.

Comestible Bases in Frosting Compositions

Highly aerated edible toppings are generally referred to as "fluffy frostings." These fluffy frostings typically have densities of less than about 0.4 g/cc. Such highly aerated compositions are generally fat-free, and depend on the whippability, of material such as egg whites, gelatins, enzymatically degraded soy whipping proteins or certain polyglyercol esters of higher fatty acids.

Such frostings can also comprise sugar, water and optionally a wide variety of foam stabilizers such as algin, gelatin or a cellulosic derivative, such as carboxymethylcellulose. Fluffy frostings have a unique desirable mouthfeel and eating quality provided by the high level of aeration.

A primary disadvantage possessed by conventional frostings is their instability over a period of time. For example, it is well known that fluffy frostings, although highly aerated and of desirable eating quality when fresh, generally lose air and/or liquid and become rubbery or marshmallow-like upon storage or while standing on a cake overnight. This disadvantage is seen in both homemade fluffy frostings and in currently available mix products. This instability characteristic also explains why "ready-to-spread" prepared fluffy frostings have not been made commercially available.

It would be desirable to have an edible topping which combined the desirable attributes of both creme icings which contain shortening such as a creamy, rich mouthfeel and also the attributes of fluffy frostings which are fat-sparing, such as their unique lightness. Thus, there is a continuing need for fat-sparing frosting compositions which provide frostings having the organoleptic properties of creme icings yet exhibiting substantially greater aeration. Accordingly, it is an object of the disclosure by *J.R. Blake; U.S. Patent 4,232,049; November 4, 1980; assigned to General Mills, Inc.* to provide frosting compositions which upon aeration realize low density fat-sparing frostings which nonetheless exhibit the eating qualities of cream icings.

It is a further object to provide frosting compositions having enhanced stability against both air loss and syneresis.

These frosting compositions are substantially free of emulsified oleaginous material. The frosting compositions comprise from about 90 to 98% by weight of a comestible base, from about 0.4 to 3% of an acid-stable whipping agent and from about 0.1 to 0.5% by weight of an acid-stable polysaccharide gum.

The comestible base which forms the major portion of the frosting compositions is prepared by the methods described in Examples 1 and 2 of U.S. Patent 4,232,053.

The frosting compositions essentially contain from about 0.4 to 3% of an acid-stable whipping agent. Better results are obtained when the compositions contain from about 1 to 2.5% of the whipping agent. By "acid-stable" it is meant herein that the employable whipping agents be able to aerate the frosting compositions, which have a pH ranging from about 2.5 to 5.5, to densities of between about 0.6 to 0.8 g/cc when the whipping agent is present within the above specified range.

Whipping agents are well known in the food art and selection of suitable materials for use herein as the acid-stable whipping agent will pose no problem to the skilled artisan. Suitable materials can be derived as protein hydrolyzates from, for example, caseinate, whey, and various vegetable proteins. The protein hydrolyzates employed herein are water soluble (i.e., soluble at least to about 20% by weight at 25°C throughout the pH range of about 2.0 to 10). The soy protein hydrolyzates are particularly effective whipping agents. These proteins may be prepared by initially chemically hydrolyzing the soy protein to a prescribed viscosity range and thereafter enzymatically hydrolyzing the soy protein with pepsin to produce a pepsin-modified hydrolyzed soy protein whipping agent.

The frosting compositions can be used as edible topping for a wide variety of foods in the same manner as any conventional fluffy frosting or creme icing. The frostings are stable against foam collapse and syneresis for extended periods.

Example 1: A frosting composition is prepared from 300 g of the comestible base prepared in Example 1 of U.S. Patent 4,232,053, 2.0 g of the whipping agent Gunther D-100WA, a water-soluble soy protein hydrolyzate (A.E. Staley Manufacturing Co.), 0.4 g xanthan gum and 0.6 g FD&C Red No. 2.

The frosting composition was prepared by mixing together the comestible base, whipping agent and the xanthan gum and color using a Hobart paddle mixer.

The frosting composition can then be aerated using a kitchen-type mixer at high speed (about 850 rpm) for three to five minutes. The frosting so prepared has a density of 0.7 g/cc. After overnight storage, loosely covered at 75°F and at 90°F, the frosting did not change in appearance or eating quality.

Frosting compositions of similar physical and organoleptic properties are realized when in the Example 1 frosting composition the xanthan gum is replaced with an equivalent amount of locust bean gum, guar gum and mixtures of the gums.

Example 2: A frosting composition is prepared from 300 g of the comestible base prepared in Example 2 of U.S. Patent 4,232,053, 20 g of strawberry puree with a 30% moisture content, 4 g of the whipping agent, Gunther D-157A, a soy protein hydrolyzate (A.E. Staley Manufacturing Co.), 0.8 g guar gum, 0.2 g locust bean gum, 0.2 g xanthan gum, 0.5 g FD&C No. 2 Red and 2 oz natural strawberry flavor. This frosting preparation is prepared in a manner similar to that of the above Example 1.

The frosting composition so prepared can be whipped in a kitchen-type mixer used at high speed to form a strawberry frosting having a density of about 0.6 g/cc. Such a frosting has a natural strawberry flavor while exhibiting the rich mouth-feel of a cream frosting and the light texture of a fluffy frosting.

Comestible Bases in Frozen Desserts

A wide variety of dessert compositions are known from which frozen desserts can be prepared. Such compositions can be divided into dairy-based composi-tions (e.g., ice cream, some sherbets, etc.) and non-dairy-based compositions (e.g., fruit ices). Non-dairy-based frozen desserts are characterized in part by an absence of any milk-derived components including milk or butter fat, nonfat milk solids or even milk-derived proteins such as whey solids or caseinate. Nondairy frozen desserts can be further divided into aerated or aeratable compositions such as fruit ices on the one hand, and nonaerated compositions such as popsicles, on the other.

It is apparent that it would be desirable to be able to make high-quality nondairy aerated frozen products from a shelf-stable mix by simply whipping with a home mixer and then statically freezing the aerated mixture in the freezing compart-ment of the home refrigerator without requiring home ice cream-making ap-paratus.

However, frozen desserts made by static freezing do not compare favorably in consistency and overall appearance to conventional frozen desserts made by the normal commercial process.

The process described by *J.R. Blake; U.S. Patent 4,244,981; January 13, 1981; assigned to General Mills, Inc.* provides a nondairy dessert composition which upon aeration and subsequent static freezing exhibits the desirable mouthfeel, body, and texture which are reminiscent of commercial ice cream.

It is a further object of the process to provide nondairy dessert compositions which are substantially free of conventional emulsifiers as well as substantially free of specifically tailored emulsifier/stabilizer systems.

The dessert compositions essentially contain from about 50 to 98% by weight of a specially prepared comestible base described in Examples 1 and 2 of U.S. Patent 4,232,053 from about 0.4 to 4% of an acid-stable whipping agent, from about 0.05 to 0.5% of an acid-stable polysaccharide gum, and from about 1 to 15% of an edible fatty triglyceride oil. The dessert compositions have moisture contents between about 48 and 65%. Upon aeration and freezing, the frozen desserts have densities of about 0.2 to 0.95 g/cc.

Example 1: A nondairy aerated frozen dessert is prepared from 300 g of the comestible base as prepared in Example 1 of U.S. Patent 4,232,053, 75 g water, 40 g of a triglyceride oil (Crisco oil from Procter and Gamble Co.), 2.0 g Gunther D-100WA whipping agent, 1.6 g vanilla flavor and 0.4 g xanthan gum.

The comestible base is mixed with the other essential and optional ingredients in a home mixer at low to medium speed for three minutes. Then, the mixture is whipped at high speed (about 850 rpm) for five minutes. The resulting aerated mixture has a density of about 0.41 g/cc. The aerated mixture is then placed in the freezing compartment of a refrigerator (0°F) for about five hours.

The resulting product is an aerated frozen dessert which has the texture and appearance of commercial ice cream. The dessert is spoonable even upon immediate withdrawal from the freezer. The moisture content of the frozen dessert is about 53%.

Dessert compositions of similar physical and organoleptic properties are realized when the xanthan gum is replaced with an equivalent amount of locust bean gum, guar gum and mixtures of the gums.

Example 2: A nondairy dessert composition is prepared from 300 g of the comestible base as prepared in Example 2 of U.S. Patent 4,232,053, 35 g water, 20 g triglyceride oil (Durkex 25 from Glidden-Durkee, Inc.), 6 g Gunther D-100WA whipping agent, 0.2 g guar gum, 0.1 g locust bean gum, 0.1 g xanthan gum, 0.1 g natural strawberry flavor and 0.1 g FD&C No. 2 Red.

Such a dessert composition is prepared in a manner similar to that described in Example 1 above. The aerated frozen dessert composition is determined to have a moisture content of about 50.8% and the texture and appearance of commercial strawberry ice cream. The product is spoonable even at 0°F.

The product is allowed to warm for one hour at room temperature with closed lids and returned to the freezer for 12-16 hours. This sequence of thermal shocking is repeated 6 times. Upon withdrawal of the product from the freezer, the product was spoonable. Taste tests indicated an absence of the formulation of ice crystals. Moreover, visual examination revealed no syneresis from the product.

YEAST AND RELATED PRODUCTS

Yeast Autolyzates as Seasonings

The yeast autolyzate is a concentrated form of product which consists of all the autolytically solubilized cellular components (such as amino acids, nucleotides, polypeptides, proteins, glycogen, trehalose, sugars, B vitamins, and many unidentified flavor compounds) which have been hydrolyzed to smaller molecules by the action of enzymes such as carbohydrases, nucleases, proteases, etc. Its primary use is as a seasoning ingredient for the preparation of sauces, gravies, soups, etc. Because it has a meaty flavor, mainly due to its high glutamic acid content, yeast autolyzate is an inexpensive substitute for meat extracts.

The process of yeast autolysis is a slow reaction. It is induced by heating an aqueous slurry of yeast cells to a temperature where the cells are killed but where the endogenous enzyme activity is still high. Typically temperatures vary between 40°-60°C, with 50°-55°C being preferred because of the correspondingly higher rate of reaction without damaging the activity of the enzymes. Even at the higher temperature, however, the reaction still requires several days to obtain a suitable degree of digestion, which is usually terminated when approximately 50% of the total nitrogen is in the form of alpha-amino nitrogen.

If an autolyzed yeast product is desired, the autolyzed slurry is pasteurized at 80°-90°C and spray-dried to give the final product. If a yeast autolyzate is the desired product, the autolyzed slurry is cooled and filtered to remove the cell debris and pasteurized. The resulting supernatant is concentrated to yield a paste having a solids content of about 70-80%.

To accelerate the autolysis reaction, it is well known to employ one of several plasmolyzing agents. The most popular plasmolyzing agent is table salt. Other plasmolyzing agents include ethyl acetate, amyl acetate, chloroform, dextrose, and ethanol.

C. Akin and R.M. Murphy; U.S. Patent 4,285,976; August 25, 1981; assigned to Standard Oil Company (Indiana) have discovered that the autolysis process can be accelerated by the presence of thiamine and/or pyridoxine at concentrations of at least 0.01 weight percent. It has also been discovered that gradually increasing the temperature of the yeast suspension to the incubation temperature over a time span of from about 20 to about 180 minutes also increases the autolysis rate. By utilizing both of these discoveries, autolysis can be accomplished within about 5 hours or less.

Example 1: Compressed baker's yeast (1.7 kg), water (1.7 kg), and salt (90 g) were mixed to form a slurry having a holding temperature below 32°C. The slurry was divided into four equal volumes in one-liter beakers. Dextrose (3.75 g) was added to each beaker and mixed. Ethanol (7.5 ml) was then added to each beaker and mixed. Thiamine was added to three of the four beakers in the amounts of 0.2 g, 0.7 g and 2.8 g. No thiamine was added to the control. The addition of the thiamine took about 20 minutes. All of the beakers were placed in a water bath and their contents heated to 55°C within about 40 minutes and maintained at that temperature during the course of the autolysis. At periodic intervals samples were taken from the incubating beakers, centrifuges, and dried to determine the dry weights (solubles).

With no thiamine added, 26% and 37% solubles were formed after 75 minutes and 165 minutes, respectively. With 0.2 g (0.0255 wt %) thiamine, the figures were slightly higher than the above control. With 0.7 g (0.0893 wt %) thiamine added, the solubles were 32% and 41%, and with 2.8 g (0.357 wt %) of thiamine added the solubles were 35% and 45% at the above time intervals respectively.

Example 2: Same as Example 1, except pyridoxine was used in place of thiamine. A similar increase in the rate and amount of soluble matter formation was observed.

Example 3: On a larger scale, a preferred process scheme would be to dissolve 550 g of salt in 10 ℓ of tap water at room temperature (not to exceed 32°C). Compressed baker's yeast (10 kg) is added to the salt solution while applying mild mixing. The yeast disperses almost immediately to form an easy flowing slurry. The dextrose (100 g) is then added, which causes carbon dioxide generation almost immediately. Ethanol (200 ml) and thiamine (20 g) are then added to the slurry. The mixture is held at room temperature for 30 minutes (the temperature should not exceed 32°C) and thereafter gradually heated to 50°-55°C over a period of 40 minutes. The slurry is maintained at this incubation temperature range for about 4 hours. The incubation temperature should not exceed 58°C at any time to prevent deactivation of the enzymes.

Upon completion of the autolysis, the autolyzed slurry contains essentially two products. One is the insoluble cell residue and the other is the solubilized cell contents. These two products can be physically separated by centrifugation and dried. Individual washing steps can optionally be included for either fraction. The residue can be suspended as a 10% slurry and spray-dried to a powder, whereas the supernatant can be spray-dried directly to yield a yeast autolyzate.

Polyporus **Mycelium as Meat Additive**

A method for obtaining mycelium from fungi (genus *Polyporus*) by culturing in depth to give products for foodstuffs having high protein content has been described by *A.K. Torev; U.S. Patent 4,212,947; July 15, 1980; assigned to DSO "HRANMASH," Bulgaria.*

This process is carried out at a temperature of 22° to 50°C, a pH from 4.5 to 7.5 and air from 0.6 m³ to 1.2 m³ per each 1 m³ of nutrient media per minute.

Preferably, the fungi are cultivated at a volume of 2 m³ in the course of a twenty-four-hour time cycle, at a volume of 20 m³ in an eighteen-hour cycle, and in three growth cycles in progressively greater volumes of nutrient media contained in separate fermentation vessels, with the proviso that the first cycle involves a nutrient media volume of 2 m³ and a 24-hour time period, the second cycle involves a nutrient media volume of 20 m³ and a time period of 18 hours, and the third cycle involves a nutrient media volume of 100 m³ and a time period of 6 hours, and with the further proviso that 50% of the volume of the cultivation media is separated and removed at the end of each cycle and the fungal mycelia being harvested at intervals during the cycles is continuously repeated every six hours, the fungi mycelium being harvested at intervals during the time cycle.

The *Polyporus* strain used can be *Polyporus squamosus* or *Polyporus brumalis*. *Polyporus squamosus* Hudo strain PS 64 is preferred. The preferred strains and variants of *Polyporus* are deposited in the State Institute of Treatment Means Control, Bulgaria, Sofia, B1. VI. Zaimov 26, under the numerals PS-64-103; Ps-24-44; and PB-33-48. The following variants can also be used: PS 24-44; PB 33-48. The strain 64-103 is characterized by its very quick growth.

The liquid nutrient media used for industrial productions in the steps below comprise 5% beet or red molasses, 0.15% ammonium nitrate or 0.1% carbamide, 0.1% potassium dihydrogen phosphate or sodium dihydrogen phosphate and 0.03% sunflower seed oil as foam suppressor.

In large-scale production, the mycelium used is grown 48 hours in an inoculator, then transferred to an intermediate apparatus where it is grown for an additional 24 hours. For the growth of the mycelium, the working apparatus can have a volume of 50-100 m^3 or more. The apparatus is washed and sterilized while empty whereupon it is loaded with prepared nutrient media (molasses). The quantity of the nutrient media is calculated so as to occupy 80% of the overall volume of the apparatus together with the sowing material from the intermediate apparatus. The sowing is carried out by using the intermediate apparatus, the mycelium being transplanted on previously sterilized lines. The technological parameters maintained in the apparatus are: temperature, $26\pm1°C$; pH 6.5 ± 0.5; air, $0.8\pm0.2m^3$ in 1 m^3 nutrient media for 1 min without agitation. The pressure is 0.5 ± 0.2 atm. The period of growth of the mycelium depends on the quantity of the inoculum. When its ratio to the nutrient media is 1:10 the growth continues for 24 hours.

After full growth of the mycelium is reached, half of the cultivation media is passed to filtration and the other half remains as sowing material. The working apparatus is filled again with prepared nutrient media and operating thus, the quantity of the sowing material to the nutrient media is in a ratio of 1:1. Full growth of the mycelium in this case is reached in 6 hours±1 hour. Then again half of the cultivation media is pumped out for filtration, fresh nutrient media is added and this process is repeated for 3 days±2 days, whereupon everything starts once more from the inoculator and the intermediate apparatus stages. The quantity of fungus mycelium obtained from 1 ℓ nutrient media is 35 ± 5 g, 9 ± 1 g dry (28% dry substance).

Separation of the mycelium is carried out by filtration in filter presses, vacuum drum filters or by centrifugation. The mycelium thus separated is washed with water in the same quantity as the cultivation media filtrate.

The chemical composition of the fungi mycelium is characterized by its high protein content (50-60%) and a complete set of unique amino acids featuring a high lysine content (8-10%). In addition, there are present mycelium "B" complex vitamins and a number of other physiologically active substances. The protein digestibility is 83%.

With regard to the essential amino acids and digestibility, (i.e. overall biological value) the fungi mycelium almost equals beef meat in value. Thus, the total content of essential amino acids in beef is 39.9% as compared to 39.0% in the mycelium; digestibility of beef is 85% as compared to 83% for the fungi mycelium.

The mycelium may be used as an addition of 10 to 20% in meat products (sausages, minced meat for meat-balls, kebapcheta, tinned meat) as a source for obtaining synthetic meat, as an addition to processed and smoked cheeses up to 20%, as an enricher for various tinned vegetables with fungus mycelium up to 25%, in preparing various dry soups (meat and vegetable) in up to 20%, and when preparing bread and other dough products in up to 10%.

Roasted Yeast as Cocoa Substitute

J.J. Liggett; U.S. Patent 4,312,890; January 26, 1982; assigned to Coors Food Products Company has produced a roasted yeast product which may be used as a cocoa extender, substitute and/or replacer.

It is preferred to utilize yeast which has been recovered from a hopped brewer's wort fermentation in a conventional commercial malt beverage fermentation process or yeast which has been propagated on a synthetic medium containing hops or hop fractions, such as acids, resins, oils, tannins and gums. However, a satisfactory product may be obtained from yeast having no prior association with brewer's wort, hops or hop fractions.

In producing the product, dried yeast is roasted by heating the yeast to a sufficient temperature, specifically between 175° and 225°C, and for a sufficient time, generally between 1 to 3 hours, if it is desired to develop the texture, color, flavor and aroma characteristics and intensities of natural cocoa powder.

Yeast roasting may be accomplished by heating the yeast in or on a suitable supporting or transporting container, bed, cylinder or the like adapted to uniformly heat and roast the yeast. Roasting may be performed on a batchwise basis or continuously such as, for example, by passing a uniform layer of yeast on a continuous belt or other flat surface through an elongated oven.

The roasting of yeast particles inherently results in a net reduction in the moisture content of the yeast. The roasted yeast product has a moisture content less than about 5%, more preferably less than about 2.5% and most preferably less than about 1.0% by weight.

Example 1: A brewer's bottom fermenting yeast, *Saccharomyces carlsbergensis,* var. Frohberg (Syn. *Saccharomyces uvarum)*, was collected routinely from a commercial beer fermentation process following a normal, vigorous, single closed vessel primary beer fermentation.

The yeast was centrifuged to separate entrained beer from the yeast and suspended solids. The yeast was slurried into water to form a suspension of approximately 20-30% solids and was then spray dried in a commercial countercurrent hot air drier to yield a finely powdered dry yeast material of about 5% by weight moisture content.

The compositions of the dried yeast before and after roasting are as follows: protein 45%/44%; fat 2%/1%; carbohydrate 40%/43%; fiber 1%/4%; ash 7%/7%; and moisture 5%/1%.

The dried yeast powder mixture was then spread loosely onto a series of flat shallow pans; placed into a preattempered, static hot air oven at 205°C (400°F); and allowed to roast at this temperature for a period of 20 minutes. During roasting, the yeast powder is periodically stirred to achieve uniform roasting. Prior to roasting, the dried yeast has a light beige color, is faintly aromatic, and is sharply and strongly bitter to taste. Following the roasting process the product has a distinct pleasant aroma and has developed a rich deep brown color best described as a chocolate color. The roasted product is removed from the oven and air cooled, while being protected throughout the process from accidental contamination.

A beverage is prepared using the roasted product to yield a beverage having the flavor qualities of "hot cocoa" or "hot chocolate," as follows. A drink base is prepared comprising 592 g granulated fine sugar, 300 g nonfat dry milk, 20 g xanthan gum, 10 g salt, 8 g vanilla, 5 g imitation cream flavor and 5 g beet powder for a total of 940 g.

To 9.4 g of the drink base is added a 0.6 g sample of the roasted yeast product as the flavoring agent replacing cocoa powder and the combined product is stirred into 100 ml of boiling water. The resulting beverage has the characteristic hot cocoa (hot chocolate) appearance, aroma, flavor and general appeal.

Example 2: The dried brewer's yeast described in Example 1 was distributed onto a series of shallow pans and roasted in a preattempered, static, hot air oven at an oven temperature of about 300°C (572°F) for an uninterrupted period of 12 minutes. During the roasting process the dried yeast develops pleasant aromatic qualities and a rich dark brown or chocolatelike color, all distinctly different from the starting material.

This roasted yeast product was prepared as a beverage as described in Example 1, resulting in an acceptable cocoa-flavored beverage comparable to beverages prepared from cocoa powder.

Yeast Extract as Meat Flavoring

It is known that the quality of the yeast extract, useful as substitute for beef extract, can be improved by addition of disodium guanosine-5'-monophosphate (GMP) or disodium inosine-5'-monophosphate (IMP) which is known to be a flavoring ingredient of a shiitake mushroom or dried skipjack.

It has been well known that a considerable amount of RNA is contained in yeast cells. But, unfortunately, a yeast extract containing a substantial amount of flavoring 5'-nucleotides such as GMP cannot be obtained because the intracellular RNA of yeast cells is usually decomposed to nonflavoring low molecular substances such as nucleosides or bases during the autolysis process. The amount of GMP, supposing GMP is formed during autolysis, is too little to improve the quality of a yeast extract. In fact, no flavoring 5'-nucleotide can be detected in commercial yeast extracts.

It is, therefore, a principal object of the process developed by *T. Tanekawa, H. Takashima and T. Hachiya; U.S. Patent 4,303,680; December 1, 1981; assigned to Ajinomoto Company, Inc., Japan* to provide a method for producing a yeast extract containing flavoring 5'-nucleotide, and having thickness or body in taste.

The method comprises the steps of:

> (1) autolyzing yeast cells in the presence of a stimulator of autolysis at a constant pH ranging from 6.0 to 6.6, more preferably 6.2 to 6.4, and a temperature of 30° to 60°C for 10 to 30 hours;

(2) heating the autolyzed suspension of yeast cells at a temperature of 90° to 100°C for 1.0 to 3.0 hours, to extract remaining RNA and thereafter performing the following steps in any convenient order;

(3) hydrolyzing the extracted RNA with 5'-phosphodiesterase converting AMP into IMP with AMP deaminase if desired, and;

(4) separating resulting clear extract from the insoluble residue.

Any of a variety of edible yeast may be employed. These include baker's yeast such as *Saccharomyces cerevisiae* CBS 1172, CBS 1234, beer yeast such as *Saccharomyces cerevisiae* CBS 1171, CBS 1230, and *Saccharomyces uvarum* CBS 1503, and *Saccharomyces carlsbergensis* IFO 2015, sake yeast such as *Saccharomyces cerevisiae* IFO 2165, IFO 2342, wine yeast such as *Saccharomyces cerevisiae* IAM 4274, and *Pichia farinosa* CBS 2004, CBS 2006.

Example: 50 ℓ nutrient medium containing 5% glucose, 1% ammonium sulfate, 0.3% KH_2PO_4, 0.1% $MgSO_4 \cdot 7H_2O$, 0.05% $CaCl_2 \cdot 2H_2O$, 0.1% corn steep liquor of pH 5.5, was put into a 70 ℓ jar fermenter, and sterilized at 110°C for 15 minutes, following which 200 ml of a seed culture broth of baker's yeast (*Saccharomyces cerevisiae* CBS 1523) cultured with aerobic shaking in the same culture medium at 30°C for 18 hours, was inoculated, and cultured aerobically at 30°C for 20 hr with vigorous aeration (½ vvm) and stirring at 500 rpm, under the pressure of 0.5 kg/cm^2.

After the cultivation, 2.75 kg of yeast cells in a form of cake with water content of 60% was obtained by centrifugation. The cake was then washed with water and water was added to the cake to prepare 8.0 ℓ of a suspension of yeast cells.

The suspension was mixed with 150 ml ethyl acetate and mixed well by stirring. Then, the suspension was divided into 1.0 ℓ suspension and the pH of the divided suspension was adjusted to from 5.5 to 7.5 with 30% NaOH. Each suspension was allowed to stand at 45°C for 18 hours and each pH was controlled with alkali during the autolysis.

After the autolysis each autolyzate suspension was heated at 90°C for 2.0 hours; thereby, the remaining enzymes were inactivated and remaining RNA was extracted.

On the other hand, to 20 kg of a commercial malt roots 20 ℓ water was added and the mixture was ground. Then, a clear extract obtained by removing the residue by filtration, was heated to 63°C for 5 minutes in order to activate 5'-nucleotidase to obtain a crude enzyme solution for use in hydrolyzing RNA.

To each heated suspension previously prepared, 200 ml of the crude enzyme solution was added and after the pH of each suspension was adjusted to 6.0, the enzymation reaction was conducted at 60°C for 5 hours. After the enzymation, each enzymation mixture was heated to 90°C for 5 minutes, and clear extract was obtained by removing the insoluble residue by centrifugation and dried under reduced pressure to prepare yeast extract in the form of powder. Each yield of solid matter and 5'-GMP forming ratio (GMP contained in yeast extract /GMP contained in the original yeast cells ratio in percentages) were measured.

It was found that both the yield of solid matter and the GMP-forming ratio are high when the autolysis is performed at a pH ranging from 6.0 to 6.6. It was found from the result of organoleptic test performed by a panel of 20 members who had been specially trained for this kind of test that the yeast extract, obtained by autolysis at pH 6.0 to 6.6, especially at pH 6.2 to 6.4 has less of unfavorable odors peculiar to yeast itself and has a thickness or body in taste which resembles that of beef extract, and that the quality of the yeast extract is excellent owing to its strong flavoring taste.

Torula Yeast as Nonstick Agent for Rice Cereal

In the preparation of puffed cereal products from milled Nato rice, it is necessary to first cook the rice in a flavoring solution which usually contains sugars, salt and the like. The processing of Nato rice in this manner has been difficult in the past because the rice tends to stick together after cooking and again after drying. The disclosure by *T.O. Martin and A.S. Clausi; U.S. Patent 4,238,514; Dec. 9, 1980; assigned to General Foods Corporation* provides a process which decreases sticking and clumping.

In preparing the puffed products, milled Nato rice is mixed with spray-dried torula yeast which has been prepared by culturing on food grade ethanol. The spray-dried torula yeast is inactive by virtue of the conditions of spray-drying, but yet provides proper functionality for the purpose of decreasing sticking and clumping after cooking and drying. The torula yeast in its preferred form is a bland-tasting cream-colored powder. It will typically have a protein content of above approximately 45% and a fat content of no greater than about 9%. It is spray-dried to a moisture content of lower than 7% under conditions effective to provide a near-neutral to slightly alkaline pH, preferably within the range of from about 7 to 8. Spray-dried torula yeast products e.g., Torutein 94, 50 and 10, have been found effective for use in this process.

Example: To prepare a batch of cereal according to this process, 634 kg of Nato rice and 6.5 kg of Torutein-94, spray-dried torula yeast is added to a rotary pressure cooker and admixed by rotating the cooker. After 5 minutes of mixing, 240 kg of an aqueous flavoring syrup solution containing about 50.3% water and about 40.7% solids is added to the cooker and mixed with the rice and spray-dried torula yeast for about 20 minutes prior to heating under pressure. The flavoring solution contains about 145 kg of aqueous sucrose containing 32.5% water, about 10 kg corn syrup containing about 19.5% water, and about 85 kg of salt brine containing about 73.5% water. At the end of 20 minutes of mixing, the cooker is heated under a pressure of 18 psig for 45 minutes with continued rotation. After this time the cooker is exhausted to 0 psig.

The cooled Nato rice is then cooled in the cooker, still under rotation, for 2 hours. The cooked and cooled Nato rice is then unloaded onto the wire mesh belt of a two-stage Proctor and Schwartz dryer. The drying air passing through both stages of the dryer is at a temperature of about 110°C and reduces the moisture content of the cooked rice from the value of about 26% at the end of the cooking, to about 18% at the end of zone 1, and about 14% at the end of zone 2. Clumps are diminished by employing a pinbreaker within the dryer and employing a vibrating grading screen at the dryer discharge. The screen has about 41% net open area with opening sizes being 0.9 x 0.9 mm. The dried, cooked Nato rice is passed to a bin wherein it is tempered for about 1 to 6 hours and is then passed between bumping rolls spaced at a distance of about 50% of the thickness of an average grain and rotated at a speed of about 160 rpm. The bumped Nato rice is then toasted in a conventional toasting oven operated at a temperature of about 240°C for about 90 seconds to finally puff and toast the flake.

The abovedescribed puffed and toasted Nato rice flakes are desirably employed to prepare an agglomerated, natural cereal product. To accomplish this, the following dry fraction ingredients are weighed and added to a multiflighted coating reel approximately 3 feet in diameter:

> Rolled oats (8% moisture)—8 kg
> Rolled wheat (8% moisture)—1.68 kg
> Torula treated Nato rice flakes (2.5% moisture)—4.12 kg
> Almonds, diced (4% moisture)—1.4 kg
> Unsweetened, dried coconut (4% moisture)—1.31 kg
> Nonfat dry milk (3% moisture)—1.04 kg

The dry fraction ingredients are mixed in a coating reel which is operated at 30-60 rpm for 5 minutes. Pure coconut oil (3.8 kg) at 43°C is poured or sprayed onto the preblended dry fraction ingredients in the rotating reel, and the resulting oil-coated dry fraction is allowed to tumble in the reel for an additional 5 minutes at the same rpm to insure distribution of oil on the surface of the ingredients and impregnation thereof.

Separately, a coating syrup having the following ingredients is prepared:

> Brown sugar (granular, 2% moisture)—5.57 kg
> Corn syrup (42 DE 80% solids)—0.54 kg
> Honey solids—0.32 kg
> Pure caramel powder (1% moisture)—0.27 kg
> Water—2.6 kg

The foregoing coating ingredients are dissolved in water at 65°C and sprayed or poured at this temperature onto the oil-coated dry ingredients fraction in the rotating reel, operated at 30-60 rpm for an additional 5 minutes of tumbling to promote uniform coating of the material and particle aggregation.

The oil/syrup coated aggregates are then removed from the coating reel at a moisture content of 13% and dried. The dried agglomerates have a final moisture content of 1.5 to 3%. The material is broken apart while still in a warm, plastic state, cooled and sized by pressing through wire screen having ¾-inch openings. The sized, cooled agglomerate is then bulk-bagged for packaging.

MISCELLANEOUS ADDITIVES

Antistick Coatings for Chewing Gum

In the production of articles of taste such as chewing gum, caramel, or fudge and so on from viscous edible materials, there has been used a powdery or liquid releasing agent to facilitate the removal of a molded article from a mold, and to facilitate the cutting of the article into pieces of an appropriate size, and also to keep the sticky surface of such piece nonadhesive to its wrapping sheet.

Most release agents in prior use have drawbacks in stability, odor or scattering of powders. *K.-I. Noborio and M. Maeda; U.S. Patent 4,208,432; June 17, 1980; assigned to Kanebo Foods Ltd., Japan* have provided an antistick which overcomes these difficulties.

This releasing agent comprises a compound selected from a group consisting of α- and β-lactoses, calcium carbonate and mixtures thereof, which are coated with saturated fatty acid monoglycerides or derivatives thereof.

Saturated fatty acid monoglycerides and derivatives thereof which are used in this process include those monoglycerides containing a saturated fatty acid moiety having at least 12 carbon atoms and mono- or diacetylation products of saturated fatty acid monoglycerides having an acid value of 2 or less and an iodine value of 2 or less.

According to this process, fine particles of a compound selected from a group consisting of α- and β-lactoses, calcium carbonate and mixtures thereof are coated with a compound selected from a group consisting of saturated fatty acid monoglycerides and derivatives thereof in a coating amount of 2 to 5% by weight, preferably 3 to 4% by weight, based on the weight of the compound which is to be coated. If the coating amount is less than 2% by weight, the lubricating and releasing abilities of resulting coated particles are unacceptably reduced. If the coating amount is higher than 5% by weight, coated particles unpreferably have an excessive flowing property so that they may nonuniformly spread on the contacting surfaces of a mold and also on the surfaces of the material charged in the mold for being shaped, adversely affecting the release of the product from the mold.

The coated particles of the above releasing agents for use in shaping food materials or chewing materials including confectionary foods such as candies and chewing gums, have a particle size of 100 mesh or less, preferably 150 to 200 mesh.

The particles of releasing agents may be produced by emulsifying a compound selected from a group consisting of α-lactose, β-lactose, calcium carbonate and mixtures thereof together with a compound selected from a group consisting of saturated fatty acid monoglycerides, derivatives thereof and mixtures thereof in water and atomizing the emulsion at an appropriate temperature to produce dried coated particles. The atomizing temperature depends upon thermal stabilities of α- or β-lactose, calcium carbonate, saturated fatty acid monoglycerides or derivatives thereof which are used. Hydrated α-lactose has a melting (decomposition) point of about 203°C and a property of gradually losing crystal water at a temperature of 120° to 130°C to produce hygroscopic α-lactose anhydride which is undesirable for use as a releasing agent. β-lactose is usually of anhydride and has a melting point of about 252°C and a property of absorbing water at room temperature to produce α-lactose hydrate.

The atomizing temperature should be preferably lower, enough to minimize the production of α-lactose anhydride. Therefore, the atomizing temperature should be lower than 120°C, preferably 50° to 120°C and more preferably 80° to 100°C. However, if only a short time period is required for atomizing the emulsion, temperatures higher than 120°C may be used during the period, provided that produced α-lactose anhydride is negligible or not so much as deleteriously affecting the properties of the product particles. An amount of water to be used for preparing the emulsion may vary in a wide range depending upon the drying effect exhibited by the atomizing device. However, it should be as little as possible to facilitate the drying of coated particles.

The coated particles of releasing agents are tasteless, odorless and harmless for one's health; they consist of cores of α-, β-lactoses and/or calcium carbonate coated with saturated fatty acid monoglycerides and/or derivatives thereof.

Example 1: One hundred parts of β-lactose and 4 parts of stearic acid monoglyceride are emulsified with 150 parts of warm water. Thereafter, the emulsion is atomized at a temperature of 100°C to produce releasing agent particles each having an average size of 150 mesh.

Example 2: The procedure of Example 1 is repeated, except that stearic acid monoglyceride is replaced by monolaurodiacetyl glyceride, to produce coated releasing agent particles.

Example 3: As control, a mixture of 50 parts of powdery sugar having a particle size of about 150 mesh and 50 parts of starch is used as a powdery releasing agent.

The products of Examples 1 and 2 showed no sticking in the Releasing Ability Test (i.e., passage through rollers) and no sticking to envelope material even after a period of two weeks. The coating also had no odor. The comparison coating of Example 3 was rated as having overall face-sticking in the above tests and had an objectionable odor.

Noncaking Coatings for Organic Acids

Water-soluble organic acids are common ingredients of food compositions where they serve as acidulants, flavoring agents and/or preservatives. Common applications for such materials are in gelatin dessert mixes and in powdered beverage mixes. Frequently utilized acids include citric, malic, tartaric and ascorbic. However, such acids have pronounced hygroscopic tendencies and absorb moisture which causes the mixes to cake upon standing. Caking renders the mixes difficult to handle, increases the time required for dissolution and has a strong adverse effect on consumer acceptability.

It is an object of the development by *P. Ciliberto and S. Kramer; U.S. Patent 4,288,460; September 8, 1981; assigned to Balchem Corporation* to provide a granular food ingredient composition which is noncaking.

It has been found that if core particles of a food ingredient such as a normally hygroscopic soluble acid are provided with a specific coating consisting essentially of 5-37% fatty acid derivative selected from the group consisting of polyoxyethylene sorbitan monooleate, polyoxyethylene sorbitan monostearate and lecithin; 15-56% propylene glycol and from 38-62% flow agent, the resulting particles are stable and noncaking over long periods of exposure to the atmosphere and are substantially instantaneously soluble when added to water.

The permissible size of the core particles may vary somewhat depending on the solubility and other characteristics of the core material. Generally, the more soluble the material, the larger the particles may be. Excessive fineness should be avoided so that problems with dusting and the like will not occur. Typical core particles will range between about 40 and about 850 microns in size.

Polyoxyethylene sorbitan monooleate, polyoxyethylene sorbitan monostearate and lecithin are well-known, commercially available surface active agents. Polyoxyethylene sorbitan monooleate and monostearate are known generally in the art as Polysorbate-80 and Polysorbate-60, respectively. Suitable material may be purchased from Atlas Chemical Industries (Tween-80 and Tween-60). Generally, the polyoxyethylene chain will be formed from about 20 mols of ethylene oxide per mol of sorbitol in order to provide the desired solubility characteristics. The use of lecithin, the oleate ester or the stearate ester in accordance with the process is critically important.

Use of other fatty acid derivatives has not been found to provide the needed solubility despite their surface active properties. While the use of any of these three materials, or mixtures thereof, will achieve the objects of the process, the coating procedure proceeds more easily with the normally liquid polyoxyethylene sorbitan monooleate. That at least equivalent performance is achieved when using a liquid coating agent is surprising.

Propylene glycol, sometimes referred to as 1,2-propanediol, also is a well-known commercially available chemical.

The coating composition also incorporates a flow promoter. Since both the operable polyoxyethylene sorbitan monoesters and the propylene glycol are liquids at ordinary ambient temperatures, the flow promoter is necessary to prevent the coated acid crystals from adhering to each other, thereby facilitating handling of the material. A preferred material is fumed silica which is usually derived from gels of colloidal silicon dioxide. Particle sizes are extremely small ranging from approximately 0.007 to approximately 0.014 micron. However, the fumed silica has an extremely high surface area on the order of 200-400 m^2/g. A particularly suitable fumed silica is commercially available (Cabot Corp., Cab-O-Sil M-5). Carboxymethylcellulose is also useful as a flow promoter.

Example 1: Granular citric acid crystals were provided with a coating consisting of 35% polyoxyethylene sorbitan monooleate, 15% propylene glycol and 50% fumed silica. Fifteen grams of the resulting coated acid crystals were added to 200 ml of water at 50°C and stirred vigorously for one minute. At the end of the one minute period, most of the granular material remained undissolved and settled out at the bottom of the container. This test demonstrates the criticality of using enough propylene glycol in the coating.

Example 2: A sample of granular citric acid crystals was provided with a coating comprising 20% polyoxyethylene sorbitan monooleate, 60% propylene glycol and 20% fumed silica. Two other samples were made, the first employing polyoxyethylene sorbitan monostearate and the second lecithin in place of the monooleate of the first sample. The resulting coated particles in all three samples were tacky and tended to agglomerate. Acid particles provided with these coatings would not flow satisfactorily and were hard to handle. This test demonstrates the importance of adding sufficient flow agent to the coating.

Example 3: Granular citric acid crystals were provided with a continuous encapsulating coating consisting essentially of 12½% polyoxyethylene sorbitan monooleate, 37½% propylene glycol and 50% fumed silica. Fifteen grams of the resulting particles were added to 200 ml of water at 50°C with stirring. After 30 seconds, no undissolved material would be observed. A slight milkiness at the surface of the liquid disappeared upon allowing the solution to stand for an additional 3 minutes. This test demonstrates the substantially instantaneous solubility of particles provided with coatings according to the process.

Osmotic Control Agents for Starch

When a concentrate containing starch is heat-processed, the starch gelatinizes in the container. When the consumer transfers the contents of the container in e.g. a bowl, and dilutes it with hot water, it is extremely difficult to disperse the gelatinized mixture to get a homogeneous product having the desired viscosity.

P.J. Anema, D.R. Haisman and R.M. Adams; U.S. Patent 4,291,066; Sept. 22, 1981; assigned to Thomas J. Lipton, Inc. have provided a process for producing a concentrate containing starch which when heat-processed will virtually not gelatinize and which on dilution with hot water will develop the desired viscosity.

To satisfy these conditions the swelling of the starch must be minimized during the preparation and heat-processing of the concentrate but it should retain sufficient swelling potential during the dilution step.

Starch granules swell at their gelatinization temperature owing to the osmotic gradient between the external environment and the interior of the starch granule. It has been found that lowering this gradient by increasing the osmotic pressure in the external environment (i.e. the aqueous liquid in which the starch is dispersed) will reduce the rate of swelling of the starch and will consequently raise the temperature at which gelatinization of the starch starts. However on dilution with hot water the concentration of osmotic-pressure-increasing agents will decrease and the starch will gelatinize at approximately its normal gelatinization temperature.

Accordingly the process for producing an ambient stable concentrate, which on dilution reconstitutes into a lump-free thick soup, sauce, gravy, or dessert involves:

    (a) producing a concentrate containing:
        (1) water,
        (2) starch,
        (3) a proportion of an appropriate osmotic-pressure-increasing agent, sufficient to substantially reduce the rate of swelling of the starch and prevent gelatinization of the starch at a temperature ranging from 60° to 100°C, and
        (4) a proportion of a water-activity depressing agent conducive to a water activity not exceeding 0.92.
    (b) pasteurizing the concentrate at a temperature ranging from 60°-100°C.

Water-activity depressing substances consist of a physiologically acceptable salt such as a sodium or calcium halide, preferably sodium chloride, a humectant such as polyhydric alcohols, preferably glycerol, protein hydrolyzates, lactose and/or glucose.

In this process, sodium chloride is preferably used as the water-activity suppressing substance in a proportion which ranges from 0.1 to 7 weight percent based on the total composition of the concentrate.

The proportion of water-activity depressing agent is so chosen that in combination with the other ingredients present in the concentrate a water activity which preferably ranges from 0.77-0.90 is achieved. Useful osmotic-pressure-increasing agents are preferably those which do not easily penetrate starch granules but remain in the liquid in which the starch is dispersed.

In the process starch hydrolyzates, particularly maltodextrins or glucose syrup solids having a DE value ranging from 10-42, hydrolyzed amylopectin and/or sucrose can be used.

Applicants have found that maltodextrin having a DE value from 10-42 and sucrose were very effective in most product situations.

Adequate proportions of the osmotic-pressure-increasing agent can be assessed in each particular case. In most instances the proportions of starch:osmotic-pressure-increasing agent:water lies within the range of 1:1-2:0.5-6.

The starch incorporated in the concentrate can be a native starch such as corn starch, potato starch, tapioca starch and the like and/or a modified starch such as starch phosphates and acetylated starch phosphates EEC No. 1410-1414, acetylated distarch adipate EEC No. 1422, hydroxy propylated distarch glycerol EEC No. 1441 or hydroxy propylated distarch phosphate EEC No. 1442.

In most instances a proportion of 10-30% starch in the concentrate will be adequate.

The pasteurization step according to step (b) of the process is carried out in containers at a temperature ranging from 60°-100°C, and preferably 70°-90°C for a period which may vary from 1-60 minutes.

The sealed containers are subsequently stored at room temperature until consumption. Before consumption the concentrates are diluted with an appropriate proportion of hot, preferably boiling water or an aqueous solution to obtain after stirring for a few seconds the desired thick product. It is of course possible to boil the diluted solution for a few minutes before consumption.

Example: A mushroom soup concentrate was prepared by mixing 17.8% of tapioca starch, 23.6% of maltodextrin (DE = 10), 13.5% of fat, 3.8% of skimmed milk powder (containing ±50% lactose), 6.3% of coffee whitener, 4.8% of salt, 1.7% of flavoring and spices, 0.2% of dried mushroom, 0.3% of onion powder, 2.5% of mushroom powder and up to 100% of water. The water activity a_w was 0.83.

The soup concentrate was filled into containers and pasteurized at 80°C for a period of 40 minutes. In the soup concentrate the starch was predominantly in the ungelatinized form. Reconstitution into a soup was achieved by diluting 41 g of concentrate with 200 ml of boiling water in a bowl and by gently stirring for 30-60 seconds. A creamy, lump-free mushroom soup was obtained having a viscosity of approximately 200 cp at a shear rate of 50 sec^{-1}.

After a storage period of 6 months at ambient temperature the concentrate did not display any sign of spoilage.

Low Calorie Coconut for Diet Foods

A process for reducing the caloric content in natural coconut, and other fatty edible items such as the edible meats of nuts, the edible seeds of grain, and the edible seeds of fruits has been described by *N.D. Kosarin and G. Finkel; U.S. Patent 4,225,624; September 30, 1980.*

The process has the advantage of extracting approximately 100% of the fat from coconut and thereby reducing the caloric content thereof approximately 70-90%. The process also has the advantage of further reducing the caloric content by removing the sugar (soluble polysaccharides) therefrom while leaving the appearance and texture of the coconut substantially unchanged as compared to the natural coconut.

A preferred method for reducing the caloric content in coconut as well as other edible items, is made up of two sequential process steps wherein the second step is optional and is effective to further reduce the caloric content of the coconut or other edible items by removing the soluble sugar therefrom.

Example 1: One part by weight shredded coconut is mixed with two parts by weight of solvent, such as petroleum ether, and is heated to a temperature of approximately 180°F (82°C). The heated mixture is maintained at approximately standard atmospheric pressure and is agitated for a period of approximately 8 hr in a Soxhlet-type extractor to separate the solvent containing the fat from the defatted coconut. As a result of this first process step, the coconut is completely defatted and the caloric content is reduced by approximately 70-90%. (Of course the above example parameters may be adjusted to remove lesser amounts of fat from the coconut.) The remaining caloric content of the coconut is contributed by the sugar retained within the defatted coconut.

It has further been discovered that the sugar may also be extracted from the defatted coconut by washing the defatted coconut in water (preferably distilled water). The washing of the defatted coconut removes up to 100% of the soluble sugar and thereby reduces the caloric content of the defatted coconut to trace amounts.

A major advantage of the aforesaid process for reducing the caloric content of coconut is that the appearance and texture of the defatted coconut remains substantially unchanged from that of the natural coconut.

It should be noted that while the aforesaid example, of reducing the caloric content of coconut, used shredded coconut, larger pieces of coconut may be equally treated by the aforesaid process. However, larger pieces will necessarily require larger amounts of time to obtain a total extraction of the fat therefrom.

The use of the defatted coconut, or other edible items, produced by the above process is seen as highly desirable in the manufacture of candies and other foods having a caloric content of less than 100 calories/ounce (353 calories/100 g). The following process is an example of using the defatted coconut to produce a candy having a relatively low calorie content while tasting and appearing to have natural shredded coconut in the center thereof.

Example 2: 63.4 g of defatted coconut (as produced in Example 1) is mixed with 263.9 g of invert syrup (71% solids), 225.8 g water, 4.2 g agar (a thickening agent) and 1 g of imitation flavor in solution. The solution is cooked until the excess water is boiled off. The cooked mixture is then poured onto a cold table where it is stirred periodically until it cools to approximately 103°F (39°C). The mixture is then cooled to room temperature and cut to the desired sizes. Subsequently, the cooled mixture is coated with sweetened chocolate so that each piece of candy comprises approximately 25% by weight of chocolate. The candy resulting from the aforesaid process contains approximately 75 calories/ounce (265 calories/100 g).

Compared with conventional chocolate bars, which contain approximately 130 calories/ounce (458 calories/100 g), and conventional chocolate-covered coconut candies, which contain approximately 175 calories/ounce (617 calories/100 g), the aforesaid chocolate-covered candy produced by the process results in an improved food product having a significant reduction in calories over the candy of the prior art.

Sorbitol as Humectant in Chewing Gum

Chewing gums generally include gum base, water-soluble sweeteners and flavoring. The water-soluble sweeteners usually include sucrose, dextrose, corn syrup and/or sodium or calcium saccharin or combinations thereof. The chewing gum is generally prepared by melting the gum base, mixing corn syrup or liquid sweetener for 3 to 5 minutes with the gum base followed by the addition of solid sweetener (for example, sugar) and flavor and mixing for 5 minutes. The chewing gum is removed from the kettle, rolled and cut to the desired shape.

In the above chewing gum, the corn syrup (which provides a substantial portion of the moisture in the gum) will be retained in the gum base as part of the oil or insoluble phase and the sucrose and/or other sweeteners will be incorporated in a water-soluble phase which is in simple admixture with the oil phase and might even be considered to be coated on the gum base. The result is that the gum base will protect the corn syrup as an internal phase thereby minimizing the amount of corn syrup on the surface of the gum (surface corn syrup causes the gum to sweat).

However, due to the equilibrium relative humidity of the chewing gum, eventual migration of moisture of the corn syrup to the gum surface and subsequent loss of moisture through evaporation is unavoidable at equilibrium relative humidities below that of the gum. Reduction in moisture content of chewing gum leads to loss of flexibility which manifests itself in increased stiffness and brittleness. The latter phenomenon is, of course, associated with stale or old chewing gum.

F. Witzel, K.W. Clark and A.I. Bakal; U.S. Patent 4,166,134; August 28, 1979; assigned to Life Savers, Inc. have provided the method for producing chewing gum having improved flexibility retention and good sweat resistance. This method includes the steps of admixing melted gum base and an aqueous softener, such as corn syrup, to provide a water-in-oil phase, admixing sweetener with the water-in-oil phase, thereafter admixing one or more humectants, such as sorbitol, and other conventional chewing gum ingredients with the gum mass comprising the sweetener-water-in-oil phase combination, to form chewing gum. The chewing gum may then be rolled and cut into the desired shape.

It has been found that chewing gum prepared in the manner described above, wherein the aqueous softener is effectively separated from the humectant, when stored at room temperature and 50% relative humidity with no wrappers, such gums are still flexible after four weeks of storage, whereas chewing gums made in a conventional manner and stored under the same conditions lose their flexibility after two weeks of storage.

In carrying out this method the gum base is first melted; thereafter, the gum base is cooled to below 200°F, and preferably below 180°F and the aqueous softener, such as corn syrup, is thoroughly mixed in the gum base to effect uniform distribution. Flavors and emulsifiers or softeners, such as lecithin, may then be added, if desired, and the mixture is mixed for from 1 to 2 minutes, and preferably for 2 minutes. At this stage, flavor oil is preferably added with mixing for from 2 to 5 minutes, preferably about 3 minutes. A particulate artificial or natural sweetener, such as sugar, free saccharin acid, sodium or calcium saccharin, cyclamate, aspartame, etc. is added to the continuous mass with mixing being continued for from 3 to 5 minutes, and preferably for from 2 to 4 minutes. The humectant is added to and mixed with the gum mass for from 1 to 5 minutes, and preferably for from 1 to 3 minutes. Thereafter, if desired, solid flavor such as gum arabic coated flavor may be added and mixed with the gum base mix to form a chewing gum which may be rolled, scored and cut into desired shapes.

Example 1: Chewing gums are prepared by this method using the ingredients listed as parts by weight (pbw).

The gum base (22 pbw) is added to a steam jacketed kettle equipped with a sigma blade mixer. The gum base is heated to about 200°F to melt same. Thereafter, the melted gum base is colored to about 180°F. The corn syrup (17 pbw), lecithin (0.2 pbw) and free saccharin acid (0.2 pbw) are added to the gum base and mixed therewith for about 2 minutes. Flavor oil (1.5 pbw) is then added and mixed therewith for about 3 minutes. The gum base mix is at this time in the form of a continuous mass.

The sugar (54 pbw) is added to and mixed with the continuous mass for about 2½ minutes. Sorbitol (5 pbw) is mixed for about 1½ minutes and then the spray dried flavor (0.5 pbw) is mixed therewith for about 1 minute.

The mass is removed from the kettle, rolled, scored and cut into chewing gum sticks.

Example 2: A chewing gum was prepared as a control by the method of Example 1 except that no sorbitol was added and 59 pbw of sugar was used.

In all cases, the tests for freshness or flexibility retention upon storage are carried out employing a Thwing-Albert Handle-O-Meter.

Using this instrument a chewing gum stick is placed under the edge and on the sample table. When the instrument is activated, the edge moves at a controlled rate, forcing the chewing gum to bend. The resistance is transferred to a meter calibrated in grams. This technique is used for all evaluations.

Test data indicated that the control sample of Example 2 which contains no sorbitol, is brittle after one week of storage at 70°F and 50% relative humidity with no wrappers. In comparison, the Example 1 sample containing sorbitol, stored under the same conditions, shows no brittleness or cracking during testing even after four weeks of storage.

Honey and Tannin for Clarification of Fruit Juices

The juice industry has used many methods of clarification for fresh and fermented juices. These methods include the enzyme clarification method such as with Klerzyme as well as treatments with bentonite, gelatin, or Sparkolloid. However, all of these methods have resulted in producing a juice product containing unwanted characteristics such as permanent hazes, off-flavors, loss of flavor, and pectin removal with loss of body. Hence, a product is produced for the consumer with a quality much less than that of its original state. In order to combat these unfavorable effects, juice processors are often forced to add sweeteners to the juice to produce a product with an acceptable taste.

R.W. Kime; U.S. Patent 4,327,115; April 27, 1982; assigned to Cornell Research Foundation, Inc. has developed a method for the clarification and removal of hazes from fresh, pasteurized, and fermented fruit juices by the use of honey as a clarifying agent. Alternatively, the addition of both tannin or a tannin derivative and honey may be used in instances where the natural fruit juice tannins are low.

According to the method, fresh, pasteurized, and fermented fruit juices are clarified by the addition of or treatment with honey. Fresh or pasteurized juices which can be clarified include, among others, apple juice, grape juice, pear juice, prune juice, and the like, and cranberry juice as well as other berry juices. Fermented fruit juices which can be clarified include those of the fresh juices listed above including both hard and sweet apple cider, apple and grape wines, and red and white wines among others.

In carrying out the clarification process, honey is added to the juice in an amount sufficient to accomplish the desired clarification. This amount is at least one-half weight percent of honey per total weight of the juice to be clarified, preferably 1 to 10 wt %, most preferably 2 to 5 wt % of the weight of the juice.

The honey and juice mixture is then agitated in a sufficient manner to promote effective mixing. The mixture is then allowed to remain at rest while precipitates are formed and settle on the bottom of the holding container.

After the juice has a clarified appearance, the precipitate is removed from the juice. Centrifugal separation as well as filtration techniques can be employed to effect separation of the precipitate from the clarified juice.

Preferred temperature for clarification is 40° to 140°F, most preferably 70° to 95°F where clarification proceeds fastest. Clarification requires a treatment time of from ½ hour to 20 hours for mixing, precipitation, and separation. This large time variance is largely dependent upon the initial clarity of the juice and the percentage of honey used, as generally the higher the percentage of honey, the less the time needed to effect clarification.

Many unclarified juices have low tannin contents. Although tannin is not a necessary component in all of the juices which are clarified by the clarification process, a small amount of tannin or a form of tannin such as tannic acid is preferably added to the juice which is low in tannin content before, during, or after the addition of honey. The tannin is preferably added to the juice prior to contact with honey. While any effective amount of tannin or its derivatives can be employed, generally between 1 to 50% by weight of a 1% tannic acid solution is employed per weight of the juice to be clarified.

Example 1: Two 100 ml samples of fermented pear juice containing 193 and 21 mg of tannin respectively were treated with 4% by weight of honey. After 1½ hours the high tannin content pear cider was completely settled out and sparkling; however, the low tannin content pear cider was not coagulated after 30 hours.

When 4% of honey and 0.08% of tannin solution was added to a low tannin content, very hazy pear cider, the cider was clarified and sparkling in 20 minutes.

Example 2: Freshly pressed apple cider was contained for 3 weeks at 32°F. Separate samples of cider were treated at 68°F with (a) 4% by weight of honey, (b) 4% of honey and 0.04% of a tannic acid, (c) 0.025% of Klerzyme (a commercially used clarifying enzyme), (d) 0.025% of Klerzyme and 4% of honey, (e) 0.025% of Klerzyme, 4% of honey and 0.04% of a tannic acid solution.

The apple cider sample treated with (a) honey alone, was clarified and sparkling after approximately 6 hours. The apple cider sample treated with (b), honey plus tannin, showed moderate coagulation after 8 hours; however, it never completely clarified having a slight haze as the tannic content was too high.

The sample treated with (c) showed only slight coagulation after 18 hr. When Klerzyme was doubled to 0.05%, the cider still did not clarify. The apple cider sample treated with (d) showed sparkling areas after 40 minutes and was completely clarified and sparkling after 90 minutes. The apple cider sample treated with (e), Klerzyme, honey, and tannin, was totally settled after 2½ hr but retained a slight haze attributed to excess tannin.

Example 3: Different samples of grape juice, grapefruit juice, and orange juice produced as reconstituted concentrates, and red and white wines were each separately treated with one of the following:

> (a) 4% of honey;
> (b) 4% of honey plus 0.04% of tannic acid solution; and
> (c) 4% of honey plus 0.08% of a solution and tannic acid.

All of the tested fruit juices showed moderate coagulation with treatment (a), (b) or (c) within 24 minutes. The grape juice samples were totally sparkling and settled within 50 minutes.

The red wine which was treated with (a) was clarified and sparkling after 5 hours while two samples of the red wine treated with (c) clarified in 1 hour and in 3 hours respectively.

The white wine tested did not clarify in (a) when no tannin was added; however, the treatment with (b) clarified in 130 minutes and with (c) clarified in 1 hour.

Elastomers for Cold-Proof Chewing Gum

In general, a chewing gum has a trend of hardening even in a mild winter season, and therefore when the chewing gum is chewed in the winter season it must be softened in a mouth to a certain degree before chewing or be chewed little by little from an edge thereof in order to facilitate softening. This hardening phenomenon has been encountered in the winter season everywhere in the world and has usually reduced the demand for the chewing gum in the cold season. Thus, the chewing gum is generally desired which has little or no trend of hardening in the cold weather and is very soft in chewing.

For obtaining such type of chewing gum, one method is to prepare the chewing gum by use of soft materials only, excluding hard types of natural resins. This method, however, makes a texture of chewing gum extremely soft and could cause a softening and melting phenomenon. Further, the method has a disadvantage of not providing a desired elasticity for the chewing gum.

A principal object of the process by *K. Ogawa, S. Tezuka, T. Maruyama and K. Kiyokawa; U.S. Patent 4,254,148; March 3, 1981; assigned to Lotte Co., Ltd., Japan* is to provide a cold-proof chewing gum base. This base comprises a 0 to 15 wt % of a natural resin, 10 to 20 wt % of a vinyl acetate resin, 6 to 16 wt % of an ester gum, 12 to 25 wt % of a rubber, 12 to 20 wt % of a wax, 14 to 20 wt % of an emulsifier and 10 to 20 wt % of a filler based on the weight of the chewing gum base.

Since this cold-proof chewing gum has a low-temperature-resistant property, a combination with an ice cream yields a frozen dessert. Since the cold-proof chewing gum maintains its ample elasticity and does not harden even when the combination is placed in a freezing temperature of about –20° to –30°C which is desired for keeping the ice cream frozen, a person who eats the frozen dessert can enjoy both the ice cream and the remaining chewing gum.

In order to prevent the hardening and keep the ample elasticity in the cold, addition of substantially larger proportions of the natural and/or synthetic rubbers and usage of extremely larger amount of the emulsifier as compared with the conventional chewing gum are particularly important in the process. The possible synthetic rubbers include general synthetic rubbers such as polybutene, polyisobutylene, isobutylene-isoprene rubber, styrene-butadiene rubber and the like.

The rubber content, which is most important in the chewing gum is in the range of 12 to 25, preferably 15 to 25% by weight, depending on the type and nature of the rubber, if for example the soft type rubber is used.

The emulsifier preferably a monoglyceride, is a very important component in the process, and it has not been clearly understood what mechanism could prevent the hardening of the chewing gum in the cold, although the considerably increased amount in the range of 14 to 20% by weight has an outstanding effect on the chewing gum, while the conventional chewing gum contains the same in an amount of 1 to 3% by weight (8 to 11% by weight for the bubble chewing gum).

The filler such as calcium carbonate may be used in an amount of 10 to 20% by weight for both the chewing gum base and the bubble gum base of the process, but desirably the amount for the bubble gum is reduced to some extent and is in the range of 10 to 15% by weight, while the amount of 15 to 20% by weight for the chewing gum base is preferred.

The foregoing amounts of materials may be kneaded by means of the conventional kneading apparatus such as a kneader to obtain the chewing gum base or the bubble gum base according to the process.

To the resulting gum base are added the conventional chewing gum additives (such as sugars, flavors and others) and if desired the softening agents (such as liquid sorbitol, glycerine and the like) and the mass is kneaded to prepare the cold-proof chewing gum or bubble gum according to the process.

Example: The composition of the gum base for the plate-type chewing gum according to the process is prepared containing (in weight percent) 5-15% natural resins, 10-13% vinyl acetate resins, 6-8% ester gums, 15-25% rubbers, 12-20% waxes, 15-20% emulsifiers and 15-20% fillers.

A conventional chewing gum base is prepared containing 20-35% natural resins, 13-17% vinyl acetate resins, 8-11% ester gums, 5-12% rubbers, 18-25% waxes, 1-3% emulsifiers and 10-13% fillers.

To the above gum bases were added the usual amounts of sugars to prepare the chewing gums. In these cases, liquid sorbitol and glycerine were used as the softening agents.

In order to measure the hardness in chewing, five samples each of the two types of chewing gums were tested in the temperature of –10°C by use of a needle-piercing meter.

Results on the conventional chewing gum averaged 4.8, while the gum of this process averaged 17.4. The higher value shows the greater softness.

Solid Elastomers in Chewing Gum

A three-step process is known for the preparation of a chewing gum base utilizing solid elastomer in which the order of mixing and blending ingredients is stated to be critical, particularly with respect to the oleaginous plasticizer and the hydrophobic plasticizer ingredients which cannot be added during mixing under high shear conditions. In this process, the hydrophobic plasticizer is added during the second step where mixing is conducted under reduced shear conditions and increased folding action; and the oleaginous plasticizer must be added in the third step wherein mixing is conducted under rapid folding action with substantially no shearing.

E.R. Koch, L.P. Abbazia and W.J. Puglia; U.S. Patent 4,187,320; February 5, 1980; assigned to Warner-Lambert Company have provided a two-stage process for the preparation of a chewing gum base utilizing solid elastomer.

According to the process, chewing gum base may be prepared in a two-stage process wherein, in the first stage, solid elastomer is initially subjected to high intensity mixing under high shear conditions to masticate the solid elastomer and obtain a substantially uniform, lump-free mass. An elastomer solvent is then added, stepwise, to the mixer containing the masticated solid elastomer, followed by the stepwise addition of an oleaginous plasticizer. Typically, these last-named ingredients are metered into the mixer containing the masticated solid elastomer. The high intensity mixing is conducted during the stepwise addition of both the elastomer solvent and the oleaginous plasticizer and continued after the addition of these ingredients has been completed until a substantially molten, uniform mass is obtained. The effluent from the high intensity mixer of stage one is a rubber compound, completely homogenous, which has no undispersed elastomeric particles and which can be stretched into a translucent film.

Solid elastomers suitable for use in the process are those normally used in chewing gum base and include synthetic gums or elastomers such as butadiene-styrene copolymer, polyisobutylene and isobutylene-isoprene copolymer; natural gums or elastomers such as chicle, natural rubber, jelutong, balata, guttapercha, lechi caspi, sorva, or mixtures thereof. Among these, butadiene-styrene copolymer, polyisobutylene, isobutylene-isoprene copolymer or mixtures thereof are preferred as the elastomer solid.

As the elastomer solvent, there may be mentioned terpene resins such as polymers of α-pinene or β-pinene; rosin derivatives such as the glycerol ester of polymerized rosin or the glycerol ester of hydrogenated rosin; and mixtures thereof.

The following oleaginous plasticizers are suitable for use in this process: hydrogenated vegetable oils, cocoa butter, natural waxes, petroleum waxes such as the polyethylene waxes, and paraffin waxes; or mixtures thereof.

The first stage of the process may be conducted without the addition of heat since heat is generated by the high intensity mixing of the solid elastomer. However, it has been found advantageous to apply heat to the mixing kettle initially (by heating the mixing kettle to a temperature range of 88°-102°C) to speed up the process. After the solid elastomer has been mixed for a period of time, heat is built up by the shearing of the elastomer so that the application of heat is no longer necessary and may be discontinued. Cooling is sometimes necessary if too much heat is built up during the shearing of the elastomer. If cooling is applied, it may then be necessary to apply additional heat to prevent the mass from adhering to the mixer. In any event, maximum temperatures built up during the mixing should not exceed the decomposition temperatures of the ingredients being mixed during a particular stage of processing. The mixing of the stage one ingredients is completed when a substantially molten, uniform mass is obtained.

The sequence of addition of the named ingredients and the high intensity mixing are critical in this first stage.

The seond stage in the process involves the stepwise addition of the remaining chewing gum base ingredients and additional high intensity mixing under high shear conditions until a uniform blend of ingredients is obtained. The second stage ingredients may include a hydrophobic plasticizer, additional oleaginous plasticizer, a nontoxic vinyl polymer, and an emulsifier. If the process is being performed in a continuous manner, the second step ingredients are added to the molten mass of stage one, stepwise, with continuous high intensity mixing. Typically, the stage two ingredients are metered into the molten mass of stage one, in sequence by decreasing order of viscosity. For example, the hydrophobic plasticizer is added first, followed by the oleaginous plasticizer, the vinyl polymer, and lastly, the emulsifier.

Example: A chewing gum base was prepared containing the following percentages (by weight) of ingredients: (1) 10% butadiene-styrene copolymer latex solid; (2) 18% of a glycol ester of hydrogenated rosin; (3) 11% of hydrogenated vegetable oil; (4) 15% calcium carbonate; (5) 10% paraffin wax; (6) 31% vinyl acetate and (7) 5% glycerol monostearate.

Preheat the high intensity mixer to 88°-102°C, add ingredient 1 and commence mixing. Continue to mix for about 20 minutes or until a lump-free mass is obtained. Discontinue heat and add ingredient 2, stepwise, and continue mixing for a period of from 1½ to 2 hours. Add 3, stepwise, and continue mixing for from ½ to ¾ of an hour. Discontinue mixing and transfer the mass to a second high intensity mixer which is jacketed to maintain a temperature of at least 85°C but no higher than 105°C. Commence mixing, add ingredient 4, stepwise, and continue mixing for approximately 20 minutes. Add ingredients 5, 6 and 7, stepwise, mixing each for at least 15 minutes before the next ingredient addition. When all ingredients have been added, an additional 15 minutes of mixing is sufficient to obtain a homogenous gum base.

Ethanol in Sprayable Vegetable Oils

The disclosure by *V.D. Sejpal; U.S. Patent 4,142,003; February 27, 1979; assigned to American Home Products Corporation* relates to vegetable oil-lecithin compositions which are suitable for dispensing from nonaerosol pump bottles or squeeze bottles in a spray form.

This may be achieved with a product formulation which contains 1 to 15% by weight of pure ethyl alcohol along with lecithin in a vegetable oil mixture. Pure ethyl alcohol is ethanol of 190° to 200° proof which conforms with USP standards for alcohol and dehydrated alcohol, respectively. The 190° proof to 200° proof (absolute) ethanols are staple articles of commerce. Ethanol of 200 U.S. proof degrees at 60°F has a specific gravity of 0.79365, is 100% by weight of ethyl alcohol and contains no water. Ethanol of 190 U.S. proof degrees at 60°F has a specific gravity of 0.81582, is 92.423% by weight of ethyl alcohol and contains 7.577% by weight of water. The 190° proof ethanol may be made by mixing 95 parts by volume of ethyl alcohol with 6.18 parts by volume of water. A shrinkage of volume occurs by mixing and results in 100 parts of 190° proof ethanol. Ethyl alcohol proof, by legal definition, is twice the percent by volume.

It has been found that the presence of 1 to 15% of pure ethyl alcohol reduces the viscosity of a vegetable oil-lecithin mixture, and provides a uniform one-phase system. The phosphatides solids in lecithin have a higher specific gravity than vegetable oil, and upon prolonged standing (1 to 2 weeks) tend to separate out on the container bottom, in the absence of ethyl alcohol. These compositions are suitable for dispensing from nonaerosol, pump or squeeze containers. The problem does not arise in aerosol-lecithin compositions because lecithin is soluble in the Freons. In the following examples, the percentages are given in percent by weight.

Example 1: The preferred formulation for use in a squeeze bottle comprises 89.0% liquid vegetable oil, 6.0% lecithin (50% phosphatides) and 5.0% pure ethyl alcohol. A pump spray formulation would comprise 84.0% liquid vegetable oil, 6.0% of the lecithin and 10.0% of pure ethyl alcohol.

The liquid vegetable oil may be soybean oil, corn oil, cottonseed oil, peanut oil, safflower oil, olive oil, and the like. Vegetable oil in the formulation may contain FDA approved antioxidants, such as BHA (butylated hydroxyanisole), propyl gallate, and TBHQ (tertiary-butylhydroxyquinone) and may also contain a crystal inhibitor, such as oxystearin, and an antifoam agent, such as methyl silicone.

Example 2: This example describes the efficacy testing of the formulations of Example 1. The results of the efficacy testing indicated that the product is a parity on antistick qualities. The test was conducted to compare the proposed product with Pam, a popular aerosol vegetable oil containing lecithin, for preventing sticking in casseroles and the like.

A clean, dry, cool Pyrex casserole was sprayed with Pam for three seconds in accord with label directions. A second casserole was sprayed and coated with proposed product (1.0 g serving size) and a third casserole was not treated in any way. The same quantity of beef stew was placed into each casserole. A pie crust was prepared and placed on top of the beef stew. The uncooked crust was then pressed against the edge of the casserole. The three casseroles were placed in a freezer and left there for one day. The casseroles were then heated at 350°F for 60 min in a gas oven. After the casseroles were heated, a panel of 5 persons observed the serving of the beef pot pie from all three casseroles. The cleaning of each was also noted by the same panelists. This test was run in duplicate.

The above test was repeated using chicken tetrazzini which was cooked, placed in the respective casseroles and then frozen as before. The casseroles were then reheated at 350°F for 60 minutes and evaluated as with the beef pot pie. This test was also run in duplicate.

In a third part of this test, a prepared macaroni and cheese (Kraft) was cooked following the label cooking directions. The macaroni and cheese was then placed in the respective casseroles and frozen as above. The next day, the casseroles were reheated at 375°F for one hour and evaluated as described above. This test was run in duplicate.

All three dishes were easily served without sticking from the casseroles treated with proposed product and with Pam. The casseroles were easily cleaned with a sponge and warm soapy water. The foods served easily from untreated casseroles; however, food stuck to the untreated casseroles which required scouring. Similar tests were conducted to compare the no-stick properties of proposed product and Pam when eggs are fried, scrambled, shirred, poached, or cooked in an omelet. In each case, the utensil was sprayed with proposed product (1.0 g serving size) and with Pam according to label directions. After the cooking was completed, the eggs were turned out of the utensil or removed with a spatula. In no case did the eggs stick, using either this product or Pam.

Example 3: This example describes the nutritional content of the formulations of this process.

The squeeze bottle and pump spray formulations (Example 1) consist of vegetable oil and lecithin solids which are the basis for the fat content and resulting caloric level. Experimental testing using an average size frying pan indicates that a 1.0 g spray portion is sufficient to achieve antistick properties and is, therefore, the basis for determining the one gram of vegetable oil-lecithin coating per portion.

The content of calories, proteins, carbohydrates, fats and cholesterol was determined from the "Composition of Foods," Agriculture Handbook No. 8.

It was found that the average spray portion contained 8 calories, 0.8 g protein, no carbohydrate, 1 g fat and no cholesterol.

It was also found that the average spray portion contains less than 2% of the U.S. recommended daily requirement for protein, vitamin A, vitamin C, thiamine, riboflavin, niacin, calcium and iron.

Ethanol in Aerosol Vegetable Oils

An improved aerosol formulation containing ethyl alcohol has also been developed by *V.D. Sejpal; U.S. Patent 4,188,412; February 12, 1980; assigned to American Home Products Corporation.*

This is achieved with a product formulation which contains 7.5 to 25% by weight of ethyl alcohol and 10 to 75% by weight of a hydrocarbon propellent along with lecithin in a vegetable oil mixture.

The ethyl alcohol which may be used in this process is the same as that described in U.S. Patent 4,142,003.

It has been found that the presence of 7.5 to 25% of ethyl alcohol reduces the viscosity of a vegetable oil, lecithin and hydrocarbon propellent mixture, provides a uniform, one-phase system, and delivers a clear, nonfoamy product from an aerosol container. The phosphatides solids in lecithin have a higher specific gravity than vegetable oil, and upon prolonged standing (1 to 2 weeks) tend to separate out on the container bottom, in the absence of ethyl alcohol or a hydrocarbon propellent. The mixture of lecithin, vegetable oil and a hydrocarbon solvent alone provides a lower viscosity, clear solution. Such a solution, however, when dispensed from an aerosol container has undesirable foamy characteristics. This composition which includes 7.5 to 25% ethyl alcohol in the product formulation results in nonfoaming, desirable, light, clear, liquid type of characteristic. In the following examples, percentages given are by weight.

Example 1: This example describes the preferred formulation comprising 60.0% liquid vegetable oil, 10.0% lecithin (50% phosphatides), 15% ethyl alcohol and 15.0% of a 60/40 isobutane/propane blend. An operable range would comprise 5-65% of the liquid vegetable oil, 3-15% of the lecithin, 7.5-25% of ethyl alcohol and 10-75% of the propellent mixture.

The liquid vegetable oil may be soybean oil, corn oil, cottonseed oil, peanut oil, safflower oil, olive oil, sesame oil, peanut oil, coconut oil, coconut butter, palm nut and other fruit pit oils, glyceryl esters of lauric, linoleic, oleic, linolenic acids including their lightly hydrogenated derivatives, and the like. Vegetable oil in the formulation may contain FDA approved antioxidants, such as BHA, propyl gallate, and TBHQ and may also contain a crystal inhibitor, such as oxystearin, and an antifoam agent, such as methyl silicone.

Example 2: The three tests described in Example 2 of U.S. Patent 4,142,003 were repeated with the preferred formulation of Example 1 of this process.

All three dishes were easily served without sticking from the casseroles treated with the preferred Example 1 formulation and with Pam. The casseroles were easily cleaned with a sponge and warm soapy water. The foods served easily from untreated casseroles; however, food stuck to the untreated casseroles which required scouring. Similar tests were conducted to compare the no-stick properties of the preferred formulation of Example 1 and Pam when eggs are fried, scrambled, shirred, poached, or cooked in an omelet. In each case, the utensil was sprayed with the formulation (1.0 g serving size) and with Pam according to label directions. After the cooking was completed, the eggs were turned out of the utensil or removed with a spatula. In no case did the eggs stick, using either the preferred formulation or Pam.

Example 3: This example describes the nutritional content of the above formulations.

The aerosol spray formulations (Example 1) consist of vegetable oil and lecithin solids which are the basis for the fat content and resulting caloric level. Experimental testing using an average size frying pan indicates that a 1.0 g spray portion is sufficient to achieve antistick properties and is, therefore, the basis for determining the 1 g of vegetable oil-lecithin coating per portion.

The content of calories, proteins, carbohydrates, fats and cholesterol was determined from the "Composition of Foods," Agriculture Handbook No. 8.

It was found that the average spray portion contained 6 calories, no protein, no carbohydrate, 0.7 g fat and no cholesterol.

It was also found that the average spray portion contains less than 2% of the U.S. recommended daily requirement for protein, vitamin A, vitamin C, thiamine, riboflavin, niacin, calcium and iron.

Urea for Controlling Gelation of Liquid Gelatin Mixtures

A gelatin concentrate is provided by *J.L. Shank; U.S. Patent 4,341,810; July 27, 1982; assigned to Dynagel Incorporated* which is ungelled at ambient temperatures and has a gel-set temperature of less than about 20°C. It contains about 10 to about 30 weight percent gelatin, urea as a lyotropic agent, and flavorant or other ingredient. Its pH value can range from about 2.5 to about 7.5, the lower values, when present, being attained through materials other than the lyotropic agent.

The ratio of gelatin-to-urea in the concentrate is about 1:0.3 to about 1:1.5. The concentrates have the property whereby dilution of the concentrate with water to a gelatin concentration of about 2 weight percent of the composition while maintaining the pH value of the concentrate produces a resulting diluted composition which has a gel-set temperature higher than that of the concentrate from which it is prepared.

The gelatin concentrates of this process gel at temperatures lower than normally expected for compositions containing an equal amount of gelatin because they contain urea, a nonacid lyotropic agent which prevents or retards gelation. While it is not desired to be bound by any particular theory or mechanism, it is believed that when the gelatin concentrate is diluted with water, the gelation retarding effect of the urea decreases more rapidly than does the gel retardation effect of lowered gelatin concentration. The usual result of these believed different rates of decreasing effects on gelling temperature is that once the concentration of gelatin is reached which is desired for a gelled product, the gel retarding effect of the urea is minimal and the product gels at a temperature near that expected for the gelatin alone.

While the gelatin concentrates of this process are normally liquid at ambient room temperatures, storage or shipment at below room temperatures, as during winter shipment, can cause the concentrates to gel. Should gelation of these concentrates occur, storage at or above ambient room temperature will generally cause the gelled concentrates to reliquify and return to their usable condition because there is usually only a difference of a few degrees, e.g. about 3°C, between their gelling and melting temperatures.

According to this process, aqueous concentrates preferably from about 10 to about 30 weight percent gelatin are prepared and utilized. More preferably, and necessarily with gelatins of high gel strength (Bloom strength), the concentrates contain from about 15 to about 25 weight percent gelatin. The amount of gelatin is calculated on the basis of the solids content of the gelatin material used; i.e. as the amount of acid salt of the protein or as the amount of free protein used. These concentrates are liquid and ungelled at ambient temperatures, rather than being gelled; i.e., the concentrates have gel-set temperatures which are preferably less than about 20°C, and are more preferably less than about 15°C.

Example: A gelatin concentrate useful in preparing the gelled food product of this process when combined with other ingredients was prepared as follows: A gelatin liquor (75.3 lb) containing 29.77 weight percent 280 Bloom gelatin-HCl was mixed with urea (22.0 lb at 98% purity) at a temperature of 65°C. The pH value of the composition was 4.3, and a total of 2.7 lb of additional water was added with mixing to form a substantially homogeneous concentrate.

The concentrate so produced contained 22.4 weight percent gelatin and had a gelatin-to-urea weight ratio of 1:0.96. This concentrate had a gel-set temperature of 8°C. When diluted to contain 2.24 weight percent gelatin, the diluted and subsequently mixed composition had a gel-set temperature of 14°C.

COMPANY INDEX

The company names listed below are given exactly as they appear in the patents, despite name changes, mergers and acquisitions which have, at times, resulted in the revision of a company name.

403

INVENTOR INDEX

U.S. PATENT NUMBER INDEX

NOTICE

Nothing contained in this Review shall be construed to constitute a permission or recommendation to practice any invention covered by any patent without a license from the patent owners. Further, neither the author nor the publisher assumes any liability with respect to the use of, or for damages resulting from the use of, any information, apparatus, method or process described in this Review.

EDIBLE OILS AND FATS
Developments Since 1978

Edited by S. Torrey

Food Technology Review No. 57

This book presents more than 225 processes for the manufacture and/or use of edible oils and fats developed since 1978. Edible oils and fats, important components of the human diet, represent a highly concentrated source of energy for the body. Weight for weight, they release about twice the energy as the same weight of carbohydrate or protein, and, in addition, they are digested slowly, thus delaying hunger sensations between meals.

The book covers a wide range of processes ranging from extraction and fractionation, by solvents or mechanical means, to purification. Modification of properties to prepare particular oils or fats for specific purposes is detailed, and there is a chapter on emulsifiers for fat-containing products. Other chapters describe formulations for margarines, low calorie spreads, cooking and salad oils, confectioners' fats, dairy products and dairy product substitutes, salad dressings, pan release agents, and meat and meat analogs.

The condensed table of contents listed below gives **chapter titles and selected subtitles.** Parenthetic numbers indicate the number of processes per topic.

ISBN 0-8155-0923-5 (1983)

Other Noyes Publications

PLANT GROWTH REGULATORS AND HERBICIDE ANTAGONISTS
Recent Advances

Edited by J.C. Johnson

Chemical Technology Review No. 212

The latest advances in plant growth regulators and herbicide antagonists (also called "safeners") are detailed here in about 240 processes. Plant growth regulators are compounds, other than nutrients, which, in relatively small amounts, inhibit, promote, or otherwise alter physiological plant processes. Herbicide antagonists are compounds applied to a plant to increase its tolerance for particular herbicides, ideally without affecting the herbicide's potency.

Many synthetic compounds have been prepared and tested for controlling one or more aspects of plant growth. In many cases, the growth control can be so drastic that the compound qualifies as a herbicide, or it can be a herbicide at higher concentration and have desirable control properties at lower application rates.

Commercial use of growth control regulators can provide increases in the size of fruits, vegetables, seeds and tubers, in the number of tillers in cereals, in the yield of corn, and in the nutritive value of the fruits, seed and vegetables produced. Other control agents may retard the growth of grasses, cotton, cereals, etc., to reduce labor in maintenance or harvest, prevent the formation of side shoots in tobacco, and reduce leaf formation in soy beans to increase seed formation.

A condensed table of contents, with **chapter headings and selected subtitles,** is given below. Parenthetic numbers indicate the number of processes per topic.

ISBN 0-8155-0915-4 (1982)

303 pages

Other Noyes Publications

PREEMERGENCE HERBICIDES
Recent Advances

Edited by S. Torrey

Chemical Technology Review No. 211

This book details about 250 of the latest advances in preemergence herbicides. Weeds are formidable problems for the world's farmers. Food crops may be infested with as many as 50 types of weeds. What farmers, and homeowners as well, want are efficient herbicides that kill broad ranges of weeds, are easy to apply and are relatively inexpensive.

Important trends in preemergence herbicides in recent years include 1) the development of more selective weedkillers, 2) the introduction of herbicides which are effective at very low application rates, and 3) the development of new types of herbicidal compounds. The latter is essential because of the ability of living organisms to develop tolerances for given poisons.

The book covers primarily preplant and preemergence effects. Preplant herbicides are worked into the soil before a crop is planted. Preemergence herbicides are usually applied a few days after a crop is planted but before weeds emerge from the ground.

Postemergence compounds are applied to weed foliage. Many of the compounds in the various processes in the book can be used for both pre- and postemergence treatment. Occasionally a compound's effects will depend on when it is applied.

Listed below is a condensed table of contents giving **chapter titles and selected subtitles**. Parenthetic numbers indicate the number of processes per topic.

ISBN 0-8155-0914-6 (1982)

MEAT, POULTRY AND SEAFOOD TECHNOLOGY
Recent Developments

by Endel Karmas, Ph.D.
Department of Food Science
Rutgers University

Food Technology Review No. 56

Over 250 processes covering the latest technological advances in the meat, poultry and seafood industries are described in this book. Foods of animal origin play an important role not only in the American diet but also in the diet of other developed countries of the world. The magnitude of total flesh food industry operations and its importance to the economy of the U.S. is emphasized by the fact that approximately one-third of the average American food dollar is spent on flesh foods.

The book is divided into three major parts covering fresh meat and meat products, poultry products, and fish and shellfish products. It is particularly interesting to note that the public apparently accepts "tailored," or reconstituted, foods more readily now than previously, provided the product is nutritionally sound; "natural" products are no longer the only acceptable foods.

Subjects under study are flavors and flavorings, coloring methods, curing operations, tenderizing treatments, injection and marination methods, and meat analogs. Processes for the separation of shellfish from their shells are of considerable interest, as are kamaboko-related products. Kamaboko is a gelled, comminuted fish product very popular in the Japanese diet.

The condensed table of contents below lists **part and chapter titles**. Parenthetic numbers indicate the number of processes per topic.

ISBN 0-8155-0887-5 (1982)

427 pages

SNACK FOOD TECHNOLOGY
Recent Developments

Edited by J.I. Duffy

Food Technology Review No. 55

Americans consume vast quantities of snack foods in a wide variety of forms ranging from potato chips and pretzels, which provide calories but little nutritional value, to high protein food bars and full meal substitutes. This book presents 121 recent processes dealing with snack foods and serves as an excellent reference on current trends in this area of food processing technology.

The public has become more sophisticated and more concerned about what should and should not be in a food product. Snack foods should be convenient, fresh tasting, flavorful, satisfying, nutritious, low in calories and sodium, and inexpensive. As a result, the industry has devised new methods to mask raw protein flavor, incorporate vitamins and minerals, limit the quantity of oil absorbed during frying, improve textures and flavors, and increase production efficiency.

Thus, as increasing numbers of the population have substituted snack foods for meals, nutritional necessity has brought challenges to the industry. Various chapters in the book cover products based on potatoes, grains, nuts, and legumes, and there is a section on the increasingly popular ethnic foods.

A condensed table of contents, with **chapter headings and selected subtitles,** is given below. Parenthetic numbers indicate the number of processes per topic.

ISBN 0-8155-0873-5 (1981)

255 pages

PROTEIN FOOD SUPPLEMENTS
Recent Advances

Edited by M.A. Maltz

Food Technology Review No. 54

The problem of increasing the world's food supply can be solved, to a degree, by the use of protein food supplements prepared from conventional and unconventional animal and vegetable sources. The products involved may be formed as primary products such as, for example, soybean curd; or they may be secondary products, prepared from processing wastes, as in the case of protein supplements derived from bone and bone marrow.

Protein food supplements may actually simulate meat, cheese, or peanut butter, as we "know" them, or they may be used in combination with or added to other foodstuffs. In addition, they may be designed to satisfy particular dietary needs where necessary.

The book, which reviews over 200 recent processes, will be of value to food technologists and researchers who must evaluate society's present nutritional problems and develop new approaches for increasing and enriching future world food supplies.

The condensed table of contents given below lists **chapter titles and selected subtitles.** Parenthetic numbers indicate the number of processes per topic.

ISBN 0-8155-0865-4 (1981)

404 pages

Other Noyes Publications

PESTICIDE MANUFACTURING AND TOXIC MATERIALS CONTROL ENCYCLOPEDIA 1980

Edited by Marshall Sittig

Chemical Technology Review No. 168
Environmental Health Review No. 3
Pollution Technology Review No. 69

This book contains a total of 514 pesticide materials arranged in an alphabetical and encyclopedic fashion by the common or generic name of each pesticide. It is a thorough revision of our previous *Pesticides Process Encyclopedia* published in 1977, plus additional material relative to toxic materials control.

The data on manufacturing processes were drawn primarily from the patent literature, while the data on product toxicity, emissions and product use came mostly from published and unpublished reports released by the Environmental Protection Agency.

This book is definitely of interest to pesticide manufacturers, chemical raw material suppliers, formulators, growers, farmers and food processors. It should also prove useful to chemists, lawyers, industrial hygienists and environmentalists.

It contains much useful extrinsic information, e.g. *allowable tolerance limits, animal and human toxicities,* and similar data not easily ascertained elsewhere.

The use of pesticides leads to healthier plants and bigger crops, and exports of pesticides could provide fast growth for U.S. producers in the coming years.

An indication of the comprehensive nature of this one-volume encyclopedia is given here:

INTRODUCTION
What Is a Pesticide?
Pesticide Manufacture
Pollution Problems
Pesticide Formulations
   Dusts & Wettable Powders
   Granules
   Liquid Formulations
   Packing & Storage
Pesticide Applications

TOXIC MATERIALS CONTROL
Safe Work Practices
Pollution Control in Manufacture
Restrictions on Exposure & Use
   Concentrations in Air/Water
   Registration
   Residue Tolerances

ENVIRONMENTALLY ACCEPTABLE ALTERNATIVES
Biodegradable Pesticides
Physical Control of Toxic Pesticides

Controlled Release Pesticides
Ultra-Low Volume Application
Undesigned Pesticides
Biological Controls
Pheromones
Integrated Pest Management

DATA ON 514 INDIVIDUAL PESTICIDES:
Acephate
Acrolein
Acrylonitrile
Alachlor
Aldicarb
Aldoxycarb
Aldrin
Allethrin
Allidochlor
Allyl Alcohol
Aluminum Phosphide
Ametryne
Aminocarb
Amitraz
AMS
Ancymidol
Anilazine
Anthraquinone
ANTU
Arsenic Acid
Asulam
Atrazine
Azinphos-Ethyl
Azinphos-Methyl
Aziprotryn
Bacillus Thuringiensis
Barban
Benazolin
Bendiocarb
Benfluralin
Benodalin
Benomyl
Bensulide
Bentazon
Benzene Hexachloride
Benzoximate
Benzoylprop-Ethyl
Benzthiazuron
S-Benzyl Di-sec-butylthiocarbamate
Bifenox
plus 474 other pesticides

RAW MATERIALS INDEX

TRADE NAMES INDEX

ISBN 0-8155-0814-X

810 pages